T0348371

PROGRESS IN

Molecular Biology
and Translational Science

Volume 107

PROGRESS IN

Molecular Biology and Translational Science

Molecular Biology of Neurodegenerative Diseases

edited by

David B. Teplow, Ph.D.

Professor of Neurology
Interim Director, Mary S. Easton Center for Alzheimer's Disease Research at UCLA
Director, Biopolymer Laboratory
David Geffen School of Medicine at UCLA
Los Angeles, CA

Volume 107

AMSTERDAM • BOSTON • HEIDELBERG • LONDON
NEW YORK • OXFORD • PARIS • SAN DIEGO
SAN FRANCISCO • SINGAPORE • SYDNEY • TOKYO
Academic Press is an imprint of Elsevier

ELSEVIER

Academic Press is an imprint of Elsevier
32 Jamestown Road, London, NW1 7BY, UK
Radarweg 29, PO Box 211, 1000 AE Amsterdam, The Netherlands
225 Wyman Street, Waltham, MA 02451, USA
525 B Street, Suite 1900, San Diego, CA 92101-4495, USA

This book is printed on acid-free paper. ∞

Library of Congress Cataloging-in-Publication Data
A catalog record for this book is available from the Library of Congress

British Library Cataloguing in Publication Data
A catalogue record for this book is available from the British Library

ISBN: 978-0-12-385883-2
ISSN: 1877-1173

For information on all Academic Press publications
visit our website at elsevierdirect.com

Printed and bound by CPI Group (UK) Ltd, Croydon, CR0 4YY

Transferred to Digital Print 2012

Working together to grow
libraries in developing countries

www.elsevier.com | www.bookaid.org | www.sabre.org

ELSEVIER   BOOK AID
International   Sabre Foundation

Dedication

The quest for scientific knowledge is an all-consuming, exhilarating, joyful experience for the scientist. However, spouses often experience this quest in quite a different way. I therefore dedicate this volume to my wife, Arden, who lovingly made the personal sacrifices necessary to enable me to pursue my scientific passions.

Contents

Tau and Tauopathies 263

Gloria Lee and Chad J. Leugers

Membrane Pores in the Pathogenesis of
Neurodegenerative Disease 295

Bruce L. Kagan

Protein Quality Control in Neurodegenerative
Disease 327

Jason E. Gestwicki and Dan Garza

Biology of Mitochondria in Neurodegenerative Diseases 355

Lee J. Martin

Fungal Prions. 417

Gemma L. Staniforth and Mick F. Tuite

Contributors

Numbers in parentheses indicate the pages on which the authors' contributions begin.

Lars Bertram, Department of Vertebrate Genomics, Max Planck Institute for Molecular Genetics, Berlin, Germany (79)

Marc I. Diamond, Hope Center for Neurological Diseases, Knight-Alzheimer Disease Research Center, Department of Neurology, Washington University in St. Louis, St. Louis, Missouri, USA (189)

Nikolay V. Dokholyan, Department of Biochemistry and Biophysics; Curriculum in Bioinformatics and Computational Biology; and Center for Computational and Systems Biology, University of North Carolina, Chapel Hill, North Carolina, USA (215)

Dan Garza, Proteostasis Therapeutics Inc., Cambridge, Massachusetts, USA (327)

Jason E. Gestwicki, Department of Pathology and the Life Sciences Institute, University of Michigan, Ann Arbor, Michigan, USA (327)

Ming Guo, Department of Neurology, and Department of Molecular and Medical Pharmacology, Brain Research Institute, David Geffen School of Medicine, University of California, Los Angeles, California, USA (125)

Bruce A. Hay, Division of Biology, MC126-29, California Institute of Technology, Pasadena, California, USA (125)

Bruce L. Kagan, Department of Psychiatry & Biobehavioral Sciences, David Geffen School of Medicine at UCLA, Semel Institute for Neuroscience and Human Behavior, Los Angeles, California, USA (295)

Gloria Lee, Department of Internal Medicine, University of Iowa Carver College of Medicine, Iowa City, Iowa, USA (263)

Chad J. Leugers, Department of Internal Medicine, University of Iowa Carver College of Medicine, Iowa City, Iowa, USA (263)

Lee J. Martin, Division of Neuropathology, Department of Pathology, The Pathobiology Graduate Program, and Department of Neuroscience, Johns Hopkins University School of Medicine, Baltimore, Maryland, USA (355)

Hironobu Naiki, Division of Molecular Pathology, Department of Pathological Sciences, Faculty of Medical Sciences, University of Fukui, Fukui, Japan (41)

Rachel L. Redler, Department of Biochemistry and Biophysics; Curriculum in Bioinformatics and Computational Biology; and Center for Computational and Systems Biology, University of North Carolina, Chapel Hill, North Carolina, USA (215)

Jean-Christophe Rochet, Department of Medicinal Chemistry and Molecular Pharmacology, Purdue University, West Lafayette, Indiana, USA (125)

Gemma L. Staniforth, Kent Fungal Group, School of Biosciences, University of Kent, Canterbury, Kent, United Kingdom (417)

Rudolph E. Tanzi, Genetics and Aging Research Unit, Massachusetts General Hospital, Harvard Medical School, Boston, USA (79)

David B. Teplow, Department of Neurology and Mary S. Easton Center for Alzheimer's Disease Research at UCLA, David Geffen School of Medicine at UCLA, Los Angeles, California, USA (101)

Paul M. Thompson, Laboratory of Neuro Imaging, David Geffen School of Medicine at UCLA & UCLA Medical Center, Los Angeles, California, USA (1)

Mick F. Tuite, Kent Fungal Group, School of Biosciences, University of Kent, Canterbury, Kent, United Kingdom (417)

Harry V. Vinters, Department of Pathology & Laboratory Medicine (Section of Neuropathology), and Department of Neurology, David Geffen School of Medicine at UCLA & UCLA Medical Center, Los Angeles, California, USA (1)

Dominic M. Walsh, Laboratory for Neurodegenerative Research, Center for Neurologic Diseases, Brigham & Women's Hospital, Harvard Institutes of Medicine, Boston, Massachusetts, USA (101)

Masahito Yamada, Department of Neurology and Neurobiology of Aging, Kanazawa University Graduate School of Medical Science, Kanazawa, Japan (41)

Zhiqiang Zheng, Hope Center for Neurological Diseases, Knight-Alzheimer Disease Research Center, Department of Neurology, Washington University in St. Louis, St. Louis, Missouri, USA (189)

Preface

The central nervous system (CNS) is the most complex system in the human body. The brain alone comprises ≈100 billion neurons and an equal or greater number of glial cells. The neurons themselves form ≈100 trillion synapses, the structures through which the electrochemical functions of the brain are mediated. The sheer number of cells, and their tremendous activity, account for the fact that the brain, which represents only ≈2% of total body weight, is estimated to consume 15% of the body's cardiac output, 20% of its oxygen, and 25% of its total glucose. The brain's complexity and its continuous operation make it especially vulnerable to injury.

The most common form of CNS injury among otherwise healthy children and young adults arises from trauma. This form of injury also has been a major cause of neurological dysfunction and death in the military and a factor in late life brain dysfunction among athletes, most notably boxers and football players. Among older adults, the predominant cause of brain dysfunction and death is stroke.

We focus here on a particularly devastating class of CNS diseases, that of neurodegenerative diseases. These diseases generally occur in mid- to late-life and progress relentlessly from subtle clinical onset to increasing dysfunction, and eventually, to death. The etiologies are variable. Sporadic cases are frequent, yet the molecular determinants of disease initiation often are unknown. Known causative factors include environmental insults, earlier physical trauma, gene mutations, age-related metabolic dysfunction, protein dysfunction or deposition, and even infection (e.g., in the infectious forms of prion diseases).

The goal of this volume is to provide the reader with a broad perspective on the molecular biology of neurodegenerative diseases, where "molecular biology" refers to the consideration of disease at numerous levels, including the organismal, genetic, anatomic, histologic, cellular, and subcellular. Historically, the study of neurological disease began with the gross and light microscopic examination of postmortem brain tissue. This practice is maintained here in the initial chapters discussing the anatomic and microscopic features of neurological diseases. In addition, the reader is provided with images obtained using state-of-the-art light and electron microscopy, magnetic resonance imaging, and positron emission tomography.

What are the mechanistic bases for the pathologies thus observed? The genetic bases for many diseases are quite clear, namely, mutations in the structural genes for specific proteins. Prominent examples include early onset familial Alzheimer's disease and cerebral amyloid angiopathy, which are associated with increased amyloid β-protein production or alterations in the primary structure of the peptide, and "triplet-repeat diseases" such as Huntington's disease and some spinocerebellar ataxias. However, the majority of AD cases, and of cases of other neurodegenerative diseases, are of late onset and complex genetic etiology.

Detailed discussion of the molecular biology of prevalent neurodegenerative diseases, including Alzheimer's, Parkinson's, Huntington's, and Lou Gehrig's (amyotrophic lateral sclerosis), are provided in individual chapters. Although distinct, these diseases share some pathologic features, often the formation of tau protein aggregates. A chapter on tau and the "tauopathies" thus also is included.

Each of the diseases discussed in this volume have distinct etiologies, yet as discussed above, there does exist overlap in some pathologic features of disease. Similarly, overlap may exist in the molecular mechanisms responsible for neuronal injury, dysfunction, and death. Three chapters are devoted to common mechanisms of cellular injury, including membrane pore formation by disease-linked proteins, perturbation of proteostasis (the ability of the cell to balance production and elimination of proteins), and mitochondrial dysfunction.

A nascent area of study in neurodegeneration is prion-like intercellular spreading of cytopathology in the CNS. Some studies have suggested that this process could be one mechanism for disease progression, a hypothesis that is being tested actively but for which answers remain elusive. In contrast, studies of non-mammalian prions, especially the fungal prions that are discussed in the final chapter, have provided a wealth of information about prion biology and transmissibility that can inform future tests of prion-like human neuropathologic processes.

It is hoped that the clinical, molecular biological, and methodological information contained within this volume will provide the reader with a comprehensive perspective on the field of neurodegeneration, a perspective that will contribute to the conception of novel theories and approaches for preventing, treating, and curing this tragic disease class.

DAVID B. TEPLOW

Pathologic Lesions in Neurodegenerative Diseases

PAUL M. THOMPSON* AND HARRY V. VINTERS[†,‡]

*Laboratory of Neuro Imaging, David Geffen School of Medicine at UCLA & UCLA Medical Center, Los Angeles, California, USA

[†]Department of Pathology & Laboratory Medicine (Section of Neuropathology), David Geffen School of Medicine at UCLA & UCLA Medical Center, Los Angeles, California, USA

[‡]Department of Neurology, David Geffen School of Medicine at UCLA & UCLA Medical Center, Los Angeles, California, USA

This chapter will discuss two of the most widely used approaches to assessing brain structure: neuroimaging and neuropathology. Whereas neuropathologic approaches to studying the central nervous system have been utilized for many decades and have provided insights into morphologic correlates of dementia for over 100 years, accurate structural imaging techniques "blossomed" with the development and refinement of computerized tomographic scanning and magnetic resonance imaging (MRI), beginning in the late 1970s. As Alzheimer disease progresses over time, there is progressive atrophy of the hippocampus and neocortex—this can be quantified and regional accentuation of the atrophy can be evaluated using quantitative MRI scanning. Furthermore, ligands for amyloid proteins have recently been developed—these can be used in positron emission tomography studies to localize amyloid proteins, and (in theory) study

Progress in Molecular Biology
and Translational Science, Vol. 107
DOI: 10.1016/B978-0-12-385883-2.00009-6

1

the dynamics of their deposition (and clearance) within the brain over time. Neuropathologic studies of the brain, using highly specific antibodies, can demonstrate synapse loss and the deposition of proteins important in AD progression—specifically ABeta and phosphor-tau. Finally, neuropathologic assessment of (autopsy) brain specimens can provide important correlation with sophisticated neuroimaging techniques.

I. Introduction

The scientific and medical challenges at hand are framed well in a recent review article [1a]: "*Neurodegenerative diseases [NDDs] are traditionally defined as disorders with progressive loss of neurons in distinct anatomical distribution (s), and accordingly different clinical phenotypes. [These diseases] are also referred to as conformational diseases. . .emphasizing the central pathogenetic role of altered protein processing.*" A defining biochemical theme in the study of many neurodegenerative disorders, including the most common CNS amyloidosis, Alzheimer disease (AD), is that of protein misfolding—the molecular basis of which is explored from various perspectives throughout this volume. However, the provenance of this theme is the "diseased brain." Medical autopsies (including examination of the brain) have been an integral part of medical teaching, research, and diagnostic work since the pioneering studies of Karl von Rokitansky (1804–1878) and Rudolf Virchow (1821–1902), though human dissection has actually been carried out since the time of Andreas Vesalius in the 16th century [1b]. Vast amounts of data have been derived from autopsy brain examinations, including patients with dementias. For this reason, we provide foundational information on the practical aspects of assessing gross, microscopic, and immunohistochemical features of some of the most common neurodegenerative disorders, how the neuropathologist approaches evaluation of an autopsy or biopsy brain specimen, and how biochemical, molecular, and genetic findings—of which there have been an explosion in recent years—inform the clinicopathologic evaluation of brains from afflicted individuals. We will discuss how the structural anatomy of progressive brain atrophy is studied in a dynamic fashion *in vivo* by new neuroimaging methods and how data imaging should be integrated with neuropathologic observations.

A. Neuropathologic Features

Neuropathologic examination of the brain—either at autopsy or, less commonly, biopsy—continues to be the gold standard for the diagnosis of AD and non-AD dementias.[2,3] This is true even as high-resolution neuroradiographic techniques are emerging that are capable of both quantifying AD-associated cerebral atrophy and detecting the amyloid β-protein (Aβ) or amyloid proteins in the brain while patients are alive (and even asymptomatic).[4-6] The obvious problem in designing studies that compare neuroimaging data with neuropathologic findings (in the same patient set) is that the most recent imaging in a given subject may, for obvious reasons (patient frailty, intercurrent illness, lack of mobility), have been carried out months or even years before necropsy examination is performed. This time period is precisely when major structural changes may occur, often at an accelerating pace, and thus neuropathology would play a pivotal role in illuminating the structures that are being imaged by structural and metabolic neuroimaging methods.[7] Careful clinicopathologic correlation, that is, attempting to explain complex neurologic symptoms in a deteriorating, often end-stage central nervous system by autopsy examination of the brain, was a central pillar of dementia research through the early 1980s, at which time structural imaging emerged and began to provide valuable information about the CNS *in vivo*. It bears re-emphasis that the starting point for important biochemical/molecular studies that have linked abnormally folded proteins to neurodegeneration was rapidly harvested (usually autopsy) brain tissue, neuropathologic features of which were subsequently correlated with the novel neurochemical data.[3,8,9]

The main neuropathologic feature of the AD brain on gross inspection is cortical atrophy, which is usually diffuse and fairly symmetrical throughout the cerebral hemispheres, rather than being accentuated in certain lobes.[3] Fresh brain weight is usually below the normal range for an adult (1200–1400g), though not necessarily so, and it may be normal. A review of fresh brain weights from 40 individuals (examined by the Neuropathology Core of the Easton AD Center at UCLA) who had confirmed AD changes (Braak stage V–VI) on microscopic examination of the CNS revealed that 25% of them had weights that were in the traditional normal range, that is, 1200–1400g, or even slightly exceeded this. When the fixed brain is cut, the cortical atrophy (manifest as thinning of the cortical ribbon) is usually accompanied by ventriculomegaly, or "hydrocephalus *ex vacuo*," and sometimes shrinkage or atrophy of the subcortical white matter (Figs. 1 and 2). The precise etiology of the white matter change is not known—it may in part represent downstream (Wallerian) degeneration secondary to cortical neuron loss, or be the manifestation of an intrinsic leukoencephalopathy. If the brain of a demented patient shows hydrocephalus out of proportion to the degree of cerebral cortical atrophy, the possibility of normal pressure hydrocephalus must be considered, though microscopic lesions of AD should still be sought by the neuropathologist in such a brain. Most experienced neuropathologists are struck (and sometimes baffled) by the variability in brain weights, cerebral cortical atrophy, and hydrocephalus *ex vacuo* among individuals who eventually have the diagnosis of AD confirmed by light microscopy and immunohistochemical study of the brain.[10]

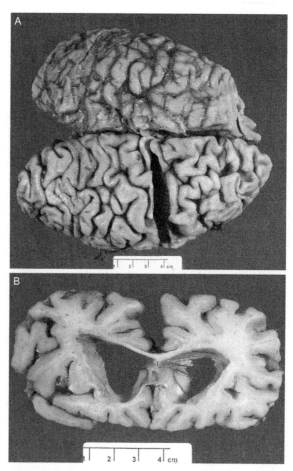

Fig. 1. View from the cerebral convexities (panel A) of fixed brain (autopsy specimen) from a patient with AD. Meninges are intact over the left cerebral hemisphere (top of panel) but have been removed from right hemisphere (bottom of panel). Cortical atrophy is more easily appreciated when leptomeninges have been removed. Panel B shows coronal section of the fixed brain, demonstrating severe cortical atrophy and marked fairly symmetrical ventriculomegaly (hydrocephalus *ex vacuo*). (For color version of this figure, the reader is referred to the Web version of this chapter.)

1. RELATION TO THE SEQUENCE OF MYELINATION

As shown by the maps of Braak and Braak (Ref. 11; Fig. 3), the trajectory of pathology in Alzheimer's disease follows a characteristic pattern. Neurofibrillary tangle (NFT) accumulation, cortical atrophy, and impaired glucose metabolism are typically found in higher-order association cortices early in the disease, with primary cortices being resistant to atrophy until very late in the disease.[12,13]

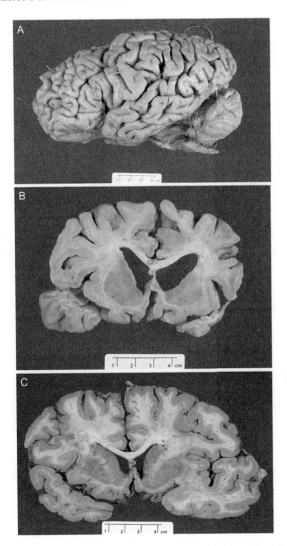

FIG. 2. Lateral view of an AD brain (fixed autopsy specimen) shows relatively minimal diffuse cortical atrophy (panel A). Panel B (coronal section of fixed brain, same specimen as in panel A) shows a slight degree of ventricular enlargement and confirms minimal cortical atrophy. Panel C shows coronal slice (fixed brain, autopsy specimen) from another subject with severe AD, but negligible cortical atrophy or ventricular enlargement. (For color version of this figure, the reader is referred to the Web version of this chapter.)

To explain this, both Braak and Braak[11] and Reisberg et al.[14] noted a process called "retrogenesis"—cortical regions that mature earliest in infancy tend to degenerate last in AD. Recent work with MRI in children has revealed a temporal

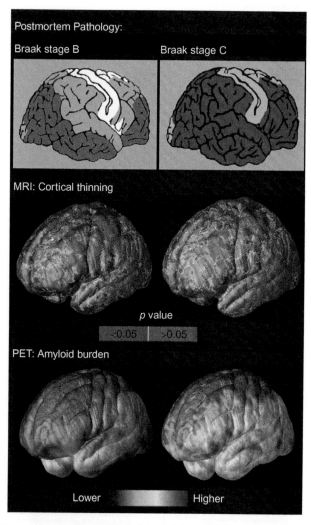

FIG. 3. Progression of AD based on neuropathology, structural MRI, and PET with the amyloid- and tau-sensitive ligand, [^{18}F] FDDNP. Neurofibrillary tangles, one of the molecular hallmarks of AD, spread in the brain in a characteristic advancing trajectory (*top row*; adapted from Ref. 11). Darker red colors denote areas with greater tangle deposition, based on histologic staining of postmortem material. On MRI, the areas with gray matter atrophy in mild AD include the temporal lobes, but in moderate AD these deficits have typically spread to involve the frontal cortices (*middle row*; adapted from a longitudinal study[12]). Finally, cerebral amyloid and neurofibrillary tangle burden estimated *in vivo* with the PET ligand FDDNP is low in controls, but higher in those with impaired cognition (*bottom row*; adapted from Ref. 13). Although individuals vary in the rate, extent, and sequence of these changes, the anatomical agreement is striking between these *in vivo* maps and the well-established *postmortem* maps for the staging of AD. In all maps, the sensorimotor cortex shows least disease-related degeneration. (Adapted, with permission from the authors and publishers). (See Color Insert.)

sequence of cortical growth that largely follows the myelination sequences observed by Conel[15] and Yakovlev and Lecours.[16] The most heavily myelinated structures, such as the primary sensorimotor and primary visual cortices, are among the earliest to develop and are also phylogenetically among the oldest, and the most resistant to AD pathology. By contrast, NFTs and neuropil threads in dendrites accumulate early in the late-myelinating heteromodal association cortices, posterior cingulate, and phylogenetically older limbic areas that remain highly plastic throughout life (Fig. 4).[17] Two theories have been proposed for the resilience of primary cortices in AD. Mesulam[17] has argued that the high neuronal plasticity of memory-related structures makes them more vulnerable to pathologic processes. Bartzokis[18] argues that the heavy myelination of the primary cortices protects them, to some degree, from the effects of plaque and tangle pathology. Regardless of the cause, there is striking agreement between pathology and neuroimaging in the sequence of cortical degeneration in AD.[12,13]

2. ALZHEIMER DISEASE: MICROSCOPIC CONFIRMATION OF AD DIAGNOSIS

The microscopic lesions that accumulate in the central nervous system (mainly cerebral cortex, but also deep central gray structures such as basal ganglia, thalamus) of individuals with AD can, when prominent, be seen on routine (hematoxylin-and-eosin stained; H&E) sections of brain (Fig. 5), but are much more easily demonstrated by the use of special stains and immunohistochemical

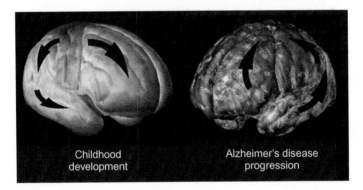

FIG. 4. Degenerative sequence in AD is the reverse of the normal developmental sequence. In a process termed *retrogenesis* (e.g., in Ref. 14), cortical regions that mature earliest in infancy tend to degenerate last in AD. The developmental sequence echoes the phylogenetic sequence in which structures evolved. The most heavily myelinated structures, with least neuronal plasticity, resist AD-related neurodegeneration. Arrows denote the childhood cortical maturation sequence (*left panel*) and the gray matter atrophy sequence in AD (*right panel*[12]). Images are from time-lapse films compiled from cortical models in subjects scanned longitudinally with MRI. These may be viewed at http://www.loni.ucla.edu/~thompson/DEVEL/dynamic.html and http://www.loni.ucla.edu/~thompson/AD_4D/dynamic.html. (See Color Insert.)

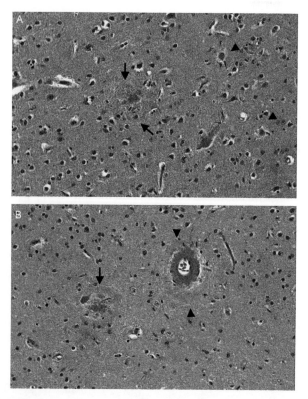

Fɪɢ. 5. H&E-stained section from an AD brain. In panel A, arrows indicate a neuritic SP, with an amyloid core surrounded by neurites, seen as a "coarsening" of the neuropil around the amyloid core. Arrowheads indicate two amyloid "cores" lacking a significant surrounding neuritic component. In panel B, also from an H&E-stained section, arrow indicates a large neuritic SP, whereas arrowheads indicate a vessel showing pronounced amyloid angiopathy (CAA). (See Color Insert.)

techniques (to be described). If key lesions (senile plaques (SPs), NFTs, amyloid angiopathy) are identifiable on routine stains, they will usually be noted to be extremely abundant using these other approaches. As the antigenic nature of AD lesions has become elucidated, immunohistochemistry using highly specific antibodies (monoclonal or polyclonal) against the components of SPs and NFTs has been used increasingly to demonstrate these lesions in the CNS and to allow for their quantification. Historically, the special stains used to demonstrate the SPs and NFTs in the cerebral cortex of an end-stage AD patient have been silver impregnation techniques (Fig. 6), usually the modified Bielschowsky, Bodian, Campbell-Switzer, and Gallyas methods—the latter two used effectively in seminal studies of SP and NFT distribution by Heiko and Eva Braak and their colleagues[11,19a,b,20]. Many of these stains, while demonstrating an intrinsic

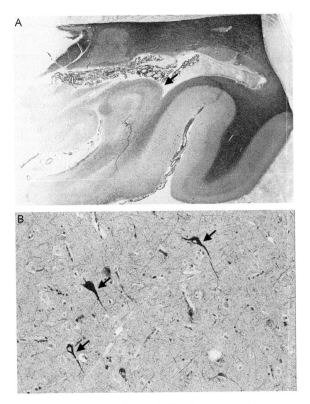

FIG. 6. Section stained with modified Bielschowsky, showing hippocampus from an AD patient—whole mount is shown in Panel A. Arrow indicates region of the pyramidal cell layer where AD lesions are seen, even at this low magnification. Panel B shows magnified view of the pyramidal cell layer, with several NFTs (arrows). (See Color Insert.)

elegance, were often capricious and resulted in annoying (and inconsistent) tissue section artifacts that tended to limit their usefulness, especially in quantitative morphometric studies.

a. Senile Plaques. SPs appear, on routine sections, as a coarsening or "dissolution" of the usually smooth and uniform neuropil (the "neuritic" component of the SP) centered on an amorphous eosinophilic *globule* of amyloid, the core of the SP (Fig. 5 and 6). The relationship between the amyloid core of a mature SP and its neuritic corona (both seen well on silver stains) has been debated for years and remains unresolved, but such mature neuritic SPs are thought to be more representative or reflective of cortical injury (thus correlate with neuronal dysfunction) than the more diffuse SPs lacking a neuritic component. Excellent reviews on the hypothesized cellular and molecular pathogenesis of SPs—emphasizing the role of secreted factors, microglia and astrocytes—delve into this issue further (one of the best is by Dickson[21]). Though SPs (especially neuritic SPs) have a neuronal component, insofar as the neurites surrounding the amyloid core represent processes emerging from presumably damaged nerve cell bodies, they are substantially extraneuronal or located within the neuropil. Although SPs are often found in elderly individuals without AD or any evidence of cognitive impairment, their density (in this context) is generally far less than that in patients with AD.[22] However, most neuropathologists have encountered autopsy brain specimens from cognitively intact elderly that contain abundant neuritic SPs. Anecdotal reports have described all neuropathologic features of AD (abundant SPs and NFTs) in cognitively normal elderly—indeed rare individuals who had been carefully examined for cognitive function shortly before death.[23]

   b. Neurofibrillary Tangles.   NFTs, a second major brain lesion of AD, are dense intraneuronal protein aggregates that include, on ultrastructural examination, characteristic paired helical filaments (also described less frequently as "bifilar helices").[24,25] NFTs (Fig. 6) are usually accompanied by "neuropil threads" in the adjacent brain parenchyma—these threads represent processes of tangle-bearing neurons.[19,26] NFTs may occur in the CNS in many non-AD neurodegenerative conditions including subacute sclerosing panencephalitis, dementia pugilistica, aluminum intoxication, postencephalitic Parkinsonism, and the Parkinsonian–amyotrophic lateral sclerosis (ALS)–dementia complex of Guam.[27] Of interest, NFT-like neuronal cytoplasmic lesions are commonly encountered within the dysmorphic and enlarged neuronal cell bodies of infants and children with cortical dysplasia or cortical tubers of tuberous sclerosis complex (TSC).[28] These NFT-like fibrillar aggregates, easily demonstrable using the same silver stains (e.g., Bielschowsky) used to highlight AD lesions, are *not* composed of paired helical filaments; rather, they show disorganized clumps and skeins of cytoplasmic neurofilaments and neurotubules.[29]

   c. Cerebral Amyloid (Congophilic) Angiopathy.   A third important lesion of AD is cerebral amyloid or congophilic angiopathy (CAA).[30] Indeed, CAA was the microscopic AD lesion from which Glenner and Wong[31a] isolated A4 protein (now designated amyloid β-protein (Aβ)).[31b] The reason that CAA is less prominently discussed (than SPs and NFTs) when considering AD neuropathologic features may be that it is extremely variable among AD patients, though found (to some extent) in an estimated 90–95% of AD brains and often in the brains of "normal aged".[30,32] CAA defines a histopathologic finding that results from a process, whereby the media of parenchymal arterioles, normally composed of smooth muscle cells (SMC), undergoes progressive loss of these SMC coincident with the accumulation of an eosinophilic hyaline material (in the vessel wall) that has the staining properties of amyloid (Fig. 7), that is, positivity for thioflavin S or T (on fluorescence microscopy), and congophilia.[25] When a brain with prominent CAA is stained with Congo red and visualized using polarized light, the walls of affected arterioles show characteristic yellow-green birefringence. (Physical properties of brain amyloids and the dyes that stain them have played an important role in the rational design of ligands that can be used to localize CNS amyloids using positron emission tomography (PET) technology; see below). CAA may also involve cortical parenchymal venules and capillaries. Some have suggested that at least a subset of SPs in the neocortex are intimately associated with capillaries and may even originate from them.[33,34] Meningeal arteries are often affected by CAA, and sometimes an amyloid-laden arteriole may be seen traversing the subarachnoid space to enter the underlying cortex. When CAA occurs in the subarachnoid space, the amyloid deposits are usually adventitial rather than medial in the walls of affected arteries and have a "chunky" appearance, suggesting they may have resulted from aggregates of Aβ in the CSF. CAA almost never involves the subcortical white matter, basal ganglia, brainstem, or spinal cord but may involve the cerebellar molecular layer and overlying meninges.[30,32] The pathogenesis of CAA is complex and probably involves overproduction of Aβ in or near the vessel wall, together with abnormal/impaired clearance of Aβ, probably along perivascular adventitial pathways of brain microvessels.[35] Neprilysin has been implicated as a major molecule protecting against CAA and Aβ-induced degeneration of cerebrovascular SMC.[36]

CAA is important as a cause of nontraumatic intracerebral hemorrhage in the elderly—including many who do not manifest overt features of a dementing illness or even cognitive impairment at the time of their stroke. A small subset of these patients has predominantly severe CAA (with a small load of SPs and NFTs) as their major neuropathologic finding.[30] CAA-related intraparenchymal hematomas are usually lobar, unlike centrencephalic hypertensive bleeds resulting from lipohyalinosis.[3,30] In some patients, multiple hematomas caused by CAA occur over months or years. These large and invariably symptomatic, sometimes fatal, hematomas occur in a relatively small proportion of those with AD and severe CAA, but CAA-related "micro-bleeds" (detectable on high-resolution MRI scanning using special sequences, e.g., susceptibility weighted imaging/SWI) are now considered a reliable biomarker for the presence of CAA within the brain.[37] More recently, severe CAA has also been associated with the occurrence of cerebral microinfarcts that contribute to the worsening of age-related cognitive impairment.[38]

    d. Other Lesions. While SPs, NFTs, and CAA are the major microscopic lesions of AD and are widely distributed throughout the cortex, two other lesions merit mention. Granulovacuolar degeneration (GVD, of Simchowicz)

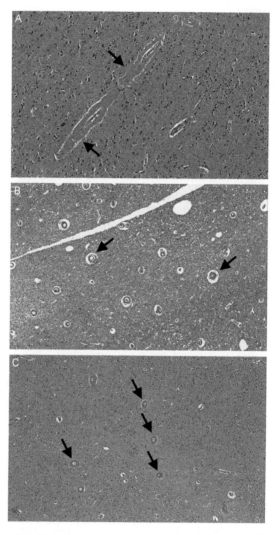

Fig. 7. Severe CAA. Panel A shows a cortical parenchymal artery cut in longitudinal section. Note thickening of the arterial walls by a glassy eosinophilic material (arrows), and complete loss of medial smooth muscle cells. Panel B shows severe CAA, several arterial profiles cut in transverse section. Arrow at left indicates an artery with a "double barrel" lumen, whereas arrow at right of the panel indicates an artery with secondary intimal thickening. Panel C shows another region with severe CAA (H&E-stained section). (All images are from sections of brain originating from autopsy of a patient with AD, severe CAA, and extensive CAA-associated microinfarcts throughout the CNS.) (See Color Insert.)

describes a neuronal cytoplasmic lesion in which the neuronal cytoplasm of hippocampal pyramidal cells is replaced by vacuoles containing small basophilic granules. Hippocampi showing prominent GVD (Fig. 8) also often show eosinophilic hyaline rod-like structures in the adjacent neuropil—described as "rod-like bodies of Hirano" or simply as "Hirano bodies." GVDs and rod-like bodies of Hirano have been the subject of limited study in terms of assessing their possible contributions to AD pathogenesis. Nevertheless, neurons showing GVD and Hirano bodies are a frequent finding in AD hippocampi but are almost never seen in neocortex.

While this section has emphasized lesions commonly seen in AD brains— either focally or diffusely in the cortex—one of the most important findings, demonstrable biochemically or by immunohistochemistry in AD brain, is synapse loss.[39,40a,b] This is shown on sections of affected cortex when they are immunostained with antibodies against synaptic proteins such as synaptophysin.[40] Interpretation (and quantification) of the immunohistochemical signal in such cases must be done by careful comparison to brain from a cognitively normal control (and using tissue that has been comparably fixed and processed), since the loss of synaptophysin protein may be subtle. AD brains also frequently show evidence of clinical comorbidity, not surprising given the many age-related diseases (e.g., tumors, cerebrovascular disease) that may impact on the aging brain.[41] Coexistent Parkinson disease (PD) changes and evidence of infarcts or hemorrhage have been noted in as many as 20–25% of AD brains.[42] The theme of comorbidity between ischemic brain lesions and AD microscopic changes—both common in the elderly—features prominently in modern dementia research, possibly because it represents a more accurate

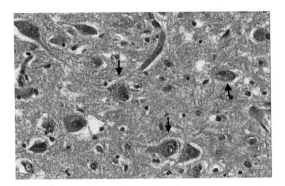

FIG. 8. Hippocampal pyramidal cell layer (H&E-stained section) from an elderly patient with trisomy 21 (Down syndrome). Arrows indicate several pyramidal neurons containing varying severity of granulovacuolar degeneration, characterized as granules within neuronal cytoplasmic vacuoles. This lesion is almost always confined to the hippocampus. (See Color Insert.)

and realistic scenario than considering AD or multi-infarct/ischemic-vascular dementia as pure entities.[43,44] Whereas cystic infarcts (regions of encephalomalacia) are easily detectable by structural imaging, and lacunar infarcts (≤ 1.0 cm diameter) are often noted, cortical and white matter microinfarcts may be difficult or impossible to detect using current methodology.

3. CHARACTERISTIC NEUROIMAGING FEATURES IN AD

Several international initiatives have been set up to use a variety of neuroimaging methods, as well as CSF biomarkers, to chart the progression of Alzheimer's disease. One of these, the Alzheimer's disease neuroimaging initiative, has evaluated the relative ability of MRI, PET, perfusion, and other methods for tracking disease progression, and predicting future decline. A major goal has been to understand which imaging features correlate best with clinical decline and with pathological measures, and which changes are found the earliest in the course of the disease. More practically, there is an effort to identify which imaging techniques require the smallest sample sizes to establish whether drug treatment significantly slows disease progression.[45]

a. MRI. Standard T1-weighted MRI scans reveal the anatomy of the brain in three dimensions (3D). A typical scan takes ≈ 8 min, with a spatial resolution of ~ 1 mm. In these scans, the gray matter is clearly visualized. The scans of patients with AD typically show widespread cortical and hippocampal atrophy, ventricular expansion, and widening of intrasulcal CSF spaces. In mild cognitive impairment (MCI[46]), which is a prodromal phase with a 15%/year risk of conversion to AD, there is usually more subtle atrophy in the medial temporal lobe, but this is harder to appreciate without computational techniques that aggregate data from multiple individuals. Although subtler than the changes in AD, amnestic MCI patients show diffuse hippocampal atrophy[47] and reduced volumes in the mesial temporal lobe including the hippocampus, entorhinal cortex, and amygdala.[48] Nonamnestic MCI patients—who show cognitive impairments in single or multiple domains other than memory— typically have greater atrophy outside of the hippocampus, specifically in multimodal association cortices.

Cortical atrophy on MRI is believed to be a reasonable index of both cell shrinkage and neuronal loss. When measures of cortical atrophy from multiple subjects are combined or compared across groups, characteristic patterns emerge. If patients are scanned longitudinally, the progression of atrophy can be mapped on the cortical mantle over time as patients progress from moderate to severe AD.[12] On average, patients with moderate AD have gray matter reductions, relative to matched healthy controls, of 15% or more in the medial temporal, posterior cingulate, temporal, and temporoparietal cortices. Gray matter loss also progresses at a rate of 3–4% per year in most of these areas.[12]

b. Relation to Cognition. The progression of cortical atrophy is also tightly linked with progressive decline in specific cognitive domains.[49] Performance on the mini-mental state examination,[50] a widely used measure of global cognition in AD, is strongly correlated with gray matter volume and with cortical thickness in the entorhinal, parahippocampal, precuneus, superior parietal, and subgenual cingulate/orbitofrontal cortices.[51] More specific correlations have also been discovered, suggesting that specific cognitive domains may be differentially affected by AD-related atrophy in specific cortical regions. For instance, cortical atrophy in the anterior cingulate and supplementary motor cortices has been associated with apathy in patients with AD.[52] Furthermore, linear regression models, fitted at each location on the cortex, detected association between the degree of language impairment and atrophy of the left temporal and parietal cortices, regions critically involved in language production and comprehension.[53] Cognitive decline in multiple domains is typical in AD. Even so, neurodevelopmental studies support the notion that motor and language performance is selectively linked with cortical morphometry in the regions subserving those tasks. As such, 3D maps of cortical thickness or gray matter density have become widely accepted as a structural correlate of functional decline in AD.

c. Hippocampal Atrophy. Mapping tissue loss is also feasible in smaller brain regions, such as the hippocampus. The combination of manual outlining of the hippocampal boundary with a digital reconstruction technique (hippocampal radial mapping[49]) has allowed better localization of tissue loss throughout the surface of the hippocampus. When architectural data from a neuroanatomical atlas are overlaid on the MRI scans,[54] the regions with earliest atrophy on MRI correspond to those cytoarchitectural regions (CA1 and subiculum) that are affected earliest by pathology. Interestingly, regions known to be relatively spared by pathology (CA2–3) show minimal atrophy on MRI.[55,56] The same regions have been found atrophic, although less severely, in patients with MCI who would later develop AD during a 3-year follow-up interval.[57] As AD progresses, subregional hippocampal atrophy spreads in a pattern that follows the known trajectory of NFT dissemination, although the time lag is unknown between the pathological sequence and the changes that can be reliably detected on MRI. Some MRI studies are now being conducted at ultra-high magnetic fields (e.g., 4–7 Tesla), in an effort to detect cortical thinning earlier in the disease, and with greater spatial resolution and statistical power.[58,59]

d. Early-Onset Alzheimer's Disease. Frisoni *et al.*[60] compared cortical atrophy between early- and late-onset AD (EOAD, LOAD) and found that EOAD affected neocortical and LOAD medial temporal areas more heavily, compared with age-matched controls. Consistent with neuropsychological findings of greater memory impairment in LOAD and impaired neocortical functions in EOAD, these different atrophy patterns suggest different predisposing or etiologic factors. At comparable levels of cognitive decline, total gray matter loss was 19.5% in EOAD but only about half as great (11.9%) in LOAD, relative to appropriate matched controls.[55,60]

   e. *Pet Scanning.* For the past 30 years, brain changes in AD have been evaluated using radiotracer imaging methods, in conjunction with 3D computed tomography. Two such methods are PET and single photon emission-computed tomography (SPECT[61]). PET was originally developed as a method to use radiolabeled tracer molecules to assess glucose metabolism and cerebral blood flow. The distribution of these tracers in the brain is recovered using a scanner that detects radiation emitted by the tracer compound, using a cylindrical detector surrounding the patient. The 3D distribution of the reporter molecule is then reconstructed using mathematical methods such as the inverse Radon transform. More recently, amyloid- and NFT-sensitive probes have been developed, which can be used in conjunction with PET scanning to study the molecular pathology of AD.

   PET and SPECT studies of AD show a consistent pattern of focally decreased cerebral metabolism and perfusion, especially in posterior cingulate and neocortical association cortex, that largely spares basal ganglia, thalamus, cerebellum, and primary sensorimotor cortex.[62] The parietal lobe deficit tends to be more sensitive to disease severity than the temporal lobe deficit. Frontal lobe hypoperfusion is also often reported, but not in the absence of parieto-temporal abnormalities. A pattern of focal cortical inhomogeneities, all accounted for by areas of infarction on MRI, implies dementia secondary to cerebrovascular disease, which also often affects cerebellum and subcortical structures. A pattern of focal cortical inhomogeneities unmatched by MRI findings is consistent with a primary neurodegenerative disorder (e.g., AD, Pick's disease, other frontotemporal dementia (FTD), dementia with Lewy bodies (DLB), dementia of PD, Huntington's disease, progressive subcortical gliosis). As a general rule, the pattern of bilateral parietotemporal hypoperfusion or hypometabolism provides good discrimination of AD patients from age-matched normal controls, and from vascular dementia or frontal lobe dementia patients. Nuclear imaging abnormalities correlate with severity and specific patterns of cognitive failure in AD patients. They also correlate with regional densities of NFTs.[63]

B. Immunohistochemical Features of AD

   The major protein component of amyloid cores of SPs and cerebral vessel walls affected by CAA is Aβ. The immunohistochemical study of AD brain lesions began shortly after the partial peptide sequence of Aβ was determined.[31] Several groups developed antibodies specific for synthetic peptides of portions of Aβ.[64] Currently, numerous commercially available antibodies to Aβ (various amino acid lengths), tau, ubiquitin, α-synuclein (to detect Lewy bodies) and TDP-43 (TAR DNA binding protein-43) are available to facilitate accurate immunohistochemical characterization of a given autopsy brain specimen. With some exceptions, SPs are more prominently immunoreactive for the 42 amino acid forms of Aβ (Aβ42), whereas CAA immunolabels better with antibodies specific for Aβ40 (Fig. 9). Diffuse SPs are stained well by anti-Aβ antibodies, as are the amyloid cores of mature SPs. The neuritic coronas of mature SPs, NFTs and neuropil threads are prominently immunolabeled with anti-phospho-tau (Fig. 10). In cases of severe CAA, gamma-trace may also be found in affected vessel walls and a heavily infiltrated arteriole may even be surrounded by a "halo" of perivascular Aβ immunoreactivity or tau-immunoreactive neurites.[65]

C. Imaging Cerebral Amyloids

To better understand how amyloid deposits (load) accumulate in the living brain, Braskie et al.[13] examined 23 subjects (10 controls, 6 amnestic MCI, 7 AD) scanned with both MRI and [18F]FDDNP, a recently developed PET ligand sensitive to plaque and tangle pathology.[66,67] They aligned parametric PET images of amyloid load to MRI scans from the same subjects, mapped the PET signals onto the cortex, and combined them across subjects. Figure 11 shows two frames from a time-lapse sequence that shows the degree of amyloid burden at different levels of cognitive impairment. The advancing pathology follows the classical Braak trajectory for NFT accumulation. Related work by Mintun et al.[4] and Rowe et al.[68] with Pittsburgh Compound B ([11C]PIB) shows frontal lobe labeling early in the degenerative sequence. The PIB progression pattern is consistent with the Braak trajectory for amyloid deposition. Unlike tangle deposition, PIB shows early increases in the basal neocortex, particularly in frontal and temporal lobes and primarily in poorly myelinated regions such as the perirhinal cortex. These PET changes may occur at a much earlier stage of the disease than cortical thinning. Amyloid PET appears to be sensitive to pathological changes earlier than gray matter measures and is also correlated with subclinical cognitive decline, even in normal subjects.[13,69] Clearly, the sensitivity and specificity of changes in each imaging modality depends on the population studied, the sample size, and details of signal reconstruction, partial volume correction, and other image processing choices. As such, it is difficult to make absolute statements as to how the trajectories of various PET ligands (e.g., FDDNP vs. PIB) relate chronologically to each other and to cortical thinning, unless all measures are compared head-to-head in the same subjects, which has not yet been done.

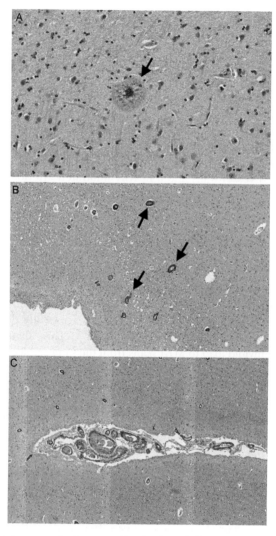

FIG. 9. ABeta-immunostained sections. Panel A (arrow) shows a neuritic SP, with a densely immunoreactive ABeta core surrounded by less prominent ABeta immunostained corona. Panel B shows prominently ABeta-immunoreactive profiles of CAA-affected arteries (arrows), predominantly in cerebral cortex. Panel C shows both meningeal and parenchymal (cortical) CAA, with prominently ABeta-immunorective staining of vessel walls. (See Color Insert.)

Nevertheless, amyloid-sensitive PET signal correlates with cognitive performance, even within the normal range, and with correlations in cortical areas that deteriorate earliest in AD, suggesting it may be useful for early diagnosis.

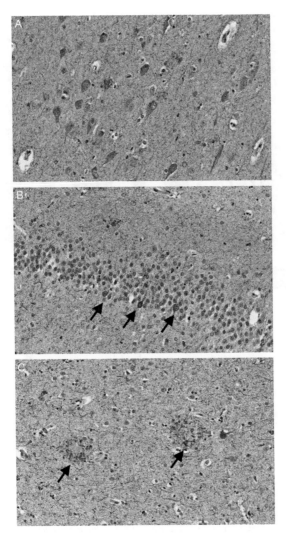

FIG. 10. All micrographs are from sections (autopsy brain) immunostained with anti-phospho-tau. Top panel shows tau-immunoreactive NFTs in cortex of an AD patient, as well as numerous surrounding neuropil threads. Note few pale-staining nonimmunoreactive neurons. Panel B shows prominently tau-immunoreactive neurons in hippocampal granule cell layer, a common finding in advanced AD. Panel C shows two SPs in which the neuritic component is prominently tau-immunoreactive. SP at left demonstrates a prominent central core which is not immunoreactive for tau but (in parallel sections) was prominently ABeta-immunoreactive. Neuropil threads are abundant in the neuropil. (See Color Insert.)

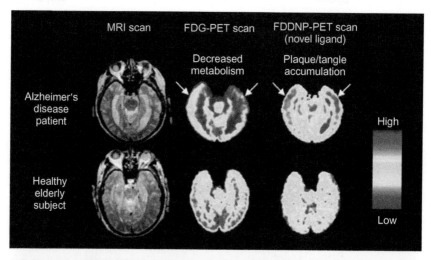

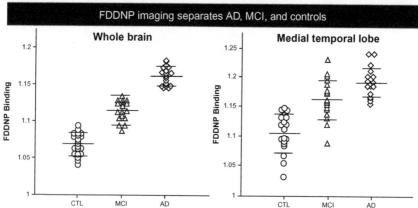

Fɪɢ. 11. Novel PET tracers for mapping Alzheimer's Disease. [^{18}F]FDDNP, a new PET tracer compound,[67,68] labels the molecular hallmarks of Alzheimer's disease—plaques and tangles—as they build up in the living brain. Representative scans (*top 2 rows*) are from an AD patient and a healthy elderly subject. Standard FDG-PET scans (*middle column*) show lower glucose metabolism in MCI and AD (*blue colors*). FDDNP scans *increase* in signal as the disease progresses, reflecting plaque and tangle accumulation (*red colors, last column*). Whole brain FDDNP binding differentiates AD, MCI and controls (*lower left*). MCI subjects already have AD-like pathology in the medial temporal lobe. (Adapted, with permission, from Ref. 67.) (See Color Insert.)

D. *Staging* AD and Quantifying Pathologic AD Lesions

Essentially, all AD lesions described above may be encountered in the cerebral cortex of cognitively normal elderly individuals, though tau-immunoreactive NFTs are usually found in the neocortex (isocortex) almost exclusively in demented subjects. It is useful to quantify these abnormalities and assess their topographic distribution. Correlations between lesion load, severity of neuropathologic findings, clinical neuropsychological symptoms, and ante-mortem neuroimaging abnormalities in a given patient are important, even essential. These correlations become somewhat problematic when the neuro-pathologist has the brain of an end-stage patient to examine, yet that patient may have experienced maximal neurologic deficit months or years before death and may have had the ultimate neuroimaging study (for reasons discussed above) months or even years prior to death.[70] When biopsies are carried out to confirm the diagnosis of AD, only a small portion of the brain is available for study. However, small clinical series have used biopsy and autopsy data from one and the same patient to monitor the progression of AD lesions over many years.[71] These studies have shown that there can be significant AD lesions in the brain of an individual who is, as judged by a reasonably high MMSE score, at a cogni-tively early stage of clinical symptoms. A more recent review of the use of brain biopsy in dementia diagnosis (from a large British tertiary referral center, The National Hospital for Neurology and Neurosurgery) showed that improved clinical and diagnostic selection criteria make such biopsies redundant on some occasions. When performed, they usually yielded a diagnosis of spongi-form encephalopathy (Creutzfeldt-Jakob disease, CJD), AD, or an inflammatory disorder (e.g., primary CNS vasculitis) or show nonspecific changes.[72] On the rare occasion when biopsy tissue becomes available from a demented patient, it should be triaged such that optimal use is made of it (e.g., small piece submitted for electron microscopy, small fragment snap frozen—assuming size is suffi-ciently large to first allow for a definitive diagnosis).

Many attempts have been made to standardize neuropathologic diagnostic criteria allowing one to distinguish AD or senile dementia of Alzheimer's type (SDAT) from normal aging.[42,73] A consensus conference in the 1980s resulted in the widely used "Khachaturian criteria" for the neuropathologic diagnosis of AD,[74] which were modified and updated by the Consortium to Establish a Registry for AD.[42,73] Braak criteria for AD severity[11,20] assume a progression of neuropathologic abnormalities (predominantly NFT and neuropil thread accu-mulation) from the transentorhinal cortex (stages I and II) to the hippocampus (III and IV), with ultimate widespread involvement of the neo-/isocortex (stages V and VI). Braak stages III and IV AD neuropathologic change is associated clinically with mild MCI but not overt dementia. However, brains of subjects with amnestic MCI (aMCI) are available for examination infre-quently and show a wide range of neuropathologic lesion density as well as coexistent heterogeneous ischemic-vascular lesions.[75,76]

In the late 1990s, the NIA-Reagan Institute Criteria for the Neuropathologic Diagnosis of AD came into widespread use and have been tested and operatio-nalized by various groups.[77] These criteria assign a "high," "intermediate," or "low" likelihood that a given individual's dementia was due to AD neuropathologic features. One study found a good correlation between a high NIA-Reagan "probability/likelihood" of AD and clinical dementia and concluded that the Khachaturian and CERAD criteria correlated fairly well with those of NIA-Reagan.[77] Occasional cases arise—especially among the "oldest old," for example, nonagenarians and centenarians—where Braak stage VI AD changes and an NIA-Reagan assessment of high likelihood of AD are clearly present in the brain of a subject who was known to be cognitively intact until shortly before death.[23] Several groups have suggested that this "clinicopathologic dissociation" (e.g., clinical dementia but minimal neuropathologic change; cognitively intact subject but abundant AD neuropathologic changes) is especially common in the oldest old.[78] Quantification of AD neuropathologic change is increasingly facilitated by the ability to produce digitized images of immunostained tissue sections, retain the images as a permanent electronic record of a given autopsy (no more worries about lost or misplaced, somewhat cumbersome glass slides!), and, if needed, use these digital images for further quantitative morphometry (Fig. 12).

E. Correlating Pathology to (Neuro)Imaging

Several unique studies have attempted to relate *postmortem* pathology, including histological and biochemical maps, to images collected *in vivo* from the same subject before their death. Mega *et al.*[79,80] created 2D histologic maps of NFT staining density using the Gallyas method, as well as lobar measures of soluble and insoluble Aβ, in a patient who had been scanned with whole brain anatomical MRI and FDG-PET before death. Using 3D cryosection imaging,[81,82] tissue was collected while preserving the capability to reconstruct histologically stained images into the original 3D conformation. After cryoprotection to allow microscopic analysis, the fixed brain was frozen and sectioned in the coronal plane in a large industrial cryomacrotome. The cryomacrotome enables image capture by a digital camera integral to the hydraulic descending blade allowing in-register capture of serial blockface images at constant magnification throughout whole-head sectioning.[81,83–85] A numbered grid overlay was placed on the cover-slipped section (Fig. 13) and each section was registered to the target section in the cryoimage dataset, and to the premortem MRI and 3D metabolic images. In one study,[79] NFT staining density, assessed using the Gallyas method, was localized to the paralimbic cortex of the basal forebrain, medial temporal, and orbital frontal regions. In the subject assessed, this was poorly correlated with hypometabolism on FDG-PET, suggesting alternate mechanisms for the metabolic defect in AD, including secondary effects of pathology elsewhere in the brain. A later study[80] found significant inverse correlations between FDG-PET measures and both soluble and

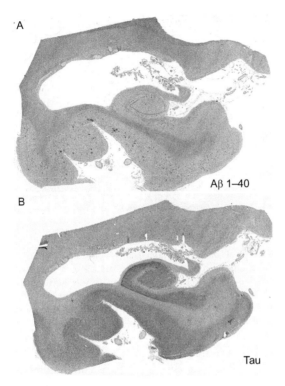

Fig. 12. Parallel sections of AD hippocampus immunostained with anti-ABeta (panel A) and anti-tau (panel B) to show differing patterns of immunoreactivity. Entire immunostained sections were digitized prior to further analysis. (See Color Insert.)

insoluble amyloid concentrations, except in temporal lobes. This suggests that pathology and metabolism are related in the AD brain at least in regions that are still relatively viable metabolically.[80]

II. Parkinsonian Signs and Symptoms ("Parkinsonism"), PD, and Diffuse Lewy Body Disease

Classic Parkinsonian signs and symptoms—a pill-rolling tremor, axial rigidity, festinating gait, etc.—may occur in a patient for many reasons, including as a complication of medication (extrapyramidal complications of phenothiazines), recreational drug use (inadvertent MPTP administration), or be a manifestation of a neurodegenerative disorder (though the "guilty" degenerative disease is not necessarily PD). Many disorders in the frontotemporal lobar degeneration (FTLD) spectrum (see below) may present with extrapyramidal

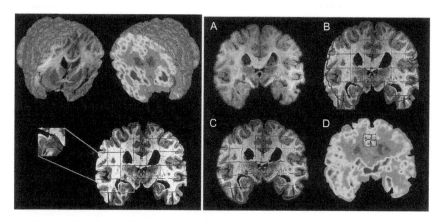

Fig. 13. Mapping biochemistry to metabolism. Here, 3D cryosection images (*top left*) are spatially aligned to match MRI and PET data from the same individual before their death [(d), *bottom right*[81,82]]. Anatomical landmarks in both datasets constrain the accurate geometric matching of the 3D reconstructed cryo-volumes to an *in vivo* MRI. Part A shows sectioned brain tissue that has been partitioned, in B, for biochemical assessment. By warping the tissue image back to the cryo blockface configuration (C), the biochemical measures can be related to *in vivo* assessments of metabolism (D). The deformed square (D) shows the location in the PET data from which a tissue sample originated. (Adapted, with permission, from Ref. 81,82.) (See Color Insert.)

features or affected individuals may manifest such signs and symptoms in the course of progression. Disorders in which this is common include corticobasal ganglionic degeneration (CBGD) and progressive supranuclear palsy (PSP) (see below). The diagnostic situation becomes even more complicated when a patient with extrapyramidal symptoms develops cognitive impairment, even dementia. The differential diagnosis of Parkinson's disease with dementia (PDD) includes coincident PD and AD, Diffuse Lewy Body Disease (DLBD), or another disorder, for example, in the FTLD spectrum. Parkinsonian clinical features correlate with nigrostriatal dopaminergic degeneration, the morphological correlate of which is neuronal loss in the substantia nigra pars compacta (SNpc).[86,87] Neuropathologic features of idiopathic PD include pigmented neuron loss (with attendant astrocytic gliosis) in the substantia nigra and other neuromelanin-containing nuclei (locus ceruleus and dorsal motor nucleus of the vagus nerve). Recently, a scheme that depicts progression of LB pathologic change in the brain, beginning in the brainstem (especially dorsal motor nucleus of the vagus and glossopharyngeal nerves and anterior olfactory nucleus) and progressing to the isocortex, has been proposed.[88] This "progression" is analogous to the accumulation of AD lesions within brain, proposed some 12 years earlier by Braak and Braak.[11]

Lewy bodies (Fig. 14) may occur in any brainstem nucleus, pigmented or not, for example, they are commonly seen in neurons of the pontine nuclei of the basis pontis. DLBD, which is almost always associated with dementia (it is

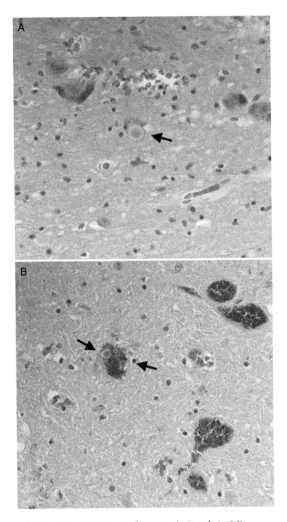

FIG. 14. Lewy bodies (LB; H&E-stained sections). Panel A: LBs are most easily seen in neuromelanin-containing neurons. Note glassy "magenta-colored" cytoplasmic inclusions, with an apparent surrounding "halo" (arrows). Neuron shown in Panel B contains two LBs (arrows). (See Color Insert.)

then sometimes better described as dementia with Lewy bodies, DLB), is clinically distinct from AD/SDAT, though the overlap between DLBD and AD neuropathologic change is striking (the authors of this chapter have rarely encountered a case of DLBD in which there was a total absence of some AD changes—more commonly the AD changes are quite advanced).[89] DLBD or DLB is characterized by fluctuating cognition with variations in attention and

alertness, neuroleptic sensitivity, recurrent visual hallucinations, and Parkinsonian features.[90] The diagnosis of PD is now refined using molecular genetic
evidence—and the knowledge that PD may be associated with LBs while others
are not. LBs and associated Lewy neurites (Fig. 15) are demonstrated by
immunohistochemistry using antibodies specific for either α-synuclein or ubiquitin. The major component of LBs is abnormally aggregated α-synuclein[91];
therefore, this is the optimal antibody to use. Anti-ubiquitin antibody has the
disadvantage of being an immunoreagent that will nonspecifically highlight AD
SPs and NFT, which, as noted, are often found in DLBD brains. Counting of
LBs is problematic and yields significant interobserver variability; therefore, a
consensus panel has recommended a semiquantitative scoring system for these
cytoplasmic inclusions using a scale from 0 (none) to 4 (abundant LBs and many
Lewy neurites).[90] Almost all individuals with DLBD have significant brainstem
pathologic changes of the type seen in idiopathic PD.

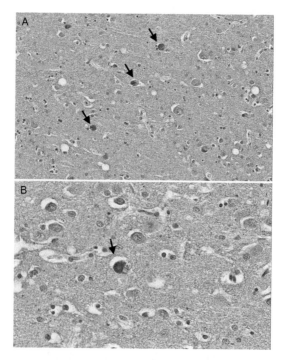

FIG. 15. LBs, alpha-synuclein immunohistochemistry. Micrographs are from sections of brain
obtained at autopsy on a subject with extensive diffuse Lewy body disease (DLBD). Note prominently immunoreactive LBs throughout the cortex, shown at low (A) and high (B) magnification.
Intraneuronal LBs are indicated by arrows. (See Color Insert.)

Dickson *et al.*[92] have recently proposed subclassifying major disorders associated with Parkinsonism using molecular pathological features, segregating them into α-synucleinopathies (e.g., PD, dementia with Lewy bodies), non-α-synucleinopathies (including tauopathies and TDP-43 proteinopathies), and nonspecific degeneration in the SNpc. α-Synuclein abnormalities are also implicated in non-DLBD/non-Parkinsonian disorders, including multiple system atrophy (MSA).[93] In MSA, α-synuclein-immunoreactive inclusions are often seen in nonneuronal cells, especially glia (*"glial cytoplasmic inclusions"* in oligodendroglia and astrocytes) in the presence of degeneration in multiple neuronal systems throughout the brain.[94] An interesting observation in recent years has been that α-synuclein abnormalities may also be prominent within the peripheral nervous system (including its autonomic components), possibly explaining some of the autonomic symptoms affected patients experience.[95]

Standard structural MRI and diffusion-weighted MRI are useful in the diagnosis of PD and related disorders, and both have been proposed for presymptomatic detection and staging of the disease. Neuronal loss in PD targets the dopamine cells in the substantia nigra. Atrophy of the *substantia nigra* on standard T1-weighted MRI correlates quite well with measures of striatal dopaminergic function using ^{18}F-DOPA PET.[96] Even so, the substantia nigra is relatively small. Accurate quantification requires high-field MRI. Brooks[97] noted that SNpc pathology is detectable using high-field MRI in the majority of idiopathic PD patients.[98]

Many imaging efforts in PD use PET scanning to assess the dopamine systems in the brain. For example, terminal dopa decarboxylase activity may be measured with ^{18}F-DOPA PET. Tropane-based PET and SPECT tracers have also been developed to measure the availability of presynaptic dopamine transporters. And finally, PET scanning with the tracer compound ^{11}C-dihydro-tetrabenazine may be used to examine vesicle monoamine transporter density in dopamine terminals.

Several ^{18}F-DOPA PET studies show lower tracer uptake in the putamen in PD, and this measure continues to decline over the course of the disease. ^{18}F-DOPA PET may also show hemispheric asymmetries in hemi-Parkinsonian patients, for whom only one side of their body is affected. There is also an anatomical sequence in the regions that show depleted tracer uptake in PD. As PD progresses, initial deficits tend to be found in the dorsal putamen but then spread to engulf the head of the caudate, as dopamine stores are progressively depleted. For clinical assessment, SPECT tracers are still widely used, as their uptake in the brain stabilizes more quickly than the PET tracers. This allows scans to be done on the same day as the tracer injection, which is logistically easier.

III. Frontotemporal Lobar Degeneration(s)

DLBD and diseases within the FTLD spectrum share clinical and even neuropathologic features with AD, that is, DLBD rarely occurs without some degree of "Alzheimerization" of the brain, and FTLDs are frequently associated with tau pathology, which is also a consistent component of the neuritic SPs and NFTs seen in AD/SDAT (see above). The morphoanatomical study of FTLD has become one of the most challenging areas of diagnostic neuropathology. The Lund and Manchester groups made seminal clinicopathologic observations and characterizations of this interesting group of entities, initially placing them under the rubric of "FTD".[99–101] Early studies of FTDs showed that many of the clinical entities described had neuropathologic similarities, even significant overlap—brain weight was slightly to moderately reduced, grossly the brain showed varying degrees of frontal or anterior temporal atrophy, neuronal loss, mild-to-moderate spongiform changes, and astrocytic gliosis were found, primarily in the first 2–3 layers of the cortex, though not in the transcortical pattern of spongiform encephalopathy, and gliosis (especially in regions of neuron loss) was easily visualized with immunohistochemical stains for glial fibrillary acidic protein. Cortical SPs and NFTs were not prominent and often were completely lacking. Subcortical structures, such as the substantia nigra, were sometimes abnormal, for example, depigmented.

FTLDs are now understood to encompass many disorders. They much more commonly have a genetic basis than does AD/SDAT, though the genes mutated vary. Conceptualizing their pathogenesis has been revolutionized by new genetic and immunohistochemical observations. Kumar-Singh and Van Broeckhoven[102] present an illuminating synthesis of these disorders, integrating details of their core clinical features and resultant syndromes (featuring various degrees of speech and extrapyramidal abnormalities), preferential regions of brain involvement, distinctive neuropathologic and related biochemical features, and genetics. There are prominent regions of clinical and neuropathologic overlap among the entities, as well as many cases that are difficult to subclassify, highlighting the importance of detailed and careful clinicopathologic correlation in patients who come to autopsy. Many FTLD patients show aphasia and behavioral abnormalities (including disinhibited behavior), extrapyramidal, and other motor disorders. Previously distinct nosologic entities incorporated into the FTLD family include FTLD with tau abnormalities (FTLD-tau, including FTD and Parkinsonism linked to chromosome 17 (FTDP-17)), Pick's disease (Fig. 16), CBGD, PSP,[103] and argyrophilic grain disease, FTLD-U with ubiquitin abnormalities, dementia lacking distinctive histology (DLDH),[104] and FTLD associated with motor neuron disease.

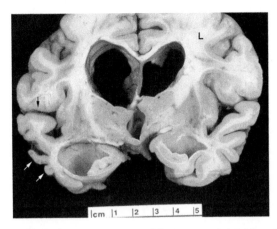

FIG. 16. Pick's disease, coronal slice of fixed brain, autopsy specimen. Note asymmetric atrophy predominantly of left middle and inferior temporal gyri (white arrows), with relative sparing of left superior temporal gyrus (black arrow) and entire right temporal lobe. There is *ex vacuo* enlargement of temporal horn of the left lateral ventricle and asymmetrical ventriculomegaly, left being substantially larger than the right. (For color version of this figure, the reader is referred to the Web version of this chapter.)

Some investigators argue for the inclusion of even more entities under this rubric, given their high frequency of tau pathology in the form of NFTs. These entities include "NFT- or tangle-predominant" AD, dementia pugilistica, multiple system tauopathy with dementia, and Parkinson–dementia complex of Guam (PDG). Genes in which mutations have been found to cause some of these disorders (especially FTDP-17) include microtubule-associated protein tau (*MAPT*). This subgroup of the FTLDs is often described using the term "tauopathies".[105,106] Other mutations described in FTLD families include those found in the genes charged multivesicular body protein 2b (*CHMP2B*), valosin-containing protein gene (*VCP*), and progranulin (*PGRN*).[107,108] *VCP* mutations encode a distinctive phenotype characterized by a frontal lobar degeneration with Paget's disease of the bone and inclusion body myopathy.

The neuropathologic work-up of these entities is incomplete without detailed immunohistochemical study. The proteins tau, ubiquitin, and (most recently) TDP-43 must be sought in brain tissue sections. TDP-43 often colocalizes with ubiquitin, and "immunopositivity" may be difficult to judge with absolute certainty as it is considered as such when positive signal is noted in the cytoplasm *rather than the nucleus* of a given neuron (Fig. 17). Some neuropathologists have suggested subclassifying all FTLDs as either tauopathies or ubiquitinopathies, depending upon which protein is detected in the brain.[109] A recent position paper has suggested a specific nomenclature for

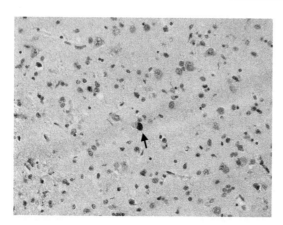

FIG. 17. TDP-43-immunoreactive cytoplasmic inclusions in a patient who had frontotemporal lobar degeneration, best characterized (using current nomenclature) as FTLD-U, with frequent ubiquitin-immunoreactive structures throughout the brain. Large TDP-43 inclusion is indicated by the arrow. (For color version of this figure, the reader is referred to the Web version of this chapter.)

neuropathologic subtypes of FTLD.[110] To many neuropathologists, the classic (though fairly rare) FTLD remains Pick disease, characterized by severe though often asymmetrical atrophy of the frontal and temporal lobes, involvement of the middle and inferior temporal gyri with sparing of a portion of the superior temporal gyrus, and intraneuronal "Pick bodies" (Fig. 16). The latter are found in abundance in the neocortex and hippocampus (including granule and pyramidal cell layer neurons). The significance of TDP-43 translocation from the neuronal nucleus to the cytoplasm (Fig. 17) as a marker for neurodegenerative disease (especially FTLD) is the subject of intense investigation.[111] TDP-43 immunoreactivity is frequently seen, for example, in the brains of patients with AD.[112]

Neuroimaging features of FTLD have been examined in a wide range of studies. In general, the most atrophic regions are the most affected by neuropathology. Differing profiles of cognitive impairments seen in patients also correspond quite well with the anatomical distribution of cortical atrophy observed on MRI.

FTLD accounts for only about 3–10% of cases of dementia,[113] and relative to Alzheimer's disease, FTLD patients show (as the name of this group of entities implies) a more focal degenerative profile targeting the frontal and anterior temporal lobes. Despite this stereotypical pattern, there are also typical neurodegenerative patterns for three of the clinical syndromes that are recognized as falling under the rubric of FTLD: FTD, progressive non-fluent aphasia (PA), and semantic dementia (SD). In a comparative, blinded study, Short et al.[114] compared the MRI scans of 59 FTLD patients and 26 AD patients and related the observed patterns of atrophy to behavioral and cognitive assessments. FTLD patients with altered personal conduct typically presented with significant bifrontal atrophy, but patients with SD also exhibited significant left temporal atrophy, in addition to frontal atrophy, as might be expected, given the dominant localization of classical language regions to the left hemisphere. Those with SD often exhibit atrophy in the middle and inferior temporal gyri. The same study also reported disinhibited behavior and hyperphagy in those with right frontal atrophy.

As in AD, profiles of frontal atrophy are relatively mild in the early stages of FTLD and may be missed on a reading of a standard T1-weighted clinical MRI scan. As a result, other functional neuroimaging methods, such as PET and SPECT, are often used to confirm the focal involvement of the frontal or temporal lobes.[115]

Other imaging studies have attempted to distinguish FTLD subtypes based on subtypes in the underlying pathology. Based on immunocytochemical staining intracellular inclusions, Kim et al.[116] examined two subgroups of patients. One group had tau-immunoreactive inclusions (FTLD-T), whereas the other had ubiquitin- and TDP-43-positive and tau-negative pathology (FTLD-U). Although the number of patients assessed was quite modest (7 with the T-subtype and 8 with the U-subtype), the FTLD-T and FTLD-U groups both showed atrophy in the frontal cortex and striatum, relative to matched controls ($N=61$). Striatal atrophy was more severe in FTLD-T. Such studies are important as they aim to distinguish between FTLD spectrum pathologic subtypes in vivo. As noted above, TDP-43 was identified as the major ubiquitinated protein associated with tau-negative FTLD.[117] Based on this distinction, Rabinovici and Miller[118] note that in patients with sporadic FTLD, SD is most often associated with TDP-43 inclusions, PA is frequently associated with one of the tau-inclusion disorders (in particular corticobasal degeneration), whereas patients with FTD are split between FTLD-T and FTLD-TDP. Distinguishing these subtypes with imaging is further complicated by the fact that 10–30% of patients presenting with an FTLD clinical syndrome are found to have AD at autopsy.[119] Typically, AD patients show posterior temporal and parietal atrophy on MRI, hypometabolism of the same regions on FDG-PET, and hypoperfusion of the same regions on SPECT scans. By contrast, FTLD patients typically show medical prefrontal changes. Although these group differences are marked after statistical averaging of imaging data from many subjects, for individual patients, the patterns are much less clear, due partly to the presence of multiple pathologies. One promising application of imaging involves the use of the Aβ-specific ligand ^{11}C-PIB to rule out the presence of Alzheimer pathology in patients who present with clinical symptoms of FTLD.

IV. Other (Miscellaneous) Disorders

We have focused on commonly encountered neurodegenerative disorders—AD/SDAT, DLBD, and touched briefly on the complex FTLD spectrum—and the "proteinopathies" that are biochemically and immunocytochemically relevant to their pathogenesis. In a neuropathology laboratory charged with characterizing and studying these disorders, use of antibodies to the proteins already mentioned will detect an estimated 90–95% of relevant diseases. One important feature that has been touched on only briefly is the need to characterize cerebrovascular comorbidity with parenchymal lesions, especially in the oldest old—a major consideration given that aging is the leading risk factor for cerebrovascular disease just as it is for SDAT.[43,44] The challenges of neuroimaging in predominantly ischemic-vascular, or mixed ischemic-vascular–parenchymal dementias, is in characterizing varying combinations of ischemic brain parenchymal abnormalities and the large arterial and microvascular diseases that cause or contribute to them. Hippocampal ischemic changes, especially common in those 80 years of age or older, have been implicated as contributing to cognitive impairment in the very old.[120] The hippocampal alterations resemble those that have been recognized for decades as being an important neuropathologic substrate of temporal lobe epilepsy (i.e., hippocampal sclerosis), with selective severe neuron loss and gliosis affecting the CA1 segment of the pyramidal cell layer, sometimes extending into the adjacent subiculum. A challenge for neuroimaging will be to develop criteria that can distinguish the resultant "collapse" of portions of the hippocampal formation from the neuron loss and atrophy that is a defining feature of hippocampal changes seen in AD/SDAT.

We also have not mentioned the importance of using antibodies to prion protein on tissue sections for confirming the diagnosis of transmissible spongiform encephalopathy (TSE, CJD) when this is suspected. When one encounters a case of possible TSE, the resources of the National Prion Disease Pathology Surveillance Center at Case Western Reserve University in Cleveland, Ohio, are also invaluable in work-up of the necropsy brain for neuropathologists and other investigators based in the USA.

Other rare neurodegenerative diseases must be considered. These include neuronal intermediate filament inclusion disease, which some now group with the FTLDs.[121] In some disorders with abnormal neuronal intranuclear inclusions, these inclusions can react with antibodies specific for a nuclear protein, "fused-in-sarcoma". Another very rare disease (never knowingly encountered by the authors) is basophilic inclusion body disease.

V. Conclusion and Future Directions

The full neuropathologic characterization of new and challenging types of neurodegenerative disease will provide "full employment" for neuropathologists in the years to come. Not only will they be charged with characterizing abnormal "shadows" and signals (including many observed with the use of novel ligands) detected by increasingly sophisticated neuroradiologists, but they will be expected to provide feedback to clinicians on how well therapies aimed at removing abnormal proteins from the brain have worked. New genes contributing to the pathogenesis of neurodegenerative diseases are almost certain to be discovered. In some cases, the relevant gene products may be of importance in understanding structural changes within the brain of a demented patient.

Radiographic-pathologic correlation will be more important than ever in drawing conclusions about which biochemical types of amyloid are being demonstrated using a given probe. A further question is: will removing abnormal brain proteins (or preventing their build-up within the brain) lead to clinical and neuropsychological improvement in patients or slow the progression of their cognitive impairment? This will be the crucible in which novel therapeutic strategies will need to be tested. Nonradiographic biomarkers (e.g., CSF levels of Aβ and phospho-tau protein) are increasingly utilized to predict the likelihood that a subject has a dementing condition. Neuropathologic observations will be useful in providing feedback to clinicians on which morphologic parameters correlate best with changes in a given biomarker.[122]

Acknowledgments

Work in HV Vinters' laboratory supported in part by PHS Grant P50 AG16570, P01 AG12435 and the Daljit S. & Elaine Sarkaria Chair in Diagnostic Medicine. Carol Appleton, Alexander Bottini, and Spencer Tung assisted with preparation of the chapter, including illustrations. Work in Paul Thompson's laboratory is supported by NIH Grants U01 AG024904, R01 EB008432, R01MH089722, P41 RR013642, R01 EB008281, and R01 EB007813.

References

1. (a)Kovacs GG, Budka H. Current concepts of neuropathological diagnostics in practice: neurodegenerative diseases. *Clin Neuropathol* 2010;**29**:271–88; (b)Wolfe DL. To see for one's self. The art of autopsy has a long history and an uncertain figure. *Am Sci* 2010;**98**: 228–35.
2. Goedert M, Ghetti B. Alois Alzheimer: his life and times. *Brain Pathol* 2007;**17**:57–62.

3. Vinters HV, Farrell MA, Mischel PS, Anders KH. *Diagnostic neuropathology.* New York: Marcel Dekker; 1998. pp. 453–507.

4. Mintun MA, LaRossa GN, Sheline YI, Dence CS, Lee SY, Mach RH, Klunk WE, Mathis CA, DeKosky ST, Morris JC. [^{11}C]PIB in a nondemented population. Potential antecedent marker of Alzheimer disease. *Neurology* 2006;**67**:446–52.

5. Small GW, Komo S, LaRue A, Saxena S, Phelps ME, Mazziotta JC, Saunders AM, Haines JL, Pericak-Vance MA, Roses AD. Early detection of Alzheimer's disease by combining apolipo-protein E and neuroimaging. *Ann N Y Acad Sci* 1996;**802**:70–8.

6. Small GW, Kepe V, Ercoli LM, Siddarth P, Bookheimer SY, Miller KJ, Lavretsky H, Burggren AC, Cole GM, Vinters HV, Thompson PM, Huang S-C, Satyamurthy N, Phelps ME, Barrio JR. PET of brain amyloid and tau in mild cognitive impairment. *N Engl J Med* 2006;**355**:2652–63.

7. Vinters HV. Imaging cerebral microvascular amyloid. *Ann Neurol* 2007;**62**:209–12.

8. Querfurth HW, LaFerla FM. Alzheimer's disease. *N Engl J Med* 2010;**362**:329–44.

9. Kovacs GG, Botond G, Budka H. Protein coding of neurodegenerative dementias: the neuropathological basis of biomarker diagnostics. *Acta Neuropathol* 2010;**119**:389–408.

10. Joachim CL, Morris JH, Selkoe DJ. Clinically diagnosed Alzheimer's disease: autopsy results in 150 cases. *Ann Neurol* 1988;**24**:50–6.

11. Braak H, Braak E. Neuropathological staging of Alzheimer related changes. *Acta Neuropathol* 1991;**82**:239–59.

12. Thompson PM, Hayashi KM, de Zubicaray G, Janke AL, Rose SE, Semple J, et al. Dynamics of gray matter loss in Alzheimer's disease. *J Neurosci* 2003;**23**(3):994–1005.

13. Braskie MN, Klunder AD, Hayashi KM, Protas H, Kepe V, Miller KJ, Huang SC, Barrio JR, Ercoli LM, Siddarth P, Satyamurthy N, Liu J, Toga AW, Bookheimer SY, Small GW, Thompson PM. Plaque and tangle imaging and cognition in normal aging and Alzheimer's disease. *Neurobiol Aging* 2010;**31**(10):1669–78. [Epub 2008 Nov 11].

14. Reisberg B, Franssen EH, Hasan SM, Monteiro I, Boksay I, Souren LE, et al. Retrogenesis: clinical, physiologic, and pathologic mechanisms in brain aging, Alzheimer's and other dementing processes. *Eur Arch Psychiatry Clin Neurosci* 1999;**249**(Suppl. 3):28–36.

15. Conel JL. *The postnatal development of the human cerebral cortex: VIII: the cortex of the six-year child.* vol. VIII. Cambridge: Harvard University Press; 1967.

16. Yakovlev PI, Lecours AR. *Regional development of the brain in early life.* Boston: Blackwell Scientific Publications; 1967.

17. Mesulam M. Brain, mind, and the evolution of connectivity. *Brain Cogn* 2000;**42**:4–6.

18. Bartzokis G, Lu PH, Mintz J. Quantifying age-related myelin breakdown with MRI: novel therapeutic targets for preventing cognitive decline and Alzheimer's disease. *J Alzheimers Dis* 2004;**6**:S53–9.

19. (a)Gogtay N, Giedd JN, Lusk L, Hayashi KM, Greenstein D, Vaituzis AC, et al. Dynamic mapping of human cortical development during childhood through early adulthood. *Proc Natl Acad Sci USA* 2004;**101**(21):8174–9; (b)Braak H, Braak E, Grundke-Iqbal I, Iqbal K. Occurrence of neuropil threads in the senile human brain and in Alzheimer's disease: a third location of paired helical filaments outside of neurofibrillary tangles and neuritic plaques. *Neurosci Lett* 1986;**65**:351–5.

20. Braak H, Duyckaerts C, Braak E, Piette F. Neuropathological staging of Alzheimer-related changes with psychometrically assessed intellectual status. In: Corain B, Iqbal K, Nicolini M, Winblad B, Wisniewski H, Zatta P, editors. *Alzheimer's disease: advances in clinical and basic research.* Chichester: Wiley; 1993. pp. 131–7.

21. Dickson DW. The pathogenesis of senile plaques. *J Neuropathol Exp Neurol* 1997;**56**:321–39.

22. Blessed G, Tomlinson BE, Roth M. The association between quantitative measures of dementia and of senile change in the cerebral grey matter of elderly subjects. *Br J Psychiatry* 1968;**117**:797–811.

23. Berlau DJ, Kahle-Wrobleski K, Head E, Goodus M, Kim R, Kawas C. Dissociation of neuropathologic findings and cognition. Case report of an apolipoprotein E epsi2/epsi2 genotype. *Arch Neurol* 2007;**64**:1193–6.

24. Dickson DW, editor. *Neurodegeneration: the molecular pathology of dementia and movement disorders*. Basel: ISN Press; 2003. 414p.

25. Vinters HV, Secor DL, Read SL, Frazee JG, Tomiyasu U, Stanley TM, Ferreiro JA, Akers MA. Microvasculature in brain biopsy specimens from patients with Alzheimer's disease: an immunohistochemical and ultrastructural study. *Ultrastruct Pathol* 1994;**18**:333–48.

26. Braak H, Braak E. Neuropil threads occur in the dendrites of tangle-bearing nerve cells. *Neuropathol Appl Neurobiol* 1988;**14**:39–44.

27. Wisniewski K, Jervis GA, Moretz RC, Wisniewski HW. Alzheimer neurofibrillary tangles in disease other than senile and presenile dementia. *Ann Neurol* 1979;**5**:288–94.

28. Mischel PS, Nguyen LP, Vinters HV. Cerebral cortical dysplasia associated with pediatric epilepsy. Review of neuropathologic features and proposal for a grading system. *J Neuropathol Exp Neurol* 1995;**54**:137–53.

29. Duong T, DeRosa MJ, Poukens V, Vinters HV, Fisher RS. Neuronal cytoskeletal abnormalities in human cerebral cortical dysplasia. *Acta Neuropathol* 1994;**87**:493–503.

30. Vinters HV. Cerebral amyloid angiopathy. A critical review. *Stroke* 1987;**18**:311–24.

31. (a)Glenner GG, Wong CW. Alzheimer's disease: initial report of the purification and characterization of a novel cerebrovascular amyloid protein. *Biochem Biophys Res Commun* 1984;**120**:885–90; (b)Sipe JD, Benson MD, Buxbaum JN, Ikeda S, Merlini G, Saraiva MJM, Westermark P. Amyloid fibril protein nomenclature: 2010 recommendations from the nomenclature committee of the International Society of Amyloidosis. *Amyloid J Protein Folding Disord* 2010;**17**:101–4.

32. Vinters HV, Gilbert JJ. Cerebral amyloid angiopathy: incidence and complications in the aging brain, II: the distribution of amyloid vascular changes. *Stroke* 1983;**14**:924–8.

33. Miyakawa T, Kimura T, Hirata S, Fujise N, Ono T, Ishizuka K, Nakabayashi J. Role of blood vessels in producing pathological changes in the brain with Alzheimer's disease. *Ann NY Acad Sci* 2000;**903**:46–54.

34. Soontornniyomkij V, Choi C, Pomakian J, Vinters HV. High-definition characterization of cerebral beta-amyloid angiopathy in Alzheimer's disease. *Hum Pathol* 2010;**41**:1601–8.

35. Weller RO, Boche D, Nicoll JAR. Microvasculature changes and cerebral amyloid angiopathy in Alzheimer's disease and their potential impact on therapy. *Acta Neuropathol* 2009;**118**:87–102.

36. Miners JS, Kehoe P, Love S. Neprilysin protects against cerebral amyloid -angiopathy and ABeta-induced degeneration of cerebrovascular smooth muscle cells. *Brain Pathol* 2011;**21**:594–605. doi: 10.1111/j.1750-3639.2011.00486.x.

37. Zhang-Nunes SX, Maat-Schieman MLC, van Duinen SG, Roos RAC, Frosch MP, Greenberg SM. The cerebral beta-amyloid angiopathies: hereditary and sporadic. *Brain Pathol* 2006;**16**:30–9.

38. Soontornniyomkij V, Lynch MD, Mermash S, Pomakian J, Badkoobehi H, Clare R, Vinters HV. Cerebral microinfarcts associated with severe cerebral beta-amyloid angiopathy. *Brain Pathol* 2010;**20**:459–67.

39. Clare R, King VG, Wirenfeldt M, Vinters HV. Synapse loss in dementias. *J Neurosci Res* 2010;**88**:2083–90.

40. (a)Terry RD, Masliah E, Salmon DP, Butters N, DeTeresa R, et al. Physical basis of cognitive alterations in Alzheimer's disease: synapse loss is the major correlate of cognitive impairment. *Ann Neurol* 1991;**30**:572–80; (b)Davidsson P, Blennow K. Neurochemical dissection of synaptic pathology in Alzheimer's disease. *Int Psychogeriatr* 1998;**10**:11–23.

41. Fu C, Chute DJ, Farag ES, Garakian J, Cummings JL, Vinters HV. Comorbidity in dementia—an autopsy study. *Arch Pathol Lab Med* 2004;**128**:32–8.

42. Gearing M, Mirra SS, Hedreen JC, Sumi SM, Hansen LA, Heyman A. The Consortium to Establish a Registry for Alzheimer's Disease (CERAD). Part X. Neuropathology confirmation of the clinical diagnosis of Alzheimer's disease. *Neurology* 1995;**45**:461–6.

43. Vinters HV, Ellis WG, Zarow C, Zaias BW, Jagust WJ, Mack WJ, Chui HC. Neuropathologic substrates of ischemic vascular dementia. *J Neuropathol Exp Neurol* 2000;**59**:931–45.

44. Selnes OA, Vinters HV. Vascular cognitive impairment. *Nat Clin Pract Neurol* 2006;**2**:538–47.

45. Beckett LA, Harvey DJ, Gamst A, Donohue M, Kornak J, Zhang H, Kuo JH. The Alzheimer's Disease Neuroimaging Initiative: annual change in biomarkers and clinical outcomes. *Alzheimers Dement* 2010;**6**(3):257–64.

46. Petersen RC, Parisi JE, Dickson DW, Johnson KA, Knopman DS, Boeve BF, et al. Neuropathologic features of amnestic mild cognitive impairment. *Arch Neurol* 2006;**63**:665–72.

47. Becker JT, Davis SW, Hayashi KM, Meltzer CC, Toga AW, Lopez OL, et al. Three-dimensional patterns of hippocampal atrophy in mild cognitive impairment. *Arch Neurol* 2006;**63**(1):97–101.

48. Bell-McGinty S, Butters MA, Meltzer CC, Greer PJ, Reynolds 3rd CF, Becker JT. Brain morphometric abnormalities in geriatric depression: long-term neurobiological effects of illness duration. *Am J Psychiatry* 2002;**159**(8):1424–7.

49. Thompson PM, Hayashi KM, de Zubicaray GI, Janke AL, Rose SE, Semple J, Hong MS, Herman DH, Gravano D, Doddrell DM, Toga AW. Mapping hippocampal and ventricular change in Alzheimer disease. *Neuroimage* 2004;**22**:1754–66.

50. Folstein MF, Folstein SE, McHugh PR. Mini-mental state. A practical method for grading the cognitive state of patients for the clinician. *J Psychiatr Res* 1975;**12**:189–98.

51. Apostolova LG, Lu PH, Rogers S, Dutton RA, Hayashi KM, Toga AW, et al. 3D mapping of mini-mental state examination performance in clinical and preclinical Alzheimer disease. *Alzheimer Dis Assoc Disord* 2006;**20**(4):224–31.

52. Apostolova LG, Akopyan GG, Partiali N, Steiner CA, Dutton RA, Hayashi KM, et al. Structural correlates of apathy in Alzheimer's disease. *Dement Geriatr Cogn Disord* 2007;**24**(2):91–7.

53. Apostolova LG, Lu PH, Rogers S, Dutton RA, Hayashi KM, Toga AW, et al. *3D mapping of language networks in clinical and pre-clinical Alzheimer's disease. Brain Lang* 2007; http://dx.doi.org/10.1016/j.bandl.2007.03.008.

54. Duvernoy HM. *The human hippocampus: an atlas of applied anatomy.* Munich: Bergmann; 1988.

55. Frisoni GB, Sabattoli F, Lee AD, Dutton RA, Toga AW, Thompson PM. In vivo neuropathology of the hippocampal formation in AD: a radial mapping MR-based study. *Neuroimage* 2006;**32**(1):104–10.

56. Csernansky JG, Wang L, Joshi S, Miller JP, Gado M, Kido D, et al. Early DAT is distinguished from aging by high-dimensional mapping of the hippocampus. Dementia of the Alzheimer type. *Neurology* 2000;**55**(11):1636–43.

57. Apostolova LG, Dutton RA, Dinov ID, Hayashi KM, Toga AW, Cummings JL, et al. Conversion of mild cognitive impairment to Alzheimer disease predicted by hippocampal atrophy maps. *Arch Neurol* 2006;**63**(5):693–9.

58. Augustinack JC, van der Kouwe AJ, Blackwell ML, Salat DH, Wiggins CJ, Frosch MP, et al. Detection of entorhinal layer II using 7Tesla [corrected] magnetic resonance imaging. *Ann Neurol* 2005;**57**(4):489–94.

59. Mueller SG, Stables L, Du AT, Schuff N, Truran D, Cashdollar N, et al. Measurement of hippocampal subfields and age-related changes with high resolution MRI at 4T. *Neurobiol Aging* 2007;**28**(5):719–26.

60. Frisoni GB, Pievani M, Testa C, Sabattoli F, Bresciani L, Bonetti M, et al. The topography of grey matter involvement in early and late onset Alzheimer's disease. *Brain* 2007;**130**(Pt. 3):720–30.

61. Silverman DHS, Devous Sr. MD. PET and SPECT imaging in evaluating Alzheimer's disease and related dementias. In: Ell PJ, Gambhir SS, editors. *Nuclear medicine in clinical diagnosis and treatment*. 3rd ed. London: Churchill-Livingstone; 2004.

62. Silverman DHS, Thompson PM. Structural and functional neuroimaging: focusing on mild cognitive impairment. *Appl Neurol* 2006;**2**(2).

63. DeCarli C, Atack JR, Ball MJ, Kay JA, Grady CL, Fewster P, Pettigrew KD, Rapoport SI, Schapiro MB. Post-mortem regional neurofibrillary tangle densities are related to regional cerebral metabolic rates for glucose during life in Alzheimer's disease patients. *Neurodegeneration* 1992;**1**:113–21.

64. Vinters HV, Pardridge WM, Yang J. Immunohistochemical study of cerebral amyloid angiopathy. Use of an antiserum to a synthetic 28-amino-acid peptide fragment of the Alzheimer's disease amyloid precursor. *Hum Pathol* 1988;**19**:214–22.

65. Vinters HV, Nishimura GS, Secor DL, Pardridge WM. Immunoreactive A4 and gamma-trace peptide co-localization in amyloidotic arteriolar lesions in the brains of patients with Alzheimer's disease. *Am J Pathol* 1990;**137**:233–40.

66. Kepe V, Huang SC, Small GW, Satyamurthy N, Barrio JR. Visualizing pathology deposits in the living brain of patients with Alzheimer's disease. *Methods Enzymol* 2006;**412**:144–60.

67. Small GW, Bookheimer SY, Thompson PM, Cole GM, Huang SC, Kepe V, Barrio JR. Current and future uses of neuroimaging for cognitively impaired patients. *Lancet Neurol* 2008;**7**(2): 161–72.

68. Rowe CC, Ng S, Ackermann U, Gong SJ, Pike K, Savage G, et al. Imaging beta-amyloid burden in aging and dementia. *Neurology* 2007;**68**(20):1718–25.

69. Jack Jr. CR, Bernstein MA, Borowski BJ, Gunter JL, Fox NC, Thompson PM, Schuff N, Krueger G, Killiany RJ, Decarli CS, Dale AM, Carmichael OW, Tosun D, Weiner MW. Alzheimer's disease neuroimaging initiative. Update on the magnetic resonance imaging core of the Alzheimer's disease neuroimaging initiative. *Alzheimers Dement* 2010;**6**(3): 212–20 Review.

70. Galasko D, Hansen LA, Katzman R, Wiederholt W, Masliah E, Terry R, Hill R, Lessin P, Thal LJ. Clinical-neuropathological correlations in Alzheimer's disease and related dementias. *Arch Neurol* 1994;**51**:888–95.

71. Di Patre PL, Read SL, Cummings JL, Tomiyasu U, Vartavarian LM, Secor DL, Vinters HV. Progression of clinical deterioration and pathological changes in patients with Alzheimer disease evaluated at biopsy and autopsy. *Arch Neurol* 1999;**56**:1254–61.

72. Schott JM, Reiniger L, Thom M, Holton JL, Grieve J, Brandner S, Warren JD, Revesz T. Brain biopsy in dementia: clinical indications and diagnostic approach. *Acta Neuropathol* 2010;**120**:327–41.

73. Mirra SS, Heyman A, McKeel D, Sumi SM, Crain BJ, Brownlee LM, Vogel FS, Hughes JP, Vanbelle G, Berg L. The Consortium to Establish a Registry for Alzheimer's Disease (CERAD). Part II. Standardization of the neuropathologic assessment of Alzheimer's disease. *Neurology* 1991;**41**:479–86.

74. Khachaturian ZS. Diagnosis of Alzheimer's disease. *Arch Neurol* 1985;**42**:1097–105.

75. Petersen RC, Parisi JE, Dickson DW, Johnson KA, Knopman DS, Boeve BF, Jicha GA, Ivnik RJ, Smith GE, Tangalos EG, Braak H, Kokmen E. Neuropathologic features of amnestic mild cognitive impairment. *Arch Neurol* 2006;**63**:665–72.

76. Vinters HV. Neuropathology of amnestic mild cognitive impairment. *Arch Neurol* 2006;**63**:645–6.

77. Newell KL, Hyman BT, Growdon JH, Hedley-Whyte ET. Application of the National Institute on Aging (NIA)-Reagan Institute criteria for the neuropathological diagnosis of Alzheimer disease. *J Neuropathol Exp Neurol* 1999;**58**:1147–55.

78. Kawas CH. The oldest old and the 90+ study. *Alzheimers Dement* 2008;**4**:S56–9.

79. Mega MS, Chen S, Thompson PM, Woods RP, Karaca TJ, Tiwari A, Vinters H, Small GW, Toga AW. Mapping pathology to metabolism: coregistration of stained whole brain sections to PET in Alzheimer's disease. *Neuroimage* 1997;**5**:147–53.

80. Mega MS, Chu T, Mazziotta JC, Trivedi KH, Thompson PM, Shah A, Cole G, Frautschy SA, Toga AW. Mapping biochemistry to metabolism: FDG-PET and beta-amyloid burden in Alzheimer's disease. *Neuroreport* 1999;**10**(14):2911–7.

81. Toga AW, Ambach KL, Quinn B, Shankar K, Schluender S. Postmortem anatomy. In: Toga AW, Mazziotta JC, editors. *Brain mapping: the methods*. San Diego: Academic Press; 1996. pp. 169–90.

82. Toga AW, Mazziotta JC, Woods RP. A high-resolution anatomic reference for PET activation studies. *Neuroimage* 1995;**2**(2).

83. Toga AW. Visualization and Warping of Multimodality Brain Imagery. In: Thatcher RW, Hallett M, Zeffiro T, John ER, Huerta M, editors. *Functional Neuroimaging: Technical Foundations* 1994. pp. 171–80.

84. Annese J, Pitiot A, Dinov ID, Toga AW. A myelo-architectonic method for the structural classification of cortical areas. *Neuroimage* 2004;**21**:15–26.

85. Annese J, Gazzaniga MS, Toga AW. Localization of the human cortical visual area MT based on computer aided histological analysis. *Cereb Cortex* 2004;**15**:1044–53.

86. Dickson DW, Fujishiro H, Orr C, DelleDonne A, Josephs KA, Frigerio R, Burnett M, Parisi JE, Klos KJ, Ahlskog JE. Neuropathology of non-motor features of Parkinson disease. *Parkinsonism Relat Disord* 2009;**15S3**:S1–5.

87. Forno LS. Neuropathology of Parkinson's disease. *J Neuropathol Exp Neurol* 1996;**55**:259–72.

88. Braak H, Del Tredici K, Rub U, de Vos RAI, Steur ENHJ, Braak E. Staging of brain pathology related to sporadic Parkinson's disease. *Neurobiol Aging* 2003;**24**:197–211.

89. Kalaitzakis ME, Pearce RKB. The morbid anatomy of dementia in Parkinson's disease. *Acta Neuropathol* 2009;**118**:587–98.

90. McKeith IG, Dickson DW, Lowe J, Emre M, O'Brien JT, Feldman H, Cummings J, Duda JE, Lippa C, Perry EK, Aarsland D, Arai H, Ballard CG, et al. Diagnosis and management of dementia with Lewy bodies. Third report of the DLB consortium. *Neurology* 2005;**65**:1863–72.

91. Maries E, Dass B, Collier TJ, Kordower JH, Steece-Collier K. The role of alpha-synuclein in Parkinsons's disease: insights from animal models. *Nat Rev Neurosci* 2003;**4**:727–38.

92. Dickson DW, Braak H, Duda JE, Duyckaerts C, Gasser T, Halliday GM, Hardy J, Leverenz JB, Del Tredici K, Wszolek ZK, Litvan I. Neuropathological assessment of Parkinson's disease: refining the diagnostic criteria. *Lancet Neurol* 2009;**12**:1150–7.

93. Trojanowski JQ, Revesz T. for the Neuropathology Working Group on MSA. Proposed neuropathological criteria for the *post mortem* diagnosis of multiple system atrophy. *Neuropathol Appl Neurobiol* 2007;**33**:615–20.

94. Apostolova LG, Klement I, Bronstein Y, Vinters HV, Cummings JL. Multiple system atrophy presenting with language impairment. *Neurology* 2006;**67**:726–7.

95. Wakabayashi K, Mori F, Tanji K, Orimo S, Takahashi H. Involvement of the peripheral nervous system in synucleinopathies, tauopathies, and other neurodegenerative proteinopathies of the brain. *Acta Neuropathol* 2010;**120**:1–12.

96. Hu MT, et al. A comparison of ^{18}F-dopa PET and inversion recovery MRI in the diagnosis of Parkinson's disease. *Neurology* 2001;**56**:1195–200.

97. Brooks DJ. Neuroimaging in Parkinson's disease. *NeuroRx* 2004;**1**(2):243–54.

98. Hutchinson M, Raff U. Structural changes of the substantia nigra in Parkinson's disease as revealed by MR imaging. *AJNR Am J Neuroradiol* 2000;**21**:697–701.

99. Snowden JS, Neary D, Mann DMA, Goulding PJ, Testa HJ. Progressive language disorder due to lobar atrophy. *Ann Neurol* 1992;**31**:174–83.

100. Neary D, Snowden JS, Northen B, Goulding P. Dementia of frontal lobe type. *J Neurol Neurosurg Psychiatry* 1988;**51**:353–61.
101. Brun A, Englund B, Gustafson L, Passant U, Mann DMA, Neary D, Snowden JS. Clinical and neuropathological criteria for frontotemporal dementia. *J Neurol Neurosurg Psychiatry* 1994;**57**:416–8.
102. Kumar-Singh S, Van Broeckhoven C. Frontotemporal lobar degeneration: current concepts in the light of recent advances. *Brain Pathol* 2007;**17**:104–13.
103. Dickson DW, Rademakers R, Hutton ML. Progressive supranuclear palsy: pathology and genetics. *Brain Pathol* 2007;**17**:74–82.
104. Knopman DS, Mastri AR, Frey WH, Sung JH, Rustan T. Dementia lacking distinctive histologic features: a common non-Alzheimer degenerative dementia. *Neurology* 1990;**40**:251–6.
105. Hernandez F, Avila J. Tauopathies. *Cell Mol Life Sci* 2007;**64**:2219–33.
106. van Swieten J, Spillantini MG. Hereditary frontotemporal dementia caused by Tau gene mutations. *Brain Pathol* 2007;**17**:63–73.
107. Mackenzie IRA, Baker M, Pickering-Brown S, Hsiung GYR, Lindholm C, Dwosh E, Gass J, Cannon A, Rademakers R, Hutton M, Feldman HH. The neuropathology of frontotemporal lobar degeneration caused by mutations in the progranulin gene. *Brain* 2006;**129**:3081–90.
108. Gass J, Cannon A, Mackenzie IR, Boeve B, Baker M, Adamson J, et al. Mutations in progranulin are a major cause of ubiquitin-positive frontotemporal lobar degeneration. *Hum Mol Genet* 2006;**15**(20):2988–3001.
109. Bigio EH. Update on recent molecular and genetic advances in frontotemporal lobar degeneration. *J Neuropathol Exp Neurol* 2008;**67**:635–48.
110. Mackenzie IRA, Neumann M, Bigio EH, Cairns NJ, Alafuzoff I, et al. Nomenclature for neuropathologic subtypes of frontotemporal lobar degeneration: consensus recommendations. *Acta Neuropathol* 2009;**117**:15–8.
111. Armstrong RA, Ellis W, Hamilton RL, Mackenzie IRA, Hedreen J, Gearing M, Montine T, Vonsattel J-P, Head E, Lieberman AP, Cairns NJ. Neuropathological heterogeneity in frontotemporal lobar degeneration with TDP-43 proteinopathy: a quantitative study of 94 cases using principal components analysis. *J Neural Transm* 2010;**117**:227–39.
112. Wilson AC, Dugger BN, Dickson DW, Wang D-S. TDP-43 in aging and Alzheimer's disease—a review. *Int J Clin Exp Pathol* 2011;**4**:147–55.
113. Heutink P. Untangling tau-related dementia. *Hum Mol Genet* 2000;**9**(6):979–86.
114. Short RA, Broderick DF, Patton A, Arvanitakis Z, Graff-Radford NR. Different patterns of magnetic resonance imaging atrophy for frontotemporal lobar degeneration syndromes. *Arch Neurol* 2005;**62**(7):1106–10.
115. Graff-Radford NR, Russell JW, Rezai K. Frontal degenerative dementia and neuroimaging. *Adv Neurol* 1995;**66**:37–47 discussion 47-50.
116. Kim EJ, Rabinovici GD, Seeley WW, Halabi C, Shu H, Weiner MW, DeArmond SJ, Trojanowski JQ, Gorno-Tempini ML, Miller BL, Rosen HJ. Patterns of MRI atrophy in tau positive and ubiquitin positive frontotemporal lobar degeneration. *J Neurol Neurosurg Psychiatry* 2007;**78**(12):1375–8 [Epub 2007 Jul 5].
117. Neumann M, Sampathu DM, Kwong LK, Truax AC, Micsenyi MC, Chou TT, Bruce J, Schuck T, Grossman M, Clark CM, McCluskey LF, Miller BL, Masliah E, Mackenzie IR, Feldman H, Feiden W, Kretzschmar HA, Trojanowski JQ, Lee VM. Ubiquitinated TDP-43 in frontotemporal lobar degeneration and amyotrophic lateral sclerosis. *Science* 2006;**314**(5796):130–3.
118. Rabinovici GD, Miller BL. Frontotemporal lobar degeneration: epidemiology, pathophysiology, diagnosis and management. *CNS Drugs* 2010;**24**(5):375–98.
119. Forman MS, Farmer J, Johnson JK, Clark CM, Arnold SE, Coslett HB, Chatterjee A, Hurtig HI, Karlawish JH, Rosen HJ, Van Deerlin V, Lee VM, Miller BL, Trojanowski JQ,

Grossman M. Frontotemporal dementia: clinicopathological correlations. *Ann Neurol* 2006;**59**(6):952–62.

120. Dickson DW, Davies P, Bevona C, van Hoeven KH, Factor SM, Grober E, Aronson MK, Crystal HA. Hippocampal sclerosis: a common pathological feature of dementia in very old (greater than or equal to 80 years of age) humans. *Acta Neuropathol* 1994;**88**:212–21.

121. Woulfe J, Gray DA, Mackenzie IRA. FUS-immunoreactive intranuclear inclusions in neuro-degenerative disease. *Brain Pathol* 2010;**20**:589–97.

122. Dubois B, Feldman HH, Jacava C, DeKosky ST, Barberger-Gateau P, Cummings J, Delacourte A, Galasko D, Gauthier S, Jicha G, Meguro K, O'Brien J, Pasquier F, Robert P, Rossor M, Salloway S, Stern Y, Visser PJ, Scheltens P. Research criteria for the diagnosis of Alzheimer's disease: revising the NINCDS-ADRDA criteria. *Lancet Neurol* 2007;**6**:734–46.

Cerebral Amyloid Angiopathy

Masahito Yamada* and
Hironobu Naiki†

*Department of Neurology and
Neurobiology of Aging, Kanazawa
University Graduate School of Medical
Science, Kanazawa, Japan

†Division of Molecular Pathology,
Department of Pathological Sciences,
Faculty of Medical Sciences, University of
Fukui, Fukui, Japan

Cerebral amyloid angiopathy (CAA) is cerebrovascular amyloid deposition. It is classified into several types according to the cerebrovascular amyloid proteins involved [amyloid β-protein (Aβ), cystatin C (ACys), prion protein (APrP), transthyretin (ATTR), gelsolin (AGel), ABri/ADan, and AL]. Sporadic Aβ-type CAA is commonly found in elderly individuals and patients with Alzheimer's disease (AD). CAA-related disorders include hemorrhagic and ischemic brain lesions and dementia. It has been proposed that cerebrovascular Aβ originates mainly from the brain and is transported to the vascular wall through a perivascular drainage pathway, where it polymerizes into fibrils on vascular basement membrane through interactions with extracellular components. CAA would be promoted by overproduction of Aβ40 (a major molecular species of

Progress in Molecular Biology
and Translational Science, Vol. 107
DOI: 10.1016/B978-0-12-385883-2.00006-0

41

cerebrovascular Aβ), a decrease of Aβ degradation, or reduction of Aβ clearance due to impairment of perivascular drainage pathway. Further understanding of the molecular pathogenesis of CAA would lead to development of disease-modifying therapies for CAA and CAA-related disorders.

I. Introduction

Amyloidosis is a pathological condition characterized by the formation of protein deposits comprising accretions of amyloid fibrils. Amyloidosis is classified as "systemic amyloidosis" and "localized amyloidosis." In systemic amyloidosis, a certain amyloid protein deposits in various organs of the body, whereas in localized amyloidosis, amyloid deposition is confined to a certain organ or tissue.

Cerebral amyloid angiopathy (CAA) is cerebrovascular amyloid deposition. CAA occurs as localized amyloidosis, for example, cerebral amyloid β-protein (Aβ) amyloidosis, or as cerebrovascular involvement of systemic amyloidosis, for example, hereditary transthyretin (TTR) amyloidosis. Although mild CAA, which is often detected incidentally at autopsy, is not associated with clinical manifestations, severe CAA causes cerebrovascular disorders such as cerebral hemorrhage and presents with dementia. Here, we describe clinical and molecular aspects of CAA and discuss key unanswered questions.

II. Clinical Aspects of CAA

A. Classification

CAA is classified according to the amyloid protein involved. Several cerebrovascular amyloid proteins have been identified, including Aβ, cystatin C (ACys for cystatin C-related amyloid), prion protein (PrP) (APrP for PrP-related amyloid), TTR (ATTR for TTR-related amyloid), gelsolin (AGel for gelsolin-related amyloid), ABri/ADan for amyloid protein of the BRI gene (*BRI*) product in familial British or Danish dementia, and immunoglobulin light chain (AL for light chain-related amyloid). Classification of CAA based on the amyloid proteins is shown in Table I. Among the several types, sporadic Aβ-type CAA is most commonly found in elderly individuals as well as in patients with Alzheimer's disease (AD). Mutations in the amyloid β-protein precursor(AβPP) gene (*AβPP*) are associated with hereditary CAA.

TABLE I
CLASSIFICATION OF CEREBRAL AMYLOID ANGIOPATHY (CAA)

Amyloid protein	Clinical phenotype

1. Amyloid β-protein (Aβ)	1. Sporadic; associated with:
	  a. Aging
	  b. Sporadic Alzheimer's disease (AD)
	  c. Other conditions, including vascular malformations, irradiation
	2. Hereditary or genetic; associated with:
	  a. Mutations in the amyloid β-protein precursor (AβPP) gene, including hereditary cerebral hemorrhage with amyloidosis—Dutch type (HCHWA-D) E693Q, E693K (Italian), E693G (Arctic), A692Q (Flemish), E694N (Iowa), L705V (Piedmont), A713T (Italian), and AβPP gene duplication
	  b. Mutations of presenilin genes
	  c. Down syndrome

2. Cystatin C (ACys)	HCHWA—Icelandic type (HCHWA-I) associated with a mutation (^{68}Leu→Gln) of cystatin C gene
3. Prion protein (PrP) (AScr)	Prion disease associated with mutations of *PRNP* gene (Y145Stop, Y163Stop, Y226Stop)
4. ABri/ADan	Familial British or Danish dementia (FBD/FDD) associated with mutations of BRI gene
5. Transthyretin (ATTR)	Meningocerebrovascular involvement of familial transthyretin (TTR) amyloidoses (familial oculoleptomeningeal amyloidosis, familial amyloid polyneuropathy) associated with mutations of TTR gene
6. Gelsolin (AGel)	Menigocerebrovascular involvement of gelsolin-related amyloidosis (familial amyloidosis, Finnish type) associated with mutations of gelsolin gene
7. AL	CAA with leukoencephalopathy due to brain-restricted monoclonal plasma cell proliferation

B. Clinical, Genetic, and Pathologic Features of CAA

1. SPORADIC Aβ-TYPE CAA

a. Epidemiology. The prevalence of CAA increases with age, occurring in about half the persons over 60 years of age.[1–3] CAA is commonly found in AD, with a prevalence of about 80–90%.[2,4] CAA-related lobar intracerebral hemorrhage (ICH) identified via a nationwide survey in Japan increased with age and presented with female predominance.[5]

b. Pathology. CAA is observed mainly in the leptomeningeal and cortical vessels of the cerebral lobes and cerebellum. For the distribution of CAA, the occipital lobe is preferentially affected.[2] On the other hand, CAA is uncommon in the basal ganglia, thalamus, brainstem, and white matter. In mild CAA, a small portion of the leptomeningeal and superficial cortical vessels are affected with amyloid deposition. In severe CAA, most small arteries and arterioles show marked amyloid deposition. Medium-sized leptomeningeal arteries are affected with amyloid deposition in the outer portion of the media to the adventitia. Vessel walls of the small arteries and arterioles are often totally replaced by amyloid deposits, except for the endothelial cells (Fig. 1A). Electron microscopically, amyloid fibrils are focally deposited within the outer basement membrane (BM) of the vessels in the initial stage of CAA. Later, a large amount of amyloid fibrils accumulate, and degeneration of smooth muscle cells of the media is observed (Fig. 1B).[6] Amyloid deposits in capillaries and, occasionally, in arterioles or small arteries have been found to infiltrate the surrounding parenchymal tissue and accompany dystrophic neurites forming plaque-like structures (perivascular plaques or drüsige Entartung (Scholz)) (Fig. 1C).[7] Recently, CAA in capillaries has been referred to as "capillary CAA (CAA-Type 1)," to distinguish it from noncapillary CAA (CAA-Type 2).[8]

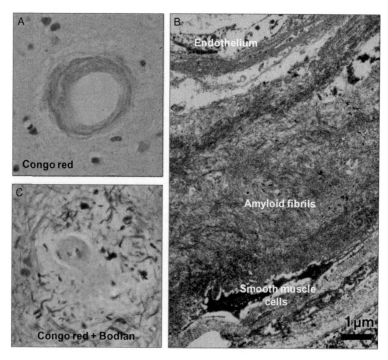

Fig. 1. Neuropathology of CAA. The vascular wall is replaced by Congo red-positive amyloid (A). Electron microscopically, a large amount of amyloid fibrils accumulate in the vessel walls with degenerative smooth muscle cells (B). Capillary amyloid deposits often infiltrate to the neuropil with degenerative neurites (perivascular plaque/drüsige Entartung) (C). (A, Congo red, original magnification 420×; B, electron micrograph, original magnification 6000×; C, Congo red-Bodian, original magnification 420×.) (See Color Insert.)

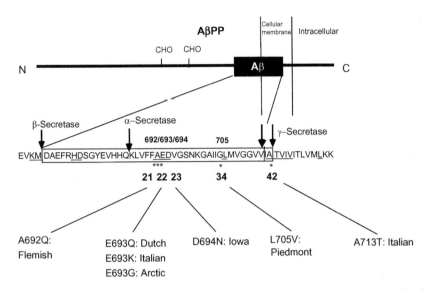

FIG. 2. Processing of AβPP to Aβ, and mutations in the AβPP gene. Sites of mutations related to familial AD or CAA are underlined. Mutations related to hereditary Aβ-type CAA are shown (°) with names, such as Dutch type.

A cerebrovascular amyloid protein was identified as Aβ in cerebrovascular amyloidosis associated with AD by Glenner and Wong.[9] Aβ is cleaved from the AβPP by β- and γ-secretase (Fig. 2). Heterogeneity of the C-terminal is present. The length of Aβ of senile plaques (SPs) is mainly 42–43 residues (Aβ42), while that of cerebrovascular Aβ is mainly 39–40 residues (Aβ40) (Fig. 3).[10,11] Tissue levels of soluble Aβ40 in the brain correlate with severity of CAA.[11] In the process of vascular Aβ deposition, Aβ42 is initially deposited, and later Aβ40 is massively accumulated.[12]

A variety of proteins other than Aβ, such as amyloid P component, apolipoprotein E (ApoE), and cystatin C, are colocalized with Aβ, some of which may have a role in the pathogenesis of CAA or CAA-related vascular changes (see Section III-C).[13,14]

SPs and neurofibrillary tangles (NFTs) are commonly present with CAA in elderly patients with or without AD, but the distribution of CAA within the brain is not directly correlated with that of SPs or NFTs.

Severe CAA is associated with vasculopathies including duplication ("double-barrel" lumen), obliterative intimal changes, hyaline degeneration, microaneurysmal dilatation, and fibrinoid necrosis (Fig. 4).[15–17] These CAA-associated vasculopathies are the pathological basis of CAA-related cerebrovascular disorders. In particular, the fibrinoid necrosis is closely associated with CAA-related hemorrhage.[15–17]

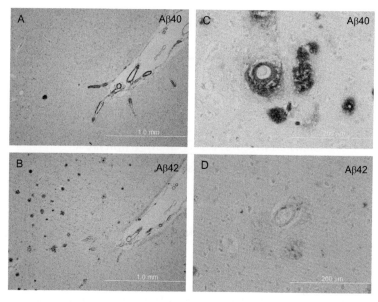

FIG. 3. Immunohistochemistry of CAA with specific antibodies to Aβ40 (A, C) and Aβ42 (B, D). Senile plaques are mainly composed of Aβ42 (B), whereas Aβ40 is a major component of CAA (A, C). (A, anti-Aβ40 antibody, original magnification 20×; B, anti-Aβ42 antibody, original magnification 20×; C, anti-Aβ40 antibody, original magnification 200×; D, anti-Aβ42 antibody, original magnification 200×). (See Color Insert.)

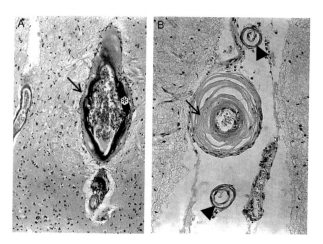

FIG. 4. CAA-associated vasculopathies. Microaneurysmal dilatation (arrow) with fibrinoid necrosis (°) (A). Thickening of the intima (arrow) and double barreling of vascular walls (arrow heads) (B). (A, Congo red, original magnification 110×; B, Congo red, original magnification 170×.)

CAA-related cerebrovascular disorders include lobar cerebral and cerebellar hemorrhages (macrohemorrhages) (Fig. 5A), leukoencephalopathy, and small cortical infarcts and hemorrhages (microhemorrhages) (Fig. 5B and C).[18–20]

Cerebral blood vessels with amyloid deposits are generally associated with an increase in number and in activation of monocyte/macrophage lineage cells. The activation of the immune system represents immune reactions against vascular amyloid deposition.[21] Furthermore, CAA may coexist with granulomatous angiitis or inflammation.[21–25]

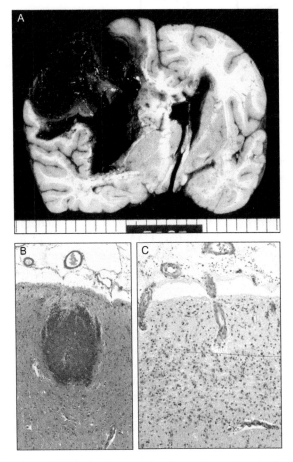

FIG. 5. Pathology of CAA-related intracerebral hemorrhage (A), cortical microhemorrhage (B), and cortical microinfarction (C). (B, H&E, original magnification 25×; C, Congo red, 25×.) (See Color Insert.)

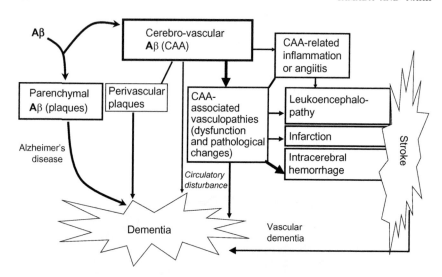

FIG. 6. Pathophysiology of CAA-related disorders.

The pathophysiology of CAA-related disorders is summarized in Fig. 6. CAA is related to both cerebrovascular disorders and dementia found in the elderly and patients with AD.

c. Risk Factors for CAA and CAA-related ICH. Genetic risk factors have been reported for sporadic Aβ-type CAA. The ε4 allele of the ApoE gene (*ApoE*), an established risk for AD, has been reported to be a risk for CAA.[26,27] Further, the ε2 allele of *ApoE* has been reported to be associated with CAA-related hemorrhage.[28] Carriers of the *ApoE* ε2 or ε4 allele, particularly, of the *ApoE* ε2/ε4 genotype were associated with early recurrence of lobar ICH in patients who survived a lobar ICH.[29] It was reported that the *ApoE* ε4 allele constituted a risk factor for capillary CAA (CAA-Type 1), while the ε2 allele did not.[8] Interestingly, the presence of CAA in head-injury cases was significantly associated with possession of an *ApoE* ε4 allele, suggesting an interaction between gene and environment in development of CAA.[30]

In addition, CAA was reported to be associated with other gene polymorphisms, including the presenilin 1 (PS1), α1-antichymotrypsin (ACT), neprilysin, transforming growth factor (TGF)-β1 (TGF-β1), low-density lipoprotein receptor related protein (LRP-1), and angiotensin-converting enzyme (ACE) genes.[4,31–39]

Regarding of risk of CAA-related ICH, the fact that lowering of blood pressure reduced risk of CAA-related ICH suggested that high blood pressure could be a factor inducing ICH in patients with CAA.[40]

There is increasing evidence that CAA could be a risk factor for ICH in thrombolytic therapies for acute myocardial infarction, pulmonary embolism, or ischemic stroke, and for ICH in warfarin therapies.[41,42] The use of anti-platelet drugs such as aspirin was related to the presence of microbleeds and to strictly lobar microbleeds suggestive of CAA.[43]

d. Clinical Manifestations.

i. Hemorrhages and Other Cerebrovascular Disorders.
Multiple and recurrent hemorrhages are common in patients with CAA. Clinical manifestations of CAA-related ICH are motor paresis, disturbance of consciousness, abnormalities in higher brain functions, such as aphasia, visual loss, and headache at the acute stage, and dementia and seizure at chronic stages.[5] Headache with meningeal signs would be caused by CAA-related subarachnoid hemorrhage (SAH).[18,44,45]

Recurrent transient neurological symptoms resembling transient ischemic attacks are found in some CAA patients,[46] and it has been suggested that these might be focal seizures due to cortical microhemorrhages or transient migraine auras caused by focal SAH.[46,47]

ii. Dementia.
Progressive dementia is frequently found in patients with CAA.[46,48] Pathomechanisms underlying the dementia are not uniform, including vascular dementia (VaD) (including Biswanger type leukoencephalopathy) due to CAA, coexistence of AD, mixed dementia of VaD and AD, and a vascular variant of AD.[19,48] It has been suggested that CAA has deleterious effects on cognition even after controlling for age and AD pathology, and that CAA was associated with white matter abnormalities and cognitive impairment.[49]

iii. CAA-Related Inflammation or Angiitis.
A subset of patients with CAA present with subacute encephalopathy characterized by cognitive symptoms, seizure, headaches, and focal neurological deficits with white matter lesions on MRI and neuropathologic evidence of CAA-associated vascular inflammation or angiitis.[22–25]

e. Laboratory Findings.
i. Imaging Studies.

(i) CT/MRI
Hemorrhagic lesions. CAA-related ICH is found in the cerebral lobes (Fig. 7). In contrast, it is usually not found in regions characteristic of hypertensive hemorrhage, such as basal ganglia, thalamus, and pons, because CAA is absent or sparse in these regions. Cerebellar hemorrhage may occur in association with CAA as well as hypertension.[18] The lobar ICH does not necessarily indicate CAA. In our pathologic study with cases of

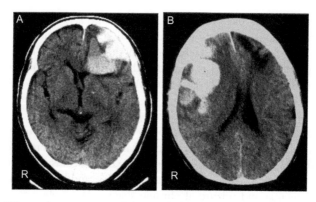

FIG. 7. CT scans showing recurrent lobar cerebral hemorrhages in an autopsy-confirmed case of sporadic Aβ-type CAA. The patient first developed a lobar hemorrhage in the left frontal lobe (A), and 3 years later, the patient had recurrent hemorrhage in the right fronto-parietal region (B).

lobar ICH, atypical hypertensive hemorrhage is most frequent, followed by CAA-related hemorrhage.[18] It is necessary to exclude causes of ICH other than CAA, such as atypical hypertensive hemorrhage, traumatic hemorrhage, systemic bleeding tendency, hemorrhages due to aneurysm or vascular malformations, and hemorrhages associated with neoplasms. CAA-related ICH in the cortico-subcortical region may accompany secondary rupture to subarachnoid space (secondary SAH) (Fig. 7).[18,44]

Multiple, recurrent hemorrhages of the lobar type strongly suggest CAA because of the tendency for CAA-related hemorrhage to recur (Fig. 7). Furthermore, cortical microhemorrhages (microbleeds) are scattered in CAA brains (Fig. 8A). Gradient-echo MRI or susceptibility-weighted MRI is useful to detect such petechial cortical hemorrhages.[50] Microbleeds in cortico-subcortical distribution are related to CAA, while those in the central gray matter to advanced hypertension.[51,52] Microbleeds in AD would be associated with CAA.[53]

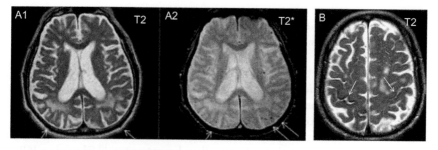

FIG. 8. MRI findings of CAA. (A) T2-(A1) and T2*-weighted images (A2) show white matter lesions and cortical microhemorrhages (arrows). (B) Focal subarachnoid hemorrhages (white arrows) are also seen on T2-weighted images.

Regarding the distribution of symptomatic CAA-related lobar ICH, the actual frequency was higher in the frontal and parietal lobes; however, after correcting for the estimated cortical volume, the parietal lobe was found to be the most frequently affected.[5] When microbleeds are included, the distribution of ICH was reported to show posterior predilection (temporal and occipital lobes), consistent with occipital predominance of CAA in pathological studies.[54] On the other hand, predilection for the parietal lobe in CAA-related microbleeds was also reported.[52]

Superficial siderosis on MRI (Fig. 8B) was suggested to be a marker of CAA in AD.[55] It was reported that CAA was a frequent cause of convexal SAH, a subtype of nonaneurysmal subarachnoid bleeding, in patients over the age of 60.[56] Superficial siderosis closely accompanied microbleeds in lobar locations, suggesting a close link between superficial siderosis and CAA.[57]

Ischemic lesions. Periventricular lucency on CT and periventricular signal hyperintensities on T2-weighted images of MRI may suggest CAA-related leukoencephalopathy (Fig. 8A). White matter damage on CT or MRI in lobar ICH is common and is associated with cognitive impairment.[58] Subjects with CAA showed a progressive increase of white matter lesions.[59]

Small cortical infarcts may also be observed. CAA was revealed to be an important risk factor for microinfarcts restricted to the watershed cortical zones in AD, suggesting a role of cerebral hypoperfusion in the pathogenesis of cortical microinfarcts.[60] Subacute, silent cortical, or subcortical infarctions were recognized on diffusion–weighted images (DWI) of MRI in patients with CAA.[61]

Reversible subacute leukoencephalopathy is a radiographic feature of CAA-related inflammation or angiitis[22,23,25]; MRI was characterized by asymmetric T2-hyperintense lesions extending to the subcortical white matter and occasionally the overlying gray matter with signal properties suggesting vasogenic edema.[25]

(ii) SPECT/PET

In a recent study, (99 m)Tc-ethylcysteinate dimer (ECD) brain perfusion SPECT revealed significantly reduced cerebral perfusion in patients with CAA, which may be related to leukoencephalopathy.[62]

In amyloid imaging with a PET ligand, (11)C-Pittsburgh compound B (PIB), it was reported that PIB binding was moderately increased in most patients with probable CAA-related ICH and that the occipital uptake was greater in CAA-related ICH compared with AD.[63,64] In an autopsy-confirmed case, PIB-PET could detect CAA as the dominant source of signal.[65]

ii. Biochemical Markers. A decrease of the cerebrospinal fluid (CSF) levels of Aβ42 level and an increase of the ratio of Aβ40 to Aβ42 in the CSF have been reported to be useful for diagnosis of AD. Recently, a significant decrease of CSF Aβ40 as well as Aβ42 was reported in patients with probable CAA, suggesting the trapping of Aβ40 as well as Aβ42 in the cerebral vasculature.[66] Plasma Aβ40 levels were reported to be associated with extent of white matter hyperintensity in AD, mild cognitive impairment, or CAA.[67]

iii. Brain Biopsy. Definite diagnosis of CAA is made by pathological confirmation. Except for autopsy, brain tissues can be obtained during hematoma evacuation or cortical biopsy in CAA-related ICH. A biopsy is recommended to demonstrate evidence of CAA-related inflammation or angiitis.

f. Diagnosis. For the clinical diagnosis of CAA-related ICH, the Boston criteria were proposed (Table II), and it was reported with the small pathologic series that the diagnosis of probable CAA can be made by the Boston criteria during life with high accuracy.[68]

TABLE II

BOSTON CRITERIA FOR DIAGNOSIS OF CAA-RELATED HEMORRHAGE[A68]

---

1. Definite CAA
Full postmortem examination demonstrating
- Lobar, cortical, or cortico-subcortical hemorrhage
- Severe CAA with vasculopathy[b]
- Absence of other diagnostic lesion

2. Probable CAA with supporting pathology
Clinical data and pathologic tissue (evacuated hematoma or cortical biopsy) demonstrating
- Lobar, cortical, or cortico-subcortical hemorrhage
- Some degree of CAA in specimen
- Absence of other diagnostic lesion

3. Probable CAA
Clinical data and MRI or CT demonstrating
- Multiple hemorrhages restricted to lobar, cortical, or cortico-subcortical regions (cerebellar hemorrhage allowed)
- Age≥55 years
- Absence of other cause of hemorrhage[c]

4. Possible CAA
Clinical data and MRI or CT demonstrating
- Single lobar, cortical, or cortico-subcortical hemorrhage
- Age≥55 years
- Absence of other cause of hemorrhage[c]

---

[a]Criteria established by the Boston Cerebral Amyloid Angiopathy Group: Steven M. Greenberg, MD, PhD, Daniel S. Kanter, MD, Carlos S. Kase, MD, and Michael S. Pessin, MD.
[b]Ref. 16
[c]Other causes of intracerebral hemorrhage: excessive warfarin (INR>3.0); antecedent head trauma or ischemic stroke; CNS tumor, vascular malformation, or vasculitis; and blood dyscrasia or coagulopathy. (INR>3.0 or other nonspecific laboratory abnormalities permitted for diagnosis of possible CAA.)

CAA may coexist with hypertension, and both CAA and hypertension may contribute to ICH. However, hypertension alone may cause lobar ICH.[18] Therefore, exclusion of advanced hypertension may increase specificity of the diagnosis, although the diagnostic sensitivity would be decreased.[5]

g. *Current Treatment and Prognosis.* Although uncontrollable peri- and postoperative hemorrhages had been reported previously, it has been suggested since the 1990s that neurosurgical procedures, especially hematoma evacuation, can be performed more safely than previously expected.[69] In our recent study with a nationwide survey in Japan,[5] neurosurgical procedures were performed without uncontrollable intraoperative or postoperative hemorrhage in 97.1% of patients. CAA-related lobar ICH recurred in 31.7% of patients during the average 35.3-month follow-up period. The mean interval between ICHs was 11.3 months and the case fatality rate was 12.2% at 1 month and 19.5% at 12 months after initial ICH.

Hypertension and possession of an *APOE* ε2 or ε4 allele were reported to be a risk factor influencing the recurrence of CAA-related lobar ICH. Lowering of blood pressure may reduce the risk of CAA-related ICH.[29,40,70]

In CAA-related inflammation including granulomatous angiitis, clinical and radiographic improvement was demonstrated after immunosuppressive treatment.[23,25]

2. OTHER TYPES OF CAA

a. Hereditary/Genetic Aβ-Type CAA.
i. *AβPP* Mutations. Mutations in the *AβPP* have been associated with familial AD or CAA (Fig. 2). Hereditary cerebral hemorrhage with amyloidosis—Dutch type (HCHWA-D) is an autosomal-dominant disorder characterized by recurrent lobar cerebral hemorrhages and leukoencephalopathy which occurs in the middle age in Dutch families. Analysis of the *AβPP* revealed a point mutation at codon 693 that causes a single amino acid substitution (Glu to Gln) at position 22 of Aβ.[71] A decrease of plasma Aβ42, but not Aβ40, levels was found in carriers of the HCHWA-D mutation.[72]

Other CAA mutations at the same codon 693 (position 22 of Aβ) of the *AβPP* as HCHWA-D were reported, including a Glu to Lys change in Italian families with multiple strokes and a Glu to Gly change in Swedish families with early-onset AD (Arctic type).[73,74]

A mutation (an Ala to Gly change) at codon 692 (position 21 of Aβ) of the *AβPP* was found in a Dutch family with presenile dementia and CAA-related hemorrhage (Flemish mutation).[75] Neuropathologically, severe CAA, vasocentric SPs, and NFTs are observed.[76]

A D694N mutation in the Aβ sequence (position 23 of Aβ) of the *AβPP* was reported in an Iowa family with autosomal-dominant dementia and severe CAA (Iowa type).[77] Selective retention of PIB in occipital cortex was reported in Iowa-type hereditary CAA.[78]

An L705V mutation in the Aβ sequence (position 34 of Aβ) of the *AβPP* was reported in a family with autosomal-dominant, recurrent ICH, and pathological examination disclosed severe CAA without parenchymal amyloid deposits or NFTs.[79]

An Italian family with autosomal-dominant dementia and multiple strokes had an A713T mutation of the *AβPP*; the disease was characterized by late-onset AD and subcortical ischemic lesions with neuropathologically confirmed CAA.[80]

Families with *AβPP* duplication showed autosomal-dominant early-onset AD with severe CAA accompanying ICH.[81]

ii. Presenilin Mutations. Familial AD caused by mutations in presenilin 1 (*PS1*) or presenilin 2 genes (*PS2*) has been reported to be frequently associated with severe CAA of Aβ type.[82–84] It was reported that cases with mutations after codon 200 of the *PS1* showed prominent CAA.[84]

iii. Down Syndrome. Patients with Down syndrome develop AD pathology in middle age with CAA of Aβ-type, because of the gene dosage effect of the trisomy of chromosome 21. CAA-related ICH was reported in some patients with Down syndrome.[85]

b. Cystatin C (ACys)-Type CAA. HCHWA-Icelandic type (HCHWA-I) or hereditary cystatin C amyloid angiopathy (HCCAA) is an autosomal-dominant disorder found in Icelanders. HCHWA-I is characterized by recurrent cerebral hemorrhages due to severe CAA before the age of 40.[86] The amyloid protein of CAA in HCHWA-I is a variant of cystatin C which is characterized by an amino acid substitution (Leu to Gln) at position 68 (L68Q) caused by a point mutation of the cystatin C gene and loss of 10 amino acid in the N-terminus.[87] Cystatin C is a cysteine protease inhibitor and is present abundantly in the CSF. The cystatin C level in the CSF is decreased in HCHWA-I.[88] The L68Q mutation was reported to destabilize the monomers and induce dimerization.[89] Cystatin C dimers as well as monomers were demonstrated in plasma and CSF from HCCAA patients.[90] The variant of cystatin C deposits mainly in brain arteries and arterioles but also to a lesser degree in tissues outside the central nervous system, such as skin and lymph nodes.[91]

c. PrP (AScr)-Type CAA. PrP Y145Stop, Y163Stop, and Y226Stop mutations were reported to show vascular deposition of PrP amyloid with or without NFTs.[92–94] Patients showed progressive dementia, but no cerebral hemorrhage or other cerebrovascular disorders were described.

d. ABri/ADan-Type CAA. An autosomal-dominant disorder in a British family showing progressive spastic paralysis, dementia, and ataxia was neuropathologically characterized by severe CAA, nonneuritic and perivascular plaques, NFTs, and ischemic leukoencephalopathy[95] and was designated familial British dementia (FBD). A novel 4K protein subunit named ABri was identified from isolated amyloid fibrils.[96] The ABri is a fragment of a putative type-II single-spanning transmembrane precursor that is encoded by a novel gene, *BRI2*, located on chromosome 13. A stop codon mutation of this gene generates a longer open reading frame, resulting in a larger, 277-residue precursor, and the release of the 34 C-terminal amino acids from the mutated precursor generates the ABri amyloid subunits.[96]

Familial Danish dementia (FDD) is an autosomal-dominant disorder characterized clinically by cataracts, deafness, progressive ataxia, and dementia, and pathologically by severe CAA, hippocampal plaques, and NFTs. FDD is associated with a decamer duplication in the 3′region of the *BRI2*, which abolishes the normal stop codon, resulting in an extended precursor protein and the release of an amyloidogenic fragment, ADan.[97]

e. TTR (ATTR)-Type CAA. Hereditary amyloidoses of TTR type *(TTR)* are autosomal-dominant disorders associated with mutations in the TTR gene. Various organs exhibit amyloid deposition derived from variant TTR (ATTR). Hereditary TTR amyloidosis with prominent peripheral and autonomic neuropathies has been reported as familial amyloid polyneuropathy (FAP) (type I) from Japan, Portugal, and other countries. Most of the families with type I FAP are associated with a *TTR* Met30Val mutation. In type I FAP, leptomeningeal and meningovascular amyloid deposition of the variant TTR may be observed pathologically, although clinical manifestations of the central nervous system are rare.

Familial cases of TTR amyloidoses with Leu12Pro, Asp18Gly, Ala25Thr, Val30Met, Val30Gly, Ala36Pro, Gly53Glu, Trp69His, and Tyr114Cys have been reported to present with prominent leptomeningeal and meningovascular amyloid deposition showing cerebral infarction, hemorrhage, hydrocephalus, paresis, and dementia (see references in Yamashita *et al.*).[98] These cases include familial oculoleptomeningeal amyloidosis, in which amyloid is deposited in vitreous bodies and the retina of eyes.

It should be noted that TTR is produced from the choroid plexus as well as liver, and TTR in the CSF derived from the choroid plexus can be deposited in the leptomeninges. Liver transplantation results in improvement of prognosis of type I FAP. There is the possibility that the patients who undergo transplantation may later develop leptomeningeal amyloidosis after a long period, because variant TTR in CSF continues to be produced from the choroid plexus. However, it was reported in patients with TTR Tyr114Cys-related CAA that liver transplantation was effective for CAA-related CNS manifestations, suggesting that the variant TTR of liver origin may contribute to the development of CAA by crossing the blood–CSF barrier.[98]

f. Gelsolin (AGel)-Type CAA. Gelsolin-related amyloidosis (familial amyloidosis, Finnish type) is a rare disorder, reported worldwide in kindreds carrying a G654A or G654T mutation of the gelsolin gene.[99] Facial palsy, mild peripheral neuropathy, and corneal lattice dystrophy associated with deposition of gelsolin-related amyloid (AGel) are characteristic. Widespread spinal, cerebral, and meningeal amyloid angiopathy with AGel deposition are found accompanying white matter lesions.[99]

g. AL-Type CAA. The cerebral vasculature, except for the choroids plexus, is not involved in systemic AL amyloidosis associated with amyloidogenic monoclonal light chains present in systemic circulation. This suggests that the monoclonal light chains cannot permeate the blood–brain barrier (BBB). Isolated AL-type CAA with widespread subcortical distribution due to brain-restricted atypical monoclonal plasma cell proliferation was reported with leukoencephalopathy.[100] This condition was distinguished from other disorders of local AL deposition such as amyloidoma and solitary intracerebral plasmacytoma with amyloid deposition.

III. Molecular Aspects of CAA

A. The Origins of Aβ Deposited in Blood Vessel Walls

In 1938, Scholz first suggested that amyloid in blood vessel walls was derived from the blood.[7] Later, it was proposed that the Aβ in CAA was mainly derived from smooth muscle cells within the artery walls.[101] It was then proposed that Aβ is produced by neurons and other parenchymal cells and then entrapped in perivascular interstitial fluid (ISF) drainage pathways in CAA in the human brain.[102,103] As we see later, support for this proposal came from transgenic mice that produce mutant human Aβ only in the brain and develop prominent CAA.[104,105]

B. The Mechanisms of Carrying Aβ to Blood Vessel Walls

Several mechanisms were proposed for the elimination of Aβ from the brain.[103]

(1) Enzymic degradation of Aβ in the brain parenchyma by proteases, including neprilysin and insulin-degrading enzyme (IDE)[106,107]

(2) Direct transcytosis of Aβ–ApoE complexes into the blood via LRP-1[108]

(3) Perivascular drainage of Aβ with other solutes and ISF along capillary and artery walls.[102,103] This is some sixfold slower than absorption into the blood via LRP, but perivascular drainage appears to compensate when the LRP mechanism is blocked or fails and when neprilysin levels in the brain are reduced.[106,108,109]

Although there are no lymphatic vessels in the mammalian brain that are comparable in structure to those in the rest of the body, the rate of drainage of ISF from the brain is estimated to be comparable to the average lymphatic drainage in the rest of the body.[110,111] Weller's group studied the details of the drainage pathway by which fluorescent tracers are eliminated from the mouse brain.[103,112] Formalin-fixable fluorescent 3kDa dextran (about the same molecular weight as Aβ) and fluorescent 40kDa ovalbumin were injected (volume 0.5 μL) into the caudate putamen of mice, and the distribution of the tracers was monitored at 5min to 24h. They found that tracers in the extracellular spaces of the brain entered the capillary BMs between the endothelial layer and the surrounding astrocytes, passed into the BMs surrounding the smooth muscle cells in the tunica media of arteries, and reached the leptomeningeal vessels on their passage out of the brain.

No fluorescent tracer was detected in cervical lymph nodes in the study reported above[112] possibly because of the very small amount of tracer injected. However, previous studies have shown that ISF and solutes drain from artery walls to cervical lymphatics at the skull base, finally reaching the cervical lymph nodes.[111] Theoretical studies suggested that the contrary (or reflection) wave that follows each pulse wave is the motive force for the perivascular transport of ISF and solutes in the reverse direction of blood flow.[113]

Several studies support the perivascular drainage model. First, biochemical studies have shown that Aβ is present in the leptomeningeal arteries in individuals aged 20–90 years.[12] Aβ was present in the walls of middle cerebral arteries and basilar arteries at the base of the brain, but no Aβ was detected in the walls of internal carotid arteries in the neck.[12] This supports the observations in experimental animals that solutes draining from the brain along perivascular pathways leave the artery walls at the base of the skull probably to drain to regional lymph nodes in the neck.[111] Second, in immunized *AβPP* transgenic mice, plaque removal is accompanied by an increase in severity of CAA.[114] Human studies have also demonstrated that CAA in patients treated with immunotherapy contains increased amounts of Aβ42 and Aβ40 due to solubilization of parenchymal Aβ lesions and that, as in transgenic animals treated with passive immunization, there is a higher density of microhemorrhages and microvascular lesions.[115,116]

C. The Mechanisms of Aβ Deposition in Blood Vessel Walls

Several factors have been proposed for the deposition of Aβ in blood vessel walls.

1. A NUCLEATION-DEPENDENT POLYMERIZATION MODEL OF Aβ AMYLOID FIBRIL FORMATION

It is widely accepted that amyloid fibrils are formed by a nucleation-dependent polymerization mechanism (Fig. 9).[117,118] This mechanism consists of two steps. The first nucleation step requires a series of association steps of monomers that are thermodynamically unfavorable. Once a nucleus is formed, the second extension step spontaneously proceeds by the consecutive association of monomers to the fibril ends until the reaction reaches an equilibrium between fibrils and monomers.

Pathways of fibril formation also produce various intermediates, including "protofibrillar" intermediates.[119] In addition, oligomeric forms of Aβ are observed, including paranuclei, ADDLs, and Aβ*56. These assemblies are

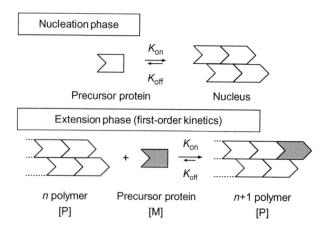

FIG. 9. Nucleation-dependent polymerization model of amyloid fibril formation.

detected both *in vitro* and *in vivo*, and these oligomers may form protofibrils or seeds leading to the formation of mature fibrils.[119] Teplow's group investigated the details of the formation mechanism of paranucleus (pentamer or hexamer) of Aβ(1–42) and oligomers of Aβ(1–40) using photochemical cross-linking, NMR, and computer simulations.[119] Glabe's group detected prefibrillar and fibrillar oligomers using conformation-dependent antioligomer antibodies, that is, A11 and OC, respectively.[120] The identification of precursor–product relationships of these species is under extensive investigation.[120,121]

Biochemical studies of Aβ in HCHWA-D and in the Iowa variant of FAD demonstrated that the amyloid deposits in CAA are composed of both variant (either E22Q or D23N) and wild-type Aβ in an ≈50:50 ratio.[93,122,123] Compared with wild-type Aβ, both the Dutch and Iowa Aβ40 synthetic peptides rapidly assemble to form amyloid fibrils *in vitro*, which are toxic to cultured human cerebrovascular endothelial cells and smooth muscle cells.[124,125] Thus, it is reasonable to consider that variant Aβ may form nuclei (seeds) first, followed by the consecutive association of both variant and wild-type Aβs to the seeds as well as to the fibril ends.

2. PATHOLOGICAL MOLECULAR INTERACTIONS FOR THE FORMATION OF Aβ AMYLOID FIBRILS

A number of amyloid-associated proteins or "pathological chaperones" codeposit with different cerebral parenchymal and cerebrovascular amyloids.[93,126,127] Such proteins are structurally and functionally diverse, and their binding to the amyloid fibrils or their precursors may be additional factors influencing the formation of toxic misfolded proteins. In Aβ-CAA, a number of amyloid-associated proteins, including complement components, serum amyloid-P component, ApoE, complement inhibitors such as apolipoprotein J and vitronectin, α1-ACT, glycosaminoglycans, and extracellular matrix proteins, are also present.[128] More specifically, capillary BMs contain collagen IV, laminin, fibronectin, and the heparan sulfate proteoglycan, perlecan. Perlecan is associated with deposits of Aβ in the brain and accelerates Aβ fibril formation *in vitro*,[129] whereas laminin binds to Aβ in BMs and is a potent inhibitor of Aβ amyloid fibril formation.[130]

Various lipid molecules have been reported to induce conformational changes of amyloid precursor proteins, as well as to initiate their amyloid fibril formation *in vitro*.[131,132] Yanagisawa's group discovered GM$_1$-ganglioside-bound Aβ in the brains of AD patients and suggested this specific form of Aβ as a seed for the formation of amyloid fibrils *in vivo*.[132] Subsequently, Matsuzaki's group reported that liposomes containing GM$_1$-ganglioside and cholesterol mimicking lipid raft microdomains strongly enhance Aβ amyloid fibril formation *in vitro*.[133]

Clusterin, α_2-macroglobulin (α_2M) and haptoglobin (Hp) are all abundant secreted glycoproteins present in human plasma and cerebrospinal fluid. Like the small heat shock proteins, all of these glycoproteins have in common the ability to protect a range of proteins from stress-induced aggregation in an ATP-independent manner and have been described as extracellular chaperones.[134] Interestingly, these glycoproteins have been found associated with amyloid deposits in AD and many other human amyloidoses. Using an array of biophysical techniques, Wilson's group established that all of these glycoproteins inhibit the formation of amyloid fibrils from a range of proteins, including Aβ, *in vitro* at substoichiometric levels, and under physiological conditions.[134–136] Together with previous findings, Wilson's group suggested that clusterin, α_2M, and Hp make up a small family of extracellular chaperones that may be an important part of an *in vivo* quality control system for extracellular proteins. They proposed that extracellular chaperones respond to misfolded and aggregated proteins in the extracellular space by binding to their exposed hydrophobic regions, maintaining the solubility of the substrate, and promoting its removal from the extracellular space via receptor-mediated endocytosis (e.g., via LRPs).

3. Reduction in Aβ Degradation or Absorption into the Blood

Although the pathways by which Aβ is generated from its precursor are largely known, Aβ catabolism under physiological and pathological conditions is only starting to be unveiled.[93] Neprilysin, endothelin-converting enzyme, IDE, β-amyloid-converting enzyme 1, plasmin, and matrix metalloproteases (MMPs) are among the major enzymes known to participate in brain Aβ catabolic pathways.[137–139] In mouse models, gene deletion of different proteases accelerates Aβ deposition.[138–140] Thus, reduced levels or catalytic activity of Aβ-degrading enzymes, as a result of age, genetic factors, or specific disease conditions, may favor Aβ accumulation. The specific association of many of these enzymes with vascular components points to their active participation in CAA pathogenesis.[141] Furthermore, the Aβ molecules that are produced by the Dutch, Flemish, Italian, and Arctic mutant *AβPP* genes are resistant to degradation by neprilysin.[142] Diversion of Aβ into perivascular drainage pathways may be one reason for the increased severity of the CAA in these disorders. Aβ is also diverted into perivascular drainage pathways when absorption of Aβ into the blood via the LRP mechanism is reduced as a result of age and genetic factors as well as specific disease conditions.[109]

4. Reduction in the Transport of Soluble Aβ along Aging Arteries

Clinical disease and severe deposition of Aβ in the brain and in CAA do not usually occur until adulthood or middle age, even in cases of familial AD and familial CAA.[103,143,144] In the sporadic forms of CAA, the prevalence increases with age in people over 60 years of age. Theoretical models suggest that perivascular drainage of ISF, and Aβ, is driven by vessel pulsations.[113] The character of pulsations in arteries changes with age as vessels become stiffer with arteriosclerosis.[145] This may be a major factor in slowing the drainage of Aβ and allowing it to form insoluble amyloid fibrils within vascular BMs. Changes in vessel tone resulting from cholinergic deafferentation in the rabbit also result in CAA.[146] From these observations, it seems that age changes and stiffening of artery walls and cholinergic deafferentation impede the perivascular drainage of Aβ, resulting in CAA.[147,148]

5. Contribution of Aβ Produced by Smooth Muscle Cells

Insoluble Aβ(1–42) appears to be deposited first in artery walls and this is followed by the much more abundant deposition of more soluble Aβ(1–40) in CAA.[149] Due to the reduction in Aβ degradation in the walls of aging arteries, Aβ(1–42) produced by smooth muscle cells may possibly provide seeds for the accumulation of the more soluble Aβ(1–40) draining from the brain in perivascular pathways.[150,151]

6. Age-Related Changes in the Constituents of Vascular BMs

Aβ is associated with various protein constituents in vascular BMs, especially heparan sulfate proteoglycan, that promote amyloid fibril formation and laminin that inhibits amyloid fibril formation.[130,152] Ultrastructural changes occur in vascular BMs with age. They become thicker and accumulate collagen.[153] Such changes may interfere with perivascular drainage of Aβ, as well as provide a suitable environment for the formation of Aβ amyloid fibrils.

7. Contribution of Apolipoprotein E

ApoE is tightly associated with Aβ in plaques in brain parenchyma as well as in vessel walls in CAA.[13,154] As described above, the ε4 allele of *ApoE* is a risk factor for the development of CAA as well as AD,[8,26,27] and the ε2 allele is associated with ICH and fibrinoid necrosis in CAA vessels.[28,155] Although the exact reasons for the association of CAA with *ApoE* polymorphisms are not clear, it is possible that ApoE is associated with fibrillogenesis of Aβ within brain tissue and in perivascular drainage pathways.[128,154] Alternatively, there is evidence to show that ApoE binds to LRP-1 and interacts with soluble and aggregated Aβ both *in vitro* and *in vivo*, influencing its conformation and clearance.[156]

D. The Mechanisms of the Various Mutations in the AβPP Gene to Determine Aβ Amyloid Deposition in Different Cerebral Compartments

Mutations in *AβPP* at or near the β- and γ-secretase sites have been shown to cause familial forms of early-onset AD.[144] These mutations increase the production of either the total Aβ or the more amyloidogenic Aβ(1–42) species. In contrast, most mutations within the Aβ domain do not result in a full range of AD pathology but characteristically result in cerebrovascular pathology.[93]

Clinical and experimental studies indicated that the Aβ40:Aβ42 ratio is an important determinant of amyloid formation in different cerebral compartments *in vivo*, that is, whether Aβ primarily deposits in blood vessel walls or brain parenchyma.[93] There are considerable differences between these two major classes of Aβ protein species. Aβ42 aggregates more readily because it nucleates and elongates more efficiently. In contrast, the more soluble Aβ40 has a lower nucleation rate, and it may also have a possible protective role with direct inhibitory effect on Aβ42 aggregation into amyloid both *in vitro* and *in vivo*.[157–159] The relationship between the Aβ40:Aβ42 ratio and the morphological phenotype is based on the different aggregation and fibrillation propensities of the two major classes of Aβ when they are in different compositions.[105] An increase in total cerebral Aβ, with an increase in both Aβ40 and Aβ42 levels, results in an increased degree of amyloid deposition in both cerebral vasculature and parenchyma. An example is the KM670/671NL Swedish double mutation, which affects the two residues located just N-terminal to the β-secretase cleavage site and results in a six- to eightfold increase of both Aβ40 and Aβ42 (Fig. 2).[160,161] The neuropathological phenotype of both human disease and its transgenic animal model is characterized by a mixed plaque and CAA-rich picture.[162,163] In contrast, mutations such as the London mutation (V717I), which are just C-terminal to the γ-secretase cleavage site of the *AβPP*, specifically increase the levels of the more insoluble and fibrillogenic Aβ42 and, as such mutations do not influence total Aβ production, there is a consequent decrease in the Aβ40 to Aβ42 ratio.[144] In such human cases and their transgenic animal models, there is significant parenchymal, but less vascular Aβ deposition.[105]

Both human and experimental data indicate that an increased Aβ40 to Aβ42 ratio, such as that found in affected members of families with HCHWA-D and the transgenic mouse model of this disease, significantly shifts Aβ deposition toward the cerebral vasculature, resulting in prominent CAA.[104] Jucker's group found that neuronal overexpression of human E693Q *AβPP* in mice caused extensive CAA, smooth muscle cell degeneration, hemorrhages, and neuroinflammation.[104] Parenchymal amyloid is nearly absent in these transgenic mice, and the few parenchymal plaques found are diffuse. In contrast, overexpression of human wild-type *AβPP* resulted in predominantly parenchymal amyloidosis, similar to that seen in AD.[104] The observation that neuronal expression of Dutch Aβ is sufficient for cerebrovascular amyloidosis, smooth muscle cell degeneration, and hemorrhage in a mouse model strongly suggests that neurons are the source of the cerebrovascular amyloid in HCHWA-D. Moreover, these results demonstrate that smooth muscle cell degeneration does not require intracellular Aβ production but can be initiated by extracellular, neuron-derived Aβ that is transported to and accumulates at the vasculature.

WT mice expressed AβPP at levels comparable to those in Dutch mice, but the former developed abundant parenchymal plaques and only sparse vascular amyloidosis, suggesting that the single E693Q amino acid substitution is sufficient to target neuron-derived Aβ to the vessel wall. Notably, the Aβ40/Aβ42 ratio was significantly lower in wt mice than in Dutch mice.[104] Thus, a straightforward explanation for why the Dutch mutation leads to CAA could be that it favors the production of Aβ40, which, in turn, is vasculotropic. To examine this hypothesis, the Aβ40/Aβ42 ratio was determined in young transgenic mice before the onset of amyloid deposition, where a twofold higher ratio of Aβ40/Aβ42 was seen in Dutch mice than in wt mice. This suggests that the Dutch mutation affects Aβ40/Aβ42 ratios at the level of Aβ production or clearance. Recent results show that Dutch Aβ40 is more resistant to proteolysis by both neprilysin and IDE and is less efficiently cleared into the blood than is wtAβ40.[142,164,165]

Familial AD-causing *PS1* mutations shift the generation of Aβ to favor Aβ42, which results in early and robust parenchymal amyloid deposition in transgenic mice that produce human wt Aβ.[166–168] Crossing the Dutch mouse with the PS45 line resulted in abundant parenchymal plaque formation at a young age, with limited CAA pathology.[104] Thus, although DutchAβ preferentially accumulates around cerebral vessels, genetically shifting the DutchAβ40/DutchAβ42 ratio to favor DutchAβ42 was sufficient to alter the distribution of the resulting amyloid pathology from the vasculature to the parenchyma. Moreover, this demonstrates that DutchAβ can form dense and congophilic plaques within the parenchyma. Therefore, parenchymal amyloid formation in Dutch mice and humans with HCHWA-D is likely to be limited by the absence of Aβ42-driven parenchymal amyloid seeding.

Yanagisawa's group showed that assembly of hereditary variant Dutch- (E22Q), Italian- (E22K) and Iowa-type (D23N) Aβs, and Flemish-type (A21G) Aβ, was accelerated by GM3 ganglioside and GD3 ganglioside, respectively.[169,170] Notably, cerebrovascular smooth muscle cells, which constitute the cerebral vessel wall at which the Dutch-, Italian- and Iowa-type Aβs deposit, exclusively express GM3, whereas GD3 is upregulated in the coculture of endothelial cells and astrocytes, which forms the cerebrovascular BM, the site of Flemish-type Aβ deposition. They also reported that the Arctic-type (E22G) Aβ, but not the Dutch-type Aβ, rapidly assembles in the presence of GM1 ganglioside (GM1),[171] suggesting that GM1 plays a critical role in the induction of Aβ deposition in the brain parenchyma, the site of Arctic-type Aβ deposition. They recently reported the profile of amyloid deposition in the brains of transgenic mice expressing a Swedish/London double mutant AβPP with a disrupted GM2 synthase gene, in which GM3 accumulates, whereas GM1 is lacking.[172] These mice showed a significantly increased level of deposited Aβ in the vascular tissues. Furthermore, they observed the severe dyshoric form of amyloid angiopathy, in which amyloid extended from the blood vessel walls deeply into the surrounding parenchyma. These results indicate that the expression of gangliosides is also a critical determinant for the amyloid pathology in AD brain.

E. The Mechanisms of the Aβ Amyloid Deposited in Blood Vessel Walls to Cause Cerebral Hemorrhages and Dementia

The initial effect of deposits of Aβ on the walls of capillaries and arteries is a change in the protein composition of BMs. There is a decrease in collagen IV, laminin, and perlecan in artery walls.[103,173] Loss of collagen IV and laminin from capillary BMs occurs at sites of Aβ deposition.[103] In contrast, the amount of fibronectin associated with the walls of capillaries is increased at sites of Aβ deposition.[103]

After Aβ amyloid fibrils are deposited in blood vessel walls, they trigger a secondary cascade of events that include, among others, release of inflammatory components, activation of the complement system, oxidative stress, alteration of BBB permeability, formation of ion-like channels, and cell toxicity.[93] Eventually, smooth muscle cells die and the artery wall becomes almost completely replaced by Aβ.[143] Experiments using a transgenic mouse expressing the Swedish double mutation have shown that degradation of extracellular matrix proteins by MMP-9 may be pivotal in the rupture of CAA arteries.[174]

Dementia in AD is the most important complication resulting from the failure of perivascular elimination of Aβ from the brain (Fig. 6).[103] In the normal young brain, Aβ is produced by neurons and other cells, diffuses through the extracellular spaces, and is either degraded by neprilysin and IDE or absorbed into the blood via LRP-1-mediated mechanisms. Some Aβ also drains with ISF along perivascular pathways in the walls of capillaries and arteries. With age, disposal of Aβ by neprilysin, IDE, and LRP-1-mediated mechanisms fails, and more Aβ is diverted to the perivascular drainage pathways. As arteries stiffen with age, perivascular drainage of Aβ becomes less efficient and ultimately fails because of blockage of the pathways by deposits of amyloid fibrils (CAA). Insoluble Aβ is deposited as plaques in the brain parenchyma, and this interferes with diffusion of Aβ and other solutes through the extracellular spaces. Eventually, perivascular drainage fails and levels of soluble Aβ and other soluble metabolites in the brain rise, resulting in disturbed homeostasis of the neuronal environment, neuronal malfunction, cognitive decline, and dementia (Fig 6).

IV. Reflections

A. Interactions of Aβ with Various Molecules Leading to Amyloid Fibril Formation

The biological membranes or other interfaces as well as the convection of the extracellular fluids in the brain may influence Aβ amyloid fibril formation *in vivo*. In an initial stage of CAA, amyloid fibrils are focally deposited within outer BM of the vessels,[6] clearly indicating that BMs provide a suitable scaffold for the formation of Aβ amyloid fibrils. Moreover, as we discussed above, the perivascular flow of ISF draining Aβ along capillary and artery walls may have a critical role for the development of CAA.[102,103] Therefore, it is essential to establish the *in vitro* experimental system where the effects of biological interfaces and fluid flow can be evaluated accurately and sensitively.

We recently investigated the effects of interfaces and convection on the aggregation behavior of Aβ *in vitro* using fluorescence spectroscopy and fluorescence microscopy with the amyloid specific dye thioflavin T.[175] Above 10 μM, Aβ monomers spontaneously nucleated, irrespective of the presence or absence of an air–water interface (AWI). In contrast, below 5 μM, the presence of appropriate hydrophobic–hydrophilic interfaces, that is, AWI or plastic surface, was essential for the nucleation. These data indicate that the effects of AWI should be carefully eliminated to precisely evaluate the effects of biological interfaces (e. g., BM) for Aβ amyloid fibril formation *in vitro*. With these elaborate *in vitro* as well as *in vivo* experimental systems, the molecular interactions among Aβ peptides, BM components, and other proteins (e.g., serum amyloid-P component, ApoE) should be investigated in the future study.

B. Pathogenesis of Macro- and Microhemorrhages

Two types of CAA-related hemorrhages, that is, microhemorrhages (microbleeds) and macrohemorrhages (macrobleeds), have been described. Is the pathogenesis different between microhemorrhages and macrohemorrhages? The fact that higher numbers, or the new appearance, of microhemorrhages predicted increased risk of symptomatic macrohemorrhage[176] suggests that microhemorrhages and macrohemorrhages could share common pathogenic mechanisms. Interestingly, however, it was reported that the volumes of hemorrhagic lesions showed a two-peak pattern corresponding to microbleeds and macrobleeds, suggesting that they are distinct entities.[177] The size, location, and pathology of vessels responsible for hemorrhages may be different between macrohemorrhages and microhemorrhages. Ruptured vascular lesions responsible for macrohemorrhages cannot be directly observed in pathological examination, because they are completely destroyed. It was suggested that increased vessel wall thickness might predispose to formation of microbleeds relative to macrobleeds.[177]

The reported risk factors for CAA-related hemorrhages included genetic polymorphisms such as *ApoE* ε2,[28,29,38] hypertension,[40] and use of thrombolytic/anticoagulation/antiplatelet therapies.[41–43,178,179] Molecular mechanisms of endothelial and smooth muscle cell injury, as well as vessel rupture, have been under extensive investigation, including toxicity of Aβ oligomers, Aβ-induced oxidative stress, and inflammation in blood vessels.[124,125,174,180] Differences in these factors may contribute to development of the two entities, that is, CAA-related macrohemorrhages or microhemorrhages.

C. What Will Be CAA-Specific Diagnostic Markers?

Currently, MRI is widely used to detect hemorrhages with a characteristic distribution of CAA for the clinical diagnosis of CAA-related hemorrhages.[50–52,56,68,181,182] However, CAA may not present with hemorrhagic lesions, but with other clinical features, such as dementia. Imaging techniques such as PIB-PET have been used to directly image cerebral amyloid deposits. However, not all PET tracers are able to distinguish Aβ from other amyloid proteins, or to distinguish CAA from SPs, though the greater uptake in the occipital lobe may be suggestive of CAA.[63–65]

Will CAA-specific imaging be possible? It was reported that clinical field-strength MRI could directly visualize amyloid plaques, as demonstrated in animal models.[183] Postmortem detection of Aβ deposition in CAA as well as AD with 7.0T MRI was reported.[184] Further progress in high-resolution MRI may enable direct visualization of CAA. Regarding amyloid-PET, resolution of amyloid deposits on PET imaging would not be enough to distinguish CAA from SPs, requiring a new tracer or technique to identify vascular amyloid.

As for biochemical markers, patients with CAA show a decrease of CSF levels of Aβ40 as well as Aβ42, in contrast with AD in which only Aβ42 levels are decreased. The decrease of CSF Aβ40 levels could be related to trapping of Aβ40 in the vasculature in the brain. Blood and brain/CSF levels of Aβ influence each other through transport across the BBB, from brain to blood and from blood to brain, using LRP-1 and receptor for advanced glycation end products (RAGE), respectively. The integrity of blood vessels is disrupted in CAA.[66,185] Changes of plasma as well as CSF levels of Aβ40 and Aβ42 need to be investigated in a longitudinal study of patients with CAA to further clarify their value.

D. Antiamyloid Therapies and CAA

Recent clinical as well as experimental studies with Aβ immunotherapies for AD have demonstrated that CAA has important implications for the future of antiamyloid therapies. Experimental studies with AD model mice indicated that Aβ immunotherapies can induce CAA-related cerebral hemorrhage, although long-term Aβ immunotherapies could decrease CAA.[114,115,186–189]

In clinical trials with Aβ42 immunization for AD (AN1792, Elan), immunized patients showed significantly higher levels of CAA and cortical microhemorrhages and microvascular lesions compared with unimmunized AD controls, although the longest living had virtually complete absence of both plaques and CAA.[116] A CAA-related macrohemorrhage was reported in an AD patient in the phase 2a AN1792 trial.[190] The findings suggest that Aβ immunization results in solubilization of plaque Aβ42, which flows out of the brain via the perivascular pathway, causing an increase of CAA and CAA-related hemorrhages. Eventually, this flow also results in Aβ clearance from the cerebral vasculature.

In the AN1792 trial, 6% of patients treated with the AN1792 suffered menigoencephalitis,[190] and there was perivascular infiltration of lymphocytes around vessels with CAA.[191] In a phase 2 trial of bapineuzumab, a humanized monoclonal anti-Aβ antibody, vasogenic edema occurred in 9.7% of patients, suggesting a vascular permeability change of CAA-laden vessels.[192] These data clearly indicate the possibilities of CAA-related vascular complications in antiamyloid therapies for AD and CAA.

It is important to establish methods for prevention of CAA-related hemorrhages in patients treated with antiamyloid therapies. Some hemorrhage-inducing factors are controllable or avoidable, including hypertension and use of thrombolytic/anticoagulation/antiplatelet therapies.[40–43,179] Future strategies for protection of vessel walls against amyloid-related vascular damage such as toxicity of Aβ oligomers, Aβ-induced oxidative stress, and inflammation need to be developed.

ACKNOWLEDGMENTS

The study was supported in part by a Grant from the Amyloidosis Research Committee from the Ministry of Health, Labor and Welfare, Japan (MY & HN), and by a Grant-in-Aid for Scientific Research from the Ministry of Education, Science, Sports and Culture, Japan (MY & HN). The authors are very grateful to Ms. Etsuko Tsujiguchi for her excellent secretarial work.

REFERENCES

1. Vinters HV, Gilbert JJ. Cerebral amyloid angiopathy: incidence and complications in the aging brain. II. The distribution of amyloid vascular changes. *Stroke* 1983;**14**:924–8.
2. Yamada M, Tsukagoshi H, Otomo E, Hayakawa M. Cerebral amyloid angiopathy in the aged. *J Neurol* 1987;**234**:371–6.

3. Masuda J, Tanaka K, Ueda K, Omae T. Autopsy study of incidence and distribution of cerebral amyloid angiopathy in Hisayama, Japan. *Stroke* 1988;**19**:205–10.

4. Yamada M. Risk factors for cerebral amyloid angiopathy in the elderly. *Ann N Y Acad Sci* 2002;**977**:37–44.

5. Hirohata M, Yoshita M, Ishida C, Ikeda SI, Tamaoka A, Kuzuhara S, et al. Clinical features of non-hypertensive lobar intracerebral hemorrhage related to cerebral amyloid angiopathy. *Eur J Neurol* 2010;**17**:823–9.

6. Yamaguchi H, Yamazaki T, Lemere CA, Frosch MP, Selkoe DJ. β Amyloid is focally deposited within the outer basement membrane in the amyloid angiopathy of Alzheimer's disease. An immunoelectron microscopic study. *Am J Pathol* 1992;**141**:249–59.

7. Scholz W. Studien zur Pathologie der Hirgefässe. II. Die drusige Entartung der Hirnarterien und -capillaren. *Z ges Neurol Psychiat* 1938;**162**:694–715.

8. Thal DR, Ghebremedhin E, Rub U, Yamaguchi H, Del Tredici K, Braak H. Two types of sporadic cerebral amyloid angiopathy. *J Neuropathol Exp Neurol* 2002;**61**:282–93.

9. Glenner GG, Wong CW. Alzheimer's disease: initial report of the purification and characterization of a novel cerebrovascular amyloid protein. *Biochem Biophys Res Commun* 1984;**120**:885–90.

10. Prelli F, Castano E, Glenner GG, Frangione B. Differences between vascular and plaque core amyloid in Alzheimer's disease. *J Neurochem* 1988;**51**:648–51.

11. Suzuki N, Iwatsubo T, Odaka A, Ishibashi Y, Kitada C, Ihara Y. High tissue content of soluble β 1–40 is linked to cerebral amyloid angiopathy. *Am J Pathol* 1994;**145**:452–60.

12. Shinkai Y, Yoshimura M, Ito Y, Odaka A, Suzuki N, Yanagisawa K, et al. Amyloid β-proteins 1-40 and 1-42(43) in the soluble fraction of extra- and intracranial blood vessels. *Ann Neurol* 1995;**38**:421–8.

13. Namba Y, Tomonaga M, Kawasaki H, Otomo E, Ikeda K. Apolipoprotein E immunoreactivity in cerebral amyloid deposits and neurofibrillary tangles in Alzheimer's disease and kuru plaque amyloid in Creutzfeldt-Jakob disease. *Brain Res* 1991;**541**:163–6.

14. Vinters HV, Nishimura GS, Secor DL, Pardridge WM. Immunoreactive A4 and γ-trace peptide colocalization in amyloidotic arteriolar lesions in brains of patients with Alzheimer's disease. *Am J Pathol* 1990;**137**:233–40.

15. Mandybur TI. Cerebral amyloid angiopathy: the vascular pathology and complications. *J Neuropathol Exp Neurol* 1986;**45**:79–90.

16. Vonsattel JP, Myers RH, Hedley-Whyte ET, Ropper AH, Bird ED, Richardson Jr. EP. Cerebral amyloid angiopathy without and with cerebral hemorrhages: a comparative histological study. *Ann Neurol* 1991;**30**:637–49.

17. Maeda A, Yamada M, Itoh Y, Otomo E, Hayakawa M, Miyatake T. Computer-assisted three-dimensional image analysis of cerebral amyloid angiopathy. *Stroke* 1993;**24**:1857–64.

18. Itoh Y, Yamada M, Hayakawa M, Otomo E, Miyatake T. Cerebral amyloid angiopathy: a significant cause of cerebellar as well as lobar cerebral hemorrhage in the elderly. *J Neurol Sci* 1993;**116**:135–41.

19. Gray F, Dubas F, Roullet E, Escourolle R. Leukoencephalopathy in diffuse hemorrhagic cerebral amyloid angiopathy. *Ann Neurol* 1985;**18**:54–9.

20. Soontornniyomkij V, Lynch MD, Mermash S, Pomakian J, Badkoobehi H, Clare R, et al. Cerebral microinfarcts associated with severe cerebral β-amyloid angiopathy. *Brain Pathol* 2010;**20**:459–67.

21. Yamada M, Itoh Y, Shintaku M, Kawamura J, Jensson O, Thorsteinsson L, et al. Immune reactions associated with cerebral amyloid angiopathy. *Stroke* 1996;**27**:1155–62.

22. Oh U, Gupta R, Krakauer JW, Khandji AG, Chin SS, Elkind MS. Reversible leukoencephalopathy associated with cerebral amyloid angiopathy. *Neurology* 2004;**62**:494–7.

23. Eng JA, Frosch MP, Choi K, Rebeck GW, Greenberg SM. Clinical manifestations of cerebral amyloid angiopathy-related inflammation. *Ann Neurol* 2004;**55**:250–6.
24. Scolding NJ, Joseph F, Kirby PA, Mazanti I, Gray F, Mikol J, et al. Aβ-related angiitis: primary angiitis of the central nervous system associated with cerebral amyloid angiopathy. *Brain* 2005;**128**:500–15.
25. Kinnecom C, Lev MH, Wendell L, Smith EE, Rosand J, Frosch MP, et al. Course of cerebral amyloid angiopathy-related inflammation. *Neurology* 2007;**68**:1411–6.
26. Greenberg SM, Rebeck GW, Vonsattel JP, Gomez-Isla T, Hyman BT. Apolipoprotein E epsilon 4 and cerebral hemorrhage associated with amyloid angiopathy. *Ann Neurol* 1995;**38**:254–9.
27. Premkumar DR, Cohen DL, Hedera P, Friedland RP, Kalaria RN. Apolipoprotein E-epsilon4 alleles in cerebral amyloid angiopathy and cerebrovascular pathology associated with Alzheimer's disease. *Am J Pathol* 1996;**148**:2083–95.
28. Nicoll JA, Burnett C, Love S, Graham DI, Dewar D, Ironside JW, et al. High frequency of apolipoprotein E epsilon 2 allele in hemorrhage due to cerebral amyloid angiopathy. *Ann Neurol* 1997;**41**:716–21.
29. O'Donnell HC, Rosand J, Knudsen KA, Furie KL, Segal AZ, Chiu RI, et al. Apolipoprotein E genotype and the risk of recurrent lobar intracerebral hemorrhage. *N Engl J Med* 2000;**342**:240–5.
30. Leclercq PD, Murray LS, Smith C, Graham DI, Nicoll JA, Gentleman SM. Cerebral amyloid angiopathy in traumatic brain injury: association with apolipoprotein E genotype. *J Neurol Neurosurg Psychiatry* 2005;**76**:229–33.
31. Yamada M, Sodeyama N, Itoh Y, Suematsu N, Otomo E, Matsushita M, et al. Association of presenilin-1 polymorphism with cerebral amyloid angiopathy in the elderly. *Stroke* 1997;**28**:2219–21.
32. Yamada M, Sodeyama N, Itoh Y, Suematsu N, Otomo E, Matsushita M, et al. Association of α1-antichymotrypsin polymorphism with cerebral amyloid angiopathy. *Ann Neurol* 1998;**44**:129–31.
33. Yamada M, Sodeyama N, Itoh Y, Takahashi A, Otomo E, Matsushita M, et al. Association of neprilysin polymorphism with cerebral amyloid angiopathy. *J Neurol Neurosurg Psychiatry* 2003;**74**:749–51.
34. Yamada M. Cerebral amyloid angiopathy and gene polymorphisms. *J Neurol Sci* 2004;**226**:41–4.
35. Hamaguchi T, Okino S, Sodeyama N, Itoh Y, Takahashi A, Otomo E, et al. Association of a polymorphism of the transforming growth factor-β1 gene with cerebral amyloid angiopathy. *J Neurol Neurosurg Psychiatry* 2005;**76**:696–9.
36. Christoforidis M, Schober R, Krohn K. Genetic-morphologic association study: association between the low density lipoprotein-receptor related protein (LRP) and cerebral amyloid angiopathy. *Neuropathol Appl Neurobiol* 2005;**31**:11–9.
37. Peila R, Yucesoy B, White LR, Johnson V, Kashon ML, Wu K, et al. A TGF-β1 polymorphism association with dementia and neuropathologies: the HAAS. *Neurobiol Aging* 2007;**28**:1367–73.
38. Domingues-Montanari S, Hernandez-Guillamon M, Fernandez-Cadenas I, Mendioroz M, Boada M, Munuera J, et al. ACE variants and risk of intracerebral hemorrhage recurrence in amyloid angiopathy. *Neurobiol Aging* 2010;**32**:551.e13–22.
39. Thal DR, Papassotiropoulos A, Saido TC, Griffin WS, Mrak RE, Kolsch H, et al. Capillary cerebral amyloid angiopathy identifies a distinct APOE epsilon4-associated subtype of sporadic Alzheimer's disease. *Acta Neuropathol* 2010;**120**:169–83.
40. Arima H, Tzourio C, Anderson C, Woodward M, Bousser MG, MacMahon S, et al. Effects of perindopril-based lowering of blood pressure on intracerebral hemorrhage related to amyloid angiopathy: the PROGRESS trial. *Stroke* 2010;**41**:394–6.

41. McCarron MO, Nicoll JA. Cerebral amyloid angiopathy and thrombolysis-related intracerebral haemorrhage. *Lancet Neurol* 2004;**3**:484–92.

42. Rosand J, Hylek EM, O'Donnell HC, Greenberg SM. Warfarin-associated hemorrhage and cerebral amyloid angiopathy: a genetic and pathologic study. *Neurology* 2000;**55**:947–51.

43. Vernooij MW, Haag MD, van der Lugt A, Hofman A, Krestin GP, Stricker BH, et al. Use of antithrombotic drugs and the presence of cerebral microbleeds: the Rotterdam Scan Study. *Arch Neurol* 2009;**66**:714–20.

44. Yamada M, Itoh Y, Otomo E, Hayakawa M, Miyatake T. Subarachnoid haemorrhage in the elderly: a necropsy study of the association with cerebral amyloid angiopathy. *J Neurol Neurosurg Psychiatry* 1993;**56**:543–7.

45. Ohshima T, Endo T, Nukui H, Ikeda S, Allsop D, Onaya T. Cerebral amyloid angiopathy as a cause of subarachnoid hemorrhage. *Stroke* 1990;**21**:480–3.

46. Greenberg SM, Vonsattel JP, Stakes JW, Gruber M, Finklestein SP. The clinical spectrum of cerebral amyloid angiopathy: presentations without lobar hemorrhage. *Neurology* 1993;**43**:2073–9.

47. Izenberg A, Aviv RI, Demaerschalk BM, Dodick DW, Hopyan J, Black SE, et al. Crescendo transient Aura attacks: a transient ischemic attack mimic caused by focal subarachnoid hemorrhage. *Stroke* 2009;**40**:3725–9.

48. Yamada M, Itoh Y, Suematsu N, Otomo E, Matsushita M. Vascular variant of Alzheimer's disease characterized by severe plaque-like β protein angiopathy. *Dement Geriatr Cogn Disord* 1997;**8**:163–8.

49. Greenberg SM, Gurol ME, Rosand J, Smith EE. Amyloid angiopathy-related vascular cognitive impairment. *Stroke* 2004;**35**:2616–9.

50. Greenberg SM, Finklestein SP, Schaefer PW. Petechial hemorrhages accompanying lobar hemorrhage: detection by gradient-echo MRI. *Neurology* 1996;**46**:1751–4.

51. Koennecke HC. Cerebral microbleeds on MRI: prevalence, associations, and potential clinical implications. *Neurology* 2006;**66**:165–71.

52. Lee SH, Kim SM, Kim N, Yoon BW, Roh JK. Cortico-subcortical distribution of microbleeds is different between hypertension and cerebral amyloid angiopathy. *J Neurol Sci* 2007;**258**:111–4.

53. Nakata-Kudo Y, Mizuno T, Yamada K, Shiga K, Yoshikawa K, Mori S, et al. Microbleeds in Alzheimer disease are more related to cerebral amyloid angiopathy than cerebrovascular disease. *Dement Geriatr Cogn Disord* 2006;**22**:8–14.

54. Rosand J, Muzikansky A, Kumar A, Wisco JJ, Smith EE, Betensky RA, et al. Spatial clustering of hemorrhages in probable cerebral amyloid angiopathy. *Ann Neurol* 2005;**58**:459–62.

55. Feldman HH, Maia LF, Mackenzie IR, Forster BB, Martzke J, Woolfenden A. Superficial siderosis: a potential diagnostic marker of cerebral amyloid angiopathy in Alzheimer disease. *Stroke* 2008;**39**:2894–7.

56. Kumar S, Goddeau Jr. RP, Selim MH, Thomas A, Schlaug G, Alhazzani A, et al. Atraumatic convexal subarachnoid hemorrhage: clinical presentation, imaging patterns, and etiologies. *Neurology* 2010;**74**:893–9.

57. Vernooij MW, Ikram MA, Hofman A, Krestin GP, Breteler MM, van der Lugt A. Superficial siderosis in the general population. *Neurology* 2009;**73**:202–5.

58. Smith EE, Gurol ME, Eng JA, Engel CR, Nguyen TN, Rosand J, et al. White matter lesions, cognition, and recurrent hemorrhage in lobar intracerebral hemorrhage. *Neurology* 2004;**63**:1606–12.

59. Chen YW, Gurol ME, Rosand J, Viswanathan A, Rakich SM, Groover TR, et al. Progression of white matter lesions and hemorrhages in cerebral amyloid angiopathy. *Neurology* 2006;**67**:83–7.

60. Suter OC, Sunthorn T, Kraftsik R, Straubel J, Darekar P, Khalili K, et al. Cerebral hypoperfusion generates cortical watershed microinfarcts in Alzheimer disease. *Stroke* 2002;**33**:1986–92.
61. Kimberly WT, Gilson A, Rost NS, Rosand J, Viswanathan A, Smith EE, et al. Silent ischemic infarcts are associated with hemorrhage burden in cerebral amyloid angiopathy. *Neurology* 2009;**72**:1230–5.
62. Chung YA, O JH, Kim JY, Kim KJ, Ahn KJ. Hypoperfusion and ischemia in cerebral amyloid angiopathy documented by 99mTc-ECD brain perfusion SPECT. *J Nucl Med* 2009;**50**:1969–74.
63. Johnson KA, Gregas M, Becker JA, Kinnecom C, Salat DH, Moran EK, et al. Imaging of amyloid burden and distribution in cerebral amyloid angiopathy. *Ann Neurol* 2007;**62**:229–34.
64. Ly JV, Donnan GA, Villemagne VL, Zavala JA, Ma H, O'Keefe G, et al. 11C-PIB binding is increased in patients with cerebral amyloid angiopathy-related hemorrhage. *Neurology* 2010;**74**:487–93.
65. Bacskai BJ, Frosch MP, Freeman SH, Raymond SB, Augustinack JC, Johnson KA, et al. Molecular imaging with Pittsburgh Compound B confirmed at autopsy: a case report. *Arch Neurol* 2007;**64**:431–4.
66. Verbeek MM, Kremer BP, Rikkert MO, Van Domburg PH, Skehan ME, Greenberg SM. Cerebrospinal fluid amyloid β (40) is decreased in cerebral amyloid angiopathy. *Ann Neurol* 2009;**66**:245–9.
67. Gurol ME, Irizarry MC, Smith EE, Raju S, Diaz-Arrastia R, Bottiglieri T, et al. Plasma β-amyloid and white matter lesions in AD, MCI, and cerebral amyloid angiopathy. *Neurology* 2006;**66**:23–9.
68. Knudsen KA, Rosand J, Karluk D, Greenberg SM. Clinical diagnosis of cerebral amyloid angiopathy: validation of the Boston criteria. *Neurology* 2001;**56**:537–9.
69. Greene GM, Godersky JC, Biller J, Hart MN, Adams Jr. HP. Surgical experience with cerebral amyloid angiopathy. *Stroke* 1990;**21**:1545–9.
70. Izumihara A, Suzuki M, Ishihara T. Recurrence and extension of lobar hemorrhage related to cerebral amyloid angiopathy: multivariate analysis of clinical risk factors. *Surg Neurol* 2005;**64**:160–4 discussion 164.
71. Levy E, Carman MD, Fernandez-Madrid IJ, Power MD, Lieburg I, van Duinen SG, et al. Mutation of the Alzheimer's disease amyloid gene in hereditary cerebral hemorrhage, Dutch type. *Science* 1990;**248**:1124–6.
72. Bornebroek M, De Jonghe C, Haan J, Kumar-Singh S, Younkin S, Roos R, et al. Hereditary cerebral hemorrhage with amyloidosis Dutch type (AβPP 693): decreased plasma amyloid-β 42 concentration. *Neurobiol Dis* 2003;**14**:619–23.
73. Bugiani O, Giaccone G, Rossi G, Mangieri M, Capobianco R, Morbin M, et al. Hereditary cerebral hemorrhage with amyloidosis associated with the E693K mutation of APP. *Arch Neurol* 2010;**67**:987–95.
74. Nilsberth C, Westlind-Danielsson A, Eckman CB, Condron MM, Axelman K, Forsell C, et al. The 'Arctic' APP mutation (E693G) causes Alzheimer's disease by enhanced Aβ protofibril formation. *Nat Neurosci* 2001;**4**:887–93.
75. Hendriks L, van Duijn CM, Cras P, Cruts M, Van Hul W, van Harskamp F, et al. Presenile dementia and cerebral haemorrhage linked to a mutation at codon 692 of the β-amyloid precursor protein gene. *Nat Genet* 1992;**1**:218–21.
76. Cras P, van Harskamp F, Hendriks L, Ceuterick C, van Duijn CM, Stefanko SZ, et al. Presenile Alzheimer dementia characterized by amyloid angiopathy and large amyloid core type senile plaques in the APP 692Ala–>Gly mutation. *Acta Neuropathol* 1998;**96**:253–60.
77. Grabowski TJ, Cho HS, Vonsattel JP, Rebeck GW, Greenberg SM. Novel amyloid precursor protein mutation in an Iowa family with dementia and severe cerebral amyloid angiopathy. *Ann Neurol* 2001;**49**:697–705.

78. Greenberg SM, Grabowski T, Gurol ME, Skehan ME, Nandigam RN, Becker JA, et al. Detection of isolated cerebrovascular β-amyloid with Pittsburgh compound B. *Ann Neurol* 2008;**64**:587–91.

79. Obici L, Demarchi A, de Rosa G, Bellotti V, Marciano S, Donadei S, et al. A novel AβPP mutation exclusively associated with cerebral amyloid angiopathy. *Ann Neurol* 2005;**58**:639–44.

80. Rossi G, Giaccone G, Maletta R, Morbin M, Capobianco R, Mangieri M, et al. A family with Alzheimer disease and strokes associated with A713T mutation of the APP gene. *Neurology* 2004;**63**:910–2.

81. Rovelet-Lecrux A, Hannequin D, Raux G, Le Meur N, Laquerriere A, Vital A, et al. APP locus duplication causes autosomal dominant early-onset Alzheimer disease with cerebral amyloid angiopathy. *Nat Genet* 2006;**38**:24–6.

82. Nochlin D, Bird TD, Nemens EJ, Ball MJ, Sumi SM. Amyloid angiopathy in a Volga German family with Alzheimer's disease and a presenilin-2 mutation (N141I). *Ann Neurol* 1998;**43**:131–5.

83. Dermaut B, Kumar-Singh S, De Jonghe C, Cruts M, Lofgren A, Lubke U, et al. Cerebral amyloid angiopathy is a pathogenic lesion in Alzheimer's disease due to a novel presenilin 1 mutation. *Brain* 2001;**124**:2383–92.

84. Mann DM, Pickering-Brown SM, Takeuchi A, Iwatsubo T. Amyloid angiopathy and variability in amyloid β deposition is determined by mutation position in presenilin-1-linked Alzheimer's disease. *Am J Pathol* 2001;**158**:2165–75.

85. Belza MG, Urich H. Cerebral amyloid angiopathy in Down's syndrome. *Clin Neuropathol* 1986;**5**:257–60.

86. Gudmundsson G, Hallgrimsson J, Jonasson TA, Bjarnason O. Hereditary cerebral haemorrhage with amyloidosis. *Brain* 1972;**95**:387–404.

87. Levy E, Lopez-Otin C, Ghiso J, Geltner D, Frangione B. Stroke in Icelandic patients with hereditary amyloid angiopathy is related to a mutation in the cystatin C gene, an inhibitor of cysteine proteases. *J Exp Med* 1989;**169**:1771–8.

88. Grubb A, Jensson O, Gudmundsson G, Arnason A, Lofberg H, Malm J. Abnormal metabolism of γ-trace alkaline microprotein. The basic defect in hereditary cerebral hemorrhage with amyloidosis. *N Engl J Med* 1984;**311**:1547–9.

89. Janowski R, Kozak M, Jankowska E, Grzonka Z, Grubb A, Abrahamson M, et al. Human cystatin C, an amyloidogenic protein, dimerizes through three-dimensional domain swapping. *Nat Struct Biol* 2001;**8**:316–20.

90. Bjarnadottir M, Nilsson C, Lindstrom V, Westman A, Davidsson P, Thormodsson F, et al. The cerebral hemorrhage-producing cystatin C variant (L68Q) in extracellular fluids. *Amyloid* 2001;**8**:1–10.

91. Palsdottir A, Snorradottir AO, Thorsteinsson L. Hereditary cystatin C amyloid angiopathy: genetic, clinical, and pathological aspects. *Brain Pathol* 2006;**16**:55–9.

92. Ghetti B, Piccardo P, Spillantini MG, Ichimiya Y, Porro M, Perini F, et al. Vascular variant of prion protein cerebral amyloidosis with tau-positive neurofibrillary tangles: the phenotype of the stop codon 145 mutation in PRNP. *Proc Natl Acad Sci USA* 1996;**93**:744–8.

93. Revesz T, Holton JL, Lashley T, Plant G, Frangione B, Rostagno A, et al. Genetics and molecular pathogenesis of sporadic and hereditary cerebral amyloid angiopathies. *Acta Neuropathol* 2009;**118**:115–30.

94. Jansen C, Parchi P, Capellari S, Vermeij AJ, Corrado P, Baas F, et al. Prion protein amyloidosis with divergent phenotype associated with two novel nonsense mutations in PRNP. *Acta Neuropathol* 2010;**119**:189–97.

95. Plant GT, Revesz T, Barnard RO, Harding AE, Gautier-Smith PC. Familial cerebral amyloid angiopathy with nonneuritic amyloid plaque formation. *Brain* 1990;**113**(Pt. 3):721–47.

96. Vidal R, Frangione B, Rostagno A, Mead S, Revesz T, Plant G, et al. A stop-codon mutation in the BRI gene associated with familial British dementia. *Nature* 1999;**399**:776–81.

97. Vidal R, Revesz T, Rostagno A, Kim E, Holton JL, Bek T, et al. A decamer duplication in the 3′ region of the BRI gene originates an amyloid peptide that is associated with dementia in a Danish kindred. *Proc Natl Acad Sci USA* 2000;**97**:4920–5.

98. Yamashita T, Ando Y, Ueda M, Nakamura M, Okamoto S, Zeledon ME, et al. Effect of liver transplantation on transthyretin Tyr114Cys-related cerebral amyloid angiopathy. *Neurology* 2008;**70**:123–8.

99. Kiuru S, Salonen O, Haltia M. Gelsolin-related spinal and cerebral amyloid angiopathy. *Ann Neurol* 1999;**45**:305–11.

100. Schroder R, Deckert M, Linke RP. Novel isolated cerebral ALlambda amyloid angiopathy with widespread subcortical distribution and leukoencephalopathy due to atypical monoclonal plasma cell proliferation, and terminal systemic gammopathy. *J Neuropathol Exp Neurol* 2009;**68**:286–99.

101. Wisniewski HM, Wegiel J. β-Amyloid formation by myocytes of leptomeningeal vessels. *Acta Neuropathol* 1994;**87**:233–41.

102. Weller RO, Massey A, Newman TA, Hutchings M, Kuo YM, Roher AE. Cerebral amyloid angiopathy: amyloid β accumulates in putative interstitial fluid drainage pathways in Alzheimer's disease. *Am J Pathol* 1998;**153**:725–33.

103. Weller RO, Subash M, Preston SD, Mazanti I, Carare RO. Perivascular drainage of amyloid-β peptides from the brain and its failure in cerebral amyloid angiopathy and Alzheimer's disease. *Brain Pathol* 2008;**18**:253–66.

104. Herzig MC, Winkler DT, Burgermeister P, Pfeifer M, Kohler E, Schmidt SD, et al. Aβ is targeted to the vasculature in a mouse model of hereditary cerebral hemorrhage with amyloidosis. *Nat Neurosci* 2004;**7**:954–60.

105. Herzig MC, Van Nostrand WE, Jucker M. Mechanism of cerebral β-amyloid angiopathy: murine and cellular models. *Brain Pathol* 2006;**16**:40–54.

106. Miners JS, Van Helmond Z, Chalmers K, Wilcock G, Love S, Kehoe PG. Decreased expression and activity of neprilysin in Alzheimer disease are associated with cerebral amyloid angiopathy. *J Neuropathol Exp Neurol* 2006;**65**:1012–21.

107. Leissring MA, Farris W, Chang AY, Walsh DM, Wu X, Sun X, et al. Enhanced proteolysis of β-amyloid in APP transgenic mice prevents plaque formation, secondary pathology, and premature death. *Neuron* 2003;**40**:1087–93.

108. Bell RD, Sagare AP, Friedman AE, Bedi GS, Holtzman DM, Deane R, et al. Transport pathways for clearance of human Alzheimer's amyloid β-peptide and apolipoproteins E and J in the mouse central nervous system. *J Cereb Blood Flow Metab* 2007;**27**:909–18.

109. Shibata M, Yamada S, Kumar SR, Calero M, Bading J, Frangione B, et al. Clearance of Alzheimer's amyloid-ss(1-40) peptide from brain by LDL receptor-related protein-1 at the blood-brain barrier. *J Clin Invest* 2000;**106**:1489–99.

110. Abbott NJ. Evidence for bulk flow of brain interstitial fluid: significance for physiology and pathology. *Neurochem Int* 2004;**45**:545–52.

111. Szentistvanyi I, Patlak CS, Ellis RA, Cserr HF. Drainage of interstitial fluid from different regions of rat brain. *Am J Physiol* 1984;**246**:F835–44.

112. Carare RO, Bernardes-Silva M, Newman TA, Page AM, Nicoll JA, Perry VH, et al. Solutes, but not cells, drain from the brain parenchyma along basement membranes of capillaries and arteries: significance for cerebral amyloid angiopathy and neuroimmunology. *Neuropathol Appl Neurobiol* 2008;**34**:131–44.

113. Schley D, Carare-Nnadi R, Please CP, Perry VH, Weller RO. Mechanisms to explain the reverse perivascular transport of solutes out of the brain. *J Theor Biol* 2006;**238**:962–74.

114. Wilcock DM, Rojiani A, Rosenthal A, Subbarao S, Freeman MJ, Gordon MN, et al. Passive immunotherapy against Aβ in aged APP-transgenic mice reverses cognitive deficits and depletes parenchymal amyloid deposits in spite of increased vascular amyloid and microhemorrhage. *J Neuroinflammation* 2004;**1**:24.

115. Pfeifer M, Boncristiano S, Bondolfi L, Stalder A, Deller T, Staufenbiel M, et al. Cerebral hemorrhage after passive anti-Aβ immunotherapy. *Science* 2002;**298**:1379.

116. Boche D, Zotova E, Weller RO, Love S, Neal JW, Pickering RM, et al. Consequence of Aβ immunization on the vasculature of human Alzheimer's disease brain. *Brain* 2008;**131**:3299–310.

117. Harper JD, Lansbury Jr. PT. Models of amyloid seeding in Alzheimer's disease and scrapie: mechanistic truths and physiological consequences of the time-dependent solubility of amyloid proteins. *Annu Rev Biochem* 1997;**66**:385–407.

118. Naiki H, Nagai Y. Molecular pathogenesis of protein misfolding diseases: pathological molecular environments versus quality control systems against misfolded proteins. *J Biochem* 2009;**146**:751–6.

119. Roychaudhuri R, Yang M, Hoshi MM, Teplow DB. Amyloid β-protein assembly and Alzheimer disease. *J Biol Chem* 2009;**284**:4749–53.

120. Glabe CG. Structural classification of toxic amyloid oligomers. *J Biol Chem* 2008;**283**:29639–43.

121. Wu JW, Breydo L, Isas JM, Lee J, Kuznetsov YG, Langen R, et al. Fibrillar oligomers nucleate the oligomerization of monomeric amyloid β but do not seed fibril formation. *J Biol Chem* 2010;**285**:6071–9.

122. Prelli F, Levy E, van Duinen SG, Bots GT, Luyendijk W, Frangione B. Expression of a normal and variant Alzheimer's β-protein gene in amyloid of hereditary cerebral hemorrhage, Dutch type: DNA and protein diagnostic assays. *Biochem Biophys Res Commun* 1990;**170**:301–7.

123. Tomidokoro Y, Rostagno A, Neubert TA, Lu Y, Rebeck GW, Frangione B, et al. Iowa variant of familial Alzheimer's disease: accumulation of posttranslationally modified AβD23N in parenchymal and cerebrovascular amyloid deposits. *Am J Pathol* 2010;**176**:1841–54.

124. Miravalle L, Tokuda T, Chiarle R, Giaccone G, Bugiani O, Tagliavini F, et al. Substitutions at codon 22 of Alzheimer's Aβ peptide induce diverse conformational changes and apoptotic effects in human cerebral endothelial cells. *J Biol Chem* 2000;**275**:27110–6.

125. Van Nostrand WE, Melchor JP, Cho HS, Greenberg SM, Rebeck GW. Pathogenic effects of D23N Iowa mutant amyloid β-protein. *J Biol Chem* 2001;**276**:32860–6.

126. Ghiso J, Frangione B. Amyloidosis and Alzheimer's disease. *Adv Drug Deliv Rev* 2002;**54**:1539–51.

127. Lashley T, Holton JL, Verbeek MM, Rostagno A, Bojsen-Moller M, David G, et al. Molecular chaperones, amyloid and preamyloid lesions in the BRI2 gene-related dementias: a morphological study. *Neuropathol Appl Neurobiol* 2006;**32**:492–504.

128. Verbeek MM, Otte-Holler I, Veerhuis R, Ruiter DJ, De Waal RM. Distribution of A β-associated proteins in cerebrovascular amyloid of Alzheimer's disease. *Acta Neuropathol* 1998;**96**:628–36.

129. Hart M, Li L, Tokunaga T, Lindsey JR, Hassell JR, Snow AD, et al. Overproduction of perlecan core protein in cultured cells and transgenic mice. *J Pathol* 2001;**194**:262–9.

130. Castillo GM, Lukito W, Peskind E, Raskind M, Kirschner DA, Yee AG, et al. Laminin inhibition of β-amyloid protein (Aβ) fibrillogenesis and identification of an Aβ binding site localized to the globular domain repeats on the laminin a chain. *J Neurosci Res* 2000;**62**:451–62.

131. Gorbenko GP, Kinnunen PK. The role of lipid-protein interactions in amyloid-type protein fibril formation. *Chem Phys Lipids* 2006;**141**:72–82.

132. Yanagisawa K. Role of gangliosides in Alzheimer's disease. *Biochim Biophys Acta* 2007;**1768**:1943–51.
133. Matsuzaki K. Physicochemical interactions of amyloid β-peptide with lipid bilayers. *Biochim Biophys Acta* 2007;**1768**:1935–42.
134. Wilson MR, Yerbury JJ, Poon S. Potential roles of abundant extracellular chaperones in the control of amyloid formation and toxicity. *Mol Biosyst* 2008;**4**:42–52.
135. Yerbury JJ, Poon S, Meehan S, Thompson B, Kumita JR, Dobson CM, et al. The extracellular chaperone clusterin influences amyloid formation and toxicity by interacting with prefibrillar structures. *FASEB J* 2007;**21**:2312–22.
136. Yerbury JJ, Kumita JR, Meehan S, Dobson CM, Wilson MR. α2-Macroglobulin and haptoglobin suppress amyloid formation by interacting with prefibrillar protein species. *J Biol Chem* 2009;**284**: 4246–4254.
137. Miners JS, Baig S, Palmer J, Palmer LE, Kehoe PG, Love S. Aβ-degrading enzymes in Alzheimer's disease. *Brain Pathol* 2008;**18**:240–52.
138. Selkoe DJ. Clearing the brain's amyloid cobwebs. *Neuron* 2001;**32**:177–80.
139. Wang YJ, Zhou HD, Zhou XF. Clearance of amyloid-β in Alzheimer's disease: progress, problems and perspectives. *Drug Discov Today* 2006;**11**:931–8.
140. Vardy ER, Catto AJ, Hooper NM. Proteolytic mechanisms in amyloid-β metabolism: therapeutic implications for Alzheimer's disease. *Trends Mol Med* 2005;**11**:464–72.
141. Morelli L, Llovera RE, Mathov I, Lue LF, Frangione B, Ghiso J, et al. Insulin-degrading enzyme in brain microvessels: proteolysis of amyloid β vasculotropic variants and reduced activity in cerebral amyloid angiopathy. *J Biol Chem* 2004;**279**:56004–13.
142. Tsubuki S, Takaki Y, Saido TC. Dutch, Flemish, Italian, and Arctic mutations of APP and resistance of Aβ to physiologically relevant proteolytic degradation. *Lancet* 2003;**361**:1957–8.
143. Revesz T, Ghiso J, Lashley T, Plant G, Rostagno A, Frangione B, et al. Cerebral amyloid angiopathies: a pathologic, biochemical, and genetic view. *J Neuropathol Exp Neurol* 2003;**62**:885–98.
144. Selkoe DJ. Alzheimer's disease: genes, proteins, and therapy. *Physiol Rev* 2001;**81**:741–66.
145. Nagasawa S, Handa H, Okumura A, Naruo Y, Moritake K, Hayashi K. Mechanical properties of human cerebral arteries. Part 1: effects of age and vascular smooth muscle activation. *Surg Neurol* 1979;**12**:297–304.
146. Beach TG, Potter PE, Kuo YM, Emmerling MR, Durham RA, Webster SD, et al. Cholinergic deafferentation of the rabbit cortex: a new animal model of Aβ deposition. *Neurosci Lett* 2000;**283**:9–12.
147. Beach TG, Kuo YM, Spiegel K, Emmerling MR, Sue LI, Kokjohn K, et al. The cholinergic deficit coincides with Aβ deposition at the earliest histopathologic stages of Alzheimer disease. *J Neuropathol Exp Neurol* 2000;**59**:308–13.
148. Weller RO, Nicoll JA. Cerebral amyloid angiopathy: pathogenesis and effects on the ageing and Alzheimer brain. *Neurol Res* 2003;**25**:611–6.
149. Van Dorpe J, Smeijers L, Dewachter I, Nuyens D, Spittaels K, Van Den Haute C, et al. Prominent cerebral amyloid angiopathy in transgenic mice overexpressing the london mutant of human APP in neurons. *Am J Pathol* 2000;**157**:1283–98.
150. Frackowiak J, Potempska A, LeVine H, Haske T, Dickson D, Mazur-Kolecka B. Extracellular deposits of Aβ produced in cultures of Alzheimer disease brain vascular smooth muscle cells. *J Neuropathol Exp Neurol* 2005;**64**:82–90.
151. Nicoll JA, Yamada M, Frackowiak J, Mazur-Kolecka B, Weller RO. Cerebral amyloid angiopathy plays a direct role in the pathogenesis of Alzheimer's disease. Pro-CAA position statement. *Neurobiol Aging* 2004;**25**:589–97 [discussion 603–604].
152. Kisilevsky R, Ancsin JB, Szarek WA, Petanceska S. Heparan sulfate as a therapeutic target in amyloidogenesis: prospects and possible complications. *Amyloid* 2007;**14**:21–32.

153. Farkas E, De Jong GI, de Vos RA, Jansen Steur EN, Luiten PG. Pathological features of cerebral cortical capillaries are doubled in Alzheimer's disease and Parkinson's disease. *Acta Neuropathol* 2000;**100**:395–402.

154. Navarro A, Del Valle E, Astudillo A, Gonzalez del Rey C, Tolivia J. Immunohistochemical study of distribution of apolipoproteins E and D in human cerebral β amyloid deposits. *Exp Neurol* 2003;**184**:697–704.

155. McCarron MO, Nicoll JA, Stewart J, Ironside JW, Mann DM, Love S, et al. The apolipoprotein E epsilon2 allele and the pathological features in cerebral amyloid angiopathy-related hemorrhage. *J Neuropathol Exp Neurol* 1999;**58**:711–8.

156. Holtzman DM. In vivo effects of ApoE and clusterin on amyloid-β metabolism and neuropathology. *J Mol Neurosci* 2004;**23**:247–54.

157. Snyder SW, Ladror US, Wade WS, Wang GT, Barrett LW, Matayoshi ED, et al. Amyloid-β aggregation: selective inhibition of aggregation in mixtures of amyloid with different chain lengths. *Biophys J* 1994;**67**:1216–28.

158. McGowan E, Pickford F, Kim J, Onstead L, Eriksen J, Yu C, et al. Aβ42 is essential for parenchymal and vascular amyloid deposition in mice. *Neuron* 2005;**47**:191–9.

159. Kim J, Onstead L, Randle S, Price R, Smithson L, Zwizinski C, et al. Aβ40 inhibits amyloid deposition in vivo. *J Neurosci* 2007;**27**:627–33.

160. Citron M, Oltersdorf T, Haass C, McConlogue L, Hung AY, Seubert P, et al. Mutation of the β-amyloid precursor protein in familial Alzheimer's disease increases β-protein production. *Nature* 1992;**360**:672–4.

161. Mullan M, Crawford F, Axelman K, Houlden H, Lilius L, Winblad B, et al. A pathogenic mutation for probable Alzheimer's disease in the APP gene at the N-terminus of β-amyloid. *Nat Genet* 1992;**1**:345–7.

162. Lannfelt L, Bogdanovic N, Appelgren H, Axelman K, Lilius L, Hansson G, et al. Amyloid precursor protein mutation causes Alzheimer's disease in a Swedish family. *Neurosci Lett* 1994;**168**:254–6.

163. Calhoun ME, Burgermeister P, Phinney AL, Stalder M, Tolnay M, Wiederhold KH, et al. Neuronal overexpression of mutant amyloid precursor protein results in prominent deposition of cerebrovascular amyloid. *Proc Natl Acad Sci USA* 1999;**96**:14088–93.

164. Morelli L, Llovera R, Gonzalez SA, Affranchino JL, Prelli F, Frangione B, et al. Differential degradation of amyloid β genetic variants associated with hereditary dementia or stroke by insulin-degrading enzyme. *J Biol Chem* 2003;**278**:23221–6.

165. Monro OR, Mackic JB, Yamada S, Segal MB, Ghiso J, Maurer C, et al. Substitution at codon 22 reduces clearance of Alzheimer's amyloid-β peptide from the cerebrospinal fluid and prevents its transport from the central nervous system into blood. *Neurobiol Aging* 2002;**23**:405–12.

166. Citron M, Westaway D, Xia W, Carlson G, Diehl T, Levesque G, et al. Mutant presenilins of Alzheimer's disease increase production of 42-residue amyloid β-protein in both transfected cells and transgenic mice. *Nat Med* 1997;**3**:67–72.

167. Borchelt DR, Ratovitski T, van Lare J, Lee MK, Gonzales V, Jenkins NA, et al. Accelerated amyloid deposition in the brains of transgenic mice coexpressing mutant presenilin 1 and amyloid precursor proteins. *Neuron* 1997;**19**:939–45.

168. Holcomb L, Gordon MN, McGowan E, Yu X, Benkovic S, Jantzen P, et al. Accelerated Alzheimer-type phenotype in transgenic mice carrying both mutant amyloid precursor protein and presenilin 1 transgenes. *Nat Med* 1998;**4**:97–100.

169. Yamamoto N, Hirabayashi Y, Amari M, Yamaguchi H, Romanov G, Van Nostrand WE, et al. Assembly of hereditary amyloid β-protein variants in the presence of favorable gangliosides. *FEBS Lett* 2005;**579**:2185–90.

170. Yamamoto N, Van Nostrand WE, Yanagisawa K. Further evidence of local ganglioside-dependent amyloid β-protein assembly in brain. *Neuroreport* 2006;**17**:1735–7.

171. Yamamoto N, Hasegawa K, Matsuzaki K, Naiki H, Yanagisawa K. Environment- and mutation-dependent aggregation behavior of Alzheimer amyloid β-protein. *J Neurochem* 2004;**90**:62–9.

172. Oikawa N, Yamaguchi H, Ogino K, Taki T, Yuyama K, Yamamoto N, et al. Gangliosides determine the amyloid pathology of Alzheimer's disease. *Neuroreport* 2009;**20**:1043–6.

173. Zhang WW, Lempessi H, Olsson Y. Amyloid angiopathy of the human brain: immunohisto-chemical studies using markers for components of extracellular matrix, smooth muscle actin and endothelial cells. *Acta Neuropathol* 1998;**96**:558–63.

174. Lee JM, Yin K, Hsin I, Chen S, Fryer JD, Holtzman DM, et al. Matrix metalloproteinase-9 in cerebral-amyloid-angiopathy-related hemorrhage. *J Neurol Sci* 2005;**229–230**:249–54.

175. Morinaga A, Hasegawa K, Nomura R, Ookoshi T, Ozawa D, Goto Y, et al. Critical role of interfaces and agitation on the nucleation of Aβ amyloid fibrils at low concentrations of Aβ monomers. *Biochim Biophys Acta* 2010;**1804**:986–95.

176. Greenberg SM, Eng JA, Ning M, Smith EE, Rosand J. Hemorrhage burden predicts recurrent intracerebral hemorrhage after lobar hemorrhage. *Stroke* 2004;**35**:1415–20.

177. Greenberg SM, Nandigam RN, Delgado P, Betensky RA, Rosand J, Viswanathan A, et al. Microbleeds versus macrobleeds: evidence for distinct entities. *Stroke* 2009;**40**:2382–6.

178. Sloan MA, Price TR, Petito CK, Randall AM, Solomon RE, Terrin ML, et al. Clinical features and pathogenesis of intracerebral hemorrhage after rt-PA and heparin therapy for acute myocardial infarction: the Thrombolysis in Myocardial Infarction (TIMI) II Pilot and Randomized Clinical Trial combined experience. *Neurology* 1995;**45**:649–58.

179. Trouillas P, von Kummer R. Classification and pathogenesis of cerebral hemorrhages after thrombolysis in ischemic stroke. *Stroke* 2006;**37**:556–61.

180. Hernandez-Guillamon M, Mawhirt S, Fossati S, Blais S, Pares M, Penalba A, et al. Matrix metalloproteinase 2 (MMP-2) degrades soluble vasculotropic amyloid-β E22Q and L34V mutants, delaying their toxicity for human brain microvascular endothelial cells. *J Biol Chem* 2010;**285**:27144–58.

181. Kim M, Bae HJ, Lee J, Kang L, Lee S, Kim S, et al. APOE epsilon2/epsilon4 polymorphism and cerebral microbleeds on gradient-echo MRI. *Neurology* 2005;**65**:1474–5.

182. Vernooij MW, van der Lugt A, Ikram MA, Wielopolski PA, Niessen WJ, Hofman A, et al. Prevalence and risk factors of cerebral microbleeds: the Rotterdam Scan Study. *Neurology* 2008;**70**:1208–14.

183. Ronald JA, Chen Y, Bernas L, Kitzler HH, Rogers KA, Hegele RA, et al. Clinical field-strength MRI of amyloid plaques induced by low-level cholesterol feeding in rabbits. *Brain* 2009;**132**:1346–54.

184. van Rooden S, Maat-Schieman ML, Nabuurs RJ, van der Weerd L, van Duijn S, van Duinen SG, et al. Cerebral amyloidosis: postmortem detection with human 7.0-T MR imaging system. *Radiology* 2009;**253**:788–96.

185. Bell RD, Zlokovic BV. Neurovascular mechanisms and blood-brain barrier disorder in Alzheimer's disease. *Acta Neuropathol* 2009;**118**:103–13.

186. Racke MM, Boone LI, Hepburn DL, Parsadainian M, Bryan MT, Ness DK, et al. Exacerbation of cerebral amyloid angiopathy-associated microhemorrhage in amyloid precursor protein transgenic mice by immunotherapy is dependent on antibody recognition of deposited forms of amyloid β. *J Neurosci* 2005;**25**:629–36.

187. Wilcock DM, Jantzen PT, Li Q, Morgan D, Gordon MN. Amyloid-β vaccination, but not nitro-nonsteroidal anti-inflammatory drug treatment, increases vascular amyloid and microhemorrhage while both reduce parenchymal amyloid. *Neuroscience* 2007;**144**:950–60.

188. Prada CM, Garcia-Alloza M, Betensky RA, Zhang-Nunes SX, Greenberg SM, Bacskai BJ, et al. Antibody-mediated clearance of amyloid-β peptide from cerebral amyloid angiopathy revealed by quantitative *in vivo* imaging. *J Neurosci* 2007;**27**:1973–80.

189. Schroeter S, Khan K, Barbour R, Doan M, Chen M, Guido T, et al. Immunotherapy reduces vascular amyloid-β in PDAPP mice. *J Neurosci* 2008;**28**:6787–93.
190. Orgogozo JM, Gilman S, Dartigues JF, Laurent B, Puel M, Kirby LC, et al. Subacute meningoencephalitis in a subset of patients with AD after Aβ42 immunization. *Neurology* 2003;**61**:46–54.
191. Nicoll JA, Wilkinson D, Holmes C, Steart P, Markham H, Weller RO. Neuropathology of human Alzheimer disease after immunization with amyloid-β peptide: a case report. *Nat Med* 2003;**9**:448–52.
192. Salloway S, Sperling R, Gilman S, Fox NC, Blennow K, Raskind M, et al. A phase 2 multiple ascending dose trial of bapineuzumab in mild to moderate Alzheimer disease. *Neurology* 2009;**73**:2061–70.

The Genetics of Alzheimer's Disease

LARS BERTRAM* AND RUDOLPH
E. TANZI[†]

*Department of Vertebrate Genomics, Max
Planck Institute for Molecular Genetics,
Berlin, Germany

[†]Genetics and Aging Research Unit,
Massachusetts General Hospital, Harvard
Medical School, Boston, USA

Genetic factors play a major role in determining a person's risk to develop Alzheimer's disease (AD). Rare mutations transmitted in a Mendelian fashion within affected families, for example, APP, PSEN1, and PSEN2, cause AD. In the absence of mutations in these genes, disease risk is largely determined by common polymorphisms that, in concert with each other and nongenetic risk factors, modestly impact risk for AD (e.g., the ε4-allele in APOE). Recent genome-wide screening approaches have revealed several additional AD susceptibility loci and more are likely to be discovered over the coming years. In this chapter, we review the current state of AD genetics research with a particular focus on loci that now can be considered established disease genes. In addition to reviewing the potential pathogenic relevance of these genes, we provide an outlook into the future of AD genetics research based on recent advances in high-throughput sequencing technologies.

Progress in Molecular Biology
and Translational Science, Vol. 107
DOI: 10.1016/B978-0-12-385883-2.00008-4

79

I. Introduction

A. The Genetic Basis of Many Neurodegenerative Disorders is Complex

For Alzheimer's disease (AD), and also for many of the other neurodegenerative disorders, familial aggregation was already recognized as a salient feature decades before any of the underlying molecular genetic and biochemical properties were known.[1] As a matter of fact, it was often only the identification of specific, disease-segregating mutations in previously unknown genes that directed the attention of molecular biologists to certain proteins and pathways that are now considered crucial in the development of the various diseases. The mutated proteins included the amyloid β-protein (Aβ) causing AD, mutations in α-synuclein causing Parkinson's disease, or mutations in microtubule-associated protein tau causing frontotemporal dementia with Parkinsonism. Another feature observed in AD and in most other common neurodegenerative diseases is the prominent dichotomy of familial (rare, often following Mendelian inheritance) versus seemingly nonfamilial (common, following non-Mendelian inheritance) forms of disease. The latter are also frequently described as "sporadic" or "idiopathic" forms, although this terminology has proved overly simplistic because a large proportion of apparently sporadic cases are actually also influenced significantly by genetic factors.

B. The Search for Novel AD Genes

Despite the previous successes and recent advances in molecular and analytic techniques, the identification of genuine risk factors for AD is complicated by several circumstances. First, while diagnostic criteria have been proposed and revised, a "definite" diagnosis of AD can only be made upon neuropathological examination. A diagnosis based on purely clinical grounds without autopsy—which is how the vast majority of AD cases are defined in typical genetic studies—can at best represent "probable" AD. Thus, such a sample may actually be a conglomerate of predominantly Alzheimer's, but also other forms of dementia (i.e., phenocopies), for example, Lewy-body dementia or frontotemporal dementia. A second and related issue is that most cases of AD manifest in old age, that is, beyond 60 or 70 years. This makes the assessment of reliable family histories, a prerequisite for genetic analyses, very difficult because a number of relatives may not have lived through the typical onset age or may suffer from other conditions that can mask or mimic the phenotype of interest. Third, AD displays a large degree of genetic and phenotypic heterogeneity. This means not only that the same phenotype can be caused or modified by a number of different genetic loci and alleles but also that mutations or polymorphisms in the same gene may lead to clinically different syndromes. Further, certain combinations of genetic and nongenetic risk factors may significantly increase the odds to develop a disease in one ethnic group or geographic area, while another set of factors may be acting together elsewhere.

Collectively, these and other factors have led to the proposition of a large number of AD susceptibility genes and environmental risk factors.[2] Until recently, substantial scientific evidence supporting the importance of these genes or risk factors have been lacking, except for a few notable exceptions. This situation has changed to some degree since the advent of massively parallel genotyping (and more recently, sequencing) techniques that now allow interrogation of the genomes of a large number of subjects at varying degrees of resolution. Currently, the most popular approach is based on genome-wide association studies (GWAS) where up to one million genetic markers are simultaneously genotyped and assessed for potential correlations with disease risk and other phenotypic variables (e.g., disease onset, progression, survival). Since 2005, the genetics community has seen a deluge of GWAS, including over a dozen in AD (Table I). While the success rate still varies from study to study, a number of well-replicated AD loci have already emerged from these projects and more are likely to be discovered in the future.

Despite its achievements, the GWAS approach is limited to studying only relatively common types of genetic variation or polymorphisms, that is, those occurring with a frequency greater than $\approx 1\%$ in the general population. It is likely, however, that a substantial portion of the genetic risk underlying common polygenic disorders is actually conferred by rare sequence variants, those occurring with a frequency $<<1\%$, in the general population. *De novo* identification of these rare variants requires resequencing in affected patients, which can now be achieved using novel high-throughput massively parallel sequencing technologies. These can reliably measure any sequence change—common or rare—allowing, for the first time in scientific history, the study of whole genomes at base-pair resolution. This approach has already led to a number of breakthrough discoveries in 2009 and 2010[6,7] and can be expected to become the mainstay of human genetics research by 2020.

Notwithstanding the current difficulties and limitations, genetic analyses have laid the foundation for understanding a wide variety of pathologic mechanisms contributing to neurodegeneration and dementia in AD and other neurodegenerative disorders. In the following sections, we outline the major genetics findings leading to both Mendelian and non-Mendelian forms of AD. More details on the molecular consequences, as well as neuropathogenic mechanisms affected by the various AD genes discussed here, can be found in other chapters of this book.

TABLE I

Overview of All Published Genome-Wide Association Studies in AD

GWAS	Design	Population	No. SNPs	No. AD GWAS (follow-up)	No. CTRL GWAS (follow-up)	"Featured" genes
Grupe (2007)[3]	Case control	USA, UK	17,343	380 (1428)	396 (1666)	*APOE*, *ACAN, BCR, CTSS, EBF3, FAM63A**, GALP, GWA 14q32.13, GWA 7p15.2, LMNA, LOC651924, MYH13, PCK1, PGBD1, TNK1, TRAK2, UBD*
Coon (2007)[82]	Case control	USA, Netherlands#	502,627	446 (415)	290 (260)	*APOE*, *GAB2*
Li (2008)[83]	Case control	Canada, UK	469,438	753 (418)	736 (249)	*APOE*, *GOLM1, GWA 15q21.2, GWA 9p24.3*
Poduslo (2009)[84]	Family based and case control	USA	489,218	9 (199)	10 (225)	*TRPC4AP*
Abraham (2008)[85]	Case control	UK‡	561,494	1082 (–)	1239 (1400)	*APOE*, *LRAT*
	Family based	USA	484,522	941 (1767)	404 (838)	*APOE*, *ATXN1, CD33, GWA 14q31*
Beecham (2009)[86]	Case control	USA^	532,000	492 (238)	496 (220)	*APOE*, *FAM113B*
Carrasquillo (2009)[42]	Case control	USA•	313,504	844 (1547)	1255 (1209)	*APOE*, *PCDH11X*
Lambert (2009)[47]	Case control	Europe‡	~540,000	2035 (3978)	5328 (3297)	*APOE*, *CLU (APOJ), CR1*
Harold (2009)[46]	Case control	USA, Europe•‡	~610,000	3941 (2023)	7848 (2340)	*APOE*, *CLU (APOJ), PICALM*
Heinzen (2009)[87] [CNV]	Case control	Beecham (2009)~	n.g.	331 (–)	368 (–)	*APOE*, *CHRNA7*
Potkin (2009)[88]	Case control	USA (ADNI)†	516,645	172 (–)	209 (–)	*APOE*, *ARSB, CAND1, EFNA5, MAGI2, PRUNE2*

| Seshadri (2010)[55] | Case control | Europe, USA•,‡,# | ~2,540,000 | 3006 (6505) | 22,604 (13,532) | **APOE*, BIN1, CLU (APOJ), EXOC3L2, PICALM** |
| Naj (2010)[89] | Case control | USA, Europe†,#,ˆ | 483,399 | 931 (1338) | 1104 (2003) | **APOE*, MTHFD1L** |

Modified after content from the AlzGene website (http://www.alzgene.org; current on September 27, 2010). Studies are listed in order of publication date (determined by PubMed-ID number). "Featured genes" are those genes or loci that were declared as "associated" in the original publication but note that criteria for declaring association may vary across studies. Genes underlined and in bold font were reported to show experiment-wide "genome-wide significant" association. Numbers of "AD cases" and "controls" refer to sample sizes used in initial GWA screening, whereas "follow-up" refers to follow-up datasets (where applicable). Please consult AlzGene website for more details on these studies. Symbols (•, ‡, #, †, ˆ) indicate sample overlap across studies with identical symbols. *In many studies, surrogate markers were used for APOE. **This locus was originally named "THEM5." Table reprinted with permission from Ref. 5.

II. Early-Onset Familial AD with Mendelian Transmission

A. Mendelian Forms of AD are Rare

Only 5% (or less) of all AD cases can be explained by early-onset familial AD (EOFAD). Despite its rarity, genetic studies of this form of AD are actually facilitated by the availability of large multigenerational pedigrees allowing genetic linkage analysis and subsequent positional cloning. This is usually not possible in LOAD families, in which fewer relatives survive the family-specific onset age and genetic information from parents is almost always lacking (see below). The search for causative mutations is expected to be greatly facilitated through massively parallel sequencing technologies that enable whole genome sequencing in one experiment at decreasing cost.[8,9] The amyloid β-protein (Aβ), which is thought to be the causative agent of AD, is produced through the sequential action of two endoproteinases, β-secretase and γ-secretase, that produce the Aβ N- and C-terminus, respectively.

B. *APP*, *PSEN1*, and *PSEN2* are Causal AD Genes

Using conventional genetics analyses, data were reported in 1987 that showed EOFAD linkage to the long arm of chromosome 21 encompassing a region that harbored the gene encoding the amyloid β-protein precursor (*APP*), a compelling candidate gene for AD.[10] In 1991, the first *APP* missense mutation in a family with EOFAD was described.[11] Since then, over 30 additional AD-causing mutations have been reported in *APP* which, in total, account for probably not more than one-tenth of all early-onset autosomal-dominant AD (Table II; for an up-to-date overview of AD mutations visit the "AD and FTD Mutation Database"; http://www.molgen.ua.ac.be/ADMutations/).[12] Interestingly, most of the *APP* mutations occur near the putative γ-secretase site between residues 714 and 717, suggesting that especially the γ-cleavage event of APP or its (dys)regulation are critical for the development of AD.

TABLE II

GENES CAUSING MENDELIAN FORMS OF ALZHEIMER'S DISEASE

Gene	Protein	Location	# Mutations[a]	Proposed molecular effects/ pathogenic relevance
APP	Amyloid β-protein precursor	21q21	32	Increase in Aβ production or $A\beta_{42}/A\beta_{40}$ ratio
PSEN1	Presenilin 1	14q24	182	Increase in $A\beta_{42}/A\beta_{40}$ ratio
PSEN2	Presenilin 2	1q31	14	Increase in $A\beta_{42}/A\beta_{40}$ ratio

[a]Number of pathogenic mutations as listed on the "AD & FTD mutation database" (current on January 15, 2011; http://www.molgen.ua.ac.be/admutations/).[12]

Recently, two additional *APP* variants were suggested to cause AD by increasing the levels of the wild-type protein. The first variant had a duplication of the *APP*-containing chromosomal segment and caused AD with cerebral amyloid angiopathy.[13] The second variant had promoter mutations thought to increase APP mRNA levels.[14] Both segregated with the disease in an autosomal-dominant fashion in several unrelated families. While more data are needed to estimate the overall contribution of these variants to the prevalence of EOFAD, these discoveries are in line with the decades-old observation that AD neuropathology almost invariably develops in patients with trisomy 21 (Down's syndrome), in which the extra copy of *APP* leads to increases in the expression of APP and in deposition of Aβ.[15]

Only 1 year after the discovery of the first *APP* mutation, a second AD linkage region, at 14q24, was reported almost simultaneously by four independent laboratories[16–18] It took three more years to clone the responsible gene (*PSEN1*) and identify the first AD-causing mutations.[19] It is now known that *PSEN1* encodes a highly conserved polytopic membrane protein, presenilin 1 (PS1) that plays an essential role in mediating intramembranous, γ-secretase processing of APP to generate Aβ from APP.[20] Even more than a decade after the original description of *PSEN1*, there are several new AD-causing mutations reported in this gene every year and the total number of mutations now is greater than 180 (Table II). Soon after the discovery of *PSEN1*, a second member of the presenilin family of proteins was identified by searching the then available databases. It displayed significant homology to *PSEN1* at the genomic, as well as at the protein, level.[21,22] This gene was named *PSEN2* (protein: PS2). The gene maps to the long arm of chromosome 1 and mutations in this gene account for the smallest fraction of all EOFAD cases (Table II). On average, mutations in *PSEN2* also display a later age of onset and slower disease progression than *APP* or *PSEN1* mutations.

In conclusion, while the currently known AD-causing mutations occur in three different genes located on three different chromosomes, they all share a common biochemical pathway, that is, the altered production of Aβ leading to a relative overabundance of the Aβ42 species, which eventually results in neuronal cell death and dementia. Collectively, these discoveries provided the essential connection between the long-known familial aggregation of early-onset AD and the increase in Aβ production observed in the brains of autopsied AD patients, which originally gave rise to the "amyloid cascade hypothesis of AD" (reviewed in Ref. 23).

C. Other Potential EOFAD Genes

Although no additional EOFAD gene has been unequivocally identified since the discovery of *PSEN2* in 1995, several lines of evidence suggest that further genetic factors remain to be identified for this form of AD. Numerous early-onset families do not show mutations in *APP*, *PSEN1*, or *PSEN2*, despite extensive sequencing efforts of open reading frames and adjacent intronic regions. In addition to APP and PS1, there are additional proteins involved in β- and γ-secretase cleavage and in other processes leading to the aggregation and deposition of Aβ (e.g., nicastrin, aph-1, pen-2, BACE). Furthermore, hyperphosphorylation of tau and the development of neurofibrillary tangles are inextricably linked to AD pathogenesis. Finally, a full-genome linkage screen performed by our group has identified at least four early-onset AD linkage regions in addition to the chromosomal location of *PSEN1* on 14q24.[24]

Recent reports have indicated the presence of disease-causing mutations in at least three additional genes, two of which are also strong biochemical candidates for an involvement in AD pathogenesis. First, a linkage study in a large and multigenerational clinically defined multiplex AD family from Belgium indicated the presence of an AD locus near the gene encoding tau on chromosome 17q (*MAPT*). Subsequently, a nonsynonymous mutation in exon 13 of *MAPT* (R406W) was reported to cosegregate with AD dementia in this family.[25] While this same mutation was also reported in at least one other family with dementia resembling AD,[26] the majority of cases affected by R406W appear to develop a syndrome fulfilling the criteria of frontotemporal dementia (FTDP-17[27]). It therefore remains to be determined whether at autopsy the clinically assessed Belgian AD family will show pathological features allowing a definitive diagnosis of AD.

The same group recently reported evidence of significant linkage with EOFAD to chromosome 7q36 in an extended multiplex AD family from the Netherlands.[28] The same ≈10cM haplotype also was found to cosegregate with AD in three additional multiplex families suggesting the presence of a disease-causing mutation in this chromosomal region. A synonymous mutation (Ala626) in the gene encoding PAX transcription activation domain-interacting protein1 (*PAXIP1*), located ≈400,000bp downstream of the shared haplotype region, was discovered in AD patients of the index family, but absent from 320 control individuals. However, since it also was absent from the three additional 7q36 haplotype-sharing families and, according to preliminary analyses, did not show evidence for functional abnormalities in mutation carriers, the overall evidence supporting *PAXIP1* as a novel EOFAD gene remains relatively weak.

Finally, a recent study reported the presence of a missense mutation (Asp90Asn) in the gene encoding one of the γ-secretase components, pen-2 (*PEN2*[29]). In addition to being a strong pathophysiological candidate, this gene is also interesting positionally as it maps close to a highly significant linkage region on chromosome 19, ≈9Mb proximal of *APOE*.[30] However, as the familial transmission of this mutation with AD could not be determined due to a lack of DNA specimens, and since preliminary functional analyses did not reveal an effect of this mutation on APP metabolism *in vitro*, this finding can be considered the least convincing of these putative novel EOFAD loci.

III. Late-Onset AD Without Mendelian Transmission

A. LOAD Shows Complex Inheritance Patterns

In contrast to EOFAD, late-onset Alzheimer's (LOAD, i.e., the onset age is typically >65 years) is characterized by a considerably more complex pattern of genetic and nongenetic factors that remains only partially understood. This complicates the detection of new loci in reasonably sized samples and continues to hamper the independent confirmation of proposed associations. This is demonstrated by the fact that more than a decade after the discovery of APOE in AD,[31] few other strong genetic risk factors have been found, despite intensive efforts in many laboratories worldwide.[2] Although almost none of the nearly 700 genes that have been tested for association with AD over the past 30 years have yielded consistent results, several lines of evidence suggest that further "gene-hunting" may indeed be worthwhile. First, there are a number of chromosomal regions showing evidence for genetic linkage from full-genome linkage studies.[32] Second, a number of loci have recently emerged using GWAS.[33] While overlap in results across independent GWAS is still small, a few new loci have emerged that may be novel AD risk genes (see below and Tables I and III). Finally, systematic meta-analyses on all published AD genetic association studies show significant summary odds ratios for a few genes when all available genotype data are summed across studies (see section below and Fig. I). While many of these may still represent false-positive findings, it is interesting to note that several of the significantly associated variants are actually nonsynonymous or regulators of gene expression, which at least indirectly implies a functional basis for the observed statistical associations.

B. Apolipoprotein E is the Single Most Important Genetic Risk Factor in AD

Until the advent of genome-wide screening technologies (see above), only one gene was an *established* LOAD risk gene, the ε4 allele of the apolipoprotein E gene (APOE).[31] In contrast to most other association-based findings from the pre-GWAS era in AD, the risk effect of APOE-ε4 had been consistently replicated in a large number of studies across many ethnic groups with odds ratios between ≈4 for heterozygous and ≈15 for homozygous carriers of the ε4 allele.[34] The three major alleles of the APOE locus, ε2, ε3, and ε4, correspond to combinations of two amino acid changes at residues 112 and 158 (ε2: Cys_{112}/Cys_{158}; ε3: Cys_{112}/Arg_{158}; ε4: Arg_{112}/Arg_{158}). In addition to the increased risk associated with the ε4 allele, several studies also have reported a weak, albeit significant, protective effect for the less common ε2 allele. Unlike the mutations in the known EOFAD genes, the APOEε4 allele is neither necessary nor sufficient to cause AD but instead operates as a genetic risk modifier by

TABLE III

EXAMPLES OF PROPOSED SUSCEPTIBILITY GENES FOR NON-MENDELIAN FORMS OF ALZHEIMER'S DISEASE

Gene	Protein	Location	% Risk change[a]	Proposed molecular effects/ pathogenic relevance[b]
APOE	Apolipoprotein E	19q13	~400	Aggregation and clearance of Aβ; cholesterol metabolism
BIN1[c]	Bridging integrator 1	2q14	~15	Production and clearance of Aβ
CD33[c]	CD33 molecule (siglec 3)	19q13.3	~10	Innate immune system response
CLU[c]	Clusterin	8p21.1	~10	Aggregation and clearance of Aβ; inflammation
CRI[c]	Complement component (3b/ 4b) receptor 1	1q32	~15	Clearance of Aβ; inflammation
PICALM[c]	Phosphatidylinositol binding Clathrin assembly protein	11q14	~15	Production and clearance of Aβ; synaptic transmission

Only genes or loci showing genome-wide significant ($p \leq 1 \times 10^{-7}$) risk effects and independent replication are included. For an up-to-date overview of these and other potential susceptibility genes, see "AlzGene" database (www.alzgene.org).

[a]Approximate change in disease risk (increase or decrease) per copy of minor allele as compared to non-carriers of minor allele.

[b]Selection of proposed effects (note that the functional evidence for these loci is often scarce; see text for more details).

[c]Indicates genes or loci originally identified by GWAS.

decreasing the age of onset in a dose-dependent manner. Even after the completion of over a dozen GWAS in AD, APOEε4 (or genetic markers highly correlated with it) remains the single most important genetic risk factor for AD, both in terms of effect size and statistical significance (Table III).

Despite its long-known and well-established genetic association, the biochemical consequences of APOEε4 in AD pathogenesis are not yet fully understood. Current hypotheses are based on the observation that Aβ accumulation is clearly enhanced in the brains of carriers as well as in transgenic mice expressing the human ε4 allele and mutant APP (reviewed in Ref. 35). Further, apolipoprotein E normally plays a role in cholesterol transport and lipid metabolism, and APOEε4 predisposes patients to vascular disease as a result of its association with increased plasma cholesterol levels. High plasma cholesterol, in turn, has been correlated with increased Aβ deposition in the brain. Interestingly, cholesterol has also been shown to both increase Aβ production and stabilize the peptide in the brains of transgenic AD mice. Thus, it is possible that APOEε4 confers risk for AD via a mechanism that is shared in common with its effect on vascular disease by increasing a carrier's risk for hypercholesterolemia, as this would also elevate accumulation of Aβ.

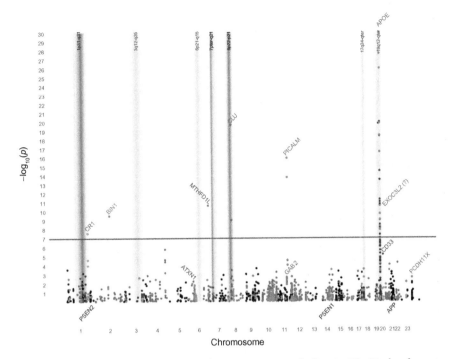

F<small>IG</small>. I. Summary of currently published genetic association findings in AD. Displayed are − log(10) *p*-values (*y*-axis) of all polymorphisms (*n*=2033) with published genetic data available on the AlzGene database (http://www.alzgene.org) on September 27, 2010, listed in genomic order (*x*-axis). Green dots represent *p*-values resulting from random effects allele-based meta-analyses of ≥4 independent datasets using either genotype summary data or effect size estimates provided in the original publications. Black/gray dots represent either single-study *p*-values or the results of meta-analyses on <4 independent datasets. Red horizontal line indicates one common threshold for genome-wide significance (*p*=1×10⁻⁷). Note that *p*-values at the *APOE* locus actually go below 1×10⁻⁵⁰ and are truncated here for display purposes. Vertical columns represent approximate locations of LOAD linkage findings (based on a "narrow definition" of diagnosing AD) as reported in a recent meta-analysis of LOAD linkage studies.[32] Dark columns represent regions that showed "genome-wide suggestive," while light columns showed "genome-wide nominal" evidence for linkage. Genes in blue font represent the approximate locations of the currently known EOFAD genes. Data from both resources were scaled to represent the NCBI36/hg18 build of the human reference genome. Reprinted with permission from Ref. 5. (See Color Insert.)

C. Genome-Wide Association Studies in LOAD

The most promising GWAS findings in AD relate to the identification of variants in or near *BIN1*, *CD33*, *CLU*, *CR1*, and *PICALM*, whose status as novel AD risk loci can now be considered *established* based on extensive independent replication data. Other potential AD susceptibility loci, such as

ATXN1, EXOC3L2, GAB2, MTHFD1L, and *PCDH11X*, should be considered provisional until further replication data become available.[5] While fine-mapping and biochemical studies are still needed to identify the actual sequence variants underlying the observed genetic associations and to confirm and characterize their presumed molecular effects, nearly all of the newly reported loci have been proposed to be linked to Aβ metabolism in one or more ways (Table III). These metabolic effects include Aβ aggregation and clearance of Aβ from the brain, either directly or indirectly, for example, through effects on the immune response to Aβ-related toxicity. However, these potential, Aβ-centered functional connections are still preliminary in most instances, and further research is needed to clarify whether or not other pathways are affected by these loci. Furthermore, it is expected that several additional AD susceptibility variants will be identified in future genome-wide efforts using higher-density microarrays in combination with substantially increased sample sizes. It remains to be seen whether these findings will reveal hitherto unrecognized, novel pathogenic mechanisms beyond those related to the metabolism of APP and Aβ. An up-to-date overview on the status of these and other potential AD candidate genes, including meta-analyses across published genetic association studies, can be found at the "AlzGene" database (URL: http://www.alzgene.org).[2]

D. Examples of Recent GWAS Signals in LOAD

The first genome-wide significant finding was reported for *GAB2* (GRB2-associated binding protein 2) by Ref. 36. This finding has been met with mixed replications and currently shows a p-value of 2.2×10^{-3} in the ongoing AlzGene meta-analyses. Functionally, GAB2 protein may be involved in the production of Aβ, as it binds to Grb2 (growth factor receptor-bound protein 2), which can bind APP and both presenilins.[37] Other data also suggest a potential involvement of Grb2 in tau phosphorylation and NFT formation.[36]

In 2008, our group reported results from a family-based GWAS that *ATXN1* (ataxin 1), *CD33* (siglec 3), and as yet uncharacterized locus on chromosome 14 (GWA 14q31.2)[4] were linked to AD. *ATXN1* causes spinocerebellar ataxia type 1, a rare, Mendelian neurodegenerative disorder caused by polyglutamine expansion in *ATXN1*. Functional genetic experiments suggest that differences in ataxin1 expression can modulate Aβ levels *in vitro*, an effect that appears to be modulated through β-secretase cleavage of APP.[38] *CD33* belongs to the family of sialic acid-binding, immunoglobulin-like lectins that are believed to promote cell–cell interactions and to regulate the functions of cells in the adaptive and innate immune systems,[39,40] both involved in contributing to the inflammatory reactions observed in the brains of AD patients. The innate immune system, in particular, has gained recent attention in AD, owing to the discovery that Aβ may function as an antimicrobial peptide.[41]

In 2009, several AD GWAS were published suggesting the presence of additional AD susceptibility genes. First, Carrasquillo *et al.* highlighted *PCDH11X* (protocadherin 11 X-linked), currently the only GWAS signal on the X-chromosome.[42] If confirmed, this could at least partially explain the well-established increase in AD prevalence in women versus men. In agreement with that notion, the associated alleles were reported to predominantly confer their risk effects (odd ratios (ORs) between 1.25 and 1.75) to female carriers, corroborated by current AlzGene meta-analyses (*p*-value of 2.5×10^{-3}). The encoded protein belongs to the cadherin superfamily, and its Y-chromosome homologue, *PCDH11Y*, is a member of the cadherin family of cell surface receptors that are involved in cell–cell adhesion and signaling, possibly in synaptic junctions.[43,44] Since some protocadherins have been proposed as γ-secretase substrates,[45] it remains to be seen whether or not PCDH11X competes with APP for γ-secretase.

Later in 2009, two large GWAS from the UK[46] and France[47] were published back-to-back highlighting three novel AD genes, that is, *CLU* (clusterin or apolipoprotein J), *CR1* (complement component (3b/4b) receptor 1), and *PICALM* (phosphatidylinositol-binding clathrin assembly protein). All three of these loci have since received overwhelming support from independent follow-up studies and currently rank at the very top of the AlzGene meta-analyses, directly following *APOE*. All three loci show genome-wide significant association in allelic meta-analyses combining all available data with *p*-values of 2.1×10^{-20} (*CLU*; rs11136000), 2.7×10^{-8} (*CR1*; rs3818361), and 1.1×10^{-16} (*PICALM*; rs3851179; Fig. I). In addition, there are several other SNPs in each of these loci that show highly significant association (*p*-values$<1 \times 10^{-5}$) with AD risk, leaving no doubt that these loci represent genuine AD susceptibility genes. Furthermore, it is interesting to note that, like *APOE*, two of these novel AD genes map in or close to previously implicated LOAD linkage regions,[32] *CLU* on chromosome 8p21 and *CR1* on 1q32.2 (Fig. I), demonstrating good agreement between family-derived and case-control data in these instances. Despite the strong statistical support for these findings, it should be emphasized that the effect sizes exerted by these loci are collectively low (allelic ORs≈1.15 for all three loci), which is much less than for *APOE* ε4 (allelic OR≈4), or other established neurodegenerative disease loci (e.g., ORs between ≈1.3 and 3 for the four established Parkinson's susceptibility loci, *SNCA, MAPT, LRRK2, GBA*; www.pdgene.org).

Functionally, the novel loci implicated by Harold *et al.* and Lambert *et al.* can be assigned to two different general groups (Table III), Aβ clearance via lipoprotein vesicles (*CLU, PICALM*) or via the regulation of inflammatory responses (*CLU, CR1*). Clusterin is an ≈75kDa chaperone molecule that is expressed in all tissues, including the CNS. The main associated SNP (rs11136000) lies deeply intronic with no known or implied functional effect.

In addition to the two main links to AD pathophysiology outlined above, clusterin has also been reported to be involved in Aβ fibrillization, regulation of brain cholesterol and lipid metabolism, and the inhibition of neuronal apoptosis/potentiation of neuroprotection.[48] PICALM plays a role in clathrin-mediated endocytosis,[49] which is important in a number of processes such as the regulation of receptor signaling, synaptic transmission, and the removal of apoptotic cells.[50] With respect to AD, it is interesting that the C-terminal fragment of APP generated by β-secretase cleavage undergoes endocytosis via clathrin-coated pits/vesicles before being cleaved by γ-secretase.[51] Thus, dysfunctional PICALM protein could interfere with this process and, ultimately, Aβ production. Furthermore, recent data suggest that brain-expressed PICALM protein is predominately present in endothelial cells, from where it could transport Aβ across vessel walls into the bloodstream.[52] Finally, CR1 is main receptor of the complement C3b protein, which originates from cleaved C3, the central component of complement and a key inflammatory protein activated in AD.[53] *In vitro* and *in vivo* experiments suggest that complement activation can protect against Aβ-induced neurotoxicity and may reduce the accumulation or promote the clearance of amyloid and degenerating neurons.[54]

In 2010, another large GWAS was published suggesting the existence of three additional AD susceptibility loci.[55] This study resulted from a large collaborative effort that also included the GWAS data from four of the five aforementioned studies (Table I). In addition to replicating the association between *CLU* and *PICALM*, this study highlighted two potential additional AD risk factors, *BIN1* (bridging integrator 1; originally implicated at sub-genome-wide significance[46]) and *EXOC3L2* (exocyst complex component 3-like 2) (or a locus nearby on chromosome 19q13.32). Combining all available data, both genes currently display highly significant association with AD risk on AlzGene with p-values $\sim 3.0 \times 10^{-10}$ and $\sim 2.1 \times 10^{-10}$, respectively (Fig. I), and allelic ORs of ~ 1.15. *BIN1* (also known as amphiphysin II) encodes several isoforms of a nucleocytoplasmic adaptor protein, most of which are highly expressed in the CNS and are involved in receptor-mediated endocytosis.[56,57] Thus, as hypothesized for *PICALM*, *BIN1* could have an effect on Aβ production (see above). The predominant mechanism of interest in AD could thus also be the clearance of Aβ from the brain. In addition, rare, homozygous mutations in *BIN1* have been found to cause recessive centronuclear myopathy, a condition characterized by muscle weakness and abnormal centralization of nuclei in muscle fibers.[58] The disease-causing effect is probably triggered by abrogating BIN1's interaction with dynamin 2 which has also been associated with risk for LOAD in candidate gene analyses, albeit inconsistently. The biological function of the protein encoded by *EXOC3L2* remains largely elusive. It should be emphasized, however, that the ~ 100kb region harboring the risk-associated variant (rs597668) on chromosome 19q13.32 contains several other genes (*NKPD1, TRAPPC6A, BLOC1S3, MARKL1*, and *MARK4*) that could also represent the functional correlates underlying this association. It is noteworthy that the associated SNP only maps ≈300 kb distal to the *APOE* region, so it remains to be seen whether these two regions are genetically or functionally related. It also remains possible that rs597668 is merely "tagging" the association with *APOE* and does not actually represent a novel AD locus in its own right.

E. Other Currently Top-Ranked LOAD Candidate Genes

Taking into account all the available association data from both GWAS and candidate gene studies, some three dozen loci show at least nominally significant association with risk for AD (Fig. I), including most of the major GWAS loci outlined above. Applying recently proposed guidelines for the cumulative assessment of genetic association data in complex disorders suggests that at least 10 of these loci show particularly strong "epidemiologic credibility" when all the available genotype data are taken into consideration. *SORL1* (sortilin-related receptor) belongs to a family of sorting receptors that contain a VPS10 (vacuolar protein sorting protein 10) domain, through which these proteins mediate a variety of intracellular sorting and trafficking functions.[59] Biochemical evidence suggests that the protein encoded by *SORL1* directly influences the production of Aβ by affecting the processing or trafficking of APP[60,61] via binding to a complement-type repeat (CR) domain in the *SORL1* protein.[62] *SORL1* shows the strongest association with AD risk.[63] The *GWA 14q32* locus was identified in one of the first GWAS published in AD[3] and has not yet been assigned to a specific gene. Until genetic experiments uncover the functional basis for the observed association(s), its potential pathogenetic relevance remains unclear. *TNK1* (tyrosine kinase, nonreceptor, 1), also originally identified in GWAS,[3] is a nonreceptor tyrosine kinase originally known as "thirty-eight-negative kinase 1" that could be linked to AD via its reported role of enabling tumor necrosis factor alpha (TNFα)-induced apoptosis.[64] Thus, TNK1 may act as a novel molecular switch that can affect TNFα signaling and, potentially, neuronal cell death. *ACE* (angiotensin converting enzyme 1) is a ubiquitously expressed zinc metalloprotease that is involved in blood pressure regulation. Of potential relevance for AD is the observation that the ACE-protein is able to degrade naturally secreted Aβ *in vitro*,[65,66] which could explain the observed increase in AD risk.[67] In addition, its risk-exerting effect also could be the result of ACE's prominent role in blood pressure regulation, as some epidemiological studies suggest that high mid-life blood pressure may increase the risk for AD in later life. *IL8* (interleukin 8, or CXCL8) is a member of the CXC chemokine family and represents a major mediator of inflammatory responses. In AD, IL8 has been reported to show increased expression in the brains of AD patients versus controls, in some studies.[68] Besides potentially playing a role in mediating inflammatory responses accompanying AD neuropathology, a more specific role has recently been suggested in a study showing that the receptor for IL8, CXCR2, may be involved in APP metabolism and Aβ production via modulation of γ-secretase activity.[69] *LDLR* (low-density lipoprotein receptor) is a membrane-spanning glycoprotein that plays a critical role in removing low-density lipoproteins (LDL and VLDL) from the blood and is the main receptor for apoE in neurons. Recent *in vitro* and *in vivo* experiments suggest that LDLR may have an important function in AD neuropathogenesis, in particular, the clearance and deposition of Aβ.[70,71] Finally, *CST3* (cystatin C), which is the most abundant extracellular inhibitor of cysteine proteases, was found to bind Aβ[72] and inhibit Aβ fibril formation in a concentration-dependent manner *in vivo*.[73] These findings, together with the observation that there is a general reduction in neuroprotection when levels of CysC are decreased *in vitro*, may be the functional correlates of the observed epidemiological association between this gene and increased risk in AD.[74]

While many of the proposed biochemical mechanisms for the potential LOAD genes outlined above appear plausible and well supported by functional genetics data, it should be added that the majority of the genes currently highlighted in AlzGene were originally tested for genetic association precisely *because* they emerged as promising functional candidates. This is even true for the top-ranking non-APOE GWAS hit, *CLU*, which owing to its functional relatedness with apoE, actually represents one of first studied candidate genes in AD.[75] At the time, however, lack of power had precluded an earlier recognition of this locus as an AD risk gene. Thus, more functional and molecular genetic experiments are necessary to decipher the precise role of the implicated sequence variants in AD pathogenesis and to determine their potential usefulness in diagnostic or pharmacogenetic settings.

IV. Outlook for Future Genetics Studies of AD

While the systematic collection and analysis of extant data clearly facilitates the interpretation and follow-up of the hitherto proposed candidate genes, the loci investigated to date represent only a fraction of the sequences currently known to exert functional roles in the human genome. The ongoing development and availability of better and more affordable genotyping and high-throughput sequencing technologies will allow for a much more systematic and efficient assessment of the remainder of the genome in the search for common and rare disease modifiers over the coming years. Whole genome sequencing, in particular, will reveal whether or not the currently observed association results all point to the same underlying disease-modifying allele (as is likely the case for ε4 in *APOE*), or whether or not they merely represent a common chromosomal background for much less common mutational events that cannot currently be interrogated by any of the available genome-wide association arrays. It is beyond the scope of this chapter to provide a detailed account of the various available approaches for the generation and analysis of large-scale resequencing data aimed at identifying rare and common variants linked to disease. Due to the vastly increased amount of data, however, the outcome of such projects will depend on careful study design, which includes choosing appropriate ascertainment strategies, maintaining a high phenotyping accuracy, devising a careful analysis plan able to handle the immense multiple testing problem, and successfully and consistently replicating the observed effects in independent samples from different populations before drawing any firm conclusions. This has been achieved in several landmark, proof-of-principle projects (reviewed in Refs. 6,7). These studies not only succeeded in identifying novel Mendelian disease-causing variants in genes previously un-linked to the specific traits,[76–78] but they also were able to "resolve" the complex patterns that typically emerge from GWAS approaches.[79–81] It does not seem unreasonable to expect that in AD as well, such efforts will revolu-tionize our understanding of the true genetic forces underlying disease suscep-tibility, possibly more so than GWAS have taken us beyond *APOE*.

V. Conclusion

Despite the great progress in the field of AD genetics that has led to the discovery and confirmation of three autosomal-dominant early-onset genes (*APP, PSEN1, PSEN1*) and one late-onset risk factor (*APOE*), strong evidence exists suggesting the presence of additional AD genes for both forms of the disease. The hunt for these genes is complicated by factors that include locus or allelic heterogeneity, small effect sizes of the underlying variants, unknown and difficult to model interaction patterns, population differences, insufficient sample sizes or sampling strategies, and linkage disequilibrium among poly-morphisms other than those initially associated with the disease. The emer-gence of more powerful and efficient genotyping and sequencing technologies (e.g., whole genome sequencing) and analysis tools (systematic and continu-ously updated meta-analyses) should enable us to better disentangle the genet-ics of AD and other complex diseases. Eventually, the insights gained from such studies will lead to a better understanding of the pathophysiological mechan-isms leading to neurodegeneration and dementia. Ultimately, this knowledge will lay the foundation for developing new treatment strategies for delaying, preventing, or even curing this devastating disease.

References

1. Bertram L, Tanzi RE. The genetic epidemiology of neurodegenerative disease. *J Clin Invest* 2005;**115**(6):1449–57.
2. Bertram L, McQueen MB, Mullin K, Blacker D, Tanzi RE. Systematic meta-analyses of Alzheimer disease genetic association studies: the AlzGene database. *Nat Genet* 2007;**39**(1):17–23.
3. Grupe A, Abraham R, Li Y, Rowland C, Hollingworth P, Morgan A, et al. Evidence for novel susceptibility genes for late-onset Alzheimer's disease from a genome-wide association study of putative functional variants. *Hum Mol Genet* 2007;**16**(8):865–73.
4. Bertram L, Lange C, Mullin K, Parkinson M, Hsiao M, Hogan MF, et al. Genome-wide association analysis reveals putative Alzheimer's disease susceptibility loci in addition to APOE. *Am J Hum Genet* 2008;**83**(5):623–32.
5. Bertram L, Lill CM, Tanzi RE. The genetics of Alzheimer disease: back to the future. *Neuron* 2010;**68**(2):270–81.
6. Manolio TA, Collins FS, Cox NJ, Goldstein DB, Hindorff LA, Hunter DJ, et al. Finding the missing heritability of complex diseases. *Nature* 2009;**461**(7265):747–53.
7. McClellan J, King M. Genetic heterogeneity in human disease. *Cell* 2010;**141**(2):210–7.
8. Tucker T, Marra M, Friedman JM. Massively parallel sequencing: the next big thing in genetic medicine. *Am J Hum Genet* 2009;**85**(2):142–54.
9. Lupski JR, Reid JG, Gonzaga-Jauregui C, Rio Deiros D, Chen DCY, Nazareth L, et al. Whole-genome sequencing in a patient with Charcot-Marie-Tooth neuropathy. *N Engl J Med* 2010;**362**(13):1181–91.
10. Tanzi RE, Gusella JF, Watkins PC, Bruns GA, St George-Hyslop P, Van Keuren ML, et al. Amyloid beta protein gene: cDNA, mRNA distribution, and genetic linkage near the Alzheimer locus. *Science* 1987;**235**(4791):880–4.
11. Goate A, Chartier-Harlin MC, Mullan M, Brown J, Crawford F, Fidani L, et al. Segregation of a missense mutation in the amyloid precursor protein gene with familial Alzheimer's disease. *Nature* 1991;**349**(6311):704–6.
12. Cruts M, Van Broeckhoven C. Molecular genetics of Alzheimer's disease. *Ann Med* 1998;**30**(6):560–5.
13. Rovelet-Lecrux A, Hannequin D, Raux G, Le Meur N, Laquerrière A, Vital A, et al. APP locus duplication causes autosomal dominant early-onset Alzheimer disease with cerebral amyloid angiopathy. *Nat Genet* 2006;**38**(1):24–6.
14. Brouwers N, Sleegers K, Engelborghs S, Bogaerts V, Serneels S, Kamali K, et al. Genetic risk and transcriptional variability of amyloid precursor protein in Alzheimer's disease. *Brain* 2006;**129**(Pt. 11):2984–91.
15. Wisniewski KE, Dalton AJ, McLachlan C, Wen GY, Wisniewski HM. Alzheimer's disease in Down's syndrome: clinicopathologic studies. *Neurology* 1985;**35**(7):957–61.
16. Mullan M, Houlden H, Windelspecht M, Fidani L, Lombardi C, Diaz P, et al. A locus for familial early-onset Alzheimer's disease on the long arm of chromosome 14, proximal to the alpha 1-antichymotrypsin gene. *Nat Genet* 1992;**2**(4):340–2.
17. Schellenberg GD, Bird TD, Wijsman EM, Orr HT, Anderson L, Nemens E, et al. Genetic linkage evidence for a familial Alzheimer's disease locus on chromosome 14. *Science* 1992;**258**(5082):668–71.
18. Van Broeckhoven C, Backhovens H, Cruts M, De Winter G, Bruyland M, Cras P, et al. Mapping of a gene predisposing to early-onset Alzheimer's disease to chromosome 14q24.3. *Nat Genet* 1992;**2**(4):335–9.
19. Sherrington R, Rogaev EI, Liang Y, Rogaeva EA, Levesque G, Ikeda M, et al. Cloning of a gene bearing missense mutations in early-onset familial Alzheimer's disease. *Nature* 1995;**375**(6534):754–60.

20. Steiner H, Fluhrer R, Haass C. Intramembrane proteolysis by gamma-secretase. *J Biol Chem* 2008;**283**(44):29627–31.
21. Levy-Lahad E, Wasco W, Poorkaj P, Romano DM, Oshima J, Pettingell WH, et al. Candidate gene for the chromosome 1 familial Alzheimer's disease locus. *Science* 1995;**269**(5226):973–7.
22. Rogaev EI, Sherrington R, Rogaeva EA, Levesque G, Ikeda M, Liang Y, et al. Familial Alzheimer's disease in kindreds with missense mutations in a gene on chromosome 1 related to the Alzheimer's disease type 3 gene. *Nature* 1995;**376**(6543):775–8.
23. Tanzi RE, Bertram L. Twenty years of the Alzheimer's disease amyloid hypothesis: a genetic perspective. *Cell* 2005;**120**(4):545–55.
24. Blacker D, Bertram L, Saunders AJ, Moscarillo TJ, Albert MS, Wiener H, et al. Results of a high-resolution genome screen of 437 Alzheimer's disease families. *Hum Mol Genet* 2003;**12**(1):23–32.
25. Rademakers R, Dermaut B, Peeters K, Cruts M, Heutink P, Goate A, et al. Tau (MAPT) mutation Arg406Trp presenting clinically with Alzheimer disease does not share a common founder in Western Europe. *Hum Mutat* 2003;**22**(5):409–11.
26. Ostojic J, Elfgren C, Passant U, Nilsson K, Gustafson L, Lannfelt L, et al. The tau R406W mutation causes progressive presenile dementia with bitemporal atrophy. *Dement Geriatr Cogn Disord* 2004;**17**(4):298–301.
27. Rosso SM, Donker Kaat L, Baks T, Joosse M, de Koning I, Pijnenburg Y, et al. Frontotemporal dementia in The Netherlands: patient characteristics and prevalence estimates from a population-based study. *Brain* 2003;**126**(Pt. 9):2016–22.
28. Rademakers R, Cruts M, Sleegers K, Dermaut B, Theuns J, Aulchenko Y, et al. Linkage and association studies identify a novel locus for Alzheimer disease at 7q36 in a Dutch population-based sample. *Am J Hum Genet* 2005;**77**(4):643–52.
29. Sala Frigerio C, Piscopo P, Calabrese E, Crestini A, Malvezzi Campeggi L, Civita di Fava R, et al. PEN-2 gene mutation in a familial Alzheimer's disease case. *J Neurol* 2005;**252**(9):1033–6.
30. Bertram L, Menon R, Mullin K, Parkinson M, Bradley ML, Blacker D, et al. PEN2 is not a genetic risk factor for Alzheimer's disease in a large family sample. *Neurology* 2004;**62**(2):304–6.
31. Strittmatter WJ, Saunders AM, Schmechel D, Pericak-Vance M, Enghild J, Salvesen GS, et al. Apolipoprotein E: high-avidity binding to beta-amyloid and increased frequency of type 4 allele in late-onset familial Alzheimer disease. *Proc Natl Acad Sci USA* 1993;**90**(5):1977–81.
32. Butler AW, Ng MYM, Hamshere ML, Forabosco P, Wroe R, Al-Chalabi A, et al. Meta-analysis of linkage studies for Alzheimer's disease—a web resource. *Neurobiol Aging* 2009;**30**(7):1037–47.
33. Bertram L, Tanzi RE. Genome-wide association studies in Alzheimer's disease. *Hum Mol Genet* 2009;**18**(R2):R137–45.
34. Farrer LA, Cupples LA, Haines JL, Hyman B, Kukull WA, Mayeux R, et al. Effects of age, sex, and ethnicity on the association between apolipoprotein E genotype and Alzheimer disease. A meta-analysis. APOE and Alzheimer Disease Meta Analysis Consortium. *JAMA* 1997;**278**(16):1349–56.
35. Vance JE, Hayashi H. Formation and function of apolipoprotein E-containing lipoproteins in the nervous system. *Biochim Biophys Acta* 2010;**1801**(8):806–18.
36. Reiman EM, Webster JA, Myers AJ, Hardy J, Dunckley T, Zismann VL, et al. GAB2 alleles modify Alzheimer's risk in APOE epsilon4 carriers. *Neuron* 2007;**54**(5):713–20.
37. Nizzari M, Venezia V, Repetto E, Caorsi V, Magrassi R, Gagliani MC, et al. Amyloid precursor protein and presenilin1 interact with the adaptor GRB2 and modulate ERK 1,2 signaling. *J Biol Chem* 2007;**282**(18):13833–44.
38. Zhang C, Browne A, Child D, Divito JR, Stevenson JA, Tanzi RE. Loss of function of ATXN1 increases amyloid beta-protein levels by potentiating beta-secretase processing of beta-amyloid precursor protein. *J Biol Chem* 2010;**285**(12):8515–26.

39. Crocker PR, Paulson JC, Varki A. Siglecs and their roles in the immune system. *Nat Rev Immunol* 2007;**7**(4):255–66.

40. von Gunten S, Simon H. Sialic acid binding immunoglobulin-like lectins may regulate innate immune responses by modulating the life span of granulocytes. *FASEB J* 2006;**20**(6):601–5.

41. Soscia SJ, Kirby JE, Washicosky KJ, Tucker SM, Ingelsson M, Hyman B, et al. The Alzheimer's disease-associated amyloid beta-protein is an antimicrobial peptide. *PLoS One* 2010;**5**(3): e9505.

42. Carrasquillo MM, Zou F, Pankratz VS, Wilcox SL, Ma L, Walker LP, et al. Genetic variation in PCDH11X is associated with susceptibility to late-onset Alzheimer's disease. *Nat Genet* 2009;**41**(2):192–8.

43. Blanco P, Sargent CA, Boucher CA, Mitchell M, Affara NA. Conservation of PCDHX in mammals; expression of human X/Y genes predominantly in brain. *Mamm Genome* 2000;**11** (10):906–14.

44. Senzaki K, Ogawa M, Yagi T. Proteins of the CNR family are multiple receptors for Reelin. *Cell* 1999;**99**(6):635–47.

45. Haas IG, Frank M, Véron N, Kemler R. Presenilin-dependent processing and nuclear function of gamma-protocadherins. *J Biol Chem* 2005;**280**(10):9313–9.

46. Harold D, Abraham R, Hollingworth P, Sims R, Gerrish A, Hamshere ML, et al. Genome-wide association study identifies variants at CLU and PICALM associated with Alzheimer's disease. *Nat Genet* 2009;**41**(10):1088–93.

47. Lambert J, Heath S, Even G, Campion D, Sleegers K, Hiltunen M, et al. Genome-wide association study identifies variants at CLU and CR1 associated with Alzheimer's disease. *Nat Genet* 2009;**41**(10):1094–9.

48. Nuutinen T, Suuronen T, Kauppinen A, Salminen A. Clusterin: a forgotten player in Alzheimer's disease. *Brain Res Rev* 2009;**61**(2):89–104.

49. Tebar F, Bohlander SK, Sorkin A. Clathrin assembly lymphoid myeloid leukemia (CALM) protein: localization in endocytic-coated pits, interactions with clathrin, and the impact of overexpression on clathrin-mediated traffic. *Mol Biol Cell* 1999;**10**(8):2687–702.

50. Harel A, Wu F, Mattson MP, Morris CM, Yao PJ. Evidence for CALM in directing VAMP2 trafficking. *Traffic* 2008;**9**(3):417–29.

51. Koo EH, Squazzo SL. Evidence that production and release of amyloid beta-protein involves the endocytic pathway. *J Biol Chem* 1994;**269**(26):17386–9.

52. Baig S, Joseph SA, Tayler H, Abraham R, Owen MJ, Williams J, et al. Distribution and expression of picalm in Alzheimer disease. *J Neuropathol Exp Neurol* 2010;**69**(10):1071–7.

53. Khera R, Das N. Complement Receptor 1: disease associations and therapeutic implications. *Mol Immunol* 2009;**46**(5):761–72.

54. Rogers J, Li R, Mastroeni D, Grover A, Leonard B, Ahern G, et al. Peripheral clearance of amyloid beta peptide by complement C3-dependent adherence to erythrocytes. *Neurobiol Aging* 2006;**27**(12):1733–9.

55. Seshadri S, Fitzpatrick AL, Ikram MA, DeStefano AL, Gudnason V, Boada M, et al. Genome-wide analysis of genetic loci associated with Alzheimer disease. *JAMA* 2010;**303**(18):1832–40.

56. Pant S, Sharma M, Patel K, Caplan S, Carr CM, Grant BD. AMPH-1/Amphiphysin/Bin1 functions with RME-1/Ehd1 in endocytic recycling. *Nat Cell Biol* 2009;**11**(12):1399–410.

57. Wigge P, Köhler K, Vallis Y, Doyle CA, Owen D, Hunt SP, et al. Amphiphysin heterodimers: potential role in clathrin-mediated endocytosis. *Mol Biol Cell* 1997;**8**(10):2003–15.

58. Nicot A, Toussaint A, Tosch V, Kretz C, Wallgren-Pettersson C, Iwarsson E, et al. Mutations in amphiphysin 2 (BIN1) disrupt interaction with dynamin 2 and cause autosomal recessive centronuclear myopathy. *Nat Genet* 2007;**39**(9):1134–9.

59. Yamazaki H, Bujo H, Kusunoki J, Seimiya K, Kanaki T, Morisaki N, et al. Elements of neural adhesion molecules and a yeast vacuolar protein sorting receptor are present in a novel

mammalian low density lipoprotein receptor family member. *J Biol Chem* 1996;**271** (40):24761–8.

60. Andersen OM, Reiche J, Schmidt V, Gotthardt M, Spoelgen R, Behlke J, et al. Neuronal sorting protein-related receptor sorLA/LR11 regulates processing of the amyloid precursor protein. *Proc Natl Acad Sci USA* 2005;**102**(38):13461–6.

61. Offe K, Dodson SE, Shoemaker JT, Fritz JJ, Gearing M, Lovcy AI, et al. The lipoprotein receptor LR11 regulates amyloid beta production and amyloid precursor protein traffic in endosomal compartments. *J Neurosci* 2006;**26**(5):1596–603.

62. Andersen OM, Schmidt V, Spoelgen R, Gliemann J, Behlke J, Galatis D, et al. Molecular dissection of the interaction between amyloid precursor protein and its neuronal trafficking receptor SorLA/LR11. *Biochemistry* 2006;**45**(8):2618–28.

63. Rogaeva E, Meng Y, Lee JH, Gu Y, Kawarai T, Zou F, et al. The neuronal sortilin-related receptor SORL1 is genetically associated with Alzheimer disease. *Nat Genet* 2007;**39** (2):168–77.

64. Azoitei N, Brey A, Busch T, Fulda S, Adler G, Seufferlein T. Thirty-eight-negative kinase 1 (TNK1) facilitates TNFalpha-induced apoptosis by blocking NF-kappaB activation. *Oncogene* 2007;**26**(45):6536–45.

65. Hu J, Igarashi A, Kamata M, Nakagawa H. Angiotensin-converting enzyme degrades Alzheimer amyloid beta-peptide (A beta); retards A beta aggregation, deposition, fibril formation; and inhibits cytotoxicity. *J Biol Chem* 2001;**276**(51):47863–8.

66. Hemming ML, Selkoe DJ. Amyloid beta-protein is degraded by cellular angiotensin-converting enzyme (ACE) and elevated by an ACE inhibitor. *J Biol Chem* 2005;**280** (45):37644–50.

67. Kehoe PG, Russ C, McIlory S, Williams H, Holmans P, Holmes C, et al. Variation in DCP1, encoding ACE, is associated with susceptibility to Alzheimer disease. *Nat Genet* 1999;**21** (1):71–2.

68. Mines M, Ding Y, Fan G. The many roles of chemokine receptors in neurodegenerative disorders: emerging new therapeutical strategies. *Curr Med Chem* 2007;**14**(23):2456–70.

69. Bakshi P, Margenthaler E, Laporte V, Crawford F, Mullan M. Novel role of CXCR2 in regulation of gamma-secretase activity. *ACS Chem Biol* 2008;**3**(12):777–89.

70. Abisambra JF, Fiorelli T, Padmanabhan J, Neame P, Wefes I, Potter H. LDLR expression and localization are altered in mouse and human cell culture models of Alzheimer's disease. *PLoS One* 2010;**5**(1):e8556.

71. Kim J, Castellano JM, Jiang H, Basak JM, Parsadanian M, Pham V, et al. Overexpression of low-density lipoprotein receptor in the brain markedly inhibits amyloid deposition and increases extracellular A beta clearance. *Neuron* 2009;**64**(5):632–44.

72. Vinters HV, Nishimura GS, Secor DL, Pardridge WM. Immunoreactive A4 and gamma-trace peptide colocalization in amyloidotic arteriolar lesions in brains of patients with Alzheimer's disease. *Am J Pathol* 1990;**137**(2):233–40.

73. Kaeser SA, Herzig MC, Coomaraswamy J, Kilger E, Selenica M, Winkler DT, et al. Cystatin C modulates cerebral beta-amyloidosis. *Nat Genet* 2007;**39**(12):1437–9.

74. Crawford FC, Freeman MJ, Schinka JA, Abdullah LI, Gold M, Hartman R, et al. A polymorphism in the cystatin C gene is a novel risk factor for late-onset Alzheimer's disease. *Neurology* 2000;**55**(6):763–8.

75. Tycko B, Feng L, Nguyen L, Francis A, Hays A, Chung WY, et al. Polymorphisms in the human apolipoprotein-J/clusterin gene: ethnic variation and distribution in Alzheimer's disease. *Hum Genet* 1996;**98**(4):430–6.

76. Bilgüvar K, Oztürk AK, Louvi A, Kwan KY, Choi M, Tatli B, et al. Whole-exome sequencing identifies recessive WDR62 mutations in severe brain malformations. *Nature* 2010;**467** (7312):207–10.

77. Gilissen C, Arts HH, Hoischen A, Spruijt L, Mans DA, Arts P, et al. Exome sequencing identifies WDR35 variants involved in Sensenbrenner syndrome. *Am J Hum Genet* 2010;**87** (3):418–23.

78. Ng SB, Bigham AW, Buckingham KJ, Hannibal MC, McMillin MJ, Gildersleeve HI, et al. Exome sequencing identifies MLL2 mutations as a cause of Kabuki syndrome. *Nat Genet* 2010;**42**(9):790–3.

79. Dickson SP, Wang K, Krantz I, Hakonarson H, Goldstein DB. Rare variants create synthetic genome-wide associations. *PLoS Biol* 2010;**8**(1):e1000294.

80. Johansen CT, Wang J, Lanktree MB, Cao H, McIntyre AD, Ban MR, et al. Excess of rare variants in genes identified by genome-wide association study of hypertriglyceridemia. *Nat Genet* 2010;**42**(8):684–7.

81. Nejentsev S, Walker N, Riches D, Egholm M, Todd JA. Rare variants of IFIH1, a gene implicated in antiviral responses, protect against type 1 diabetes. *Science* 2009;**324** (5925):387–9.

82. Coon KD, Myers AJ, Craig DW, Webster JA, Pearson JV, Lince DH, et al. A high-density whole-genome association study reveals that APOE is the major susceptibility gene for sporadic late-onset Alzheimer's disease. *J Clin Psychiatry* 2007;**68**(4):613–8.

83. Li H, Wetten S, Li L. St Jean PL, Upmanyu R, Surh L, et al. Candidate single-nucleotide polymorphisms from a genomewide association study of Alzheimer disease. Arch Neurol 2008;**65**(1):45–53.

84. Poduslo SE, Huang R, Huang J, Smith S. Genome screen of late-onset Alzheimer's extended pedigrees identifies TRPC4AP by haplotype analysis. *Am J Med Genet B Neuropsychiatr Genet* 2009;**150B**(1):50–5.

85. Abraham R, Moskvina V, Sims R, Hollingworth P, Morgan A, Georgieva L, et al. A genome-wide association study for late-onset Alzheimer's disease using DNA pooling. *BMC Med Genomics* 2008;**1**:44.

86. Beecham GW, Martin ER, Li Y-J, Slifer MA, Gilbert JR, Haines JL, et al. Genome-wide association study implicates a chromosome 12 risk locus for late-onset Alzheimer disease. *Am J Hum Genet* 2009;**84**(1):35–43.

87. Heinzen EL, Need AC, Hayden KM, Chiba-Falek O, Roses AD, Strittmatter WJ, et al. Genome-wide scan of copy number variation in late-onset Alzheimer's disease. *J Alzheimers Dis* 2010;**19**(1):69–77.

88. Potkin SG, Guffanti G, Lakatos A, Turner JA, Kruggel F, Fallon JH, et al. Hippocampal atrophy as a quantitative trait in a genome-wide association study identifying novel susceptibility genes for Alzheimer's disease. *PLoS ONE* 2009;**4**(8):e6501.

89. Naj AC, Beecham GW, Martin ER, Gallins PJ, Powell EH, Konidari I, et al. Dementia revealed: novel chromosome 6 locus for late-onset Alzheimer disease provides genetic evidence for folate-pathway abnormalities. PLoS Genet [Internet]. 2010 [cited 2011 Feb 21];**6**(9). Available from: http://www.ncbi.nlm.nih.gov/pubmed/20885792.

Alzheimer's Disease and the Amyloid β-Protein

Dominic M. Walsh* and
David B. Teplow[†]

*Laboratory for Neurodegenerative
Research, Center for Neurologic Diseases,
Brigham & Women's Hospital, Harvard
Institutes of Medicine, Boston,
Massachusetts, USA

[†]Department of Neurology and Mary S.
Easton Center for Alzheimer's Disease
Research at UCLA, David Geffen School
of Medicine at UCLA, Los Angeles,
California, USA

Alzheimer's disease is a devastating disorder that is estimated to affect more than 25 million people worldwide and for which there are no preventive, disease-modifying, or curative therapies. Substantial evidence indicates that the amyloid β-protein (Aβ) is a seminal factor in disease causation and may be a tractable therapeutic target. The ability of Aβ to self-associate to form oligomeric assemblies appears to underlie the early toxic events that lead to memory impairment and subsequent neurodegeneration. We review here research on Aβ folding, self-assembly, and toxicity, highlighting areas critical for the development of efficacious Aβ-directed therapeutics.

I. Clinical Features of Alzheimer's Disease

Alzheimer's disease (AD) is a progressive brain disorder that slowly destroys memory and thinking. It is the most common human dementing illness and currently affects an estimated 25 million people, including almost half of those over the age of 85.[1–3] The onset of AD is often said to be imperceptible because the initial disease signs are difficult to differentiate from cognitive changes that frequently accompany normal aging. The earliest symptoms typically are subtle and intermittent deficits in episodic memory, but after many months of gradually progressive impairment of first declarative and then nondeclarative memory, other cognitive symptoms appear and slowly worsen. Over a further period of years, a profound dementia develops that affects multiple cognitive and behavioral spheres.

The diagnosis of AD may involve a detailed medical examination, a review of the patient's history, use of one or more neuropsychological testing measures, interviews of family members, assessment of plasma or CSF markers, or brain imaging. However, in many countries, the use of brain imaging and biomarkers is not common and diagnosis of probable AD is largely based on neuropsychological test performance and the exclusion of other disorders. Tests such as the mini-mental state examination, clinical dementia rating scale, and the Cambridge cognition examination assess cognitive performance over a range of domains.[4–6] The accuracy of such assessments in diagnosing pathologically confirmed AD is >80%,[7,8] but a positive diagnosis of AD currently requires both clinical confirmation of dementia and postmortem detection of neurofibrillary tangles and amyloid plaques in the neocortex of the brain.[9–11] Neurofibrillary tangles are intraneuronal filaments composed of hyperphosphorylated tau protein. Amyloid plaques are extracellular proteinaceous structures that derive their name from the Greek "amylon," meaning starch, bind the histological dye Congo red.[12] However, they are not composed of carbohydrate but contain an array of proteins, the most abundant of which is the amyloid β-protein (Aβ).

II. Genetics and Molecular Biology of Alzheimer's Disease

Many of the current theories about the molecular basis of AD have emerged from the study of plaques and tangles. In 1983, the protein core of senile plaques was isolated and its amino acid composition determined.[13] Shortly thereafter, George Glenner and Caine Wong isolated the principal protein constituent of cerebrovascular amyloid, first from AD brain and then from the brain of an individual with Down's syndrome (DS), and determined the amino acid sequence of the first 24 residues.[14,15] The partial sequence obtained did not match any known protein and led Glenner to name this new protein "β-protein," reflecting the earlier discovery that fibrils extracted from liver or spleen amyloid had a cross-β X-ray diffraction pattern.[16,17] This ≈4-kDa protein is now known as Aβ. The N-terminal sequence of Aβ isolated from AD and DS neuritic plaques was determined a year later.[18] The amino acid composition of the plaque- and cerebral amyloid angiopathy (CAA)-derived proteins was highly similar, and the N-terminal sequences were virtually identical, with the plaque-derived protein exhibiting more N-terminal hetero-geneity than the CAA protein.[14,15,18] The determination of the N-terminal amino acid sequence of Aβ enabled the cloning of the gene encoding Aβ. This gene encodes a large precursor protein, the amyloid β-protein precursor (AβPP, typically referred to as APP).[19,20]

AD-like neuropathology is invariably seen in DS (trisomy of chromosome 21)[21–23] and the location of *APP* to chromosome 21 validated Glenner's earlier prediction "that the genetic defect in Alzheimer's disease (whether acquired or inherited) is localized to chromosome 21."[14,15] These findings supported the hypothesis that increased production of APP, and hence of Aβ, was responsible for the AD-type neuropathology and age-related cognitive deficits so common in DS and that typify AD. The idea that simple overproduction of APP is sufficient to precipitate AD-type dementia has received support from two additional discoveries. First, in a case of DS where the *APP* gene in chromo-some 21 was not triplicated, no signs of dementia or significant amyloid deposition were observed at the time of death (age 78).[24] Second, duplication of the *APP* gene has been detected in more than 10 different families with early onset AD or CAA.[25–29]

The *APP* gene comprises 19 exons.[30] In humans, alternative splicing gives rise to three major transcripts, resulting in proteins of 695, 751, or 770 amino acids. All three isoforms are type 1 transmembrane glycoproteins that under-go extensive posttranslational endoproteolytic processing in the context of two distinct pathways (Fig. 1A). Both pathways involve initial production of a large, N-terminal ectodomain. α-Secretase cleaves APP between K16 and L17 of the Aβ domain[31] generating APPsα and C83. Cleavage at this position precludes formation of Aβ and is mediated by members of the ADAM ("a disintegrin and metalloproteinase") family of proteases, the most relevant of which appear to be ADAM 10 and ADAM 17.[32–35] Proteolysis leading to Aβ production is initiated by an aspartyl protease, β-amyloid cleaving enzyme-1 (BACE). BACE acts immediately N-terminal to the Aβ domain, simulta-neously generating APPsβ and C99 (Fig. 1A).[36,37] BACE also cuts at a second less-favored site 11 residues further C-terminal, producing C89 and a slightly longer APPsβ.

Each of the three C-terminal fragments (CTFs) resulting from α- or β-secretase cleavage, C99, C89, or C83, are substrates for γ-secretase, a unique aspartyl protease complex, the active site of which exists within the presenilin (PS) protein.[38] γ-Secretase is a member of a family of unusual proteases that are able to cleave peptide bonds of protein segments within the plasma membrane.[39] γ-Secretase cleaves at multiple sites.[40] When acting upon α-secretase-cleaved APP, γ-secretase produces a peptide referred to as "p3," based on its molecular weight. The most common forms of both Aβ and p3 terminate at V40 or A42.[41,42]

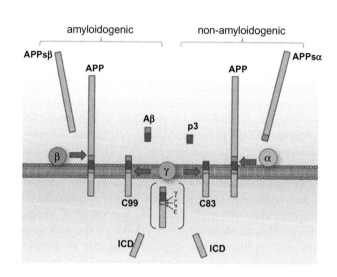

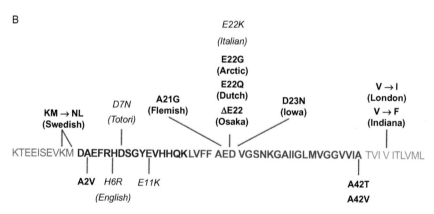

Fig. 1. (A) APP is cleaved by two activities. Proteolysis at the β-site is mediated by β-amyloid cleaving enzyme (BACE) and occurs predominantly between M671 and D672 (APP$_{770}$ numbering) releasing a 596–671-amino acid long secreted fragment (APP$_{sβ}$) and generating a 99-residue membrane-retained C-terminal fragment (CTF, C99). Cleavage at the α-site (between residues 687K and 688L) can be achieved by a number of ADAM ("a disintegrin and metalloprotease") proteases, releasing a secreted fragment, APP$_{sα}$, which is 16 residues longer than APP$_{sβ}$, and a CTF, C83, which is 16 amino acids shorter than C99. The CTFs generated by α- and β-ectodomain shedding serve as substrates for the aspartyl protease, γ-secretase, which cleaves each at multiple sites within the CTF transmembrane domain. γ-Processing of C99 leads to the production of the ≈4-kDa Aβ and the intracellular domain (ICD), and proteolysis of C83 leads to the production of the ≈3-kDa p3 fragment and the corresponding ICD. Extensive evidence suggests that Aβ is centrally involved in AD, but whether p3 and ICD contribute to disease is uncertain. Color coding is described below. (B) The amino acid sequence of Aβ$_{42}$. Orange, APP N-terminal to Aβ; purple, Aβ region N-terminal to the α-secretase cleavage site; red, C-terminus of Aβ after α-secretase

In addition to releasing Aβ or p3 peptides, γ-secretase also produces APP intracellular domains (ICDs). Interestingly, the N-termini of these ICDs start 9–10 amino acids away from the Aβ or p3 C-termini.[43] The explanation for this observation is thought to be that γ-secretase actually cleaves APP CTFs at three or more positions. The first, referred to as the ε-site, occurs 9–10 residues C-terminal to V40 of Aβ, giving rise to a 50–51 residue long ICD.[43,44] A second position, referred to as the ζ-site, exists six residues more N-termini.[40,45] The third site is the γ-site that produces the Aβ or p3 peptides.

Elucidation of this complex proteolytic cleavage process provided mechanistic insights into the causes of familial forms of AD (FAD). Most notably, more than 20 point mutations associated with early onset AD were identified in the *APP* gene (http://www.molgen.ua.ac.be/ADMutations), all of which either occur within or adjacent to the Aβ region of APP (Fig. 1B). Mutations immediately N-terminal to the Aβ domain identified in a Swedish kindred with early onset AD act to increase the amount of Aβ produced throughout life.[46] These mutations, of residues K670–M671 to N670–L671, facilitate BACE cleavage, and because BACE cleavage appears to be the rate-limiting step in Aβ production, the mutations result in increased production of Aβ.[47–49] Mutations C-terminal to the Aβ domain, which occur at more than 10 different sites, all increase the total Aβ production or the $A\beta_{42}/A\beta_{40}$ concentration ratio.[50] As we discuss later, this ratio appears to strongly influence the formation of toxic Aβ assemblies.[51,52] Mutations producing amino acid substitutions within or adjacent to the central hydrophobic cluster of Aβ (L17–A21; Fig. 1B) cause FAD or CAA.[53] The substitutions also can alter the biophysical properties of Aβ, including aggregation pathway, aggregation kinetics, and protease sensitivity.[53,54] Similarly, mutations near the C- and N-termini of the Aβ domain act to increase Aβ production or to enhance aggregation.[55,56] Recently, an E11K substitution was identified in an individual with probable FAD.[57] Expression of this mutant APP in cultured cells caused a reduction of BACE cleavage at position 11 and a compensatory increase in proteolysis at D1.[57]

The fact that *APP* mutations cause AD has focused attention on the gene product Aβ. However, triplication of, and mutations in, *APP* accounts for <20% of early onset FAD cases and <1% of all AD cases.[58,59] In contrast, mutations in the *presenilin 1* and *2* genes account for the vast majority of FAD cases, which represent 1–2% of all AD cases.[58,59] Presenilin (PS) 1 and 2 are multipass transmembrane proteins that possess the active site of γ-secretase (Fig. 1B).[60,61]

---

cleavage; green, APP C-terminal to Aβ. One mutation immediately N-terminal to the Aβ domain of APP, 10 mutations within the Aβ sequence, and 17 mutations (not all shown) C-terminal to the Aβ domain cause familial AD or cerebral amyloid angiopathy (CAA). Italics denote mutations reported in at least one AD case but not confirmed as causative through further genetic analysis. (See Color Insert.)

Mutations in PS-1 and PS-2 vary in how they affect total $A\beta$ production, causing modest increases, no effects, or small decreases. Irrespective of their effects on total $A\beta$ levels, all PS mutations that have been examined *in vitro* act to increase the amount of $A\beta_{42}$ relative to $A\beta_{40}$.[62] Thus, mutations in PS-1 and PS-2 indicate that $A\beta_{42}$ may play a critical role in AD. It should be noted that changes in the $A\beta_{42}/A\beta_{40}$ concentration ratio are widely accepted as an indicator of AD and are being considered as biomarkers to assess the efficacy of putative therapeutics.[63]

The vast majority of late-onset AD is not monogenic in nature. Considerable efforts have been made to identify susceptibility genes, but, whereas many candidates have been suggested, the sole consensus genetic risk factor is *APOE*, the gene encoding apolipoprotein E (ApoE). ApoE is a 299-residue protein involved in lipid transport and metabolism. Three common allelic variants exist that differ only by the presence of either Cys (C) or Arg (R) residues at positions 122 and 158. These are ApoE2 (C/C), ApoE3 (C/R), and ApoE4 (R/R).[64,65] Population prevalence of each genotype is $APO\epsilon3 > APO\epsilon4 > APO\epsilon2$. Among patients with late-onset AD, the incidence of the $\epsilon4$ allele is much higher, and that of the $\epsilon2$ much lower, than expected.[66–68] Moreover, $APOE\epsilon4$ carriers suffering from AD have a higher amyloid burden compared to other AD patients,[64,67,68] and the course of their disease appears more aggressive.[69–71] At a molecular level, ApoE appears to affect $A\beta$ aggregation and clearance in an isoform-specific fashion, with ApoE4 accelerating aggregation[72] and decreasing $A\beta$ clearance relative to ApoE2 and ApoE3.[73] The advent of genome-wide association studies likely will reveal new susceptibility genes[58,59], contributing to a deeper understanding of AD etiology.

III. The Amyloid Cascade Hypothesis

The centrality of $A\beta$ in AD was predicted by Glenner and Wong even before the identification of the *APP* gene. However, this prediction gained strong support when the first disease-causing mutations in APP were discovered.[74,75] This led some to posit that because mutations in APP cause AD, and the proteolytic product of APP, $A\beta$, is found invariably in AD patients in the form of amyloid plaques, $A\beta$ plaques must be central to the disease process.[76–79] Amyloid plaques were considered the key pathogenetic factor in AD, one that led to a "cascade" of events culminating in the neuronal compromise and death characterizing AD. Perturbations in $A\beta$ homeostatis that increased steady-state levels of $A\beta$ thus would be expected to increase aggregation of $A\beta$ into amyloid plaques. Once formed, plaques would cause activation of microglia and astrocytes and local neuritic damage, which, in turn, would lead to the formation of NFTs, further neuronal compromise, depletion of neurotransmitters, and, finally, all the clinical signs of the dementing illness AD.

Over the past two decades, researchers have built up a detailed understanding of APP and Aβ biology and, consequently, additional nongenetic evidence has emerged further supporting a seminal role for Aβ in AD. For example, forms of Aβ are toxic to neurons[80] and can trigger disease-relevant changes in tau.[81–87] AD-causing mutations have not been detected in the gene encoding tau, and the finding that Aβ can induce pathologically relevant changes in tau aggregation and phosphorylation seems to confirm that changes in Aβ precede alterations in tau.

IV. The Amyloid β-Protein—Lessons from *In Vitro* Studies

There is a general consensus that Aβ is a seminal pathologic agent in AD, yet how Aβ mediates its toxic effect(s) remains incompletely understood. Specifically, what forms of Aβ cause toxicity and what the toxic pathways activated by Aβ are remain unresolved. Here, we will focus on the former question, which itself comprises two interrelated parts, namely, Aβ primary structure and Aβ aggregation. The best-studied aspect of Aβ primary sequence is the length of the C-terminal domain. Aβ terminating at V40 is the principal form of Aβ detected in plasma and CSF, but Aβ terminating at A42, although present in humans in $\sim$10-fold lower concentration,[88] appears to be particularly important in disease. $A\beta_{42}$ deposits in the parenchyma before $A\beta_{40}$,[89–91] and a decreased CSF $A\beta_{42}/A\beta_{40}$ concentration ratio is diagnostic of AD.[92,93] $A\beta_{42}$ aggregates faster *in vitro* than does $A\beta_{40}$,[94] suggesting that toxicity and the ability to aggregate are related. Indeed, numerous studies have found that aggregation of Aβ is a prerequisite for toxicity.[95–99] However, the actual assembly form or forms of Aβ that cause toxicity were not identified, and the relative contribution of different ratios of $A\beta_{40}$ and $A\beta_{42}$ to toxicity have only recently been addressed.[51,52]

A major outcome of research in the first decade following the determination of the Aβ sequence was the demonstration that aggregation leading to the formation of amyloid fibrils was associated with toxicity, whereas aggregation of Aβ into so-called amorphous assemblies did not produce toxicity.[98,100] A reasonable interpretation of these data was that amyloid fibrils were the proximate toxic agent in AD. This interpretation was consistent with the fact that amyloid plaques were primarily composed of amyloid fibrils similar to those formed *in vitro*.[101–103] However, the presence of abundant amyloid in the brains of some cognitively normal individuals, and the fact that the amount of brain amyloid does not correlate as strongly with clinical state as do levels of tau deposition[104–106] conflicted directly with the amyloid hypothesis.

Continued hypothesis testing has weakened the original "amyloid cascade hypothesis," leading to what has been called the "oligomer cascade hypothesis."[107] An important trigger that led to the hypothesis reappraisal process was finding that in multiple APP transgenic mouse lines, cognitive impairment was observed prior to frank Aβ deposition.[108–114] Importantly, many different forms of nonfibrillar Aβ assemblies mediate neuronal dysfunction.[54] Rigorous structure–activity relationships (SAR) can be determined only if the structural characterization of a putative active entity is sufficiently detailed. The majority of SAR studies extant have not achieved this goal. Assembly state and structure generally have been poorly defined. This is a natural consequence of the fact that Aβ is an intrinsically disordered protein that possesses great conformational diversity and exhibits substantial assembly polydispersity and metastability.[54] These factors have led some to suggest that it is the *process* of polymerization, rather than a discrete Aβ assembly that causes neuronal death.[52,115] More recent studies, however, have achieved the goal of establishing formal oligomer size–neurotoxicity relationships.[107] A key aspect of these studies was the ability to stabilize the different oligomers species so that their structures were constant during biological assays. In these studies, and in others that do not involve stable assemblies, it is essential that the duration of the biological assay is sufficiently short so as to preclude or minimize changes in the structures of the assemblies under investigation.[80,116] We now turn to some of the most commonly encountered and better-characterized Aβ preparations, giving special attention to studies using rapid bioassays.

Protofibrils (PFs) are polydisperse curvilinear structures of ≈5 nm diameter that generally do not exceed 150–200 nm in length. PFs have been characterized using size exclusion chromatography (SEC), quasielastic light scattering (QLS), multiangle light scattering (MALS), analytical ultracentrifugation (AUC), electron microscopy (EM), and atomic force microscopy (AFM). Temporal studies of PF assembly have shown that they are precursors to mature amyloid-type fibrils. For this reason, they were termed *proto*fibrils to indicate that they are the first or original fibrillar assembly observed during Aβ self-association. PFs share some physical properties with amyloid fibrils, including significant β-sheet content, binding of thioflavin T and Congo red, and neurotoxic activity.[117–122] PF formation is dependent on concentration, pH, ionic strength, and cosolvent content (e.g., fluorinated alcohols, free fatty acids, lipids). Under conditions where there appeared to be little conversion of PFs to fibrils, addition of PFs to rat cortical cultures caused a time-dependent decrease in neuronal viability as measured by LDH release or an increase in Hoechst staining.[123,124] Given the aforementioned caveats about establishing SAR with specific Aβ preparations, it is important to note that PFs were found to cause a dose-dependent inhibition of MTT metabolism by primary neuronal cells that was detectable after only 2h of incubation.[117] In addition, an almost immediate enhancement of electrical activity of neurons[123] and a robust block of LTP[125,126] have also been observed. It has also been demonstrated that certain naturally occurring lipids cause disassembly of fibrils into protofibrils that can spread rapidly throughout the mouse brain and alter learned behavior after intraventricular injection.[127] Compelling evidence for PF involvement in AD comes from the "Arctic" form of FAD, which is caused by an E22G amino acid substitution that increases PF formation.[128]

Annular structures with external and internal diameters of ~8–14 and ~2–4nm, respectively, have been detected.[122,129–133] These types of assemblies form readily and seem to be particularly well populated in preparations of the Arctic Aβ peptide.[129,134] Recent data suggest that annular assemblies cannot directly form fibrils and that annular structures highly similar to those produced *in vitro* can be immunoprecipitated from water-soluble extracts of human brain.[135] It has been speculated that annuli can form pores on neuronal membranes and disrupt cellular homeostasis in a manner analogous to bacterial pore-forming toxins.[121,136,137]

Aβ-derived diffusible ligands (ADDLs) refer to a mixture of different sized Aβ species defined by their method of preparation.[138] Originally, ADDLs were described in AFM studies as relatively homogeneous imperfect spheres ~5–6 nm in diameter and with molecular weights consistent with dodecamers. However, subsequent analyses using SEC, light scattering, and AUC indicated that the classic "Klein ADDL preparation" contains a mixture of different species, including Aβ assemblies that are not well detected by EM or AFM.[139–141] ADDL preparations contain species ranging from monomer to assemblies with molecular weights approaching 1 million.[139] SEC analysis of such material reveals two major species: a peak that elutes in the void (of both Superdex 75 and 200 columns) with a size distribution in the range of 100–1000 kDa, and an included peak with an estimated size consistent with that of a monomer.[139,140] AUC analysis of unfractionated ADDLs also revealed a mixture of different sized species, including Aβ monomer and a mixture of species with calculated masses ranging from 90 to 400kDa.[141] The relative abundance of the high molecular weight species and monomer appears to vary. For instance, some groups found monomer to be the major species in their ADDL preparations,[139,142–144] whereas others found the high molecular weight species to predominate.[140,141] The ability of ADDLs to form fibrils has not been determined *in vitro*. However, recent *in vivo* experiments have shown that, when biotinylated ADDLs were injected into the brains of APP transgenic mice, these ADDLs were incorporated into growing plaques and could seed the formation of new plaques. These results suggest that ADDLs can assemble into amyloid fibrils.[144] ADDLs and PFs share many physical similarities, and thus it may be that ADDLs are simply lower-molecular-weight forms of PFs.

Like PFs, ADDLs have also been shown to be cytotoxic. ADDLs cause neuronal death, block LTP,[138,145] inhibit reduction of MTT by neural cells,[145–147] and initiate synaptic and dendritic degeneration.[148,149] When incubated with organotypic mouse brain slice cultures for 24h, low concentrations of ADDLs (5nM) caused ≈20% loss in cell number. At higher concentration (500nM) and brief incubation periods (45–60min), cell loss was not evident but a near-complete abrogation of LTP was observed.[138,145] Consistent with their synaptotoxic activity, ADDLs have been shown to avidly bind and decorate dendritic arbors of certain cultured neurons[140,148,150] and to mediate dendritic spine loss.[149] Interestingly, as with Aβ derived from human brain,[141,151] the synaptotoxic effects of ADDLs can be blocked with certain anti-PrP antibodies.[140,141] Further support for the *in vivo* relevance of ADDLs comes from studies using antibodies specific for ADDLs. Antibodies M93 and M94 prevented brain-derived Aβ binding to the surface of cultured hippocampal neurons,[150,152] and M93 dot blot analysis of extracts from AD and control brains revealed dramatically higher immunoreactivity in AD brain extracts.[150] Similarly, another "anti-ADDL" antibody, ACU-954, stained amyloid deposits throughout the laminae of AD cortex, with labeling seen in and around neuritic plaques, but staining was not evident in control brain.[144]

A variety of other nonfibrillar preparations have been found to be neurotoxic.[126,153–158] For example, Deshpande *et al.* examined the effects of high molecular weight oligomers (formed as described in Ref. 159), ADDLs, and fibrillar Aβ. All three were toxic to primary human cortical neurons, but the extent and mechanism of toxicity differed.[160] Low concentrations (5µM) of high molecular weight synthetic oligomers caused widespread death within 24h, whereas similar concentrations of ADDLs took five times longer to cause cell loss, and fourfold higher concentrations of fibrillar Aβ took 10 days to induce only modest cell death. High molecular weight oligomers and ADDLs both bound rapidly and avidly to synaptic contacts. High molecular weight oligomers caused activation of the mitochondrial death pathway, but activation of this pathway also occurred when *sublethal* levels of the same oligomers were used, suggesting that such changes may underlie defective synaptic activity in neurons that are still viable. Whether the differences in overt neuronal loss observed using these different assemblies reflects innate differences in toxic activity, or disparities in the extent of ongoing polymerization, is unclear[115,122] and serves to further underscore the difficulty in ascribing cytotoxic activity to a discrete species. Together, these *in vitro* studies demonstrate that Aβ toxicity is a complex and multifaceted phenomenon that may be induced by multiple assembly forms of Aβ and that it can result in a variety of effects ranging from reversible changes in synaptic form and function all the way to frank neuronal loss. If, as seems likely, a similar situation exists *in vivo*, Aβ may mediate a variety of different toxic effects that could occur simultaneously and that will make therapeutic targeting of manifold downstream Aβ-initiated pathways difficult.

V. The Amyloid β-Protein — Lessons from the Human Brain

Early research focused on the elucidation of Aβ primary structure and the association between plaques and AD-type dementia, but surprisingly few studies have dealt with Aβ extracted from human brain and its activity or relationship with disease severity. This is in part due to the limited tools available to study Aβ assemblies in complex mixtures such as brain extracts. A commonly used approach to distinguish between plaque-associated Aβ and other forms of Aβ is based on the known physical properties of amyloid plaques— that is, they are insoluble in aqueous solutions and can be removed from suspension by high-speed centrifugation.[13–15,18,161] Aβ that remains in solution following centrifugation is referred to as soluble,[1] whereas Aβ that is pelleted from aqueous solution but that can be solubilized in formic acid is taken to represent plaque Aβ and is often referred to as "insoluble."[162] While the origin of insoluble Aβ appears obvious, that of soluble Aβ is ambiguous. Extraction of Aβ from brain invariably involves homogenization and consequent cell fracture; thus the extracted pool will include truly soluble extracellular Aβ, extracellular Aβ loosely associated or in equilibrium with plaques, and a portion of intracellular Aβ. To date, most studies of soluble cerebral Aβ have employed ELISA methods, which cannot reveal the aggregation state of the species detected and for the most part appear to preferentially detect Aβ monomer.[163–167]

Crude fractionation approaches have revealed important information about nonfibrillar soluble forms of Aβ. Lue *et al.* found a strong correlation between soluble Aβ and synaptic loss and that the increase in soluble $A\beta_{40}$ better distinguished high pathology control cases from AD than did other parameters.[168] Although this latter finding is controversial,[169] McLean *et al.* also reported a link between soluble Aβ levels and synaptic loss.[170] Synaptic loss is the most direct pathological correlate of Alzheimer-type dementia, and therefore these results suggest that soluble Aβ levels correlate with severity of dementia. Unfortunately, to our knowledge, a direct comparison between the

---

[1]Solubility in the context of amyloid proteins should be distinguished from solubility in the strictly chemical sense (e.g., concentration of solute at saturation). All Aβ species, from monomers through large fibrils, are soluble. However, certain species may or may not be precipitable. This depends on buoyant density, not solubility.

cerebral levels of soluble Aβ and the clinical indicators of AD-type dementia
has not been made. Naslund *et al.* have posited a central role for nonplaque
forms of Aβ in AD. Using a single formic acid extract (which recovered in
excess of 90% of the Aβ extractable in water, formic acid, and hexafluoroiso-
propanol (HFIP) combined), they measured the levels of $A\beta_{40}$ and $A\beta_{42}$ in 79
brains and found that both were elevated early in dementia and both correlated
with the degree of cognitive decline.[171] Approximately half of the study subjects lacked plaques quantifiable by immunohistochemistry but had readily
detectable levels of Aβ, thus suggesting that the Aβ measured in the formic
acid extract was not just derived from amyloid plaques but also came from
soluble forms of Aβ and small deposits not detectable by microscopy. These
findings may explain why previous histopathology studies failed to demonstrate
a strong link between Aβ and dementia. The findings also highlight the potential importance of "preamyloid" Aβ assemblies.

Efforts to better understand the aggregation states of Aβ *in vivo*, although
limited, have revealed similarities to states observed in *in vitro* studies with
synthetic Aβ peptides. A distribution of Aβ species from monomer up to oligomers in excess of 100kDa has been detected in water-soluble extracts by SEC
and ultrafiltration techniques.[162,172] Interestingly, analysis of soluble and insoluble Aβ by SDS-PAGE consistently revealed the presence of two major species,
a ≈4-kDa monomer and an ≈8-kDa SDS-stable dimer.[18,161,170,172–174] A recent
study found that Aβ dimer was specifically detected in both the water-soluble
and detergent-soluble fraction of AD brain.[175] In each case, monomer and dimer
were detected by antibodies to the N- and C-termini of Aβ and both Aβ40 and
Aβ42 were shown to be present. Analysis under native conditions confirmed that
at least a portion of soluble Aβ exists in brain extracts that elutes from SEC as an
authentic dimer, but that SDS-stable dimers can also contribute to larger assemblies.[172] Moreover, the disease relevance of the dimer-containing species is
supported by the finding that SEC-isolated fractions of *assemblies* containing
SDS-stable Aβ dimers disrupt synaptic form and function *in vitro* and perturb
memory consolidation in the rat.[172] SDS-stable dimers have also been detected
in human CSF and formic acid extracts of AD brain,[176–179] and their identity
confirmed as dimers by mass spectrometry.[176,177]

Recent *in vitro* studies using covalently linked dimers or mutant monomers
that readily form dimers indicate that dimers rapidly assemble to form kineti-
cally trapped prefibrillar aggregates similar to ADDLs and protofibrils.[125,180,181] Importantly, oligomerization of monomeric synthetic Aβ has
been observed in real-time mass spectrometrically using the technique of ion
mobility spectrometry.[182] In these studies, $A\beta_{40}$ readily formed dimers and
tetramers. Trimers and hexamers were not observed. In contrast, $A\beta_{42}$ formed
dimers, tetramers, hexamers, and dodecamers. These results were consistent
with other *in vitro* studies demonstrating distinct oligomerization behaviors for

$A\beta_{40}$ and $A\beta_{42}$.[183] It is tempting to speculate that such assemblies may form *in vivo* and mediate important aspects of $A\beta$ toxicity. Consequently, further physical characterization of brain-derived $A\beta$ oligomers is warranted, as is the assessment of their relationship to early cognitive changes. Such studies may lead to exciting new diagnostic and therapeutic procedures.

VI. Testing the Amyloid Hypothesis in Humans

Data extant suggest strongly that $A\beta$ plays an important pathogenetic role in AD. $A\beta$ may mediate adverse effects by directly altering neuronal function or through the formation of plaques that induce inflammatory responses that subsequently decrease neuronal activity and viability. However, what forms of $A\beta$ are involved in disease is still unclear. Much attention has focused on $A\beta$ assembly molecular weight, but this level of understanding is insufficient to establish mechanistic insights into disease causation. Secondary, tertiary, and quaternary structure must influence the interaction with, and effects upon, cells in the brain, including not only neurons but also glial cells. An important goal of future studies must be to characterize more rigorously the complex conformational dynamics and equilibrium relationships comprising the $A\beta$ system. Basic physical principles can be determined *in vitro*, but these results then must be translated into the *in vivo* realm, which will be a much more difficult proposition because of its complexity. In the case of the *ex vivo* material, it will be essential to purify $A\beta$ from the water-soluble fraction of human brain to enable *in vitro* structural characterization of this material. The major caveat will be the effect of the isolation procedures on the assembly state and structure of the brain-derived material. An interesting observation has been the difference in neurotoxic potential between $A\beta$ species produced through chemical synthesis and those isolated directly from biological sources.[141,184] This observation suggests that structural differences must exist between the two $A\beta$ preparations. These differences could be in the primary structure, the presence of posttranslational modifications (cross-linking, oxidation, isomerization), or the association of other biomolecules with the $A\beta$ species. Obtaining sufficient amounts of pure brain $A\beta$ species to allow high-resolution structural analysis is unlikely; therefore, it will be necessary to produce assemblies of synthetic $A\beta$ with structures as similar as possible to those isolated from the brain.

If the $A\beta$ hypothesis is indeed true, then targeting the formation of, or neutralizing, toxic $A\beta$ assemblies is a reasonable approach toward prevention or treatment of early AD and would probably be most useful for preventing cognitive decline in patients with prodromal AD, MCI, or early stage AD. The more difficult therapeutic challenge is that of moderate to severe AD, disease stages at which substantial neuronal loss already has occurred.[185,186] Indeed, a number of clinical trials of anti-amyloid therapeutics have been completed.[187] None has proved significantly more beneficial than current palliative treatments.

Although disappointing, the lack of success in clinical trials may be due more to issues of trial design and execution than to the failure of the Aβ hypothesis. Most clinical trials involved patients with mild to moderate AD, and because the symptoms in these patients likely are driven by factors downstream of Aβ, targeting Aβ at this stage would not be anticipated to be effective. Second, there remain serious challenges in accurately diagnosing the disease. AD is not a single entity but rather could be considered, much like autism, as a "spectrum disorder." If trial selection criteria do not incorporate this notion, then the "AD" cohorts used in clinical trials likely include significant numbers of patients suffering non-AD disorders, mixed AD disorders, and classical AD. The resulting data thus could not be interpreted correctly. This problem underscores the urgent need to develop reliable biomarkers for antemortem diagnosis of AD so as to ensure the integrity of patient populations enrolled in clinical trials. Many biomarkers, including various aggregated forms of Aβ, are currently being assessed for diagnostic usefulness. Finally, none of the anti-amyloid agents that failed in phase III trials was an optimized agent or showed disease-modifying activity in phase II trials.[185] For instance, R-fluoriprofen was advanced to phase III trial as a γ-secretase modulator, but this compound exhibited low potency and poor brain penetrance.[188] Alzhemed, a putative inhibitor of Aβ aggregation, entered phase III trial despite being only a weak aggregation inhibitor and showing little evidence of efficacy in transgenic mice.[189]

It is essential that only well-validated candidates are advanced to phase III trials and that the patient populations tested comprise those at earlier disease stages (ideally preclinical or MCI stages). The DIAN (Dominantly Inherited Alzheimer Network) study offers this possibility because the gene mutations responsible for disease in these families are fully penetrant and the age-of-onset is predictable. In contrast, efforts to treat mild to moderate AD should employ a combination of therapeutics designed to intervene in different aspects of the disease process. In addition to targeting Aβ, such combinatorial treatments should include agents designed to modulate inflammation, to enhance synaptic activity and neurite regrowth, and to suppress pathways leading to tau misprocessing, hyperphosphorylation, and aggregation. Undoubtedly, treating later stages of AD will be extremely challenging, and the development of combinatorial therapy will require greater definition of the downstream targets on which Aβ acts. Nonetheless, since the partial elucidation of the amino acid sequence of vascular Aβ amyloid in 1984,[14,15] considerable progress has been made such that we now have a clear perspective on the principal therapeutic targets likely to allow treatment of this devastating disorder.

Acknowledgment

We thank Carlo Sala Frigerio for assistance with figures.

References

1. Ferri CP, Prince M, Brayne C, Brodaty H, Fratiglioni L, Ganguli M, et al. Global prevalence of dementia: a Delphi consensus study. *Lancet* 2005;**366**:2112–7.
2. Association. 2011. Generation Alzheimers: the Defining Disease of the Baby Boomers [Pamphlet].
3. Prince AC, Brooks SJ, Stahl D, Treasure J. Systematic review and meta-analysis of the baseline concentrations and physiologic responses of gut hormones to food in eating disorders. *Am J Clin Nutr* 2009;**89**:755–65.
4. Blessed G, Tomlinson BE, Roth M. The association between quantitative measures of dementia and senile change in the cerebral grey matter of elderly subjects. *Br J Psychiatry* 1968;**114**:797–811.
5. Folstein MF, Folstein SE, McHugh PR. "Mini-mental state". A practical method for grading the cognitive state of patients for the clinician. *J Psychiatr Res* 1975;**12**:189–98.
6. Hughes CP, Berg L, Danziger WL, Coben LA, Martin RL. A new clinical scale for the staging of dementia. *Br J Psychiatry* 1982;**140**:566–72.
7. Burke WJ, Miller JP, Rubin EH, Morris JC, Coben LA, Duchek J, et al. Reliability of the Washington University Clinical Dementia Rating. *Arch Neurol* 1988;**45**:31–2.
8. Knopman DS, DeKosky ST, Cummings JL, Chui H, Corey-Bloom J, Relkin N, et al. Practice parameter: diagnosis of dementia (an evidence-based review). Report of the Quality Standards Subcommittee of the American Academy of Neurology. *Neurology* 2001;**56**:1143–53.
9. Mirra S, Heyman A, McKeel D, Sumi S, Crain B, Brownless L, et al. The Consortium to Establish a Registry for Alzheimer's disease (CERAD). Part II. Standardization of the neuropathologic assessment of Alzheimer's disease. *Neurology* 1991;**41**:479–86.
10. Ball M, Braak H, Coleman P, Dickson D, Duyckaerts C, Gambetti P, et al. Consensus recommendations for the postmortem diagnosis of Alzheimer's disease. *Neurobiol Aging* 1997;**18**:S1–2.
11. Ball MJ, Murdoch GH. Neuropathological criteria for the diagnosis of Alzheimer's disease—are we really ready yet. *Neurobiol Aging* 1997;**18**:S3–S12.
12. Pepys MB. Amyloidosis. *Annu Rev Med* 2006;**57**:223–41.
13. Allsop D, Landon M, Kidd M. The isolation and amino acid composition of senile plaque core protein. *Brain Res* 1983;**259**:348–52.
14. Glenner GG, Wong CW. Alzheimer's disease: initial report of the purification and characterization of a novel cerebrovascular amyloid protein. *Biochem Biophys Res Commun* 1984;**120**:885–90.
15. Glenner GG, Wong CW. Alzheimer's disease and Down's syndrome: sharing of a unique cerebrovascular amyloid fibril protein. *Biochem Biophys Res Commun* 1984;**122**:1131–5.
16. Geddes AJ, Parker KD, Atkins ED, Beighton E. "Cross-β" conformation in proteins. *J Mol Biol* 1968;**32**:343–58.
17. Bonar L, Cohen AS, Skinner MM. Characterization of the amyloid fibril as a cross-β protein. *Proc Soc Exp Biol Med* 1969;**131**:1373–5.
18. Masters CL, Simms G, Weinman NA, Multhaup G, McDonald BL, Beyreuther K. Amyloid plaque core protein in Alzheimer disease and Down syndrome. *Proc Natl Acad Sci USA* 1985;**82**:4245–9.

19. Kang J, Lemaire HG, Unterbeck A, Salbaum JM, Masters CL, Grzeschik KH, et al. The precursor of Alzheimer's disease amyloid A4 protein resembles a cell-surface receptor. *Nature* 1987;**325**:733–6.

20. Goldgaber D, Lerman MI, McBridge OW, Saffiotti V, Gajdusek DC. Characterization and chromosomal localization of a cDNA encoding brain amyloid of Alzheimer's disease. *Science* 1987;**235**:877–80.

21. Olson MI, Shaw CM. Presenile dementia and Alzheimer's disease in mongolism. *Brain* 1969;**92**:147–56.

22. Mann DM, Yates PO, Marcyniuk B. Alzheimer's presenile dementia, senile dementia of Alzheimer type and Down's syndrome in middle age form an age related continuum of pathological changes. *Neuropathol Appl Neurobiol* 1984;**10**:185–207.

23. Motte J, Williams RS. Age-related changes in the density and morphology of plaques and neurofibrillary tangles in Down syndrome brain. *Acta Neuropathol* 1989;**77**:535–46.

24. Prasher VP, Farrer MJ, Kessling AM, Fisher EM, West RJ, Barber PC, et al. Molecular mapping of Alzheimer-type dementia in Down's syndrome. *Ann Neurol* 1998;**43**:380–3.

25. Guyant-Marechal I, Berger E, Laquerriere A, Rovelet-Lecrux A, Viennet G, Frebourg T, et al. Intrafamilial diversity of phenotype associated with app duplication. *Neurology* 2008;**71**:1925–6.

26. McNaughton D, Knight W, Guerreiro R, Ryan N, Lowe J, Poulter M, et al. Duplication of amyloid precursor protein (APP), but not prion protein (PRNP) gene is a significant cause of early onset dementia in a large UK series. *Neurobiol Aging* 2010; **33**:426.e13–426.e21.

27. Sleegers K, Brouwers N, Gijselinck I, Theuns J, Goossens D, Wauters J, et al. APP duplication is sufficient to cause early onset Alzheimer's dementia with cerebral amyloid angiopathy. *Brain* 2006;**129**:2977–83.

28. Rovelet-Lecrux A, Hannequin D, Raux G, Le Meur N, Laquerriere A, Vital A, et al. APP locus duplication causes autosomal dominant early-onset Alzheimer disease with cerebral amyloid angiopathy. *Nat Genet* 2006;**38**:24–6.

29. Rovelet-Lecrux A, Frebourg T, Tuominen H, Majamaa K, Campion D, Remes AM. APP locus duplication in a Finnish family with dementia and intracerebral haemorrhage. *J Neurol Neurosurg Psychiatry* 2007;**78**:1158–9.

30. Yoshikai S, Sasaki H, Doh-ura K, Furuya H, Sakaki Y. Genomic organization of the human-amyloid beta-protein precursor gene corrigendum. *Gene* 1991;**102**:291–2.

31. Esch FS, Keim PS, Beattie EC, Blacher RW, Culwell AR, Oltersdorf T, et al. Cleavage of amyloid β-peptide during constitutive processing of its precursor. *Science* 1990;**248**:1122–4.

32. Parvathy S, Karran EH, Turner AJ, Hooper NM. The secretases that cleave angiotensin converting enzyme and the amyloid precursor protein are distinct from tumour necrosis factor-alpha convertase. *FEBS Lett* 1998;**431**:63–5.

33. Lammich S, Kojro E, Postina R, Gilbert S, Pfeiffer R, Jasionowski M, et al. Constitutive and regulated alpha-secretase cleavage of Alzheimer's amyloid precursor protein by a disintegrin metalloprotease. *Proc Natl Acad Sci USA* 1999;**96**:3922–7.

34. Slack BE, Ma LK, Seah CC. Constitutive shedding of the amyloid precursor protein ectodomain is up-regulated by tumour necrosis factor-alpha converting enzyme. *Biochem J* 2001;**357**:787–94.

35. Hartmann D, de Strooper B, Serneels L, Craessaerts K, Herreman A, Annaert W, et al. The disintegrin/metalloprotease ADAM 10 is essential for Notch signalling but not for alpha-secretase activity in fibroblasts. *Hum Mol Genet* 2002;**11**:2615–24.

36. Vassar R, Bennett BD, Babu-Khan S, Kahn S, Mendiaz EA, Denis P, et al. Beta-secretase cleavage of Alzheimer's amyloid precursor protein by the transmembrane aspartic protease BACE. *Science* 1999;**286**:735–41.

37. Cai H, Wang Y, McCarthy D, Wen H, Borchelt DR, Price DL, et al. BACE1 is the major beta-secretase for generation of Abeta peptides by neurons. *Nat Neurosci* 2001;**4**:233–4.

38. Kopan R, Ilagan MX. Gamma-secretase: proteasome of the membrane? *Nat Rev Mol Cell Biol* 2004;**5**:499–504.
39. De Strooper B. Loss-of-function presenilin mutations in Alzheimer disease. Talking Point on the role of presenilin mutations in Alzheimer disease. *EMBO Rep* 2007;**8**:141–6.
40. Zhao G, Mao G, Tan J, Dong Y, Cui MZ, Kim SH, Xu X. Identification of a new presenilin-dependent zeta-cleavage site within the transmembrane domain of amyloid precursor protein. *J Biol Chem* 2004;**279**:50647–50.
41. Haass C, Koo EH, Mellon A, Hung AY, Selkoe DJ. Targeting of cell-surface beta-amyloid precursor protein to lysosomes: alternative processing into amyloid-bearing fragments. *Nature* 1992;**357**:500–3.
42. Haass C, Schlossmacher MG, Hung AY, Vigo-Pelfrey C, Mellon A, Ostaszewski BL, et al. Amyloid beta-peptide is produced by cultured cells during normal metabolism. *Nature* 1992;**359**:322–5.
43. Gu Y, Misonou H, Sato T, Dohmae N, Takio K, Ihara Y. Distinct intramembrane cleavage of the beta-amyloid precursor protein family resembling gamma-secretase-like cleavage of Notch. *J Biol Chem* 2001;**276**:35235–8.
44. Weidemann A, Eggert S, Reinhard FB, Vogel M, Paliga K, Baier G, et al. A novel epsilon-cleavage within the transmembrane domain of the Alzheimer amyloid precursor protein demonstrates homology with Notch processing. *Biochemistry* 2002;**41**:2825–35.
45. Kakuda N, Funamoto S, Yagishita S, Takami M, Osawa S, Dohmae N, et al. Equimolar production of amyloid beta-protein and amyloid precursor protein intracellular domain from beta-carboxyl-terminal fragment by gamma-secretase. *J Biol Chem* 2006;**281**:14776–86.
46. Mullan M, Crawford F, Houlden H, Axelman K, Lilius L, Winblad B, et al. A pathogenic mutation for probable Alzheimer's disease in the APP gene at the N-terminus of β-amyloid. *Nat Genet* 1992;**1**:345–7.
47. Citron M, Oltersdorf T, Haass C, McConlogue L, Hung AY, Seubert P, et al. Mutation of the β-amyloid precursor protein in familial Alzheimer's disease increases β-protein production. *Nature* 1992;**360**:672–4.
48. Turner 3rd RT, Koelsch G, Hong L, Castanheira P, Ermolieff J, Ghosh AK, et al. Subsite specificity of memapsin 2 (beta-secretase): implications for inhibitor design. *Biochemistry* 2001;**40**:10001–6.
49. Sala Frigerio C, Fadeeva JV, Minogue AM, Citron M, Van Leuven F, Staufenbiel M, et al. Beta-secretase cleavage is not required for generation of the intracellular C-terminal domain of the amyloid precursor family of proteins. *FEBS J* 2010;**277**:1503–18.
50. De Strooper B. Proteases and proteolysis in Alzheimer disease: a multifactorial view on the disease process. *Physiol Rev* 2010;**90**:465–94.
51. Kuperstein I, Broersen K, Benilova I, Rozenski J, Jonckheere W, Debulpaep M, et al. Neurotoxicity of Alzheimer's disease Abeta peptides is induced by small changes in the Abeta42 to Abeta40 ratio. *EMBO J* 2010;**29**:3408–20.
52. Jan A, Gokce O, Luthi-Carter R, Lashuel HA. The ratio of monomeric to aggregated forms of Abeta40 and Abeta42 is an important determinant of amyloid-beta aggregation, fibrillogenesis, and toxicity. *J Biol Chem* 2008;**283**:28176–89.
53. Betts V, Leissring ML, Dolios G, Wang R, Selkoe DJ, Walsh DM. Aggregation and catabolism of disease-associated intra-Abeta mutations: reduced proteolysis of AbetaA21G by neprilysin. *Neurobiol Dis* 2008;**31**:442–50.
54. Roychaudhuri R, Yang M, Hoshi MM, Teplow DB. Amyloid β-protein assembly and Alzheimer disease. *J Biol Chem* 2009;**284**:4749–53.
55. Jones CT, Morris S, Yates CM, Maffoot A, Brock DJ, St. Clair D. Mutation in codon 713 of the beta-amyloid precursor protein gene presenting with schizophrenia. *Nat Genet* 1992;**1**:306–9.

56. Di Fede G, Catania M, Morbin M, Rossi G, Suardi S, Mazzoleni G, et al. A recessive mutation in the APP gene with dominant-negative effect on amyloidogenesis. *Science* 2009;**323**:1473–7.

57. Zhou L, Brouwers N, Benilova I, Vandersteen A, Mercken M, Van Laere K, et al. Amyloid precursor protein mutation E682K at the alternative beta-secretase cleavage beta'-site increases Abeta generation. *EMBO Mol Med* 2011;**3**:291–302.

58. Guerreiro RJ, Baquero M, Blesa R, Boada M, Bras JM, Bullido MJ, et al. Genetic screening of Alzheimer's disease genes in Iberian and African samples yields novel mutations in presenilins and APP. *Neurobiol Aging* 2010;**31**:725–31.

59. Guerreiro RJ, Gustafson DR, Hardy J. The genetic architecture of Alzheimer's disease: beyond APP, PSENs and APOE. *Neurobiol Aging* 2010;**33**:437–56.

60. Imbimbo BP, Del Giudice E, Colavito D, D'Arrigo A, Dalle Carbonare M, Villetti G, et al. 1-(3′,4′-Dichloro-2-fluoro[1,1′-biphenyl]-4-yl)-cyclopropanecarboxylic acid (CHF5074), a novel gamma-secretase modulator, reduces brain beta-amyloid pathology in a transgenic mouse model of Alzheimer's disease without causing peripheral toxicity. *J Pharmacol Exp Ther* 2007;**323**:822–30.

61. Wolfe MS. When loss is gain: reduced presenilin proteolytic function leads to increased Abeta42/Abeta40. Talking Point on the role of presenilin mutations in Alzheimer disease. *EMBO Rep* 2007;**8**:136–40.

62. Bentahir M, Nyabi O, Verhamme J, Tolia A, Horre K, Wiltfang J, et al. Presenilin clinical mutations can affect gamma-secretase activity by different mechanisms. *J Neurochem* 2006;**96**:732–42.

63. Dubois B, Feldman HH, Jacova C, Cummings JL, Dekosky ST, Barberger-Gateau P, et al. Revising the definition of Alzheimer's disease: a new lexicon. *Lancet Neurol* 2010;**9**:1118–27.

64. Schmechel DE, Saunders AM, Strittmatter WJ, Crain BJ, Hulette CM, Joo SH, et al. Increased amyloid β-peptide deposition in cerebral cortex as a consequence of apolipoprotein E genotype in late-onset Alzheimer disease. *Proc Natl Acad Sci USA* 1993;**90**:9649–53.

65. Mahley RW. Apolipoprotein E: cholesterol transport protein with expanding role in cell biology. *Science* 1988;**240**:622–30.

66. Corder EH, Saunders AM, Strittmatter WJ, Schmechel DE, Gaskell PC, Small GW, et al. Gene dose of apolipoprotein E type 4 allele and the risk of Alzheimer's disease in late onset families. *Science* 1993;**261**:921–3.

67. Strittmatter WJ, Saunders AM, Schmechel D, Pericak-Vance M, Enghild J, Salvesen GS, et al. Apolipoprotein E: high-avidity binding to β-amyloid and increased frequency of type 4 allele in late-onset familial Alzheimer disease. *Proc Natl Acad Sci USA* 1993;**90**:1977–81.

68. Strittmatter WJ, Weisgraber KH, Huand D, Dong L-M, Salvesen GS, Pericak-Vance M, et al. Binding of human apolipoprotein E to synthetic amyloid β peptide: isoform specific effects and implications for late-onset Alzheimer disease. *Proc Natl Acad Sci USA* 1993;**90**:8098–102.

69. Tiraboschi P, Hansen LA, Masliah E, Alford M, Thal LJ, Corey-Bloom J. Impact of APOE genotype on neuropathologic and neurochemical markers of Alzheimer disease. *Neurology* 2004;**62**:1977–83.

70. Tiraboschi P, Hansen LA, Thal LJ, Corey-Bloom J. The importance of neuritic plaques and tangles to the development and evolution of AD. *Neurology* 2004;**62**:1984–9.

71. Tiraboschi P, Sabbagh MN, Hansen LA, Salmon DP, Merdes A, Gamst A, et al. Alzheimer disease without neocortical neurofibrillary tangles: "a second look" *Neurology* 2004;**62**:1141–7.

72. Bales KR, Liu F, Wu S, Lin S, Koger D, DeLong C, et al. Human APOE isoform-dependent effects on brain beta-amyloid levels in PDAPP transgenic mice. *J Neurosci* 2009;**29**:6771–9.

73. Sharman MJ, Morici M, Hone E, Berger T, Taddei K, Martins IJ, et al. APOE genotype results in differential effects on the peripheral clearance of amyloid-beta42 in APOE knock-in and knock-out mice. *J Alzheimers Dis* 2010;**21**:403–9.

74. Levy E, Carman MD, Fernandez-Madrid IJ, Power MD, Lieberburg I, van Duinen SG, et al. Mutation of the Alzheimer's disease amyloid gene in hereditary cerebral hemorrhage, Dutch-type. *Science* 1990;**248**:1124–6.

75. Goate A, Chartier-Harlin M-C, Mullan M, Brown J, Crawford F, Fidani L, et al. Segregation of a missense mutation in the amyloid precursor protein gene with familial Alzheimer's disease. *Nature* 1991;**349**:704–6.

76. Hardy J, Allsop D. Amyloid deposition as the central event in the aetiology of Alzheimer's disease. *Trends Pharmacol Sci* 1991;**12**:383–8.

77. Selkoe DJ. The molecular pathology of Alzheimer's disease. *Neuron* 1991;**6**:487–98.

78. Selkoe DJ. Amyloid protein and Alzheimer's disease. *Sci Am* 1991;**265**:68–78.

79. Hardy JA, Higgins GA. Alzheimer's disease: the amyloid cascade hypothesis. *Science* 1992;**256**:184–5.

80. Walsh DM, Hartley DM, Selkoe DJ. The many faces of Aβ: structures and activity. *Curr Med Chem Immunol Endocrinol Metab Agents* 2003;**3**:277–91.

81. Gotz J, Chen F, van Dorpe J, Nitsch RM. Formation of neurofibrillary tangles in P301l tau transgenic mice induced by Abeta 42 fibrils. *Science* 2001;**293**:1491–5.

82. Rapoport M, Dawson HN, Binder LI, Vitek MP, Ferreira A. Tau is essential to beta-amyloid-induced neurotoxicity. *Proc Natl Acad Sci USA* 2002;**99**:6364–9.

83. Park SY, Ferreira A. The generation of a 17 kDa neurotoxic fragment: an alternative mechanism by which tau mediates beta-amyloid-induced neurodegeneration. *J Neurosci* 2005;**25**:5365–75.

84. Hurtado DE, Molina-Porcel L, Iba M, Aboagye AK, Paul SM, Trojanowski JQ, et al. A(beta) accelerates the spatiotemporal progression of tau pathology and augments tau amyloidosis in an Alzheimer mouse model. *Am J Pathol* 2010;**177**:1977–88.

85. Zempel H, Thies E, Mandelkow E, Mandelkow EM. Abeta oligomers cause localized Ca(2+) elevation, missorting of endogenous Tau into dendrites, Tau phosphorylation, and destruction of microtubules and spines. *J Neurosci* 2010;**30**:11938–50.

86. Jin M, Shepardson N, Yang T, Chen G, Walsh D, Selkoe DJ. Soluble amyloid beta-protein dimers isolated from Alzheimer cortex directly induce Tau hyperphosphorylation and neuritic degeneration. *Proc Natl Acad Sci USA* 2011;**108**:5819–24.

87. Lewis J, Dickson DW, Lin WL, Chisholm L, Corral A, Jones G, et al. Enhanced neurofibrillary degeneration in transgenic mice expressing mutant tau and APP. *Science* 2001;**293**:1487–91.

88. Naslund J, Schierhorn A, Hellman U, Lannfelt L, Roses AD, Tjernberg LO, et al. Relative abundance of Alzheimer A beta amyloid peptide variants in Alzheimer disease and normal aging. *Proc Natl Acad Sci USA* 1994;**91**:8378–82.

89. Iwatsubo T, Odaka A, Suzuki N, Mizusawa H, Nukina N, Ihara Y. Visualization of A beta 42 (43) and A beta 40 in senile plaques with end-specific A beta monoclonals: evidence that an initially deposited species is A beta 42(43). *Neuron* 1994;**13**:45–53.

90. Iwatsubo T, Mann DM, Odaka A, Suzuki N, Ihara Y. Amyloid β protein (Aβ) deposition: Aß42 (43) precedes Aβ40 in Down syndrome. *Ann Neurol* 1995;**37**:294–9.

91. Lemere CA, Blustzjan JK, Yamaguchi H, Wisniewski T, Saido TC, Selkoe DJ. Sequence of deposition of heterogeneous amyloid β-peptides and Apo E in Down syndrome: implications for initial events in amyloid plaque formation. *Neurobiol Dis* 1996;**3**:16–32.

92. Hansson O, Zetterberg H, Buchhave P, Andreasson U, Londos E, Minthon L, et al. Prediction of Alzheimer's disease using the CSF Abeta42/Abeta40 ratio in patients with mild cognitive impairment. *Dement Geriatr Cogn Disord* 2007;**23**:316–20.

93. Lewczuk P, Esselmann H, Bibl M, Beck G, Maler JM, Otto M, et al. Tau protein phosphorylated at threonine 181 in CSF as a neurochemical biomarker in Alzheimer's disease: original data and review of the literature. *J Mol Neurosci* 2004;**23**:115–22.

94. Jarrett JT, Berger EP, Lansbury Jr. PT. The carboxy terminus of the beta amyloid protein is critical for the seeding of amyloid formation: implications for the pathogenesis of Alzheimer's disease. *Biochemistry* 1993;**32**:4693–7.

95. Pike CJ, Walencewicz AJ, Glabe CG, Cotman CW. In vitro aging of β-amyloid protein causes peptide aggregation and neurotoxicity. *Brain Res* 1991;**563**:311–4.

96. Busciglio J, Lorenzo A, Yankner B. Methological variables in the assessment of b-amyloid neurotoxicity. *Neurobiol Aging* 1992;**13**:609–12.

97. May PC, Gitter BD, Waters DC, Simmons LK, Becker GW, Small JS, et al. β-Amyloid peptide in vitro toxicity: lot-to-lot variability. *Neurobiol Aging* 1992;**13**:605–12.

98. Pike CJ, Burdick D, Walencewicz AJ, Glabe CG, Cotman CW. Neurodegeneration induced by β-amyloid peptides in vitro: the role of peptide assembly state. *J Neurosci* 1993;**13**:1676–87.

99. Pike CJ, Walencewicz AJ, Glabe CG, Cotman CW. Aggregation-related toxicity of synthetic beta-amyloid protein in hippocampal cultures. *Eur J Pharmacol* 1991;**207**:367–8.

100. Lorenzo A, Yankner B. β-Amyloid neurotoxicity requires fibril formation and is inhibited by Congo red. *Proc Natl Acad Sci USA* 1994;**91**:12243–7.

101. Terry RD, Gonatas NK, Weiss M. Ultrastructural studies in Alzheimer's presenile dementia. *Am J Pathol* 1964;**44**:269–97.

102. Kidd M. Alzheimer's disease—an electron microscopical study. *Brain* 1964;**87**:307–20.

103. Narang HK, Chandler RL, Anger HS. Further observations on particulate structures in scrapie affected brain. *Neuropathol Appl Neurobiol* 1980;**6**:23–8.

104. Terry RD, Masliah E, Salmon DP, Butters N, DeTeresa R, Hill R, et al. Physical basis of cognitive alterations in Alzheimer's disease: synapse loss is the major correlate of cognitive impairment. *Ann Neurol* 1991;**30**:572–80.

105. Price JL, Morris JC. Tangles and plaques in nondemented aging and "preclinical" Alzheimer's disease. *Ann Neurol* 1999;**45**:358–68.

106. Terry RD. The pathogenesis of Alzheimer disease: an alternative to the amyloid hypothesis. *J Neuropathol Exp Neurol* 1996;**55**:1023–5.

107. Ono K, Condron MM, Teplow DB. Structure-neurotoxicity relationships of amyloid β-protein oligomers. *Proc Natl Acad Sci USA* 2009;**106**:14745–50.

108. Mucke L, Masliah E, Johnson WB, Ruppe MD, Alford M, Rockenstein EM, et al. Synapto-trophic effects of human amyloid beta protein precursors in the cortex of transgenic mice. *Brain Res* 1994;**666**:151–67.

109. D'Hooge R, Nagels G, Westland CE, Mucke L, De Deyn PP. Spatial learning deficit in mice expressing human 751-amino acid beta-amyloid precursor protein. *Neuroreport* 1996;**7**:2807–11.

110. Mucke L, Masliah E, Yu GQ, Mallory M, Rockenstein EM, Tatsuno G, et al. High-level neuronal expression of abeta 1-42 in wild-type human amyloid protein precursor transgenic mice: synaptotoxicity without plaque formation. *J Neurosci* 2000;**20**:4050–8.

111. Moechars D, Dewachter I, Lorent K, Reversé D, Baekelandt V, Naidu A, et al. Early phenotypic changes in transgenic mice that overexpress different mutants of amyloid precursor protein in brain. *J Biol Chem* 1999;**274**:6483–92.

112. Games D, Buttini M, Kobayashi D, Schenk D, Seubert P. Mice as models: transgenic approaches and Alzheimer's disease. *J Alzheimers Dis* 2006;**9**:133–49.

113. Jacobsen JS, Wu CC, Redwine JM, Comery TA, Arias R, Bowlby M, et al. Early-onset behavioral and synaptic deficits in a mouse model of Alzheimer's disease. *Proc Natl Acad Sci USA* 2006;**103**:5161–6.

114. Hsia AY, Masliah E, McConlogue L, Yu GQ, Tatsuno G, Hu K, et al. Plaque-independent disruption of neural circuits in Alzheimer's disease mouse models. *Proc Natl Acad Sci USA* 1999;**96**:3228–33.

115. Wogulis M, Wright S, Cunningham D, Chilcote T, Powell K, Rydel RE. Nucleation-dependent polymerization is an essential component of amyloid-mediated neuronal cell death. *J Neurosci* 2005;**25**:1071–80.
116. Editorial. State of aggregation. *Nat Neurosci* 2011;**14**:399.
117. Walsh DM, Hartley DM, Kusumoto Y, Fezoui Y, Condron MM, Lomakin A, et al. Amyloid beta-protein fibrillogenesis. Structure and biological activity of protofibrillar intermediates. *J Biol Chem* 1900;**274**:25945–52.
118. Harper JD, Wong SS, Lieber CM, Lansbury Jr. PT. Observation of metastable Aβ amyloid protofibrils by atomic force microscopy. *Chem Biol* 1997;**4**:119–25.
119. Harper JD, Wong SS, Lieber CM, Lansbury PT. Assembly of Aβ amyloid protofibrils: an *in vitro* model for a possible early event in Alzheimer's disease. *Biochemistry* 1999;**38**:8972–80.
120. Walsh DM, Tseng BP, Schlossmacher MG, Growdon JH, Podlisny MB, Selkoe DJ. Aβ oligomers are present in human CSF and accumulate in cultured brain cells. *Soc Neurosci Abstr* 1999;**25**(2) 720.10, 1805.
121. Lashuel HA, Hartley DM, Petre BM, Wall JS, Simon MN, Walz T, et al. Mixtures of wild-type and a pathogenic (E22G) form of Abeta40 in vitro accumulate protofibrils, including amyloid pores. *J Mol Biol* 2003;**332**:795–808.
122. Jan A, Hartley DM, Lashuel HA. Preparation and characterization of toxic Abeta aggregates for structural and functional studies in Alzheimer's disease research. *Nat Protoc* 2010;**5**:1186–209.
123. Hartley DM, Walsh DM, Ye CP, Diehl T, Vasquez S, Vassilev PM, et al. Protofibrillar intermediates of amyloid β-protein induce acute electrophysiological changes and progressive neurotoxicity in cortical neurons. *J Neurosci* 1999;**19**:8876–84.
124. Isaacs AM, Senn DB, Yuan M, Shine JP, Yankner BA. Acceleration of amyloid beta-peptide aggregation by physiological concentrations of calcium. *J Biol Chem* 2006;**281**:27916–23.
125. O'Nuallain B, Freir DB, Nicoll AJ, Risse E, Ferguson N, Herron CE, et al. Amyloid beta-protein dimers rapidly form stable synaptotoxic protofibrils. *J Neurosci* 2010;**30**:14411–9.
126. Hartley DM, Zhao C, Speier AC, Woodard GA, Li S, Li Z, et al. Transglutaminase induces protofibril-like amyloid beta-protein assemblies that are protease-resistant and inhibit long-term potentiation. *J Biol Chem* 2008;**283**:16790–800.
127. Martins R, Morais A, Dias A, Soares I, Rolao C, Ducla-Soares JL, et al. Early modification of sickle cell disease clinical course by UDP-glucuronosyltransferase 1A1 gene promoter polymorphism. *J Hum Genet* 2008;**53**:524–8.
128. Nilsberth C, Westlind-Danielsson A, Eckman CB, Condron MM, Axelman K, Forsell C, et al. The 'Arctic' APP mutation (E693G) causes Alzheimer's disease by enhanced Abeta protofibril formation. *Nat Neurosci* 2001;**4**:887–93.
129. Lashuel HA, Hartley D, Petre BM, Walz T, Lansbury Jr. PT. Neurodegenerative disease: amyloid pores from pathogenic mutations. *Nature* 2002;**418**:291.
130. Mina EW, Lasagna-Reeves C, Glabe CG, Kayed R. Poloxamer 188 copolymer membrane sealant rescues toxicity of amyloid oligomers in vitro. *J Mol Biol* 2009;**391**:577–85.
131. Glabe CG. Structural classification of toxic amyloid oligomers. *J Biol Chem* 2008;**283**: 29639–43.
132. Lasagna-Reeves CA, Castillo-Carranza DL, Guerrero-Muoz MJ, Jackson GR, Kayed R. Preparation and characterization of neurotoxic tau oligomers. *Biochemistry* 2010;**49**:10039–41.
133. Lasagna-Reeves CA, Clos AL, Midoro-Hiriuti T, Goldblum RM, Jackson GR, Kayed R. Inhaled insulin forms toxic pulmonary amyloid aggregates. *Endocrinology* 2010;**151**:4717–24.
134. Lashuel HA, Hartley DM, Balakhaneh D, Aggarwal A, Teichberg S, Callaway DJ. New class of inhibitors of amyloid-beta fibril formation. Implications for the mechanism of pathogenesis in Alzheimer's disease. *J Biol Chem* 2002;**277**:42881–90.
135. Lasagna-Reeves CA, Glabe CG, Kayed R. Amyloid-beta annular protofibrils evade fibrillar fate in Alzheimer disease brain. *J Biol Chem* 2011;**286**:22122–30.

136. Lashuel HA, Lansbury Jr. PT. Are amyloid diseases caused by protein aggregates that mimic bacterial pore-forming toxins? *Q Rev Biophys* 2006;**39**:167–201.

137. Diaz JC, Linnehan J, Pollard H, Arispe N. Histidines 13 and 14 in the Abeta sequence are targets for inhibition of Alzheimer's disease Abeta ion channel and cytotoxicity. *Biol Res* 2006;**39**:447–60.

138. Lambert MP, Barlow AK, Chromy BA, Edwards C, Freed R, Iosatos M, et al. Diffusible, nonfribrillar ligands derived from $A\beta_{1-42}$ are potent central nervous system neurotoxins. *Proc Natl Acad Sci* 1998;**95**:6448–53.

139. Hepler RW, Grimm KM, Nahas DD, Breese R, Dodson EC, Acton P, et al. Solution state characterization of amyloid beta-derived diffusible ligands. *Biochemistry* 2006;**45**:15157–67.

140. Lauren J, Gimbel DA, Nygaard HB, Gilbert JW, Strittmatter SM. Cellular prion protein mediates impairment of synaptic plasticity by amyloid-beta oligomers. *Nature* 2009;**457**:1128–32.

141. Freir DB, Nicoll AJ, Klyubin I, Panico S, Mc Donald JM, Risse E, et al. Interaction between prion protein and toxic amyloid beta assemblies can be therapeutically targeted at multiple sites. *Nat Commun* 2011;**2**:336.

142. Chromy BA, Nowak RJ, Lambert MP, Viola KL, Chang L, Velasco PT, et al. Self-assembly of Abeta(1-42) into globular neurotoxins. *Biochemistry* 2003;**42**:12749–60.

143. Chromy LR, Pipas JM, Garcea RL. Chaperone-mediated in vitro assembly of Polyomavirus capsids. *Proc Natl Acad Sci USA* 2003;**100**:10477–82.

144. Shughrue PJ, Acton PJ, Breese RS, Zhao WQ, Chen-Dodson E, Hepler RW, et al. Anti-ADDL antibodies differentially block oligomer binding to hippocampal neurons. *Neurobiol Aging* 2010;**31**:189–202.

145. Wang HW, Pasternak JF, Kuo H, Ristic H, Lambert MP, Chromy B, et al. Soluble oligomers of β amyloid (1-42) inhibit long-term potentiation but not long-term depression in rat dentate gyrus. *Brain Res* 2002;**924**:133–40.

146. Dahlgren KN, Manelli AM, Stine Jr. WB, Baker LK, Krafft GA, LaDu MJ. Oligomeric and fibrillar species of amyloid-β peptides differentially affect neuronal viability. *J Biol Chem* 2002;**277**:32046–53.

147. Kim HJ, Chae SC, Lee DK, Chromy B, Lee SC, Park YC, et al. Selective neuronal degeneration induced by soluble oligomeric amyloid beta protein. *FASEB J* 2003;**17**:118–20.

148. Lacor P, Buniel MC, Chang L, Fernandez SJ, Gong Y, Viola KL, et al. Synaptic targeting by Alzheimer's-related amyloid beta oligomers. *J Neurosci* 2004;**24**:1091–200.

149. Lacor P, Buniel MC, Furlow PW, Clemente AS, Velasco PT, Wood M, et al. Abeta oligomer-induced abberrations in synapse composition, shape, and density provide a molecular basis for loss of connectivity in Alzheimer's disease. *J Neurosci* 2007;**27**:796–807.

150. Shughrue PJ, Acton PJ, Breese RS, Zhao W-Q, Chen-Dodson E, Hepler RW, et al. Anti-ADDL antibodies differentially block oligomer binding to hippocampal neurons. *Neurobiol Aging* 2010;**31**:189–202.

151. Barry AE, Klyubin I, Mc Donald JM, Mably AJ, Farrell MA, Scott M, et al. Alzheimer's disease brain-derived amyloid-beta-mediated inhibition of LTP in vivo is prevented by immunotargeting cellular prion protein. *J Neurosci* 2011;**31**:7259–63.

152. Lambert MP, Viola KL, Chromy BA, Chang L, Morgan TE, Yu J, et al. Vaccination with soluble Abeta oligomers generates toxicity-neutralizing antibodies. *J Neurochem* 2001;**79**:595–605.

153. Kayed R, Head E, Thompson JL, McIntire TM, Milton SC, Cotman CW, et al. Common structure of soluble amyloid oligomers implies common mechanism of pathogenesis. *Science* 2003;**300**:486–9.

154. Maloney MT, Minamide LS, Kinley AW, Boyle JA, Bamburg JR. Beta-secretase-cleaved amyloid precursor protein accumulates at actin inclusions induced in neurons by stress or amyloid beta: a feedforward mechanism for Alzheimer's disease. *J Neurosci* 2005;**25**:11313–21.

155. Kelly BN, Howard BR, Wang H, Robinson H, Sundquist WI, Hill CP. Implications for viral capsid assembly from crystal structures of HIV-1 Gag(1-278) and CA(N)(133-278). *Biochemistry* 2006;**45**:11257–66.

156. Barghorn S, Zheng-Fischhofer Q, Ackmann M, Biernat J, von Bergen M, Mandelkow EM, et al. Structure, microtubule interactions, and paired helical filament aggregation by tau mutants of frontotemporal dementias. *Biochemistry* 2000;**39**.11714–21.

157. Whalen BM, Selkoe DJ, Hartley DM. Small non-fibrillar assemblies of amyloid beta-protein bearing the Arctic mutation induce rapid neuritic degeneration. *Neurobiol Dis* 2005;**20**:254–66.

158. Hoshi M, Sato M, Matsumoto S, Noguchi A, Yasutake K, Yoshida N, et al. Spherical aggregates of beta-amyloid (amylospheroid) show high neurotoxicity and activate tau protein kinase I/glycogen synthase kinase-3beta. *Proc Natl Acad Sci USA* 2003;**100**:6370–5.

159. Demuro A, Mina E, Kayed R, Milton SC, Parker I, Glabe CG. Calcium dysregulation and membrane disruption as a ubiquitous neurotoxic mechanism of soluble amyloid oligomers. *J Biol Chem* 2005;**280**:17294–300.

160. Deshpande A, Mina E, Glabe C, Busciqlio J. Different conformations of amyloid beta induce neurotoxicity by distinct mechanisms in human cortical neurons. *J Neurosci* 2006;**26**:6011–8.

161. Masters CL, Multhaup G, Simms G, Pottigiesser J, Martins RN, Beyreuther K. Neuronal origin of a cerebral amyloid: neurofibrillary tangles of Alzheimer's disease contain the same protein as the amyloid of plaque cores and blood vessels. *EMBO J* 1985;**4**:2757–63.

162. Kuo Y-M, Emmerling MR, Vigo-Pelfrey C, Kasunic TC, Kirkpatrick JB, Murdoch GH, et al. Water-soluble Aβ (N-40, N-42) oligomers in normal and Alzheimer disease brains. *J Biol Chem* 1996;**271**:4077–81.

163. Morishima-Kawashima M, Ihara Y. The presence of amyloid β-protein in the detergent-insoluble membrane compartment of human neuroblastoma cells. *Biochemistry* 1998;**37**:15247–53.

164. Enya M, Morishima-Kawashima M, Yoshimura M, Shinkai Y, Kusui K, Khan K, et al. Appearance of sodium dodecyl sulfate-stable amyloid β-protein (Aβ) dimer in the cortex during aging. *Am J Pathol* 1999;**154**:271–9.

165. Funato H, Enya M, Yoshimura M, Morishima-Kawashima M, Ihara Y. Presence of sodium dodecyl sulfate-stable amyloid β-protein in the hippocampus CA1 not exhibiting neurofibrillary tangle formation. *Am J Pathol* 1999;**155**:23–8.

166. Stenh C, Englund H, Lord A, Johansson AS, Almeida CG, Gellerfors P, et al. Amyloid-beta oligomers are inefficiently measured by enzyme-linked immunosorbent assay. *Ann Neurol* 2005;**58**:147–50.

167. Englund H, Sehlin D, Johansson AS, Nilsson LN, Gellerfors P, Paulie S, et al. Sensitive ELISA detection of amyloid-beta protofibrils in biological samples. *J Neurochem* 2007;**103**:334–45.

168. Lue LF, Kuo YM, Roher AE, Brachova L, Shen Y, Sue L, et al. Soluble amyloid beta peptide concentration as a predictor of synaptic change in Alzheimer's disease. *Am J Pathol* 1999;**155**:853–62.

169. Wang J, Dickson DW, Trojanowski JQ, Lee VM. The levels of soluble versus insoluble brain Abeta distinguish Alzheimer's disease from normal and pathologic aging. *Exp Neurol* 1999;**158**:328–37.

170. McLean CA, Cherny RA, Fraser FW, Fuller SJ, Smith MJ, Beyreuther K, et al. Soluble pool of Abeta amyloid as a determinant of severity of neurodegeneration in Alzheimer's disease. *Ann Neurol* 1999;**46**:860–6.

171. Näslund J, Haroutunian V, Mohs R, Davis K, Davies P, Greengard P, et al. Correlation between elevated amyloid β-peptide in the brain and cognitive decline. *JAMA* 2000;**283**:1571–7.

172. Shankar GM, Li S, Mehta TH, Garcia-Munoz A, Shepardson NE, Smith I, et al. Amyloid-beta dimers isolated directly from Alzheimer's disease brains impair synaptic plasticity and memory. *Nat Med* 2008;**14**:837–42.

173. Roher AE, Lowenson JD, Clarke S, Wolkow C, Wang R, Cotter RJ, et al. Structural alterations in the peptide backbone of β-amyloid core protein may account for its deposition and stability in Alzheimer's disease. *J Biol Chem* 1993;**268**:3072–83.

174. Kalback W, Watson MD, Kokjohn TA, Kuo YM, Weiss N, Luehrs DC, et al. APP transgenic mice Tg2576 accumulate Abeta peptides that are distinct from the chemically modified and insoluble peptides deposited in Alzheimer's disease senile plaques. *Biochemistry* 2002;**41**:922–8.

175. Mc Donald JM, Savva GM, Brayne C, Welzel AT, Forster G, Shankar GM, et al. The presence of sodium dodecyl sulphate-stable Abeta dimers is strongly associated with Alzheimer-type dementia. *Brain* 2010;**133**:1328–41.

176. Vigo-Pelfrey C, Lee D, Keim PS, Lieberburg I, Schenk D. Characterization of β-amyloid peptide from human cerebrospinal fluid. *J Neurochem* 1993;**61**:1965–8.

177. Roher AE, Chaney MO, Kuo Y-M, Webster SD, Stine WB, Haverkamp LJ, et al. Morphology and toxicity of Aβ-(1-42) dimer derived from neuritic and vascular amyloid deposits of Alzheimer's disease. *J Biol Chem* 1996;**271**:20631–5.

178. Fukumoto S, Miner JH, Ida H, Fukumoto E, Yuasa K, Miyazaki H, et al. Laminin alpha5 is required for dental epithelium growth and polarity and the development of tooth bud and shape. *J Biol Chem* 2006;**281**:5008–16.

179. Walsh DM, Tseng BP, Rydel RE, Podlisny MB, Selkoe DJ. Detection of intracellular oligomers of amyloid β-protein in cells derived from human brain. *Biochemistry* 2000;**39**:10831–9.

180. Sandberg A, Luheshi LM, Sollvander S, Pereira de Barros T, Macao B, Knowles TP, et al. Stabilization of neurotoxic Alzheimer amyloid-beta oligomers by protein engineering. *Proc Natl Acad Sci USA* 2010;**107**:15595–600.

181. Yamaguchi T, Yagi H, Goto Y, Matsuzaki K, Hoshino M. A disulfide-linked amyloid-beta peptide dimer forms a protofibril-like oligomer through a distinct pathway from amyloid fibril formation. *Biochemistry* 2010;**49**:7100–7.

182. Bernstein SL, Dupuis NF, Lazo ND, Wyttenbach T, Condron MM, Bitan G, et al. Amyloid-β protein oligomerization and the importance of tetramers and dodecamers in the aetiology of Alzheimer's disease. *Nat Chem* 2009;**1**:326–31.

183. Bitan G, Kirkitadze MD, Lomakin A, Vollers SS, Benedek GB, Teplow DB. Amyloid beta-protein (Abeta) assembly: Abeta 40 and Abeta 42 oligomerize through distinct pathways. *Proc Natl Acad Sci USA* 2003;**100**:330–5.

184. Wang Q, Walsh DM, Rowan MJ, Selkoe DJ, Anwyl R. Block of long-term potentiation by naturally secreted and synthetic amyloid beta-peptide in hippocampal slices is mediated via activation of the kinases c-Jun N-terminal kinase, cyclin-dependent kinase 5, and p38 mitogen-activated protein kinase as well as metabotropic glutamate receptor type 5. *J Neurosci* 2004;**24**:3370–8.

185. Golde TE, Schneider LS, Koo EH. Anti-abeta therapeutics in Alzheimer's disease: the need for a paradigm shift. *Neuron* 2011;**69**:203–13.

186. Citron M. Alzheimer's disease: strategies for disease modification. *Nat Rev Drug Discov* 2010;**9**:387–98.

187. Extance A. Alzheimer's failure raises questions about disease-modifying strategies. *Nat Rev Drug Discov* 2010;**9**:749–51.

188. Galasko DR, Graff-Radford N, May S, Hendrix S, Cottrell BA, Sagi SA, et al. Safety, tolerability, pharmacokinetics, and Abeta levels after short-term administration of R-flurbiprofen in healthy elderly individuals. *Alzheimer Dis Assoc Disord* 2007;**21**:292–9.

189. Gervais F, Paquette J, Morissette C, Krzywkowski P, Yu M, Azzi M, et al. Targeting soluble Abeta peptide with Tramiprosate for the treatment of brain amyloidosis. *Neurobiol Aging* 2007;**28**:537–47.

Molecular Insights into Parkinson's Disease

JEAN-CHRISTOPHE ROCHET,[*]
BRUCE A. HAY,[†] AND MING GUO[‡,§]

[*]*Department of Medicinal Chemistry and Molecular Pharmacology, Purdue University, West Lafayette, Indiana, USA*

[†]*Division of Biology, MC126-29, California Institute of Technology, Pasadena, California, USA*

[‡]*Department of Neurology, Brain Research Institute, David Geffen School of Medicine, University of California, Los Angeles, California, USA*

[§]*Department of Molecular and Medical Pharmacology, Brain Research Institute, David Geffen School of Medicine, University of California, Los Angeles, California, USA*

Progress in Molecular Biology
and Translational Science, Vol. 107
DOI: 10.1016/B978-0-12-385883-2.00011-4

125

Mutations in *SNCA, PINK1, parkin*, and *DJ-1* are associated with autosomal-dominant or autosomal-recessive forms of Parkinson's disease (PD), the second most common neurodegenerative disorder. Studies on the structural and functional properties of the corresponding gene products have provided significant insights into the molecular underpinnings of familial PD and the much more common sporadic forms of the disease. Here, we review recent advances in our understanding of four PD-related gene products: α-synuclein, parkin, PINK1, and DJ-1. In Part 1, we review new insights into the role of α-synuclein in PD. In Part 2, we summarize the latest developments in understanding the role of mitochondrial dysfunction in PD, emphasizing the role of the *PINK1/parkin* pathway in regulating mitochondrial dynamics and mitophagy. The role of *DJ-1* is also discussed. In Part 3, we point out converging pathways and future directions.

I. Introduction

Parkinson's disease (PD) is a neurodegenerative disorder manifested by resting tremor, bradykinesia (slowness of movement), rigidity, and postural instability.[1–3] Some of these symptoms are attributed in large part to a loss of dopaminergic neurons in the substantia nigra of the midbrain. A defining neuropathological feature of PD brain is the presence in some surviving neurons of Lewy bodies, which are cytosolic inclusions enriched with fibrillar forms of the presynaptic protein α-synuclein (αSyn), encoded by the *SNCA* gene (also known as *PARK1*).[4,5] Autosomal-dominant mutations in *SNCA* have been discovered in patients with early onset familial PD,[6–10] and evidence suggests that these mutations promote αSyn aggregation. Thus, these neuropathological and genetic data suggest that αSyn aggregation is involved in PD pathogenesis.

A second characteristic feature of PD pathogenesis is an impairment of mitochondrial function. Biochemical studies have revealed a defect of mitochondrial complex I in the postmortem brains of PD patients.[11,12] The decrease in complex I activity is predicted to cause an accumulation of reactive oxygen species (ROS) that damage proteins, lipids, and DNA.[13,14] Dopaminergic neurons in the substantia nigra are hypothesized to be particularly susceptible to a buildup of ROS because they have high levels of oxidative stress (even under basal conditions) resulting from dopamine metabolism and auto-oxidation.[13,15]

Additional evidence that mitochondrial deficits play a role in PD stems from the observation that three proteins mutated in autosomal-recessive early onset PD (DJ-1, Parkin, PINK1) regulate mitochondrial functions.[16–19]

In Part 1 of this chapter, we provide an overview of current knowledge relating to the role of αSyn aggregation in PD. Questions that are addressed include (i) Which species are formed on the αSyn self-assembly pathway? (ii) Which of these species are responsible for neurotoxicity? (iii) How is αSyn aggregation modulated by cellular perturbations such as oxidative stress and membrane binding? (iv) How are αSyn aggregation and toxicity impacted by antioxidants and molecular chaperones? In Part 2, we review molecular mechanisms by which mitochondrial dysfunction elicits cellular defects in PD, with an emphasis on cellular pathways relating to *PINK1*, *parkin*, and *DJ-1*. Questions addressed include (i) What is the evidence that *PINK1* and *parkin* function in a common pathway to regulate mitochondrial integrity? (ii) What are cytoplasmic substrates for PINK1 and Parkin that mediate their neuroprotective functions? (iii) How do PINK1 and Parkin identify damaged mitochondria and mediate their removal? (iv) What is the relationship between *DJ-1* and the *PINK1/parkin* pathway? In Part 3, we highlight recent findings suggesting that αSyn aggregation and mitochondrial dysfunction act as "co-conspirators" to trigger dopaminergic cell death in PD. Throughout this chapter we raise key questions that need to be answered to better understand how αSyn aggregation and mitochondrial dysfunction contribute to PD pathogenesis, and we suggest potential strategies to target these two toxic phenomena in patients.

II. Role of αSynuclein Aggregation in PD

A. Physiological Role of αSyn

αSyn is a member of the "synuclein" family that also includes β-synuclein (βSyn) and γ-synuclein (γSyn).[20] αSyn is expressed as multiple isoforms spanning 98-, 112-, 126-, and 140-amino acid residues as a result of alternative pre-mRNA splicing.[21] The 140-residue (14-kDa) isoform has been characterized much more extensively than the other splice variants.

The sequence of the 14-kDa isoform of αSyn can be subdivided into three domains: (i) an N-terminal domain (residues 1–67), encompassing six repeats of the highly conserved hexamer sequence "KTK(E/Q)GV"; (ii) a central domain (residues 61–95; also referred to as the "non-Aβ component of AD amyloid" (NAC) domain), characterized by a high content of hydrophobic residues; and (iii) a C-terminal domain (residues 96–140), characterized by a high content of proline, aspartate, and glutamate residues. Analysis of aqueous solutions of αSyn by circular dichroism (CD) or nuclear magnetic resonance (NMR) reveals a lack of stable secondary structure, and under these conditions the protein is referred to as "natively unfolded."[22,23] The N-terminal lysine-rich repeats are similar to lipid-binding motifs in amphipathic helical domains of exchangeable apolipoproteins, suggesting that the normal function of αSyn involves binding to phospholipid membranes.[24] Consistent with this idea, the N-terminal repeat region of αSyn binds anionic phospholipid vesicles and adopts an amphipathic α-helical structure as a result of its interaction with the membrane.[24–31] In addition, αSyn is thought to play a role in regulating neurotransmission via interactions with synaptic vesicles[32–38] or by regulating the SNARE complex assembly.[39] Data recently reported by Selkoe and colleagues[40] suggest that αSyn exists as an α-helical tetramer that remains intact when purified from mouse cortex or mammalian cell lines under nondenaturing conditions. A subsequent study showed that αSyn with an N-terminal 10-residue leader sequence derived from GST can be purified from a bacterial expression system as an apparent tetramer with some degree of helical structure.[41] These intriguing findings (currently being validated by other groups) raise the possibility that the α-helical structure of αSyn can be stabilized by contacts between neighboring subunits in an αSyn oligomer, and not just by interactions with phospholipid membranes.

B. Effects of Familial Mutations on αSyn Self-assembly

Two types of αSyn gene mutations have been identified in patients with early onset familial PD: (i) substitution mutations encoding the αSyn variants A30P, E46K, and A53T (Fig. 1)[6–8]; and (ii) mutations that increase the copy

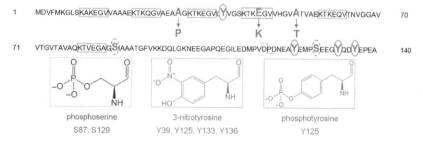

FIG. 1. Amino acid sequence of α-Syn in one-letter code. The lysine-rich repeats are enclosed in rectangular boxes. The diagram highlights the three substitutions associated with familial PD (A30P, E46K, and A53T) and posttranslational modifications identified in postmortem human brain: serine phosphorylation, tyrosine nitration, tyrosine phosphorylation, and C-terminal truncation (represented by a boundary line between residues 119 and 120, one of the several proteolytic cleavage sites in the C-terminal region[42]). (See Color Insert.)

number of the wild-type gene, including a duplication and a triplication.[9,10] The observation that Lewy bodies are enriched with fibrillar αSyn has led to the hypothesis that αSyn gene mutations cause early onset PD by promoting the formation of neurotoxic aggregates. In the case of the duplication and triplication mutants, the increased αSyn expression levels resulting from these mutations would be expected to favor aggregation of the protein via mass action. To address whether the three substitution mutations promote αSyn self-assembly, a common strategy has been to monitor the aggregation of αSyn variants in cell-free systems.

Upon prolonged incubation at 37 °C, recombinant wild-type and mutant αSyn form fibrils with characteristic features of classic amyloid deposits, similar to fibrillar αSyn isolated from Lewy bodies.[43–47] αSyn fibrillization does not occur as a simple two-step process, but rather involves the formation of pre-fibrillar intermediates referred to as "protofibrils."[48,49] The results of far-UV CD, atomic force microscopy (AFM), and electron microscopy (EM) analyses indicate that protofibrils consist of spheres, chains, and rings enriched with β-sheet secondary structure.[45,46,48–52] In addition, αSyn protofibrils are transient intermediates that accumulate to a maximum of ~15% of the total protein before being consumed by the formation of amyloid-like fibrils.[45,46,52]

A study by Lansbury and colleagues[45] revealed that A53T formed fibrils more rapidly than wild-type αSyn, whereas A30P formed fibrils less rapidly than the wild-type protein. In contrast, A53T and A30P both formed protofibrils more rapidly than the wild-type protein.[45] Similar results have been published by other groups.[53–55] These findings suggest that the neurotoxic effects of the A30P and A53T substitutions result from enhanced protofibril formation rather than accelerated fibrillization. Consistent with this idea, compounds that promote the conversion of αSyn from small aggregates to larger inclusions alleviate the protein's neurotoxicity in cellular and animal models relevant to PD.[56]

αSyn protofibrils, but not the monomeric or fibrillar protein, bind phospholipid vesicles with high affinity[50,57] and trigger membrane permeabilization.[50,51,58,59] Other groups have reported that oligomeric αSyn increases the conductance across a phospholipid bilayer, although it is unclear whether this effect involves the formation of pore-like structures.[60,61] Additional evidence suggests that oligomeric αSyn can perturb ion homeostasis by forming conducting membrane channels in cells.[62–64] These observations provide a rationale for why αSyn protofibrils may be toxic (perhaps even more so than the fibrillar protein): namely, they may trigger dopaminergic cell death by permeabilizing lipid membranes, thus causing a disruption of ion gradients necessary for neuronal homeostasis.[48,50,58] A30P and A53T have a greater membrane permeabilization activity per mole of protein than wild-type αSyn, and this property may contribute to the enhanced neurotoxicity of these two familial mutants (in addition to their increased propensity to form protofibrils).[58]

Similar to A53T, E46K forms amyloid-like fibrils more rapidly than wild-type αSyn.[54,65,66] However, in contrast to A30P and A53T, E46K does not have an enhanced propensity to form protofibrils, and protofibrillar E46K exhibits a decreased specific membrane permeabilization activity.[66] One way to interpret these findings is to infer that E46K elicits neurotoxicity via a mechanism that does not involve the formation of membrane-disrupting protofibrils. Alternatively, if we presume that E46K conforms to the "toxic protofibril hypothesis," then cellular perturbations that are neglected in current cell-free systems (e.g., posttranslational modifications) must increase the ability of this variant to form protofibrils and/or permeabilize membranes *in vivo.*

C. Modulation of αSyn Aggregation by Long-Range Interactions

The solution structure of human wild-type αSyn has been characterized extensively using NMR methods that combine measurements of paramagnetic relaxation enhancement (PRE) and/or residual dipolar couplings (RDCs) with ensemble molecular dynamics simulations.[67–69] The results indicate that the protein adopts an ensemble of conformations stabilized by long-range interactions between the C-terminal region and the N-terminal and NAC domains. In turn, the long-range interactions may result in inhibition of αSyn self-assembly via an auto-inhibitory mechanism involving the "shielding" of hydrophobic residues in the N-terminal and NAC regions by the C-terminal domain.[55,67,68] One would predict that a loss of long-range interactions should lead to an increased rate of fibrillization. Consistent with this model, C-terminally truncated αSyn variants form fibrils more rapidly than the full-length protein.[55,70–73] In addition, mouse αSyn, a variant with weaker long-range interactions compared to those of the human wild-type protein,[74] forms fibrils more rapidly than human wild-type αSyn or A53T.[46]

On the basis of PRE and RDC data, Zweckstetter and colleagues[75] reported that the A30P and A53T substitutions destabilize long-range interactions between the C-terminal region and the hydrophobic NAC domain, and they inferred that this perturbation might account for the enhanced ability of the familial mutants to form oligomers compared to the wild-type protein. In contrast, Eliezer and colleagues[76] failed to observe a loss of long-range interactions in A30P and A53T, and in fact they found that C-to-N contacts were *stronger* in E46K compared to wild-type αSyn. Accordingly, these investigators concluded that parameters such as net charge or secondary structure propensity are more important than the strength of long-range interactions in determining relative rates of self-assembly of wild-type and mutant αSyn.

D. Posttranslational Modifications of αSyn in Diseased Brains and Synucleinopathy Models

Various posttranslational modifications are associated with aggregated αSyn in patients with synucleinopathy disorders. Examples of these modifications include tyrosine nitration,[77] phosphorylation of serine 129 (S129),[42,78–80] ubiquitylation,[42,81] and C-terminal truncations resulting from the removal of approximately 20–40 residues (Fig. 1).[42,72,73,82] In contrast, phosphorylation of tyrosine 125 (Y125) was detected in the brains of aged, nondiseased individuals, but not in the brains of patients with dementia with Lewy bodies (DLB).[83]

A number of posttranslational modifications of αSyn have been detected in cellular and animal models relevant to PD and other synucleinopathy disorders, including (i) oxidation of methionine residues 116 or 127 to methionine sulfoxide (MetSO) or methionine sulfone[84] ; (ii) nitration of tyrosine residues,[85–88] including Y39[89,90] in the N-terminal domain and Y125, Y133, and Y136 in the C-terminal domain (Strathearn et al., unpublished observations)[84,89]; (iii) phosphorylation of S129 (Strathearn et al., unpublished observations)[79,80,88,91–100]; (iv) phosphorylation of Y125[83,84]; and (v) C-terminal truncation resulting from the cleavage of approximately 20–40 residues.[72,73,101,102] One group has reported the presence of αSyn isoforms phosphorylated on serine 87 (S87) in the brains of synucleinopathy patients or transgenic mouse models of synucleinopathy disorders,[103] although another group failed to detect this modification in human patients or transgenic mice.[104]

E. Effects of Posttranslational Modifications on αSyn Self-assembly and Neurotoxicity

A central question driving research in the field of synucleinopathy disorders is whether posttranslational modifications are a *cause* of enhanced αSyn aggregation and neurotoxicity. To address this question, a number of studies have been carried out in cell-free systems and cellular and animal models. The key findings from these studies are summarized below.

1. αSyn Oxidation

Data obtained from studies in cell-free systems indicate that αSyn oligomer formation is promoted by H_2O_2 and Fe^{2+} or Cu^{2+}[105,106] and by dityrosine crosslink formation involving Y125 under conditions of oxidative stress.[107–109] Modification of αSyn by metal-catalyzed oxidation[110] or oxidation by the lipid peroxidation product 4-hydroxy-2-nonenal[111] results in inhibition of fibrillization and a buildup of soluble oligomers. Another lipid peroxidation product, acrolein, also stimulates αSyn oligomer formation.[112] Oxidized cholesterol metabolites promote the self-assembly of recombinant αSyn to protofibrils *and* fibrils, apparently via a mechanism involving noncovalent interactions.[113]

Oxidation of all four methionine residues of αSyn (M1, M5, M116, and M127) to MetSO results in a nearly complete suppression of fibrillizaton at neutral pH,[114,115] and the degree of inhibition increases with the number of oxidized methionine residues.[116] The methionine-oxidized protein regains its ability to form fibrils under conditions that favor neutralization of C-terminal negative charges – notably, when incubated in the presence of various metal ions (e.g., Ti^{3+}, Zn^{2+}, Al^{3+}, Pb^{2+})[117] or at low pH (pH = 3).[115] These observations suggest that methionine oxidation interferes with αSyn fibrillization by favoring repulsive intermolecular interactions and/or auto-inhibitory long-range interactions involving the C-terminal domain.[115] In contrast to the inhibitory effect of MetSO on αSyn fibrillization, soluble oligomers are found to accumulate in mixtures of methionine-oxidized and unoxidized αSyn[114,116,118,119] or in pure solutions of the methionine-oxidized protein.[115]

2. αSyn–DA Interactions

A number of groups have shown that αSyn reacts with oxidized derivatives of DA, including indole-5,6-quinone, 5,6-dihydroxyindole, and dihydroxyphenylacetic acid (DOPAC), and dopamine-modified αSyn loses the ability to form amyloid-like fibrils and instead accumulates as soluble oligomers.[118,120–125] Data from other studies suggest that DA oxidation products stimulate αSyn oligomer formation, block fibril formation, and destabilize preformed fibrils by interacting with the protein noncovalently.[122,123,126–128] DA may also promote αSyn oligomerization and inhibit αSyn fibrillization via oxidation of the protein's four methionine residues to MetSO (see above).[129] Some αSyn oligomers formed in the presence of DA appear similar to protofibrils on the basis of their size and morphology determined by AFM or EM[120,127] or their elution behavior during gel filtration or SDS-PAGE,[49,50,118,121,130] whereas others are smaller cross-linked multimers that lack a stable secondary structure.[49,118,120,121,124,125,130,131] The results of molecular modeling studies involving docking of DA into αSyn conformations determined by solution NMR suggested that DA interacts with two sites on the protein: (i) the peptide sequence 125-YEMPS-129, via hydrophobic interactions; and (ii) residue E83, via electrostatic interactions.[132] Consistent with these binding sites, the 125-YEMPS-129 segment was previously shown to play an important role in DA-mediated suppression of αSyn fibrillization via a noncovalent mechanism,[123,127] and an αSyn mutant in which E83 was replaced with alanine (E83A) was found to be resistant to the inhibitory effects of DA on αSyn fibril formation.[132]

In support of the above findings from studies in cell-free systems, Ischiropou-los and colleagues[133] showed that inclusion formation by αSyn A53T is suppressed in a neuroblastoma cell line engineered to produce high levels of intracellular DA via TH overexpression. The inhibitory effect of DA on αSyn aggregation in cell culture was dependent on the presence of the 125-YEMPS-129 segment.[123] Moreover, detergent-insoluble αSyn aggregates and soluble αSyn oligomers were found to be less abundant and more abundant (respectively) in nigral tissue compared to cortical tissue isolated from αSyn transgenic mice.[123] From these results, the authors inferred that (i) αSyn oligomers formed in the presence of DA are nontoxic, and (ii) a loss of DA in the substantia nigra may enhance αSyn neurotoxicity in this region by promoting the protein's conversion to amyloid-like fibrils. The degree to which this model relates to the pathogenesis of human PD is unclear, however, given that increased levels of αSyn expression are associated with increased dopaminergic cell death in the substantia nigra in the brains of patients[9,10] but not in the brains of αSyn transgenic mice.[133] In addition, multiple lines of evidence suggest that interactions among cytosolic DA, αSyn, and Ca^{2+} ions contribute to preferential dopaminergic cell death in PD.[134–136] Nevertheless, the apparent lack of dopaminergic cell death in response to a buildup of αSyn oligomers in the substantia nigra of transgenic mice is an important observation because it suggests that some αSyn assemblies formed in the presence of DA are not intrinsically toxic. Alternatively, a set of conditions in the substantia nigra of αSyn transgenic mice may prevent dopaminergic cell death which would normally be triggered by toxic αSyn oligomers (and thus identification of these conditions might reveal new strategies for treating PD).

3. αSYN NITRATION

In the presence of peroxynitrite, αSyn undergoes dimerization as a result of dityrosine formation or nitration at one or more tyrosine residues.[107,108,137] Nitrated αSyn has an increased propensity to form soluble oligomers but a decreased ability to form amyloid-like fibrils.[137,138] Moreover, nitrated αSyn oligomers (but not nitrated monomers or dimers) suppress fibrillization of the unmodified protein.[138,139] To address whether nitration plays a role in αSyn neurotoxicity, He and colleagues[140] generated a fully nitrated variant of αSyn112 which was fused to the TAT signal peptide to enable transport across cell membranes. Nitrated TAT–αSyn112 was found to elicit greater dopami-nergic cell death and more pronounced motor deficits compared to the corresponding non-nitrated fusion protein after unilateral infusion into rat substantia nigra, suggesting that nitrated αSyn112 is more toxic than the unmodified protein. Although these findings revealed important new insights, they also raised questions about whether the toxicity of the nitrated fusion protein was affected by (i) the presence of the TAT peptide, which may perturb the subcellular distribution of its αSyn "cargo" compared to that of the unfused protein[141,142]; and (ii) molecular properties of αSyn112 that are distinct from those of αSyn140. One way to address these issues would be to test the effects of ablating one or more nitration sites via site-directed mutagenesis on the neurotoxicity of αSyn140 expressed in a cell-culture model. Recently, we found that an αSyn140 variant in which Y125, Y133, and Y136 were replaced with phenylalanine ("3YF") was substantially less toxic than the wild-type protein in a primary midbrain culture model, suggesting that one or more of the above tyrosine residues plays a role in dopaminergic cell death (Strathearn and Rochet, unpublished observations). A limitation of this approach is that one cannot be certain whether the reduced toxicity of the mutant protein results from the disruption of nitration, dityrosine formation (see above), or tyrosine phosphorylation (see below).

4. SERINE PHOSPHORYLATION

In an early study, Iwatsubo and colleagues[78] found that recombinant human αSyn phosphorylated by casein kinase 2 (CK2), an enzyme that phosphorylates αSyn at S129 and to a lesser degree S87,[143] formed amyloid-like fibrils more rapidly than the unphosphorylated protein. In contrast, a detailed biochemical analysis of αSyn phosphorylated uniquely at S129 (αSyn-pS129), formed by treating the S87A mutant with CK1, revealed that S129 phosphorylation disrupted long-range interactions in the natively unfolded monomer and interfered with the protein's ability to undergo oligomerization or fibrillization.[144] It is unclear why serine phosphorylation was found to affect the rate of αSyn fibrillization differently in the two studies, although even small differences in the experimental conditions can have pronounced effects on the kinetics of αSyn self-assembly.[145] Importantly, αSyn mutants in which S129 was replaced with aspartate (S129D) or glutamate (S129E) differed substantially from αSyn-pS129 in terms of their conformations in solution and fibrillization rates (S129D and S129E had unperturbed long-range interactions and a similar ability to form fibrils as wild-type αSyn).[144] These findings suggest that the S129D and S129E substitutions are poor mimics of S129 phosphorylation, at least in the context of the purified recombinant protein.

Considerable research efforts have focused on elucidating how S129 phosphorylation affects αSyn aggregation and neurotoxicity in animal models of synucleinopathy disorders. One approach has been to investigate the impact of replacing S129 with (i) alanine, which cannot be phosphorylated; or (ii) aspartate, which may serve as an *in vivo* phosphoserine mimic. In one study, Chen and Feany[93] showed that the S129A mutant had a greater propensity to form inclusions but a reduced ability to trigger dopaminergic cell death than wild-type αSyn in a transgenic *Drosophila* model. In contrast, flies expressing S129D exhibited more pronounced neurodegeneration than flies expressing the wild-type protein. Contrary to these results, two groups reported that S129A and S129D exhibited increased and decreased neurotoxicity (respectively) compared to wild-type αSyn when expressed from a recombinant adeno-associated virus (rAAV) vector injected in rat substantia nigra.[146,147] In one of these studies, S129A was also shown to produce more abundant amyloid-like (thioflavin S-positive) nigral inclusions than wild-type αSyn or S129D.[147] A third group reported that wild-type αSyn, S129A, and S129D were essentially indistinguishable in terms of their neurotoxic effects in a rat rAAV model (the outcomes of this rAAV study may have differed from those of the other two outlined above because of differences in experimental conditions—e.g., rAAV serotype, rat strain, and duration of study).[148]

Together, these findings indicate that S129 substitutions have opposite effects on αSyn neurotoxicity in transgenic *Drosophila* versus rAAV-infected rats: the rank order of toxicities is S129D>wild-type αSyn>S129A in flies, whereas it is S129A>wild-type αSyn>S129D in rats. Because the S129A mutant has an increased propensity to form amyloid-like fibrils compared to wild-type αSyn in cell-free systems,[144] it is not a reliable variant to determine the consequences of ablating S129 phosphorylation on αSyn neurotoxicity (in order to serve as a meaningful negative control, such a variant should have the same fibrillization propensity as the wild-type protein). It is unclear why the expression of S129D had opposite effects in rats versus flies. As one possibility, interactions between the human and rat proteins may modulate αSyn neurotoxicity in the rat rAAV model, whereas these interspecies effects are absent in *Drosophila* because flies do not express an endogenous αSyn homolog.[46,147] It should also be noted that S129D (or S129E) may not faithfully reproduce the conformational properties or aggregation behavior of αSyn-pS129 *in vivo*, which is a limitation that has been demonstrated in cell-free systems (see above).[144]

Kinases potentially involved in phosphorylating αSyn at S129 include casein kinase 2,[95,97,98,104,143,149–151] G-protein coupled receptor kinases 2 and 5 (GRK2, GRK5),[152,153] and polo-like kinases (PLKs),[151,154,155] whereas αSyn-pS129 dephosphorylation is mediated by phosphoprotein phosphatase 2A (PP2A).[156] Experiments designed to modulate the activity or expression level of these enzymes may provide insight into the impact of S129 phosphorylation on αSyn aggregation or neurotoxicity. In one study, Chen and Feany[93] reported that the neurotoxicity of wild-type αSyn was enhanced upon coexpression of GRK2 in their transgenic fly model. In another study, expression of PLK2 was found to mitigate αSyn-mediated dopaminergic cell death in a *Caenorhabditis elegans* model and in rat primary midbrain cultures.[157] A third study revealed that activation of PP2A with eicosanoyl-5-hydroxytryptamide, an inhibitor of PP2A demethylation, interfered with S129 phosphorylation, αSyn aggregation, dendritic degeneration, glial activation, and motor dysfunction in αSyn transgenic mice.[156] A caveat in interpreting the results of these three studies is that modulation of kinase or phosphatase activity may affect αSyn aggregation or neurotoxicity via mechanisms independent of S129 phosphorylation.

Finally, αSyn-pS87 was recently shown to have a decreased ability to form oligomers or fibrils compared to wild-type αSyn.[103] The S87D and S87E variants had similar aggregation propensities as αSyn-pS87, suggesting that aspartate and glutamate are better phosphoserine mimics at position 87 than at position 129. Phosphorylation of S87 (but not S129) lowers the affinity of αSyn for phospholipid membranes and alters the conformation of the membrane-bound protein, suggesting that this modification may interfere with the protein's normal functions (e.g., modulation of neurotransmission).[103] Kinases implicated in S87 phosphorylation include CK1[103,143] and the dual-specificity tyrosine-regulated kinase DYRK1A.[158]

5. Tyrosine Phosphorylation

An early study revealed that the nonreceptor tyrosine kinase p72^{syk} suppressed αSyn aggregation in a cell-free system by phosphorylating residues Y125, Y133, and Y136 in the C-terminal tail.[159] In contrast, phosphorylation of Y125 alone by Lyn kinase had no effect on αSyn self-assembly. More recently, Feany and colleagues[83] reported that Y125F exhibited enhanced oligomerization and neurotoxicity compared to wild-type αSyn in transgenic flies, whereas coexpression of the tyrosine kinase shark (a *Drosophila* homolog of Syk) interfered with the ability of wild-type αSyn and S129D to form oligomers or elicit neurodegeneration in this model. The authors also showed that levels of αSyn-pY125 were lower in postmortem brains from old versus young individuals or from DLB patients versus age-matched controls. From these data, the authors inferred that (i) phosphorylation of Y125 suppresses the formation of neurotoxic αSyn oligomers by antagonizing the pathologic effects of S129 phosphorylation; and (ii) a reduction in this protective effect of Y125 phosphorylation in older individuals may contribute to the increased risk of PD with aging. Our recent finding that the "3YF" mutant exhibits reduced neurotoxicity compared to wild-type αSyn in primary midbrain cultures (Strathearn and Rochet, unpublished observations) is inconsistent with the enhanced neurotoxicity of Y125F in transgenic *Drosophila*. The reasons for this discrepancy are unclear but may relate to obvious differences between the two experimental systems (e.g., presence of human and rat αSyn in midbrain cultures versus only human αSyn in fly brain; mutation of Y125/Y133/Y136 in cell-culture model versus mutation of just Y125 in *Drosophila* model).

6. C-Terminal Truncation

Recombinant C-terminal truncation mutants spanning residues 1–110 or 1–120 of wild-type αSyn, similar to truncated αSyn variants identified in Lewy bodies,[42,72,73,82] undergo fibrillization and/or oligomerization more rapidly than the full-length protein.[47,55,70–73,160] In addition, substoichiometric levels of the truncation mutants accelerate the aggregation of full-length αSyn, and this seeding effect is more pronounced in the case of truncated A53T than the truncated wild-type protein (paired with full-length A53T and wild-type αSyn, respectively).[72,73] The stimulatory effect of C-terminal truncation on αSyn self-assembly is thought to involve the disruption of auto-inhibitory long-range interactions, a mechanism that may also account for the enhanced aggregation of full-length αSyn upon binding of metal ions or polyamines to the negatively charged C-terminal region.[55,67,68]

A question of central importance to the field is whether αSyn neurotoxicity is enhanced as a result of C-terminal cleavage. In one study, Feany and colleagues[161] showed that transgenic *Drosophila* expressing a truncation mutant of human αSyn spanning residues 1–120 (αSyn_{1-120}) underwent a more pronounced accumulation of αSyn oligomers and inclusions and a more severe loss of dopaminergic neurons than flies expressing the full-length protein. In another study, Spillantini and colleagues[162] showed that expression of αSyn_{1-120} in transgenic mice lacking the endogenous mouse protein triggered the formation of fibrillar αSyn inclusions, swelling of striatal neurites, loss of striatal dopamine, and motor deficits. Similarly, expression of αSyn_{1-119} in a conditional transgenic mouse model induced a loss of striatal dopamine and dopamine metabolites.[163] Transgenic mice expressing human αSyn_{1-130} exhibited more extensive dopaminergic cell death in the substantia nigra and more striking behavioral deficits than mice expressing the full-length protein.[164] Coexpression of full-length human αSyn and αSyn_{1-110} from AAV vectors stereotactically injected into rat substantia nigra (each at a multiplicity of infection below the toxic threshold determined for that virus alone) resulted in a pronounced neurodegenerative phenotype.[165] In contrast, stereotactic injection of an AAV vector encoding $A53T_{1-93}$, a product of matrix metalloproteinase-3 (MMP-3) cleavage, in rat substantia nigra resulted in a similar degree of neurodegeneration as injection of AAV encoding full-length A53T.[102] Collectively, these results imply that C-terminally cleaved αSyn isoforms generally elicit greater neurotoxicity than full-length αSyn, and/or they may enhance the toxicity of the full-length protein. However, the effects of truncation on αSyn-mediated neurodegeneration are likely to vary with the site of cleavage (e.g., residue 110, 120, or 130 versus residue 93) and the αSyn sequence context (e.g., wild-type versus A53T sequence).

C-terminal truncation may enhance αSyn-mediated neurodegeneration by promoting fibrillization (e.g., perhaps by disfavoring interactions with DA via removal of the 125-YEMPS-129 segment[123,133]) and/or the formation of potentially toxic oligomers.[160] If C-terminally cleaved αSyn variants are indeed more toxic than the full-length protein, then one would infer that proteases responsible for cleaving αSyn may be reasonable therapeutic targets for PD and other synucleinopathy disorders. Several proteases are known to cleave αSyn in the C-terminal region, including calpain,[166–168] the 20S proteasome,[73,160] cathepsin D,[169] and MMP-3.[102]

F. Effects of Phospholipids on αSyn Self-assembly

Because αSyn readily associates with cellular membranes, a high priority in the field of synucleinopathy diseases is to understand how the protein's interactions with phospholipids affect its self-assembly behavior. In an important study, Lee and colleagues[170] showed that αSyn formed SDS-resistant oligomers more rapidly in incubated membrane fractions versus cytosolic fractions of rat brain homogenates via a mechanism dependent on oxidative stress. The formation of membrane-bound αSyn oligomers was stimulated by the addition of cytosolic αSyn to membrane fractions, implying that the cytosolic protein was recruited into membrane-bound assemblies. αSyn was also shown to readily form oligomers in cells enriched with lipid droplets[171] and to undergo accelerated oligomerization and fibrillization in the presence of long-chain polyunsaturated fatty acids.[172–174] In addition, αSyn fibril formation was stimulated in the presence of anionic detergent micelles, phospholipid vesicles, or synaptosomal membranes.[175–178] αSyn self-assembly occurred readily in these systems at a high protein–lipid ratio, whereas aggregation could be suppressed in the presence of excess phospholipids.[176,179]

Upon interacting with phospholipid membranes, αSyn adopts a conformation in which the N-terminal repeat region is folded into an amphipathic α-helical domain (either bent or extended) that binds the membrane surface, whereas the C-terminal region is disordered and exposed to the aqueous environment.[24–29,31,176,180–184] We and others have hypothesized that αSyn aggregation may be stimulated by interactions among α-helical molecules bound to the membrane surface.[49,185–188] The underlying rationale is that neighboring αSyn conformers may interact more readily in a two-dimensional space at the membrane surface compared to the less geometrically constrained three-dimensional space of the bulk solution.[189,190] Consistent with this hypothesis, we showed that αSyn formed clusters upon binding to a supported lipid bilayer consisting of a mixture of the anionic phospholipid phosphatidylglycerol (PG) and the zwitterionic phospholipid phosphatidylcholine (PC).[191–193] The ability of αSyn to form membrane-bound clusters increased with increasing anionic lipid (PG) content and/or protein concentration, and regions on the bilayer where αSyn was clustered were also enriched in PG. From these data, we inferred that (i) αSyn induces separation of phospholipids into regions enriched in anionic and zwitterionic lipids in order to neutralize charges on helical αSyn and/or relieve unfavorable lipid–lipid interactions; and (ii) αSyn forms clusters associated with regions enriched with anionic lipids on the membrane surface in order to bury hydrophobic residues exposed when the protein adopts a helical structure.

An elegant series of NMR analyses by Bax and colleagues[182,194] revealed that vesicle-bound αSyn adopts multiple conformations in which the N-terminal α-helical domain spans segments of different lengths along the polypeptide chain. Two prominent long-lived conformations—referred to as the "SL1" and "SL2" states—had immobilized, membrane-bound helical domains spanning residues 1–25 and 1–97, respectively. Data reported by Beyer and colleagues[195] suggested that αSyn initially binds the membrane in the SL1 state, which then converts to the SL2 state via a mechanism in which the N terminal helical domain spanning residues 1–25 nucleates the helical folding of the segment spanning residues 26–100. Consistent with these findings, residues 2–11 (and in particular, an aspartate residue at position 2) were found to play an important role in αSyn membrane binding and toxicity in a yeast model.[196] Because the SL1 state has a more extensive disordered region (spanning residues 26–140, including the hydrophobic NAC domain), this membrane-bound conformer should have a higher propensity to engage in intermolecular contacts compatible with the formation of β-sheet-rich, potentially toxic oligomers. The proportion of wild-type αSyn molecules existing in the SL1 state increased with increasing protein/lipid ratios,[182] and all three familial PD mutants (A30P, E46K, and A53T) populated the SL1 conformation to a greater extent than the wild-type protein.[194]

From these observations, we propose a model in which membranes enriched with anionic phospholipids promote the formation of neurotoxic αSyn oligomers. An initial step in this process involves the clustering of neighboring αSyn molecules via helix–helix contacts in membrane domains enriched with anionic phospholipids. Next, nonhelical segments of clustered αSyn molecules (particularly SL1 conformers) interact to form assemblies with increased β-sheet content and stability. Environmental perturbations may trigger αSyn self-assembly at the membrane surface by intervening at various stages of the pathway outlined above. For example, an increase in relative levels of anionic phospholipids in the brains of PD patients[197] may promote the clustering of membrane-bound helical conformers during the early stages of self-assembly.[191] Nitration of the C-terminal tyrosine residues (Y125, Y133, and Y136), which was recently shown to lower the affinity of αSyn for phospholipid vesicles via an allosteric mechanism,[198] could potentially promote the formation of membrane-bound αSyn oligomers by increasing the proportion of αSyn molecules in the SL1 state,[194] or by increasing the lifetime of this conformation.[195] Interactions between non-helical segments of membrane-bound αSyn may be favored by the presence of Ca^{2+} ions, which promote interactions of the C-terminal tail with the bilayer and induce an increase in overall β-sheet content.[199] The aggregation of membrane-associated αSyn may also be triggered by C-terminal truncation, based on evidence that the three familial mutants (and to a lesser extent the wild-type protein) undergo accelerated oligomerization when incubated in the presence of liposomes and the 20S proteasome, but not either agent alone.[160] Finally, strategies to interfere with the clustering of helical membrane-bound αSyn conformers or subsequent interactions between nonhelical segments may prove beneficial for the treatment of PD and other synucleinopathy disorders.

G. Characterization of αSyn Fibrillar Structure

The structural properties of αSyn amyloid-like fibrils were closely examined using various biophysical methods, including electron paramagnetic resonance spectroscopy,[200,201] hydrogen exchange-mass spectrometry,[202] and solid-state NMR (SS-NMR).[203–205] These studies revealed that fibrillar αSyn consists of a structured core domain encompassing residues ~30–110, flanked by an N-terminal, conformationally heterogeneous region and a C-terminal disordered region. The core region of fibrillar αSyn comprises five β-strands arranged as strand–loop–strand motifs that project along the fibril axis.[200–204] This arrangement yields a five-layered structure in which each layer is composed of an extended β-sheet with parallel, in-register strands. The five-layered structure constitutes a protofilament, and two protofilaments are aligned in a straight or twisted fashion to form a fibril.[204] Data obtained from a recent SS-NMR study suggested that the core region of fibrillar αSyn adopts a different structure consisting of two repeats of a motif encompassing a long β-strand followed by two shorter β-strands (Fig. 2).[205] Together, these structural insights may have a profound impact on drug discovery efforts by stimulating the design of αSyn fibrillization inhibitors using structure-based methods.[206–208]

H. Structure-Based Approaches to Assess Relative Toxicities of αSyn Oligomers and Fibrils

A central problem in the field of PD and other synucleinopathy disorders is whether αSyn neurotoxicity is mediated by prefibrillar intermediates (oligomers, protofibrils) or mature amyloid-like fibrils. To address this question, two groups have used the innovative approach of characterizing the neurotoxic properties of αSyn variants with a high propensity to form oligomers or protofibrils but not fibrils.[209,210] In both cases, the αSyn variants were designed using a rational approach based on the structure of fibrillar αSyn described above. In one study, the engineered αSyn mutants A56P and A30P/A56P/A76P ("TP") were found to undergo less rapid fibrillization than wild-type αSyn, and SS-NMR analysis of fibrils formed by the mutant proteins revealed a decrease in β-sheet

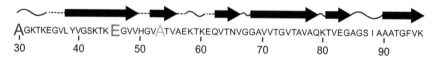

AGKTKEGVL YVGSKTK EGVVHGVATVAEKTKEQVTNVGGAVVTGVTAVAQKTVEGAGS I AAATGFVK

| | | | | | | |
| 30 | 40 | 50 | 60 | 70 | 80 | 90 |

FIG. 2. Structural model of fibrillar α-Syn determined from a recent solid-state NMR study. The core domain spans residues 38–96 and consists of two repeats of a β-sheet motif encompassing one long strand followed by two shorter strands. The three familial substitutions (A30P, E46K, and A53T) are shown in a larger font size. This figure (adapted from Ref. 205) was generously provided by Dr. Chad Rienstra (University of Illinois-Urbana/Champaign).

content.[209] In addition, both proline variants (and especially the "TP" mutant) had an enhanced propensity to accumulate as oligomers in a cell-free system and exhibited increased neurotoxicity compared to wild-type αSyn when expressed in primary cell cultures, *C. elegans*, or *Drosophila*. In the second study, the αSyn variants E35K and E57K, designed on the basis of the rationale that these substitutions might disrupt salt bridges involved in stabilizing β-sheet structure in fibrillar αSyn, were found to have a decreased propensity to form amyloid-like fibrils but an increased ability to form oligomers, including ring-like structures.[210] Strikingly, E35K and E57K exhibited enhanced neurotoxicity compared to wild-type αSyn and A53T when expressed from a lentiviral construct in rat substantia nigra, and both variants formed SDS-resistant oligomers (particularly trimers) which were detected by immunoblot analysis of membrane fractions from rat midbrain homogenates. In contrast, an engineered αSyn mutant spanning residues 30–110 with an enhanced fibrillization propensity exhibited substantially reduced neurotoxicity in the rat lentiviral model.[210] Collectively, these findings support the hypothesis that αSyn oligomers, rather than amyloid-like fibrils, are the major αSyn species involved in neurodegeneration in PD.

I. A Role for Cell-to-Cell Transmission of αSyn in PD Pathogenesis

Emerging evidence over the past 5 years has led to the intriguing hypothesis that the spread of neuropathology in PD involves cell-to-cell transmission of αSyn.[211] A number of groups have shown that monomeric and oligomeric forms of αSyn are secreted into the conditioned media of mammalian cell cultures, apparently via a nonclassical mechanism that depends on the presence of intracellular Ca^{2+} and involves the release of exosomes derived from multivesicular bodies.[212–217] αSyn secretion is stimulated by various cellular stresses, including proteasomal inhibition and lysosomal impairment, and secreted αSyn has higher levels of oxidative damage than the intracellular protein.[214,215] Oligomeric αSyn is internalized by mammalian cells via an endocytic mechanism and eliminated from recipient cells via lysosomal clearance pathways.[218] Additional studies have revealed that αSyn is transmitted between cells in culture or from host neurons to engrafted neuronal cells in αSyn transgenic mice, resulting in inclusion formation and activation of apoptosis in recipient cells.[216,217,219] These findings are important because they suggest that cell-to-cell transmission of αSyn may contribute to the spread of neuropathology in PD described by Braak and colleagues.[220] Moreover, the activation of astrocytes or microglia by extracellular αSyn (via mechanisms involving receptor binding and/or internalization) may contribute to neuroinflammatory pathways characteristic of the disease.[221–225]

J. Cellular Mechanisms to Suppress αSyn Aggregation and Neurotoxicity

Various cellular mechanisms are in place to inhibit the accumulation of neurotoxic, aggregated forms of αSyn. Examples of proteins involved in these "surveillance" mechanisms include antioxidant proteins, molecular chaperones, and proteins involved in cellular clearance mechanisms (Fig. 3). Each of these types of proteins is reviewed in greater detail below.

1. UPREGULATION OF ANTIOXIDANT RESPONSES

Evidence that oxidative stress favors the formation of potentially toxic αSyn aggregates has prompted the hypothesis that αSyn-mediated neurodegeneration may be alleviated by proteins with antioxidant activity.[226,227] In support of

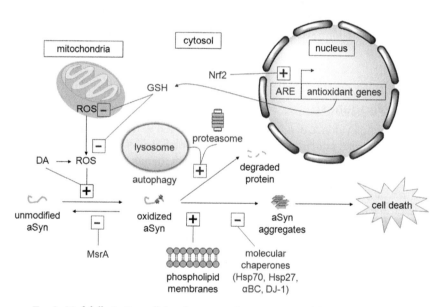

FIG. 3. Model illustrating cellular phenomena that promote or inhibit α-Syn self-assembly. A loss of mitochondrial function (e.g., impairment of complex I) or an increase in cytosolic dopamine levels triggers a buildup of ROS in dopaminergic neurons. In turn, ROS and/or dopamine oxidation products react with α-Syn, converting the protein to oxidized forms with a high propensity to aggregate. α-Syn self-assembly is promoted under some conditions by binding of the protein to phospholipid membranes. α-Syn aggregation and neurotoxicity may be mitigated by (i) cellular clearance mechanisms, including the 26S proteasome and lysosomal autophagy; (ii) cellular antioxidant responses, including Nrf2-mediated transcription, resulting in increased glutathione (GSH) synthesis, and MsrA-dependent repair of oxidized α-Syn; and (iii) molecular chaperones, including Hsp70, Hsp27, αB-crystallin, and DJ-1. (See Color Insert.)

this hypothesis, expression of the antioxidant repair enzyme methionine sulfoxide reductase A (MsrA) was shown to inhibit αSyn aggregation and neurotoxicity in cellular and animal models relevant to PD (Fig. 3).[228,229] MsrA reduces the S-stereoisomer of MetSO, including protein-bound MetSO, back to methionine. MsrA carries out its antioxidant function by (i) repairing oxidatively damaged proteins, thereby helping to preserve their activities; and (ii) depleting ROS by participating in cycles of methionine oxidation and reduction.[230,231] In one study, overexpression of bovine MsrA was found to alleviate αSyn-mediated dopaminergic cell death and cause a decrease in the abundance of αSyn oligomers in primary midbrain cultures.[229] In contrast, the ROS scavenger N-acetyl cysteine had a much less pronounced inhibitory effect, suggesting that MsrA interferes with αSyn aggregation and neurotoxicity by repairing oxidized αSyn rather than by depleting free radicals. Consistent with this interpretation, MsrA was shown to repair methionine-oxidized αSyn in a cell-free system.[229] In another study, coexpression of bovine MsrA was found to rescue dopaminergic cell death and motor deficits induced by the expression of human αSyn in a transgenic Drosophila model.[228] A similar protective effect was observed when the flies were fed S-methyl-L-cysteine (SMLC), a methionine analog that presumably triggered ROS depletion by undergoing cycles of oxidation and reduction catalyzed by endogenous MsrA in the fly brain.[228]

Collectively, these findings suggest that MsrA prevents a buildup of oxidized αSyn isoforms that readily form toxic oligomers in PD models.[228,229] The neuroprotective effect of MsrA may involve direct repair of αSyn and/or ROS scavenging (the relative importance of each of these mechanisms may vary from one model to another). Cycles of oxidation and MsrA-catalyzed repair of membrane-bound αSyn are likely to play an important role in suppressing the oxidation of unsaturated membrane lipids.[232] Because MsrA is abundant throughout the brain including the substantia nigra,[233] these observations imply that the enzyme may be involved in protecting nigral dopaminergic neurons against PD-related insults. If this is true, then the reported age-dependent decrease in MsrA activity[234] may contribute to the increased risk of PD with aging.

The antioxidant enzyme Cu/Zn superoxide dismutase (SOD1) has also been shown to alleviate dopaminergic cell death and motor deficits when coexpressed in αSyn transgenic Drosophila.[235] Moreover, activation of the Nrf2/Keap1 signaling pathway, resulting in increased expression of genes encoding enzymes involved in the cellular antioxidant response,[236–238] rescues motor deficits and dopaminergic neuronal loss in αSyn-expressing flies (Fig. 3).[239] These observations further substantiate the idea that αSyn neurotoxicity can be alleviated by the expression of proteins with antioxidant activity.

2. Upregulation of Chaperone Function

Molecular chaperones contribute to protein quality control by facilitating the refolding of misfolded polypeptides, interfering directly with protein aggregation, or directing misfolded or aggregated polypeptides to cellular clearance pathways (e.g., ubiquitin-proteasome pathway (UPP), lysosomal autophagy).[227,240] Heat shock proteins (e.g., Hsp27, Hsp40, Hsp70, and Hsp90 in eukaryotes) are a major class of chaperones upregulated in response to elevated temperatures and other stresses that cause a buildup of misfolded polypeptides. Examples of molecular chaperones that have been characterized in terms of their ability to suppress αSyn aggregation or toxicity include Hsp70, Hsp27, αB-crystallin, and DJ-1 (Fig. 3).

2a. Hsp70

Hsp70 was found to inhibit αSyn fibril formation in cell-free systems via a mechanism involving binding of the chaperone to various species on the αSyn self-assembly pathway, including early oligomers and higher order protofibrils.[241–243] The interaction of αSyn with Hsp70 led to the formation of soluble amorphous aggregates that were devoid of permeabilizing activity or cytotoxicity.[241,242,244] Dobson and colleagues[244] reported that Hsp70 and oligomeric αSyn formed a more compact complex in the presence of ATP or ADP, yielding toxic protofibrils that sequestered the chaperone via coaggregation. The stimulatory effects of nucleotides on αSyn protofibril formation and Hsp70 depletion were abrogated by co-incubation with Hip (ST13), a co-chaperone that is downregulated in PD patients,[245] and siRNA-mediated knockdown of Hip induced an Hsp70-dependent increase in αSyn-YFP inclusions in a transgenic *C. elegans* model. In another study, stable β-sheet-rich αSyn oligomers were found to inhibit Hsp70/Hsp40-mediated unfolding/refolding of protein substrates, apparently via weak hydrophobic interactions with the Hsp40 co-chaperone.[246] Together, these findings suggest that Hsp70-mediated suppression of αSyn neurotoxicity involves a balance between productive and nonproductive interactions among Hsp70, its co-chaperones, and various αSyn intermediates, resulting in the conversion of αSyn into nontoxic aggregates or in inhibition of the chaperone machinery, respectively.

Hsp70 has also been shown to modulate αSyn self-assembly and neurotoxicity in cellular and animal models. Groundbreaking studies by Bonini and colleagues[247,248] revealed that Hsp70 overexpression or upregulation alleviated αSyn neurotoxicity without altering the degree of inclusion formation in αSyn transgenic flies, whereas inactivation of the constitutive Hsp70 isoform Hsc4 resulted in accelerated dopaminergic cell death in this model. These results suggested that Hsp70 facilitates the conversion of αSyn to nontoxic aggregates in the fly brain. Hsp70 expression was also shown to inhibit αSyn oligomerization and toxicity in H4 human neuroglioma cells[249,250] and primary midbrain cultures.[251] Hsp70 was also found to suppress αSyn aggregation in the brains of αSyn transgenic mice,[249] although data from another study suggested that the chaperone had no impact on motor deficits or αSyn oligomerization in transgenic mice expressing A53T αSyn.[252] Finally, Hsp70 was recently shown to suppress the accumulation of extracellular αSyn oligomers in the H4 neuroglioma cell-culture model, potentially via a mechanism involving Hsp70 secretion.[216] Collectively, these data suggest that Hsp70 plays a major role in mitigating αSyn aggregation and toxicity in synucleinopathy disorders.

2b. αB-Crystallin and Hsp27

The small heat shock proteins αB-crystallin and Hsp27 are composed of low molecular weight subunits assembled into large oligomeric complexes. αB-crystallin was found to inhibit the fibrillization of recombinant αSyn via a mechanism in which the chaperone binds prefibrillar forms of αSyn (e.g., partially folded monomeric species) and promotes their conversion to amorphous aggregates.[253,254] A recent study revealed that αB-crystallin also interacts with amyloid-like αSyn fibrils, thereby blocking their elongation and promoting their dissociation to monomeric subunits.[255] The inhibitory effect of αB-crystallin on αSyn fibrillization is mimicked by peptides derived from the conserved α-crystallin core domain[256] and is augmented by phosphorylation of the chaperone at three serine residues (S19, S45, and S59).[257] Finally, evidence suggests that αB-crystallin and Hsp27 attenuate αSyn aggregation and/or toxicity in various cell-culture models.[258,259] These findings, together with neuropathological data showing that αB-crystallin and Hsp27 are present in synucleinopathy inclusions,[259,260] suggest that both small heat shock proteins may play a role in mitigating αSyn aggregation and toxicity in the brain.

2c. DJ-1

Mutations in the gene encoding DJ-1 (*PARK7*) cause very rare autosomal-recessive forms of PD.[261–266] DJ-1 appears to function as a homodimer of ~20-kDa subunits, each of which has an α−β fold.[267–271] The protein is a member of the large DJ-1/PfpI superfamily and has orthologs in most organisms.[272,273] Many activities have been proposed for DJ-1 (outlined in greater detail in Part 2 of this chapter), and there is currently no consensus on what its functions are.[274,275]

The crystal structure of human DJ-1 reveals the presence of a readily oxidized cysteine residue (cysteine 106) located at the subunit interface, and the protein exhibits a decrease in pI under conditions of oxidative stress due to the conversion of cysteine 106 to the sulfinic acid.[271,276,277] DJ-1 has a conserved domain that is also shared by the heat shock protein Hsp31, and several groups have observed that DJ-1 exhibits molecular chaperone activity in cell-free systems.[269,278–280] In one study, wild-type DJ-1, but not the familial mutant

L166P, was found to suppress the fibrillization and heat-induced oligomerization of αSyn.[278] DJ-1 exhibited a markedly diminished chaperone activity with the substrate citrate synthase after preincubation with the reductant dithiothreitol (DTT) or as a result of replacing cysteine 53 with alanine, although the effects of DTT treatment and C53 substitution on the ability of DJ-1 to suppress αSyn self-assembly were not investigated. A study by Fink, Petsko, and colleagues[279] revealed that human wild-type DJ-1 inhibits αSyn fibrillization via a mechanism driven by the "2O" form of DJ-1, in which cysteine 106 is oxidized to the sulfinic acid. DJ-1 isoforms oxidized to a greater or lesser extent than the 2O form had a substantially decreased ability to suppress αSyn fibril formation. Another group reported that two mutant forms of DJ-1 associated with early onset PD, M26I and A104T, had a decreased ability to inhibit αSyn fibrillization compared to wild-type DJ-1, and the chaperone activity of the wild-type and mutant proteins was augmented by introducing a disulfide bond at the subunit interface.[280]

Collectively, these findings suggest that homodimeric DJ-1 in the "2O" state suppresses αSyn fibrillization, and this activity may be at least partly responsible for DJ-1-mediated neuroprotection. Presumably, DJ-1 must exist as a dimer to carry out this redox-sensitive chaperone function because (i) a hydrophobic patch that is predicted to interact with αSyn and other protein substrates is located at the dimer interface,[269] and (ii) an "active site" pocket that appears to favor the oxidation of C106 to the sulfinic acid is lined with polar residues from both subunits.[270,271,276,281] The DJ-1 chaperone activity is expected to be disrupted by primary structure alterations that destabilize the native dimeric structure, including mutations involved in early onset PD[280,282–288] and oxidative modifications associated with sporadic PD and aging.[286,289,290] Surprisingly, attempts to demonstrate a physical interaction between 2O DJ-1 and monomeric αSyn using a variety of biophysical approaches have been unsuccessful.[279] However, DJ-1 may carry out its chaperone function by forming a complex with oligomeric or protofibrillar αSyn (Hulleman and Rochet, unpublished observations), similar to the mechanism proposed for Hsp70[241,242] (see above).

Finally, we and others have reported that DJ-1 suppresses αSyn neurotoxicity and aggregation by inducing an increase in Hsp70 expression[251,291-292] (and DJ-1 has been shown to interact with mitochondrial Hsp70)[293] in cell-culture models. Recent evidence that DJ-1 promotes the folding of a chaperone-dependent fluorescent biosensor in a human neuroblastoma cell line[294] is consistent with the idea that DJ-1 can enhance the function of other chaperones, in addition to carrying out its own chaperone function.

3. UPREGULATION OF PROTEIN DEGRADATION

Cellular clearance systems play a key role in protein quality control by ensuring the removal of misfolded or damaged polypeptides.[227,240] Two clearance systems have been extensively characterized in terms of their ability to modulate intracellular αSyn levels: proteasomal degradation and lysosomal autophagy, including chaperone-mediated autophagy (CMA) and macroauto-phagy. In general, αSyn overexpression results in inhibition of each of these catabolic pathways. Conversely, upregulation of these pathways results in enhanced αSyn degradation and thus suppression of αSyn aggregation (Fig. 3).

3a. PROTEASOMAL DEGRADATION

A number of studies have revealed that αSyn accumulates in cells exposed to proteasome inhibitors,[295–298] suggesting that the proteasome plays an important role in αSyn clearance. More recently, the proteasome was shown to preferentially degrade αSyn-pS129.[98,299] Because polyubiquitylated αSyn species were not detected in cells subjected to proteasome impairment, it was inferred that αSyn degradation occurred via a ubiquitin-independent mechanism in these models.[296,299] Other groups have reported that αSyn levels are *not* affected by interfering with proteasome function.[300–302] A potential explanation for these discrepancies is that soluble αSyn may be eliminated by the proteasome, whereas aggregated αSyn may be degraded by macroautophagy, and the distribution of the protein between soluble and aggregated forms may vary in different cell-culture models.[298] αSyn overexpression results in disruption of the UPP in various cell lines,[301,303,304] and oligomeric αSyn interferes with the enzymatic activity of the 26S proteasome.[305–308] αSyn neurotoxicity was found to be attenuated by overexpression of components of the UPP, including (i) parkin, an E3 ubiquitin ligase (described in greater detail in Part 2 of this chapter) in mouse primary midbrain cultures[301] and (ii) ubiquitin variants compatible with the formation of lysine-48 linkages in a *Drosophila* model.[309]

3b. AUTOPHAGY

In a landmark paper, Cuervo and colleagues[310] reported that αSyn is a substrate for CMA in primary neurons and in a cell-free system containing purified lysosomes. Consistent with these data, a later study revealed that αSyn turnover was disrupted in mammalian cell lines and primary neuronal cultures by introducing two amino acid substitutions (D98A and Q99A) that disrupt the αSyn CMA recognition motif ($_{95}$VKKDQ$_{99}$) or via RNAi-mediated knockdown of the lysosomal LAMP2A receptor, a central player in the CMA pathway.[302] CMA has also been implicated in the removal of αSyn from the brains of transgenic mice.[311] The familial mutants A30P and A53T and dopamine-modified wild-type αSyn have been shown to elicit CMA impairment, apparently via interactions with the LAMP2A receptor that interfere with lysosomal uptake and degradation of αSyn and other substrates.[310,312,313] The inhibitory effects of the mutant and dopamine-modified proteins on CMA may be related to the high propensity of these variants to form prefibrillar oligomers.[45,118]

Several lines of evidence suggest that αSyn is also a substrate for macro-autophagy. The macroautophagy inhibitor 3-methyladenine induces αSyn accumulation in dopaminergic PC12 cells and in cultured neurons.[298,302] A recent *in vivo* study showed that macroautophagy plays an important role in αSyn clearance when the protein is expressed at high levels in transgenic mouse brain, whereas at lower expression levels (i.e., in nontransgenic animals) the protein is eliminated primarily via proteasomal degradation.[314] Data obtained by the Rubinsztein group[315] suggest that αSyn overexpression inhibits macroautophagy by disrupting the formation of AP precursors termed "omega-somes." Another group reported that αSyn overexpression results in a buildup of the AP marker protein LC3-II,[316] suggesting that αSyn may also disrupt macroautophagy downstream of AP formation (e.g., perhaps at the AP–lyso-some fusion step). Importantly, αSyn clearance is stimulated in cells and transgenic mice exposed to pharmacological inducers of macroautophagy, including rapamycin,[298,317–321] suggesting that stimulation of this clearance mechanism may be a reasonable strategy to alleviate αSyn aggregation and neurotoxicity in PD.

III. Role of Mitochondrial Dysfunction in PD

A. Parkin, A Multifunctional E3 Ubiquitin Ligase

Mutations in *PARKIN*, the product of the *PARK2* locus, were first discov-ered in Japan in multiple families with a syndrome known as "autosomal-recessive juvenile parkinsonism" (AR-JP).[322,323] Mutations in *PARKIN* may also play an important role in some sporadic cases of PD.[324,325] Parkin is primarily localized to the cytoplasm, but as discussed further below, it also translocates to the mitochondria in response to various signals, where it plays important roles in mitochondrial homeostasis. Parkin has an N-terminal ubiquitin-like domain (Ubl), followed by a recently described zinc-coordinating motif termed RING0, and two RING-finger motifs separated from each other by a cysteine-rich in-between-RING (IBR) motif.

Ubiquitin conjugation to proteins is usually effected by a series of enzymes: an E1 that activates ubiquitin, an E2 that accepts the activated ubiquitin and works with a third enzyme, and the E3-ubiquitin ligase that provides specificity in determining which substrates will be ubiquitinated.[326] Parkin is a RING family E3 ligase.[327–329] It undergoes autoubiquitination and can ubiquitinate a number of different substrates. Pathogenic mutations occur throughout the protein-coding region and many have effects on ligase activity (reviewed in Refs. 330–332). Parkin ligase activity can also be inactivated by oxidative stress, and following nitrosylation or covalent modification by dopamine.[333–337]

Inactivation may occur through modification of residues (such as cysteines) essential for Parkin's ligase activity. Alternatively, and/or in addition, modification may result in Parkin becoming insoluble, leading indirectly to a loss of ligase activity on some or all substrates.[333,335,338] α-Syn overexprerssion has also been shown to promote Parkin insolubility.[339] Much of the time, Parkin inacti vation through these mechanisms may be pathological; but it is also possible that these modifications serve under some circumstances to modulate Parkin activity for normal physiological ends, an area that remains relatively unexplored. Since there is evidence that each of the above insults can play a role in PD pathogenesis in sporadic cases, it is likely that disruption of Parkin ligase activity plays a role in at least some cases of sporadic PD (reviewed in Ref. 331). Needless to say, an important goal in the field has been to identify substrates of this activity.

B. Cytoplasmic Parkin Substrates

Many potential Parkin substrates have been identified. In the section below, we discuss substrates in the cytoplasm. In a later section, we discuss roles for Parkin ligase activity at the mitochondrial membrane. RING domain ligases such as Parkin can catalyze K48-mediated polyubiquitination,[327–329] which targets substrates for proteasome-dependent degradation (reviewed in Ref. 326). Thus, one important hypothesis in the field has been that loss of *parkin* results in the aberrant accumulation of toxic proteins. Important predictions of this model are that the substrate should accumulate in *parkin* knockout mice and in patients carrying *parkin* mutations, and perhaps in other contexts in which *parkin* has been inactivated, such as following MPTP intoxication and/or in sporadic PD cases. Additional predictions of this model are that decreasing expression levels of the substrate in a *parkin* mutant background should prevent cell death, while overexpression should result in toxicity. A number of Parkin binding partners and substrates have been identified, but most have not been characterized in sufficient detail to know whether they meet these criteria (reviewed in Refs. 330–332). Several intriguing exceptions that meet at least several criteria are aminoacyl-tRNA synthetase (AIMP2)[340,341] and the far upstream element binding protein 1 (FBP-1).[342] AIMP2 is present in the Lewy bodies, and Parkin is able to promote its degradation, presumably via polyubiquitination and proteasome-dependent degradation. AIMP2 also accumulates in the brains of *parkin*-null mice, patients with *PARKIN* mutations, and in some sporadic PD patients, and is toxic to dopaminergic neurons when overexpressed. FBP-1 also accumulates in *parkin* knockout mice, patients with mutations in *PARKIN* sporadic cases, and animal models of MPTP intoxication. How these proteins mediate their toxic functions is unknown and warrants further investigation.

Paris, a third protein recently identified as a Parkin substrate, is interesting because a number of lines of evidence suggest that its expression is both necessary and sufficient to mediate loss of dopaminergic neurons in cells lacking *parkin*.[343] Paris and Parkin can co-immunoprecipitate from cells and brains, and Paris is ubiquitinated through a K48 polyubiquitin linkage in a Parkin-dependent manner. Paris levels are also increased in the brains of *PARKIN* patients and patients with sporadic PD, consistent with the hypothesis that Parkin-dependent ubiquitination of Paris results in its degradation. Finally, silencing of *paris* expression prevents the loss of dopaminergic neurons observed in a conditional *parkin* knockout, while overexpression of *paris* results in loss of DN neurons, which can be suppressed through coexpression of *paris*. These latter observations are particularly important because they suggest that PD pathology critically depends on the levels of Paris.

How does upregulation of Paris lead to pathology? Paris binds the *PGC-1-alpha* promoter and represses its transcription, as well as the transcription of a *PGC-1-alpha* target, *NRF-1*. These interactions are likely to be significant because overexpression of *PGC-1-alpha* prevents the death associated with upregulation of Paris, as well as death induced in other PD models.[344] Downregulation of *PGC-1-alpha* target gene expression has also been observed in *PARKIN* patients[345] and dopaminergic neurons from sporadic cases.[344] How might downregulation of *PGC-1-alpha* and its target gene *NRF-1* (and perhaps other genes as well) cause pathology? A major effect of *PGC-1-alpha* expression is to promote mitochondrial biogenesis and regulate the metabolism of ROS.[346-349] As discussed below, Parkin also participates in a process that removes damaged mitochondria. Therefore, it is tempting to propose that stress-dependent stimulation of Parkin's ligase activity promotes the removal of damaged mitochondria and, through downregulation of *paris*, promotes a compensatory increase in mitochondrial biogenesis. In such a model, inactivation of *parkin* would result in a particularly toxic situation in which damaged mitochondria (and their damaged genomes) are retained and compensatory biogenesis fails to occur. Conversely, silencing of *paris* and/or activation of *PGC-1-alpha* through other means may provide a therapeutic route to maintaining mitochondrial function to some extent, even if the Parkin-dependent removal of damaged mitochondria is unable to occur or occurs less efficiently.

C. Parkin's Cytoprotective Ability Involves Multiple Forms of Ubiquitination

Parkin expression has also been shown to be neuroprotective in response to a variety of stresses (reviewed in Refs. 330,332). This activity is generally thought to require Parkin's ligase activity—as implied by the fact that ligase-dead forms of the protein show little or no activity—but the exact mechanism of action is unknown. K48-dependent polyubiquitination may be important in some contexts, as discussed above. However, Parkin can also mediate mono-ubiquitination, K63-linked polyubiquitination (reviewed in Ref. 330) as well as a very recently described K27-linked ubiquitination,[350] and perhaps even E2-independent ubiquitination.[351] These modifications can influence cellular processes such as signal transduction, transcriptional regulation, and protein and membrane trafficking, without promoting substrate degradation.[352,353] Finally, Parkin has also been suggested to bind directly to the 19S regulatory domain of the proteasome,[354,355] activating the proteasome in an E3 ligase activity-independent manner by increasing the affinity of 19S subunits for each other.[355] The fact that Parkin expression can provide protection from death associated with proteasome inhibition[334,356] suggests that at least some component of Parkin's neuroprotective activity when overexpressed—which may or may not be the same activity lost in the absence of *parkin*—involves processes other than K48-linked ubiquitin and proteasome-dependent protein degradation. We return to some of these points in the sections below.

D. PINK1, A Serine/Threonine Kinase with Multiple Forms and Localizations

In parallel with the above work, a number of observations demonstrate that Parkin, together with the kinase PINK1, plays an important role at the mitochondrial membrane in the regulation of mitochondrial homeostasis and quality control. In the following section we outline this pathway and its significance for PD. Mutations in *PINK1* (the *PARK6* locus) were first described in Spanish and Italian families with a syndrome of AR-JP.[357] As with *PARKIN*, a variety of mutations have been reported in patients (though patients with *PINK1* mutations are less common), and the one family characterized showed Lewy body pathology,[358] as observed in some, but not all patients homozygous or trans-heterozygous for mutations in *PARKIN*.[322,359,360] *PINK1* encodes a putative serine/threonine kinase. The N-terminus contains a mitochondrial targeting sequence. This is followed by a hydrophobic transmembrane domain, the kinase domain, and a putative C-terminal regulatory domain. Expression of PINK1 is cytoprotective in some contexts, and this activity often appears to require kinase activity (reviewed in Ref. 361). The majority of mutations associated with familial PD disrupt kinase activity, while others affect other aspects of PINK1 function (see below). Multiple forms of PINK1 protein are generated, and some, but not all PINK1, localize to mitochondria with the kinase domain facing the cytoplasm. These features of PINK1 biology are discussed further below.

E. The PINK1/Parkin Pathway of Mitochondrial Homeostasis and Quality Control in *Drosophila* and Beyond

Links between *parkin* and *PINK1* were first identified in *Drosophila* (reviewed in Refs. 17,362,363). Flies lacking *parkin* exhibit dramatic mitochondrial defects—swollen mitochondria that have severely fragmented cristae—in several energy-intensive tissues, including the male germline and adult flight muscle.[364,365] The flight muscles ultimately die, and their death shows features of apoptosis.[364] Flies lacking *parkin* also display a small but significant degeneration of a subset of dopaminergic neurons.[366] Although severe defects in mitochondrial morphology are not observed in *parkin* knockout mice, these animals do display mitochondrial functional defects including reduced mitochondrial respiratory activity.[367]

Flies lacking the *Drosophila* homolog of *PINK1* show phenotypes very similar to those of flies lacking *parkin*: mutants are viable but exhibit increased stress sensitivity and mitochondrial morphological defects in testes and muscle.[368–370] *PINK1* mutants also show reduced ATP levels and mitochondrial DNA (mtDNA) content. Flies lacking endogenous *PINK1* function but expressing PD-associated mutant forms of *PINK1*[370,371] show phenotypes similar to those of the *PINK1* null mutant, consistent with the *PINK1*-associated disease being the result of a loss of *PINK1* function.[370,371] As in *parkin* mutant flies, mitochondria in *PINK1* mutant flight muscle are swollen with fragmented cristae, and these cells ultimately undergo apoptotic death.[368–370] Mitochondria within dopaminergic neurons in *PINK1* mutants also display aberrant morphology, and there is a small but statistically significant loss of a subset of these neurons with age.[369,370]

What is the relationship between *PINK1* and *parkin*? Stringent genetic studies in *Drosophila* allow one to construct a genetic pathway. *parkin* overexpression suppresses all *PINK1* mutant phenotypes tested, while *PINK1* overexpression does not compensate for loss of *parkin* function.[368–370] In addition, double mutants lacking both *PINK1* and *parkin* have phenotypes identical to, rather than stronger than, either single mutant.[368,369] Together, these observations indicate that *PINK1* and *parkin* act in a linear pathway to regulate mitochondrial integrity, with *parkin* functioning downstream of *PINK1*.

Observations on *PINK1* and *parkin* function in flies are relevant to humans for several reasons. First, expression of human *PINK1*[368,370] or *PARKIN* in *Drosophila* suppresses phenotypes caused by loss of function of *PINK1* or *parkin*, respectively, indicating that the human and fly proteins are functionally conserved. Second, as noted above, PD patients who harbor mutations in *PINK1* or *PARKIN* are clinically indistinguishable,[372] and mice lacking both *PINK1* and *parkin* show phenotypes no worse than those of the single mutants,[373] consistent with the hypothesis that these genes function in a common genetic pathway. Third, cells from patients and/or mouse knockout models of *PINK1* or *parkin* also show defects in mitochondrial morphology and/or mitochondrial respiration, particularly in complex I activity in a variety of cell types.[364,367,374–388]

Important clues as to the mechanism by which *PINK1* and *parkin* regulate mitochondrial function came from the study of mitochondrial morphology in *Drosophila* mutants. Mitochondria are continually undergoing cycles of fission and fusion. This allows them to change shape and share components. It also plays an important role in facilitating recruitment to specific cellular compartments such as synapses where ATP or Ca^{2+} buffering demands are high. Not surprisingly, dysfunction of mitochondrial fission/fusion has been linked to the pathogenesis of a number of neurodegenerative diseases (reviewed in Ref. 389). Fusion is promoted by *mitofusin* (*mfn*), which is required for outer membrane fusion, and *opa-1*, which is required for inner membrane fusion, while fission is promoted by Drp1, a predominantly cytoplasmic protein recruited to mitochondria during fission. Recruitment and/or function of Drp-1 depends on mitochondrial outer membrane proteins such as Fis1 and Mff through mechanisms still being explored.[389–391] During *Drosophila* spermatogenesis, mitochondria undergo significant morphological changes. Early spermatids undergo mitochondrial aggregation and fusion, creating a spherical structure known as the nebenkern, which is composed of two intertwined mitochondria.[392] During subsequent spermatid elongation, the nebenkern unfurls, yielding two mitochondrial derivatives that are maintained throughout subsequent stages of spermatogenesis. In both *PINK1*[393] and *parkin*[393,394] mutants, only one leaf blade is seen, suggesting a defect in mitochondrial fission or an overabundance of fusion.[393] Several other pieces of evidence support this hypothesis. First, when mitochondria in *PINK1* or *parkin* mutants are visualized with mitochondrially targeted GFP, they are clumped in large aggregates in both dopaminergic neurons and flight muscle. These phenotypes, as well as others, such as the degeneration of flight muscle and a decrease in the levels of dopamine in fly heads, can be suppressed by increasing the levels of Drp1 or Fis1, and/or decreasing levels of *mfn* or Opa1.[380,393,395,396] Finally, heterozygosity for *drp1* is lethal in a *PINK1* mutant background, consistent with the idea that *PINK1* and *drp1* work in the same direction to promote fission.[380,393]

PINK1 and *parkin* also regulate mitochondrial morphology in mammalian systems. However, in contrast to the story in *Drosophila*, which is consistent across cell types and labs, in mammalian systems various effects have been observed. Observations consistent with the *Drosophila* work have been obtained in some systems. Thus, enlarged mitochondria have been observed in *PINK1* striatal neurons[378] and COS7 cells in which *PINK1* was silenced using RNAi. In the latter system, this phenotype was suppressed by overexpressing *Fis1* or *Drp1*, as in *Drosophila*.[395] Cultured fibroblasts from *PARKIN* patients also contained longer and more branched mitochondria than wild-type controls.[397] Most recently, Yu and colleagues found that expression of *PINK1* or *parkin* in hippocampal neurons resulted in an increase in mitochondria number and a decrease in size, while silencing of *PINK1* resulted in elongated mitochondria.[398] As expected, expression of *Drp1* or silencing of *Opa1* suppressed these phenotypes. Midbrain dopaminergic neurons responded similar to manipulations of *PINK1* and *parkin*. In contrast, others have observed that loss of *PINK1* results in fission, with decreased levels of *Drp1* resulting in suppression of this phenotype.[377,382,383,399] The reasons for these differences are unknown, but likely to be interesting. One possibility is that loss of *PINK1* or *Parkin* results in an increase in the amount of mitochondrial damage, and that this leads, through other pathways, to mitochondrial fragmentation. Screens in yeast[400,401] and *C. elegans*[402] have shown that disruption of many genes leads to changes in mitochondrial morphology, including fragmentation or elongation. Thus, it is likely that the final mitochondrial morphology phenotype observed in any particular cell type with respect to the presence or absence of *PINK1/parkin* will depend on multiple variables. In any case, what is most important is not the specific morphology observed, but the functional state of the mitochondrial population and the ways in which this is influenced by *PINK1* and *parkin*, discussed below.

How could defective mitochondrial fission resulting from mutation of *PINK1* or *parkin* lead to defects in cell physiology? A major function of mitochondria is to oxidize dietary reducing equivalents to generate ATP and heat through oxidative phosphorylation, a process in which electron transport through the respiratory chain creates a potential difference across the inner mitochondrial membrane, which is used to produce ATP. Unfortunately, inefficiencies in electron transport result in the production of ROS, making mitochondria the major source of free-radical production in the cell. These radicals damage mitochondrial proteins, lipids, and DNA, which impair mitochondrial function and can lead to further increases in free-radical production, ultimately compromising cellular function and/or leading to cell death. Mitochondrial cycles of fusion and fission allow content mixing, which presumably functions to make the cellular population of mitochondria relatively homogeneous in protein content, stabilizing overall mitochondrial function over the short term. However, because mitochondrial genomes undergo mutation at high frequency, nonselective cycles of fusion and fission do nothing to prevent (and may promote) the accumulation of defective genomes (there is no selective pressure to remove them and genomes with deletions often accumulate), resulting in a time-dependent decrease in overall cellular mitochondrial function. Therefore, mechanisms must exist to specifically eliminate defective mitochondria and, presumably, any associated defective mitochondrial genomes. A number of observations suggest that this removal occurs through the process of mitophagy, a specialized form of autophagy in which cellular components are degraded following engulfment by autophagosomes (reviewed in Ref. 403). Importantly for the purposes of this review, the process of mitophagy is intimately linked with changes in mitochondrial size and shape brought about through fission and fusion (reviewed in Refs. 404,405). As we will see, loss of pink1 and/or parkin inhibits the removal of damaged mitochondria.

How are defective mitochondria selected for removal? Live cell imaging of mitochondrial dynamics in mammalian cells shows that, while many products of mitochondrial fission rapidly fuse again with the mitochondrial network, others, which also exhibited a depolarized membrane potential, showed a decreased probability of fusion and were often targeted for degradation through mitophagy.[406,407] These and other observations (reviewed in Refs. 404,408) suggest a general model of mitochondrial quality control in which fission allows the segregation of mitochondrial components, including proteins, lipids, and genomes, into separate mitochondrial units. These units are then "tested" for their ability to acquire a hyperpolarized membrane potential, an honest signal of organelle and genome health since many normal mitochondrial activities ultimately function to make the inner mitochondrial compartment hyperpolarized so as to generate ATP. Those mitochondria that fail the test are kept segregated (through mechanisms to be discussed shortly) and are ultimately targeted for degradation through mitophagy. Importantly, decreased mitophagy results in the accumulation of oxidized proteins and decreased cellular respiration, strongly suggesting that the end result of this process is the selective removal of damaged mitochondria.[407]

How are dysfunctional mitochondria identified and targeted for mitophagy? Youle and colleagues showed that Parkin is selectively recruited to mitochondria whose membrane is depolarized, resulting in their elimination through mitophagy. Parkin recruitment to mitochondria is independent of drp1, while mitophagy requires drp1, indicating that Parkin-dependent mitophagy requires fission. Work from a number of labs has gone on to show that recruitment of Parkin to mitochondria is PINK1 dependent,[350,409–415] including in neurons differentiated from PINK1 mutant patient cells.[416] PINK1 is a membrane protein with its C-terminus facing the cytoplasm.[417] PINK1 protein levels are specifically upregulated on damaged mitochondria.[410,411,413] In healthy mitochondria, PINK1 is constitutively cleaved, releasing its C-terminal kinase domain into the intermembrane space and/or cytoplasm where it is degraded in a proteasome-dependent manner. In damaged mitochondria that have lost their membrane potential, full-length PINK1 remains anchored to the membrane. Outer mitochondrial anchorage of PINK1 is all that is required to recruit Parkin, because tethering of PINK1 to the mitochondrial membrane through other methods is sufficient to recruit Parkin.[410]

How is PINK1 cleavage regulated? Early work in *Drosophila* showed that PINK1 cleavage required the inner mitochondrial membrane protease rhomboid 7.[418] This, coupled with an earlier observation that the *rhomboid-7* mutant phenotype includes a loss of mitochondrial fusion,[419] suggested a model in which Pink1's profission activity is negatively regulated through cleavage by rhomboid-7. Early experiments in mammalian systems excluded the rhomboid protease PARL as the protease cleaving PINK1, on the basis of the results of RNAi knockdown.[410] However, more recent observation by the same group[420] and others[421–423] clearly demonstrate that PINK1 is cleaved by PARL and that this activity is regulated by mitochondrial membrane potential. In healthy mitochondria, PINK1 is guided to the inner mitochondrial membrane using the general mitochondrial import machinery. Here it is cleaved by PARL, resulting in release of a cytoplasmic form that is degraded by an MG132-sensitive protease. In contrast, when the mitochondrial membrane potential is depolarized, newly synthesized full-length PINK1 is somehow shunted onto the outer mitochondrial membrane, where it is able to recruit Parkin. An unrelated yeast protein shows a similar mitochondrial membrane potential-dependent shift in localization, suggesting that alternations in the mitochondrial membrane potential may be more generally used to signal energy or health status to the rest of the cell through differential protein localization.[424]

How PINK1 stabilization on the outer mitochondrial membrane recruits Parkin is unknown. There is also evidence that recruitment of Parkin results in an activation of its ligase activity, also through an unknown mechanism.[409] What is clear is that, once recruited and activated, Parkin has a number of effects on the mitochondria-associated proteome. Many mitochondrial proteins are ubiquitinated and/or degraded in a Parkin-dependent manner. These include Mfns,[413,425–430] Milton and Miro,[428] Bcl-2,[431] which are components of the TOM mitochondrial protein import complex,[428,432] VDAC,[350,433] and Drp-1.[434] Most recently, Chan and colleagues carried out a comprehensive analysis in HeLa cells of proteins associated with depolarized mitochondria to which Parkin had been recruited.[428] They confirmed many previous findings and discovered much that is new. In particular, they found that Parkin recruitment to the mitochondrial membrane results in the recruitment of the proteasome and the K-48 and K63-linked ubiquitination, and in some cases degradation, of a number of mitochondrial proteins. Components of the autophagy pathway were also recruited. Chan *et al.*,[428] Tanaka *et al.*,[427] and to a lesser extent Yoshii *et al.*[432] also found that inhibition of the proteasome, or specifically K48-linked ubiquitination,[428] suppressed mitophagy, but not recruitment of Parkin, arguing that protein degradation downstream of Parkin recruitment is essential for mitophagy, at least in these contexts.

What are the key substrates of Parkin in this context? As noted above, a number of groups have shown that Parkin recruitment results in the degradation of Mfns. However, mitophagy still occurs in *mfn* null cells.[428] Therefore, the degradation of the *mfns* presumably serves primarily to segregate dysfunctional mitochondria by preventing fusion. Voltage-dependent cleavage and inactivation of Opa1 by the mitochondrial protease Oma1 probably serves a similar purpose with respect to the inner mitochondrial membrane.[435] Recruitment of Parkin to mitochondria also results in the mitochondrial accumulation of p97, a AAA-ATPase[427] that participates in a number of processes including endoplasmic reticulum-associated degradation in which p97 provides the driving force required to extrude ubiquitinated membrane proteins from the ER so that they can be degraded in the cytosol.[436] Mfn degradation and mitophagy require p97, suggesting a model in which p97 may be similarly required to promote the extrusion into the cytoplasm and degradation of critical substrates. VDAC, an abundant mitochondrial outer membrane protein, is ubiquitinated through a K63 linkage by Parkin and has been suggested to be important for mitophagy.[350] However, VDAC null cells still undergo mitophagy, indicating that its ubiquitination is not absolutely required.[433] It has also been suggested that K63-linked ubiquitination of mitochondrial proteins such as VDAC promotes mitophagy through recruitment of ubiquitin-binding adaptors such as HDAC6 and the autophagy adaptor p62.[350,437,438] However, as with *VDAC*, *p62* null cells still undergo Parkin-dependent mitophagy, indicating that p62 ubiquitination may play other roles, perhaps involving altering mitochondria location within the cell to facilitate access to the autophagy machinery.[433,439]

To summarize, recruitment of Parkin to the depolarized mitochondria results in the ubiquitination, and in some cases degradation, of a number of proteins. In the case of the Mfns, Milton and Miro, proteins involved in mitochondrial trafficking along microtubules, degradation probably serves to sequester damaged organelles away from the mitochondrial network and promote their localization to sites of mitophagy, respectively. Recruitment of the autophagy p62 may also function to facilitate localization of damaged mitochondria to sites of mitophagy. The proteasome and p97 are required for Parkin-dependent mitophagy, but exactly why they are required is unknown. They may be needed to remove inhibitors of mitophagy. If this is the case, it may be possible to identify these substrates as genes whose silencing allows mitophagy to occur in *mfn* null cells exposed to proteasome inhibitors. Alternatively, or in addition, p97 and the proteasome may have effects on mitophagy through action on targets in other cellular compartments. With respect to PD, at least for those with *PINK1* or *PARKIN* mutations, PD may result from a failure of mitophagy—an important aspect of mitochondrial quality control.

As attractive as the above model of *PINK1/parkin*-dependent clearance of damaged mitochondria is, it is important to note that thus far mitophagy has only been demonstrated in cultured cells, not *in vivo*. Thus, an important goal in the field is to provide evidence supporting the role of *PINK1/parkin* in mediating mitophagy to clear damaged mitochondria *in vivo*. Also, it has been reported that Parkin translocation to damaged mitochondria occurs only in nonneuronal cells, not in neurons.[415] Whether mitophagy occurs in some tissues but not others awaits future studies.

F. Other Roles for PINK1

Loss of *PINK1* results in defects in mitochondrial calcium handling. The mitochondrial matrix accumulates Ca^{2+} to higher basal levels, as a result of a decreased efflux capacity. Mitochondria from cell lacking *PINK1* are therefore less able to buffer cytoplasmic Ca^{2+} increases in response to stimuli.[440–442] This can lead to increased ROS, decreased ATP production, and opening of the mitochondrial permeability transition pore, which can lead to death. Loss of *PINK1* is also associated with increased expression of cytokines in response to LPS but decreased expression of cytokine-induced NF-κB, which promotes cell survival.[441] It is possible that some of these effects occur downstream of the accumulation of defective mitochondria because of loss of mitophagy. However, it is also possible that PINK1 has other important targets in mitochondria and other cellular compartments that are important for maintaining mitochondrial homeostasis. For example, at the mitochondria, PINK1 has been reported to phosphorylate the TNF receptor-associated protein 1 (TRAP1), a molecular chaperone, and phosphorylation of TRAP1 by PINK1 is important for preventing oxidative-stress-induced release of cytochrome *c*, which promotes apoptosis.[443]

A number of groups have found some fraction of PINK1 localized to the cytoplasm.[418,444–448] Importantly, expression of an N-terminally deleted version of PINK1 lacking its mitochondrial targeting sequence protects neurons from death induced by MPTP, and this activity requires putative PINK1 kinase activity. This raises the possibility that, at least when overexpressed, *PINK1* can have pro survival functions outside the mitochondria.[447] Recent work suggests that one pathway through which this may work involves phosphorylation of the mammalian Target of Rapamycin Complex 2 (TORC2) component Rictor by PINK1, which promotes TORC2-dependent phosphorylation of the prosurvival kinase AKT.[449] Interestingly, it was recently reported that expression of DJ-1 also promotes the oxidative-stress-induced phosphorylation of AKT and that DJ-1 promotes AKT's ability to protect cells from MPTP-induced stress.[450] It will be interesting to determine whether DJ-1's activity in this assay requires PINK1 kinase activity or physical interaction. Interactions between PINK1/

Parkin and DJ-1 in the cytoplasm have been identified by one group,[451] but not by a second group,[452] leaving the relationships between these proteins in the cytoplasm unclear.

G. Links Between the PINK1/Parkin Pathway and Other PARK Loci

Mutations in *DJ-1* cause very rare autosomal-recessive forms of PD.[262] DJ-1 has also been shown to have oncogenic activity.[453] The DJ-1 protein is small ($\sim$20kDa) and probably functions as a homodimer. DJ-1 is a member of the large DJ-1/PfpI superfamily and has members in almost all organisms. Many activities have been proposed for DJ-1, and there is currently no consensus on what its functions are. Suggested functions include a redox-regulated chaperone, a cysteine protease, a transcription coactivator, an RNA-binding protein, and a regulator of survival signaling through interactions with Daxx or the kinase ASK1 (reviewed in Refs. 275,454). What is clear is that DJ-1 is cytoprotective in a number of different contexts; this activity requires a conserved cysteine (Cys106 in mammals). Oxidation of this cysteine to cysteine sulfinic and cysteine sulfonic acids occurs under oxidative stress conditions, and oxidation of this residue to the sulfinic acid is critical for cytoprotection in a number of systems. Mutants that mimic the sulfinic acid form of DJ-1 but that lack Cys106 are still cytoprotective, indicating that DJ-1 is unlikely to be cytoprotective simply because oxidation of Cys106 consumes ROS.[455,456] Importantly for the purposes of this review, while DJ-1 is predominantly cytoplasmic,[457] oxidative stress enhances its association with mitochondria.[276,455,458–460] In addition, versions of DJ-1 targeted specifically to mitochondria by fusing a mitochondrial localization sequence to its N-terminus show enhanced cytoprotective functions, suggesting that the mitochondria is one (of perhaps many) important site of DJ-1 action.[461]

Is there a relationship between *DJ-1* and the *PINK1/parkin* pathway? Lymphoblastoid cells from *DJ-1* PD patients have an increased percentage of fragmented mitochondria. *DJ-1* mutant mice show defects in muscle mitochondrial function,[462] and the loss of *DJ-1* in a mouse neuroblastoma cell line, cortical neurons, or embryonic fibroblasts from *DJ-1* null embryos leads to mitochondrial membrane depolarization, accumulation of ROS, fragmentation of mitochondria, accumulation of markers of autophagy around mitochondria,[452,463,464] and, in the case of the *DJ-1* null cells, increased autophagic flux. In both studies, antioxidants reversed these effects and, in the neuroblastoma cells, expression of *DJ-1* blocked mitochondrial fragmentation in response to rotenone, which are results consistent with the idea that *DJ-1* functions to maintain mitochondrial health in the face of oxidative stress. Expression of *PINK1* or *parkin* could restore mitochondrial connectivity in cells lacking *DJ-1*,[452,464] and expression of *DJ-1* could restore the connectivity of mitochondria in cells lacking *PINK1* exposed to rotenone[452] but not in *PINK1*-deficient cells not exposed to rotenone.[377] In short, the ability of PINK1/Parkin or DJ-1 to protect mitochondrial function seems, most of the time, to be independent of the presence of the other. As noted above, there are conflicting reports as to whether PINK1, Parkin, and DJ-1 form a complex,[451,452] and it is intriguing that both PINK1 and DJ-1 promote the activation of AKT. But, as yet, epistasis experiments have not been carried out to determine if these effects on AKT activity occur through the same pathway. Finally, it is worth noting that triple knockout mice lacking *PINK1*, *parkin*, and *DJ-1* fail to show nigral degeneration even with aging.[373] This somewhat surprising result could reflect action in a common pathway as with *PINK1* and *parkin*. Alternatively, it could reflect developmental compensation, the presence of which is strongly hinted at in experiments utilizing conditional adult-specific *parkin* knockout mice[343] (compare with Refs. 465,466). If compensation is involved, it would of course be very interesting to know how it is brought about.

In summary, the above observations tentatively suggest that *DJ-1* acts in parallel to *PINK1* and *parkin*, not through them. A similar conclusion is suggested by experiments in *Drosophila*. While *Drosophila DJ-1* mutants show mitochondrial defects, these are in many ways distinct from those associated with loss of *PINK1* or *parkin*.[462] In addition, while expression of *DJ-1* is capable of rescuing muscle defects due to loss of *PINK1*, it cannot rescue identical-looking muscle defects due to loss of *parkin*. Expression of either *PINK1* or *parkin* in a *DJ-1* mutant also has the surprising effect of causing organismal lethality. All of these results suggest complex, not linear, interactions.

IV. Convergent Pathways and Future Directions

Several lines of evidence suggest possible links between α-Syn and mitochondrial biology. Mice lacking *SNCA* show increased resistance to MPTP[467] and reductions in the levels of the mitochondrial lipid cardiolipin.[468] Mice overexpressing α-Syn also show mitochondrial damage.[87,469–472] α-Syn has been shown to localize to mitochondria,[87,473–476] with the levels of mitochondrially localized α-Syn increasing in the brains of PD patients.[477] Recently, two groups have shown that expression of wild-type α-Syn results in increased fragmentation of the mitochondrial network, while decreased levels of α-Syn resulted in increased tubulation of the network.[475,476] Both studies suggest that α-Syn promotes these effects through direct interactions with mitochondrial lipids. Kamp and colleagues suggest that α-Syn prevents fusion, while

Nakamura and colleagues suggest that it promotes fission, an issue that needs to be resolved. Regardless of the exact mechanism, do these observations help us understand a possible basis for α-Syn toxicity? This awaits future studies.

Mitochondrial fusion allows the sharing of proteins, lipids, and small molecules, which helps to preserve the overall mitochondrial and cellular function. In the absence of fusion, individual mitochondria are likely to diverge in terms of their requirements for these factors depending on where they are and the metabolic activities they are engaged in. This would be expected to lead, over time, to increasing numbers of dysfunctional mitochondria as individual units stochastically enter states that result in a decreased ability to maintain a hyperpolarized membrane potential, the loss of which results in increased dysfunction and/or increased ROS production, leading ultimately to cell dysfunction and death.[478] In support of such a model, loss of mitochondrial fusion through mutation or loss of *mfns* results in mitochondrial depolarization and cell death (reviewed in Ref. 479). Seen in this light, α-Syn's ability to promote mitochondrial fission may tip the balance toward mitochondrial and cellular dysfunction by limiting opportunities for mitochondria to fuse with and complement each other's defects. In such a model, *PINK1* and *parkin* would still promote the removal of dysfunctional mitochondrial units as they arose, as observed by Kamp and colleagues.[476] However, if compensatory biogenesis does not keep up with this removal, and/or the *PINK1Parkin* pathway becomes saturated or energy-limited in its ability to promote mitophagy (see Ref. 415 for a possible example of energy limitation), mitochondrial damage may accumulate over time, leading to cellular dysfunction and death. An important prediction of this model is that inhibitors of mitochondrial fission, which tip the balance back more toward the fused state, should act as suppressors of α-Syn toxicity. It will be interesting to determine whether inhibitors of *drp-1* recently identified[480] have therapeutic benefit in situations in which α-Syn is either overexpressed or mutated.

Finally, we note that both mitochondrial quality control and α-Syn removal involve use of the cells' two major protein degradative systems, namely, the proteasome and autophagy. Importantly, inhibition of autophagy can compromise degradation of ubiquitin-proteasome substrates, and inhibition of the proteasome inhibits some autophagic processes (such as *PINK1/parkin*-dependent mitophagy) and places increased demand on the autophagic degradation system to remove misfolded proteins and to provide essential amino acids. As noted above, expression of α-Syn can disrupt proteasome function and autophagy. In consequence, increased levels of α-Syn may, through this mechanism alone, compromise the removal of damaged mitochondria. Conversely, preventing the generation and/or removal of damaged mitochondria through inhibition of *PINK1*, *parkin*, or *DJ-1* results in cells experiencing increased levels of cellular ROS-dependent protein and lipid damage, and decreased

162

levels of ATP. This results in increased demand for proteasome and autophagy-dependent degradation, which are energy-dependent processes, in a context in which energy may be limiting. As a result, effective removal of α-Syn (the dose dependence of which with respect to disease generation is exquisite) may be compromised, resulting in the accumulation of toxic forms of the protein. Given this interrelationship, and the fact that α-Syn accumulation and/or damage to *PINK1/parkin* pathway components are probably present in many forms of PD, therapies designed to inhibit α-Syn-dependent toxicity or defects in mitophagy may have broad therapeutic benefit.

Acknowledgments

We apologize to those whose work has not been cited because of space constraints. This work is supported by grants and funds from the National Institutes of Health (R03, R21, R01) and the Michael J. Fox Foundation to J.-C.R., and from the National Institutes of Health (R01, K02), the Glenn Family Foundation, the Esther A. and Joseph Klingenstein Fellowship, and the McKnight Foundation of Neuroscience to M.G.

References

1. Dauer W, Przedborski S. Parkinson's disease: mechanisms and models. *Neuron* 2003;**39**:889–909.
2. Dawson TM, Dawson VL. Molecular pathways of neurodegeneration in Parkinson's disease. *Science* 2003;**302**:819–22.
3. Shulman JM, De Jager PL, Feany MB. Parkinson's disease: genetics and pathogenesis. *Annu Rev Pathol* 2011;**6**:193–222.
4. Forno LS. Neuropathology of Parkinson's disease. *J Neuropathol Exp Neurol* 1996;**55**:259–72.
5. Spillantini MG, Schmidt ML, Lee VM-Y, Trojanowski JQ, Jakes R, Goedert M. α-Synuclein in Lewy bodies. *Nature* 1997;**388**:839–40.
6. Polymeropoulos MH, Lavedan C, Leroy E, Ide SE, Dehejia A, Dutra A, et al. Mutation in the α-synuclein gene identified in families with Parkinson's disease. *Science* 1997;**276**:2045–7.
7. Kruger R, Kuhn W, Muller T, Woitalla D, Graeber M, Kosel S, et al. Ala30Pro mutation in the gene encoding α-synuclein in Parkinson's disease. *Nat Genet* 1998;**18**:106–8.
8. Zarranz JJ, Alegre J, Gomez-Esteban JC, Lezcano E, Ros R, Ampuero I, et al. The new mutation, E46K, of alpha-synuclein causes Parkinson and Lewy body dementia. *Ann Neurol* 2004;**55**:164–73.
9. Singleton AB, Farrer M, Johnson J, Singleton A, Hague S, Kachergus J, et al. Alpha-synuclein locus triplication causes Parkinson's disease. *Science* 2003;**302**:841.
10. Chartier-Harlin MC, Kachergus J, Roumier C, Mouroux V, Douay X, Lincoln S, et al. Alpha-synuclein locus duplication as a cause of familial Parkinson's disease. *Lancet* 2004;**364**:1167–9.
11. Betarbet R, Sherer TB, MacKenzie G, Garcia-Osuna M, Panov AV, Greenamyre JT. Chronic systemic pesticide exposure reproduces features of Parkinson's disease. *Nat Neurosci* 2000;**3**:1301–6.

12. Orth M, Schapira AH. Mitochondrial involvement in Parkinson's disease. *Neurochem Int* 2002;**40**:533–41.
13. Jenner P. Oxidative stress in Parkinson's disease. *Ann Neurol* 2003;**53**(Suppl. 3):S26–36 [discussion S36–S28].
14. Banerjee R, Starkov AA, Beal MF, Thomas B. Mitochondrial dysfunction in the limelight of Parkinson's disease pathogenesis. *Biochim Biophys Acta* 2009;**1792**:651–63.
15. Graham DG. Oxidative pathways for catecholamines in the genesis of neuromelanin and cytotoxic quinones. *Mol Pharmacol* 1978;**14**:633–43.
16. Abou-Sleiman PM, Muqit MM, Wood NW. Expanding insights of mitochondrial dysfunction in Parkinson's disease. *Nat Rev Neurosci* 2006;**7**:207–19.
17. Dodson MW, Guo M. Pink1, Parkin, DJ-1 and mitochondrial dysfunction in Parkinson's disease. *Curr Opin Neurobiol* 2007;**17**:331–7.
18. Bogaerts V, Theuns J, van Broeckhoven C. Genetic findings in Parkinson's disease and translation into treatment: a leading role for mitochondria? *Genes Brain Behav* 2008;**7**:129–51.
19. Vila M, Ramonet D, Perier C. Mitochondrial alterations in Parkinson's disease: new clues. *J Neurochem* 2008;**107**:317–28.
20. Goedert M. Alpha-synuclein and neurodegenerative diseases. *Nat Rev Neurosci* 2001;**2**:492–501.
21. Beyer K, Domingo-Sabat M, Lao JI, Carrato C, Ferrer I, Ariza A. Identification and characterization of a new alpha-synuclein isoform and its role in Lewy body diseases. *Neurogenetics* 2008;**9**:15–23.
22. Weinreb PH, Zhen W, Poon AW, Conway KA, Lansbury Jr. PT. NACP, a protein implicated in Alzheimer's disease and learning, is natively unfolded. *Biochemistry* 1996;**35**:13709–15.
23. Bussell Jr. R, Eliezer D. Residual structure and dynamics in Parkinson's disease-associated mutants of α-synuclein. *J Biol Chem* 2001;**276**:45996–6003.
24. Davidson WS, Jonas A, Clayton DF, George JM. Stabilization of α-synuclein secondary structure upon binding to synthetic membranes. *J Biol Chem* 1998;**273**:9443–9.
25. Jo E, McLaurin J, Yip CM, St. George-Hyslop P, Fraser PE. α-Synuclein membrane interactions and lipid specificity. *J Biol Chem* 2000;**275**:34328–34.
26. Perrin RJ, Woods WS, Clayton DF, George JM. Interaction of human α-synuclein and Parkinson's disease variants with phospholipids. Structural analysis using site-directed mutagenesis. *J Biol Chem* 2000;**275**:34393–8.
27. Eliezer D, Kutluay E, Bussell Jr. R, Browne G. Conformational properties of α-synuclein in its free and lipid-associated states. *J Mol Biol* 2001;**307**:1061–73.
28. Bussell Jr. R, Eliezer D. A structural and functional role for 11-mer repeats in α-synuclein and other exchangeable lipid binding proteins. *J Mol Biol* 2003;**329**:763–78.
29. Chandra S, Chen X, Rizo J, Jahn R, Sudhof TC. A broken α-helix in folded α-synuclein. *J Biol Chem* 2003;**278**:15313–8.
30. Zhu M, Li J, Fink AL. The association of α-synuclein with membranes affects bilayer structure, stability and fibril formation. *J Biol Chem* 2003;**278**:40186–97.
31. Jao CC, Der-Sarkissian A, Chen J, Langen R. Structure of membrane-bound alpha-synuclein studied by site-directed spin labeling. *Proc Natl Acad Sci USA* 2004;**101**:8331–6.
32. Abeliovich A, Schmitz Y, Farinas I, Choi-Lundberg D, Ho WH, Castillo PE, et al. Mice lacking α-synuclein display functional deficits in the nigrostriatal dopamine system. *Neuron* 2000;**25**:239–52.
33. Murphy DD, Rueter SM, Trojanowski JQ, Lee VM-Y. Synucleins are developmentally expressed, and α-synuclein regulates the size of the presynaptic vesicular pool in primary hippocampal neurons. *J Neurosci* 2000;**20**:3214–20.

34. Cabin DE, Shimazu K, Murphy D, Cole NB, Gottschalk W, McIlwain KL, et al. Synaptic vesicle depletion correlates with attenuated synaptic responses to prolonged repetitive stimulation in mice lacking α-synuclein. *J Neurosci* 2002;**22**:8797–807.

35. Larsen KE, Schmitz Y, Troyer MD, Mosharov E, Dietrich P, Quazi AZ, et al. Alpha-synuclein overexpression in PC12 and chromaffin cells impairs catecholamine release by interfering with a late step in exocytosis. *J Neurosci* 2006;**26**:11915–22.

36. Cheng F, Vivacqua G, Yu S. The role of alpha-synuclein in neurotransmission and synaptic plasticity. *J Chem Neuroanat* 2011;**42**:242–8.

37. Nemani VM, Lu W, Berge V, Nakamura K, Onoa B, Lee MK, et al. Increased expression of alpha-synuclein reduces neurotransmitter release by inhibiting synaptic vesicle reclustering after endocytosis. *Neuron* 2010;**65**:66–79.

38. Venda LL, Cragg SJ, Buchman VL, Wade-Martins R. Alpha-synuclein and dopamine at the crossroads of Parkinson's disease. *Trends Neurosci* 2010;**33**:559–68.

39. Chandra S, Gallardo G, Fernandez-Chacon R, Schluter OM, Sudhof TC. Alpha-synuclein cooperates with CSPalpha in preventing neurodegeneration. *Cell* 2005;**123**:383–96.

40. Bartels T, Choi JG, Selkoe DJ. Alpha-synuclein occurs physiologically as a helically folded tetramer that resists aggregation. *Nature* 2011;**477**:107–10.

41. Wang W, Perovic I, Chittuluru J, Kaganovich A, Nguyen LT, Liao J, et al. A soluble alpha-synuclein construct forms a dynamic tetramer. *Proc Natl Acad Sci USA* 2011;**108**:17797–802.

42. Anderson JP, Walker DE, Goldstein JM, de Laat R, Banducci K, Caccavello RJ, et al. Phosphorylation of Ser-129 is the dominant pathological modification of alpha-synuclein in familial and sporadic Lewy body disease. *J Biol Chem* 2006;**281**:29739–52.

43. El-Agnaf OM, Bodles AM, Guthrie DJ, Harriott P, Irvine GB. The N-terminal region of non-A beta component of Alzheimer's disease amyloid is responsible for its tendency to assume beta-sheet and aggregate to form fibrils. *Eur J Biochem* 1998;**258**:157–63.

44. Narhi L, Wood SJ, Steavenson S, Jiang Y, Wu GM, Anafi D, et al. Both familial Parkinson's disease mutations accelerate α-synuclein aggregation. *J Biol Chem* 1999;**274**:9843–6.

45. Conway KA, Lee S-J, Rochet J-C, Ding TT, Williamson RE, Lansbury Jr. PT. Acceleration of oligomerization, not fibrillization, is a shared property of both α-synuclein mutations linked to early-onset Parkinson's disease: implications for pathogenesis and therapy. *Proc Natl Acad Sci USA* 2000;**97**:571–6.

46. Rochet JC, Conway KA, Lansbury Jr. PT. Inhibition of fibrillization and accumulation of prefibrillar oligomers in mixtures of human and mouse α-synuclein. *Biochemistry* 2000;**39**:10619–26.

47. Serpell LC, Berriman J, Jakes R, Goedert M, Crowther RA. Fiber diffraction of synthetic α-synuclein filaments shows amyloid-like cross-β conformation. *Proc Natl Acad Sci USA* 2000;**97**:4897–902.

48. Volles MJ, Lansbury Jr. PT. Zeroing in on the pathogenic form of α-synuclein and its mechanism of neurotoxicity in Parkinson's disease. *Biochemistry* 2003;**42**:7871–8.

49. Rochet JC, Outeiro TF, Conway KA, Ding TT, Volles MJ, Lashuel HA, et al. Interactions among alpha-synuclein, dopamine, and biomembranes: some clues for understanding neurodegeneration in Parkinson's disease. *J Mol Neurosci* 2004;**23**:23–34.

50. Volles MJ, Lee S-J, Rochet J-C, Shtilerman MD, Ding TT, Kessler JC, et al. Vesicle permeabilization by protofibrillar α-synuclein: implications for the pathogenesis and treatment of Parkinson's disease. *Biochemistry* 2001;**40**:7812–9.

51. Ding TT, Lee S-J, Rochet J-C, Lansbury Jr. PT. Annular α-synuclein protofibrils are produced when spherical protofibrils are incubated in solution or bound to brain-derived membranes. *Biochemistry* 2002;**41**:10209–17.

52. Lashuel HA, Petre BM, Wall J, Simon M, Nowak RJ, Walz T, et al. α-Synuclein, especially the Parkinson's disease-associated mutants, forms pore-like annular and tubular protofibrils. *J Mol Biol* 2002;**322**:1089–102.

53. Li J, Uversky VN, Fink AL. Effect of familial Parkinson's disease point mutations A30P and A53T on the structural properties, aggregation, and fibrillation of human α-synuclein. *Biochemistry* 2001;**40**:11604–13.

54. Choi W, Zibaee S, Jakes R, Serpell LC, Davletov B, Crowther RA, et al. Mutation E46K increases phospholipid binding and assembly into filaments of human alpha-synuclein. *FEBS Lett* 2004;**576**:363–8.

55. Hoyer W, Cherny D, Subramaniam V, Jovin TM. Impact of the acidic C-terminal region comprising amino acids 109-140 on alpha-synuclein aggregation in vitro. *Biochemistry* 2004;**43**:16233–42.

56. Outeiro TF, Kontopoulos E, Altman S, Kufareva I, Strathearn KE, Amore AM, et al. Sirtuin 2 inhibitors rescue alpha-synuclein-mediated toxicity in models of parkinson's disease. *Science* 2007;**317**:516–9.

57. Smith DP, Tew DJ, Hill AF, Bottomley SP, Masters CL, Barnham KJ, et al. Formation of a high affinity lipid-binding intermediate during the early aggregation phase of alpha-synuclein. *Biochemistry* 2008;**47**:1425–34.

58. Volles MJ, Lansbury Jr. PT. Vesicle permeabilization by protofibrillar α-synuclein is sensitive to Parkinson's disease-linked mutations and occurs by a pore-like mechanism. *Biochemistry* 2002;**41**:4595–602.

59. Giehm L, Svergun DI, Otzen DE, Vestergaard B. Low-resolution structure of a vesicle disrupting α-synuclein oligomer that accumulates during fibrillation. *Proc Natl Acad Sci USA* 2011;**108**:3246–51.

60. Kayed R, Sokolov Y, Edmonds B, McIntire TM, Milton SC, Hall JE, et al. Permeabilization of lipid bilayers is a common conformation-dependent activity of soluble amyloid oligomers in protein misfolding diseases. *J Biol Chem* 2004;**279**:46363–6.

61. Quist A, Doudevski I, Lin H, Azimova R, Ng D, Frangione B, et al. Amyloid ion channels: a common structural link for protein-misfolding disease. *Proc Natl Acad Sci USA* 2005;**102**: 10427–32.

62. Danzer KM, Haasen D, Karow AR, Moussaud S, Habeck M, Giese A, et al. Different species of alpha-synuclein oligomers induce calcium influx and seeding. *J Neurosci* 2007;**27**:9220–32.

63. Tsigelny IF, Bar-On P, Sharikov Y, Crews L, Hashimoto M, Miller MA, et al. Dynamics of alpha-synuclein aggregation and inhibition of pore-like oligomer development by beta-synuclein. *FEBS J* 2007;**274**:1862–77.

64. Feng LR, Federoff HJ, Vicini S, Maguire-Zeiss KA. Alpha-synuclein mediates alterations in membrane conductance: a potential role for alpha-synuclein oligomers in cell vulnerability. *Eur J Neurosci* 2010;**32**:10–7.

65. Greenbaum EA, Graves CL, Mishizen-Eberz AJ, Lupoli MA, Lynch DR, Englander SW, et al. The E46K mutation in alpha-synuclein increases amyloid fibril formation. *J Biol Chem* 2005;**280**:7800–7.

66. Fredenburg RA, Rospigliosi C, Meray RK, Kessler JC, Lashuel HA, Eliezer D, et al. The impact of the E46K mutation on the properties of alpha-synuclein in its monomeric and oligomeric states. *Biochemistry* 2007;**46**:7107–18.

67. Bertoncini CW, Jung YS, Fernandez CO, Hoyer W, Griesinger C, Jovin TM, et al. Release of long-range tertiary interactions potentiates aggregation of natively unstructured alpha-synuclein. *Proc Natl Acad Sci USA* 2005;**102**:1430–5.

68. Dedmon MM, Lindorff-Larsen K, Christodoulou J, Vendruscolo M, Dobson CM. Mapping long-range interactions in alpha-synuclein using spin-label NMR and ensemble molecular dynamics simulations. *J Am Chem Soc* 2005;**127**:476–7.

69. Salmon L, Nodet G, Ozenne V, Yin G, Jensen MR, Zweckstetter M, et al. NMR characterization of long-range order in intrinsically disordered proteins. *J Am Chem Soc* 2010;**132**:8407–18.

70. Crowther RA, Jakes R, Spillantini MG, Goedert M. Synthetic filaments assembled from C-terminally truncated alpha-synuclein. *FEBS Lett* 1998;**436**:309–12.

71. Murray IV, Giasson BI, Quinn SM, Koppaka V, Axelsen PH, Ischiropoulos H, et al. Role of alpha-synuclein carboxy-terminus on fibril formation in vitro. *Biochemistry* 2003;**42**:8530–40.

72. Li W, West N, Colla E, Pletnikova O, Troncoso JC, Marsh L, et al. Aggregation promoting C-terminal truncation of alpha-synuclein is a normal cellular process and is enhanced by the familial Parkinson's disease-linked mutations. *Proc Natl Acad Sci USA* 2005;**102**:2162–7.

73. Liu CW, Giasson BI, Lewis KA, Lee VM, Demartino GN, Thomas PJ. A precipitating role for truncated alpha-synuclein and the proteasome in alpha-synuclein aggregation: implications for pathogenesis of Parkinson disease. *J Biol Chem* 2005;**280**:22670–8.

74. Wu KP, Kim S, Fela DA, Baum J. Characterization of conformational and dynamic properties of natively unfolded human and mouse alpha-synuclein ensembles by NMR: implication for aggregation. *J Mol Biol* 2008;**378**:1104–15.

75. Bertoncini CW, Fernandez CO, Griesinger C, Jovin TM, Zweckstetter M. Familial mutants of alpha-synuclein with increased neurotoxicity have a destabilized conformation. *J Biol Chem* 2005;**280**:30649–52.

76. Rospigliosi CC, McClendon S, Schmid AW, Ramlall TF, Barre P, Lashuel HA, et al. E46K Parkinson's-linked mutation enhances C-terminal-to-N-terminal contacts in alpha-synuclein. *J Mol Biol* 2009;**388**:1022–32.

77. Giasson BI, Duda JE, Murray IV, Chen Q, Souza JM, Hurtig HI, et al. Oxidative damage linked to neurodegeneration by selective α-synuclein nitration in synucleinopathy lesions. *Science* 2000;**290**:985–9.

78. Fujiwara H, Hasegawa M, Dohmae N, Kawashima A, Masliah E, Goldberg MS, et al. α-Synuclein is phosphorylated in synucleinopathy lesions. *Nat Cell Biol* 2002;4:160–4.

79. Neumann M, Kahle PJ, Giasson BI, Ozmen L, Borroni E, Spooren W, et al. Misfolded proteinase K-resistant hyperphosphorylated α-synuclein in aged transgenic mice with locomotor deterioration and in human α-synucleinopathies. *J Clin Invest* 2002;**110**:1429–39.

80. Kaneko H, Kakita A, Kasuga K, Nozaki H, Ishikawa A, Miyashita A, et al. Enhanced accumulation of phosphorylated alpha-synuclein and elevated beta-amyloid 42/40 ratio caused by expression of the presenilin-1 deltaT440 mutant associated with familial Lewy body disease and variant Alzheimer's disease. *J Neurosci* 2007;**27**:13092–7.

81. Hasegawa M, Fujiwara H, Nonaka T, Wakabayashi K, Takahashi H, Lee VM, et al. Phosphorylated alpha-synuclein is ubiquitinated in alpha-synucleinopathy lesions. *J Biol Chem* 2002;**277**:49071–6.

82. Baba M, Nakajo S, Tu P-H, Tomita T, Nakaya K, Lee VM-Y, et al. Aggregation of α-synuclein in Lewy bodies of sporadic Parkinson's disease and dementia with Lewy bodies. *Am J Pathol* 1998;**152**:879–84.

83. Chen L, Periquet M, Wang X, Negro A, McLean PJ, Hyman BT, et al. Tyrosine and serine phosphorylation of alpha-synuclein have opposing effects on neurotoxicity and soluble oligomer formation. *J Clin Invest* 2009;**119**:3257–65.

84. Mirzaei H, Schieler JL, Rochet J-C, Regnier F. Identification of rotenone-induced modifications in α-synuclein using affinity pull-down and tandem mass spectrometry. *Anal Chem* 2006;**78**:2422–31.

85. Przedborski S, Chen Q, Vila M, Giasson BI, Djaldatti R, Vukosavic S, et al. Oxidative post-translational modifications of alpha-synuclein in the 1-methyl-4-phenyl-1,2,3,6-tetrahydropyridine (MPTP) mouse model of Parkinson's disease. *J Neurochem* 2001;**76**:637–40.

86. Giasson BI, Duda JE, Quinn SM, Zhang B, Trojanowski JQ, Lee VM. Neuronal α-synucleinopathy with severe movement disorder in mice expressing A53T human α-synuclein. *Neuron* 2002;**34**:521–33.

87. Martin LJ, Pan Y, Price AC, Sterling W, Copeland NG, Jenkins NA, et al. Parkinson's disease alpha-synuclein transgenic mice develop neuronal mitochondrial degeneration and cell death. *J Neurosci* 2006;**26**:41–50.

88. McCormack AL, Mak SK, Shenasa M, Langston WJ, Forno LS, Di Monte DA. Pathologic modifications of alpha-synuclein in 1-methyl-4-phenyl-1,2,3,6-tetrahydropyridine (MPTP)-treated squirrel monkeys. *J Neuropathol Exp Neurol* 2008;**67**:793–802.

89. Paxinou E, Chen Q, Weisse M, Giasson BI, Norris EH, Rueter SM, et al. Induction of alpha-synuclein aggregation by intracellular nitrative insult. *J Neurosci* 2001;**21**:8053–61.

90. Danielson SR, Held JM, Schilling B, Oo M, Gibson BW, Andersen JK. Preferentially increased nitration of alpha-synuclein at tyrosine-39 in a cellular oxidative model of Parkinson's disease. *Anal Chem* 2009;**81**:7823–8.

91. Takahashi M, Kanuka H, Fujiwara H, Koyama A, Hasegawa M, Miura M, et al. Phosphorylation of alpha-synuclein characteristic of synucleinopathy lesions is recapitulated in alpha-synuclein transgenic Drosophila. *Neurosci Lett* 2003;**336**:155–8.

92. Yamada M, Iwatsubo T, Mizuno Y, Mochizuki H. Overexpression of alpha-synuclein in rat substantia nigra results in loss of dopaminergic neurons, phosphorylation of alpha-synuclein and activation of caspase-9: resemblance to pathogenetic changes in Parkinson's disease. *J Neurochem* 2004;**91**:451–61.

93. Chen L, Feany MB. Alpha-synuclein phosphorylation controls neurotoxicity and inclusion formation in a Drosophila model of Parkinson disease. *Nat Neurosci* 2005;**8**:657–63.

94. Shults CW, Rockenstein E, Crews L, Adame A, Mante M, Larrea G, et al. Neurological and neurodegenerative alterations in a transgenic mouse model expressing human alpha-synuclein under oligodendrocyte promoter: implications for multiple system atrophy. *J Neurosci* 2005;**25**:10689–99.

95. Smith WW, Margolis RL, Li X, Troncoso JC, Lee MK, Dawson VL, et al. Alpha-synuclein phosphorylation enhances eosinophilic cytoplasmic inclusion formation in SH-SY5Y cells. *J Neurosci* 2005;**25**:5544–52.

96. Eslamboli A, Romero-Ramos M, Burger C, Bjorklund T, Muzyczka N, Mandel RJ, et al. Long-term consequences of human alpha-synuclein overexpression in the primate ventral midbrain. *Brain* 2007;**130**:799–815.

97. Wakamatsu M, Ishii A, Ukai Y, Sakagami J, Iwata S, Ono M, et al. Accumulation of phosphorylated alpha-synuclein in dopaminergic neurons of transgenic mice that express human alpha-synuclein. *J Neurosci Res* 2007;**85**:1819–25.

98. Chau KY, Ching HL, Schapira AH, Cooper JM. Relationship between alpha synuclein phosphorylation, proteasomal inhibition and cell death: relevance to Parkinson's disease pathogenesis. *J Neurochem* 2009;**110**:1005–13.

99. Schell H, Hasegawa T, Neumann M, Kahle PJ. Nuclear and neuritic distribution of serine-129 phosphorylated alpha-synuclein in transgenic mice. *Neuroscience* 2009;**160**:796–804.

100. Riedel M, Goldbaum O, Wille M, Richter-Landsberg C. Membrane lipid modification by docosahexaenoic acid (DHA) promotes the formation of alpha-synuclein inclusion bodies immunopositive for SUMO-1 in oligodendroglial cells after oxidative stress. *J Mol Neurosci* 2011;**43**:290–302.

101. Tofaris GK, Razzaq A, Ghetti B, Lilley KS, Spillantini MG. Ubiquitination of alpha-synuclein in Lewy bodies is a pathological event not associated with impairment of proteasome function. *J Biol Chem* 2003;**278**:44405–11.

102. Choi DH, Kim YJ, Kim YG, Joh TH, Beal MF, Kim YS. Role of matrix metalloproteinase 3-mediated alpha-synuclein cleavage in dopaminergic cell death. *J Biol Chem* 2011;**286**:14168–77.

103. Paleologou KE, Oueslati A, Shakked G, Rospigliosi CC, Kim HY, Lamberto GR, et al. Phosphorylation at S87 is enhanced in synucleinopathies, inhibits alpha-synuclein oligomerization, and influences synuclein-membrane interactions. *J Neurosci* 2010;**30**:3184–98.

104. Waxman EA, Giasson BI. Specificity and regulation of casein kinase-mediated phosphorylation of alpha-synuclein. *J Neuropathol Exp Neurol* 2008;**67**:402–16.
105. Hashimoto M, Hsu LJ, Xia Y, Takeda A, Sisk A, Sundsmo M, et al. Oxidative stress induces amyloid-like aggregate formation of NACP/α-synuclein in vitro. *Neuroreport* 1999;**10**:717–21.
106. Paik SR, Shin HJ, Lee JH. Metal-catalyzed oxidation of α-synuclein in the presence of Copper (II) and hydrogen peroxide. *Arch Biochem Biophys* 2000;**378**:269–77.
107. Souza JM, Giasson BI, Chen Q, Lee VM, Ischiropoulos H. Dityrosine cross-linking promotes formation of stable alpha-synuclein polymers. Implication of nitrative and oxidative stress in the pathogenesis of neurodegenerative synucleinopathies. *J Biol Chem* 2000;**275**:18344–9.
108. Takahashi T, Yamashita H, Nakamura T, Nagano Y, Nakamura S. Tyrosine 125 of alpha-synuclein plays a critical role for dimerization following nitrative stress. *Brain Res* 2002;**938**:73–80.
109. Krishnan S, Chi EY, Wood SJ, Kendrick BS, Li C, Garzon-Rodriguez W, et al. Oxidative dimer formation is the critical rate-limiting step for Parkinson's disease α-synuclein fibrillogenesis. *Biochemistry* 2003;**42**:829–37.
110. Cole NB, Murphy DD, Lebowitz J, Di Noto L, Levine RL, Nussbaum RL. Metal-catalyzed oxidation of alpha synuclein: helping to define the relationship between oligomers, protofilaments and filaments. *J Biol Chem* 2005;**280**:9678–90.
111. Qin Z, Hu D, Han S, Reaney SH, Di Monte DA, Fink AL. Effect of 4-hydroxy-2-nonenal modification on alpha-synuclein aggregation. *J Biol Chem* 2007;**282**:5862–70.
112. Shamoto-Nagai M, Maruyama W, Hashizume Y, Yoshida M, Osawa T, Riederer P, et al. In parkinsonian substantia nigra, alpha-synuclein is modified by acrolein, a lipid-peroxidation product, and accumulates in the dopamine neurons with inhibition of proteasome activity. *J Neural Transm* 2007;**114**:1559–67.
113. Bosco DA, Fowler DM, Zhang Q, Nieva J, Powers ET, Wentworth Jr. P, et al. Elevated levels of oxidized cholesterol metabolites in Lewy body disease brains accelerate alpha-synuclein fibrilization. *Nat Chem Biol* 2006;**2**:249–53.
114. Uversky VN, Yamin G, Souillac PO, Goers J, Glaser CB, Fink AL. Methionine oxidation inhibits fibrillation of human α-synuclein in vitro. *FEBS Lett* 2002;**517**:239–44.
115. Zhou W, Long C, Reaney SH, Di Monte DA, Fink AL, Uversky VN. Methionine oxidation stabilizes non-toxic oligomers of alpha-synuclein through strengthening the auto-inhibitory intra-molecular long-range interactions. *Biochim Biophys Acta* 2010;**1802**:322–30.
116. Hokenson MJ, Uversky VN, Goers J, Yamin G, Munishkina LA, Fink AL. Role of individual methionines in the fibrillation of methionine-oxidized alpha-synuclein. *Biochemistry* 2004;**43**:4621–33.
117. Yamin G, Glaser CB, Uversky VN, Fink AL. Certain metals trigger fibrillation of methionine-oxidized α-synuclein. *J Biol Chem* 2003;**278**:27630–5.
118. Conway KA, Rochet J-C, Bieganski RM, Lansbury Jr. PT. Kinetic stabilization of the α-synuclein protofibril by a dopamine-α-synuclein adduct. *Science* 2001;**294**:1346–9.
119. Glaser CB, Yamin G, Uversky VN, Fink AL. Methionine oxidation, alpha-synuclein and Parkinson's disease. *Biochim Biophys Acta* 2005;**1703**:157–69.
120. Cappai R, Leck SL, Tew DJ, Williamson NA, Smith DP, Galatis D, et al. Dopamine promotes alpha-synuclein aggregation into SDS-resistant soluble oligomers via a distinct folding pathway. *FASEB J* 2005;**19**:1377–9.
121. Li HT, Lin DH, Luo XY, Zhang F, Ji LN, Du HN, et al. Inhibition of alpha-synuclein fibrillization by dopamine analogs via reaction with the amino groups of alpha-synuclein. Implication for dopaminergic neurodegeneration. *FEBS J* 2005;**272**:3661–72.
122. Bisaglia M, Mammi S, Bubacco L. Kinetic and structural analysis of the early oxidation products of dopamine. Analysis of the interactions with alpha-synuclein. *J Biol Chem* 2007;**282**:15597–605.

123. Mazzulli JR, Armakola M, Dumoulin M, Parastatidis I, Ischiropoulos H. Cellular oligomerization of alpha-synuclein is determined by the interaction of oxidized catechols with a C-terminal sequence. *J Biol Chem* 2007;**282**:31621–30.
124. Pham CL, Leong SL, Ali FE, Kenche VB, Hill AF, Gras SL, et al. Dopamine and the dopamine oxidation product 5,6-dihydroxylindole promote distinct on-pathway and off-pathway aggregation of alpha synuclein in a pH-dependent manner. *J Mol Biol* 2009;**387**:771–85.
125. Bisaglia M, Tosatto L, Munari F, Tessari I, de Laureto PP, Mammi S, et al. Dopamine quinones interact with alpha-synuclein to form unstructured adducts. *Biochem Biophys Res Commun* 2010;**394**:424–8.
126. Li J, Zhu M, Manning-Bog AB, Di Monte DA, Fink AL. Dopamine and L-dopa disaggregate amyloid fibrils: implications for Parkinson's and Alzheimer's disease. *FASEB J* 2004;**18**:962–4.
127. Norris EH, Giasson BI, Hodara R, Xu S, Trojanowski JQ, Ischiropoulos H, et al. Reversible inhibition of alpha-synuclein fibrillization by dopaminochrome-mediated conformational alterations. *J Biol Chem* 2005;**280**:21212–9.
128. Follmer C, Romao L, Einsiedler CM, Porto TC, Lara FA, Moncores M, et al. Dopamine affects the stability, hydration, and packing of protofibrils and fibrils of the wild type and variants of alpha-synuclein. *Biochemistry* 2007;**46**:472–82.
129. Leong SL, Pham CL, Galatis D, Fodero-Tavoletti MT, Perez K, Hill AF, et al. Formation of dopamine-mediated alpha-synuclein-soluble oligomers requires methionine oxidation. *Free Radic Biol Med* 2009;**46**:1328–37.
130. Conway KA, Harper JD, Lansbury Jr. PT. Fibrils formed *in vitro* from α-synuclein and two mutant forms linked to Parkinson's disease are typical amyloid. *Biochemistry* 2000; **39**:2552–63.
131. Rekas A, Knott RB, Sokolova A, Barnham KJ, Perez KA, Masters CL, et al. The structure of dopamine induced alpha-synuclein oligomers. *Eur Biophys J* 2010;**39**:1407–19.
132. Herrera FE, Chesi A, Paleologou KE, Schmid A, Munoz A, Vendruscolo M, et al. Inhibition of alpha-synuclein fibrillization by dopamine is mediated by interactions with five C-terminal residues and with E83 in the NAC region. *PLoS One* 2008;**3**:e3394.
133. Mazzulli JR, Mishizen AJ, Giasson BI, Lynch DR, Thomas SA, Nakashima A, et al. Cytosolic catechols inhibit alpha-synuclein aggregation and facilitate the formation of intracellular soluble oligomeric intermediates. *J Neurosci* 2006;**26**:10068–78.
134. Xu J, Kao SY, Lee FJ, Song W, Jin LW, Yankner BA. Dopamine-dependent neurotoxicity of α-synuclein: a mechanism for selective neurodegeneration in Parkinson disease. *Nat Med* 2002;**8**:600–6.
135. Caudle WM, Richardson JR, Wang MZ, Taylor TN, Guillot TS, McCormack AL, et al. Reduced vesicular storage of dopamine causes progressive nigrostriatal neurodegeneration. *J Neurosci* 2007;**27**:8138–48.
136. Mosharov EV, Larsen KE, Kanter E, Phillips KA, Wilson K, Schmitz Y, et al. Interplay between cytosolic dopamine, calcium, and alpha-synuclein causes selective death of substantia nigra neurons. *Neuron* 2009;**62**:218–29.
137. Norris EH, Giasson BI, Ischiropoulos H, Lee VM. Effects of oxidative and nitrative challenges on α-synuclein fibrillogenesis involve distinct mechanisms of protein modifications. *J Biol Chem* 2003;**278**:27230–40.
138. Yamin G, Uversky VN, Fink AL. Nitration inhibits fibrillation of human α-synuclein in vitro by formation of soluble oligomers. *FEBS Lett* 2003;**542**:147–52.
139. Hodara R, Norris EH, Giasson BI, Mishizen-Eberz AJ, Lynch DR, Lee VM, et al. Functional consequences of alpha-synuclein tyrosine nitration: diminished binding to lipid vesicles and increased fibril formation. *J Biol Chem* 2004;**279**:47746–53.

140. Yu Z, Xu X, Xiang Z, Zhou J, Zhang Z, Hu C, et al. Nitrated alpha-synuclein induces the loss of dopaminergic neurons in the substantia nigra of rats. *PLoS One* 2010;**5**:e9956.

141. Ross MF, Filipovska A, Smith RA, Gait MJ, Murphy MP. Cell-penetrating peptides do not cross mitochondrial membranes even when conjugated to a lipophilic cation: evidence against direct passage through phospholipid bilayers. *Biochem J* 2004;**383**:457–68.

142. Al-Taei S, Penning NA, Simpson JC, Futaki S, Takeuchi T, Nakase I, et al. Intracellular traffic and fate of protein transduction domains HIV-1 TAT peptide and octaarginine. Implications for their utilization as drug delivery vectors. *Bioconjug Chem* 2006;**17**:90–100.

143. Okochi M, Walter J, Koyama A, Nakajo S, Baba M, Iwatsubo T, et al. Constitutive phosphorylation of the Parkinson's disease associated alpha-synuclein. *J Biol Chem* 2000;**275**:390–7.

144. Paleologou KE, Schmid AW, Rospigliosi CC, Kim HY, Lamberto GR, Fredenburg RA, et al. Phosphorylation at Ser-129 but not the phosphomimics S129E/D inhibits the fibrillation of alpha-synuclein. *J Biol Chem* 2008;**283**:16895–905.

145. Hoyer W, Antony T, Cherny D, Heim G, Jovin TM, Subramaniam V. Dependence of alpha-synuclein aggregate morphology on solution conditions. *J Mol Biol* 2002;**322**:383–93.

146. Gorbatyuk OS, Li S, Sullivan LF, Chen W, Kondrikova G, Manfredsson FP, et al. The phosphorylation state of Ser-129 in human alpha-synuclein determines neurodegeneration in a rat model of Parkinson disease. *Proc Natl Acad Sci USA* 2008;**105**:763–8.

147. Azeredo da Silveira S, Schneider BL, Cifuentes-Diaz C, Sage D, Abbas-Terki T, Iwatsubo T, et al. Phosphorylation does not prompt, nor prevent, the formation of alpha-synuclein toxic species in a rat model of Parkinson's disease. *Hum Mol Genet* 2009;**18**:872–87.

148. McFarland NR, Fan Z, Xu K, Schwarzschild MA, Feany MB, Hyman BT, et al. Alpha-synuclein S129 phosphorylation mutants do not alter nigrostriatal toxicity in a rat model of Parkinson disease. *J Neuropathol Exp Neurol* 2009;**68**:515–24.

149. Takahashi M, Ko LW, Kulathingal J, Jiang P, Sevlever D, Yen SH. Oxidative stress-induced phosphorylation, degradation and aggregation of alpha-synuclein are linked to upregulated CK2 and cathepsin D. *Eur J Neurosci* 2007;**26**:863–74.

150. Sugeno N, Takeda A, Hasegawa T, Kobayashi M, Kikuchi A, Mori F, et al. Serine 129 phosphorylation of alpha-synuclein induces unfolded protein response-mediated cell death. *J Biol Chem* 2008;**283**:23179–88.

151. Kragh CL, Lund LB, Febbraro F, Hansen HD, Gai WP, El-Agnaf O, et al. α-Synuclein aggregation and Ser-129 phosphorylation-dependent cell death in oligodendroglial cells. *J Biol Chem* 2009;**284**:10211–22.

152. Pronin AN, Morris AJ, Surguchov A, Benovic JL. Synucleins are a novel class of substrates for G protein-coupled receptor kinases. *J Biol Chem* 2000;**275**:26515–22.

153. Arawaka S, Wada M, Goto S, Karube H, Sakamoto M, Ren CH, et al. The role of G-protein-coupled receptor kinase 5 in pathogenesis of sporadic Parkinson's disease. *J Neurosci* 2006;**26**:9227–38.

154. Inglis KJ, Chereau D, Brigham EF, Chiou SS, Schobel S, Frigon NL, et al. Polo-like kinase 2 (PLK2) phosphorylates alpha-synuclein at serine 129 in central nervous system. *J Biol Chem* 2009;**284**:2598–602.

155. Mbefo MK, Paleologou KE, Boucharaba A, Oueslati A, Schell H, Fournier M, et al. Phosphorylation of synucleins by members of the Polo-like kinase family. *J Biol Chem* 2010;**285**:2807–22.

156. Lee KW, Chen W, Junn E, Im JY, Grosso H, Sonsalla PK, et al. Enhanced phosphatase activity attenuates {alpha}-synucleinopathy in a mouse model. *J Neurosci* 2011;**31**:6963–71.

157. Gitler AD, Chesi A, Geddie ML, Strathearn KE, Hamamichi S, Hill KJ, et al. Alpha-synuclein is part of a diverse and highly conserved interaction network that includes PARK9 and manganese toxicity. *Nat Genet* 2009;**41**:308–15.

158. Kim EJ, Sung JY, Lee HJ, Rhim H, Hasegawa M, Iwatsubo T, et al. Dyrk1A phosphorylates alpha-synuclein and enhances intracellular inclusion formation. *J Biol Chem* 2006; **281**:33250–7.

159. Negro A, Brunati AM, Donella-Deana A, Massimino ML, Pinna LA. Multiple phosphorylation of α-synuclein by protein tyrosine kinase Syk prevents eosin-induced aggregation. *FASEB J* 2002;**16**:210–2.

160. Lewis KA, Yaeger A, DeMartino GN, Thomas PJ. Accelerated formation of alpha-synuclein oligomers by concerted action of the 20S proteasome and familial Parkinson mutations. *J Bioenerg Biomembr* 2010;**42**:85–95.

161. Periquet M, Fulga T, Myllykangas L, Schlossmacher MG, Feany MB. Aggregated alpha-synuclein mediates dopaminergic neurotoxicity in vivo. *J Neurosci* 2007;**27**:3338–46.

162. Tofaris GK, Garcia Reitbock P, Humby T, Lambourne SL, O'Connell M, Ghetti B, et al. Pathological changes in dopaminergic nerve cells of the substantia nigra and olfactory bulb in mice transgenic for truncated human alpha-synuclein(1-120): implications for Lewy body disorders. *J Neurosci* 2006;**26**:3942–50.

163. Daher JP, Ying M, Banerjee R, McDonald RS, Hahn MD, Yang L, et al. Conditional transgenic mice expressing C-terminally truncated human alpha-synuclein (alphaSyn119) exhibit reduced striatal dopamine without loss of nigrostriatal pathway dopaminergic neurons. *Mol Neurodegener* 2009;**4**:34.

164. Wakamatsu M, Ishii A, Iwata S, Sakagami J, Ukai Y, Ono M, et al. Selective loss of nigral dopamine neurons induced by overexpression of truncated human alpha-synuclein in mice. *Neurobiol Aging* 2008;**29**:574–85.

165. Ulusoy A, Febbraro F, Jensen PH, Kirik D, Romero-Ramos M. Co-expression of C-terminal truncated alpha-synuclein enhances full-length alpha-synuclein-induced pathology. *Eur J Neurosci* 2010;**32**:409–22.

166. Mishizen-Eberz AJ, Guttmann RP, Giasson BI, Day 3rd GA, Hodara R, Ischiropoulos H, et al. Distinct cleavage patterns of normal and pathologic forms of alpha-synuclein by calpain I in vitro. *J Neurochem* 2003;**86**:836–47.

167. Mishizen-Eberz AJ, Norris EH, Giasson BI, Hodara R, Ischiropoulos H, Lee VM, et al. Cleavage of alpha-synuclein by calpain: potential role in degradation of fibrillized and nitrated species of alpha-synuclein. *Biochemistry* 2005;**44**:7818–29.

168. Dufty BM, Warner LR, Hou ST, Jiang SX, Gomez-Isla T, Leenhouts KM, et al. Calpain-cleavage of alpha-synuclein: connecting proteolytic processing to disease-linked aggregation. *Am J Pathol* 2007;**170**:1725–38.

169. Sevlever D, Jiang P, Yen SH. Cathepsin D is the main lysosomal enzyme involved in the degradation of alpha-synuclein and generation of its carboxy-terminally truncated species. *Biochemistry* 2008;**47**:9678–87.

170. Lee H-J, Choi C, Lee S-J. Membrane-bound α-synuclein has a high aggregation propensity and the ability to seed the aggregation of the cytosolic form. *J Biol Chem* 2002;**277**:671–8.

171. Cole NB, Murphy DD, Grider T, Rueter S, Brasaemle D, Nussbaum RL. Lipid droplet binding and oligomerization properties of the Parkinson's disease protein α-synuclein. *J Biol Chem* 2002;**277**:6344–52.

172. Perrin RJ, Woods WS, Clayton DF, George JM. Exposure to long chain polyunsaturated fatty acids triggers rapid multimerization of synucleins. *J Biol Chem* 2001;**276**:41958–62.

173. Sharon R, Bar-Joseph I, Frosch MP, Walsh DM, Hamilton JA, Selkoe DJ. The formation of highly soluble oligomers of α-synuclein is regulated by fatty acids and enhanced in Parkinson's disease. *Neuron* 2003;**37**:583–95.

174. De Franceschi G, Frare E, Pivato M, Relini A, Penco A, Greggio E, et al. Structural and morphological characterization of aggregated species of {alpha}-synuclein induced by docosahexaenoic acid. *J Biol Chem* 2011;**286**:22262–74.

175. Necula M, Chirita CN, Kuret J. Rapid anionic micelle-mediated alpha-synuclein fibrillization in vitro. *J Biol Chem* 2003;**278**:46674–80.

176. Zhu M, Fink AL. Lipid binding inhibits α-synuclein fibril formation. *J Biol Chem* 2003;**278**:16873–7.

177. Jo E, Darabie AA, Han K, Tandon A, Fraser PE, McLaurin J. alpha-Synuclein-synaptosomal membrane interactions: implications for fibrillogenesis. *Eur J Biochem* 2004; **271**:3180–9.

178. Giehm L, Oliveira CL, Christiansen G, Pedersen JS, Otzen DE. SDS-induced fibrillation of alpha-synuclein: an alternative fibrillation pathway. *J Mol Biol* 2010;**401**:115–33.

179. Narayanan V, Scarlata S. Membrane binding and self-association of alpha-synucleins. *Biochemistry* 2001;**40**:9927–34.

180. Bisaglia M, Tessari I, Pinato L, Bellanda M, Giraudo S, Fasano M, et al. A topological model of the interaction between alpha-synuclein and sodium dodecyl sulfate micelles. *Biochemistry* 2005;**44**:329–39.

181. Ulmer TS, Bax A, Cole NB, Nussbaum RL. Structure and dynamics of micelle-bound human alpha-synuclein. *J Biol Chem* 2005;**280**:9595–603.

182. Bodner CR, Dobson CM, Bax A. Multiple tight phospholipid-binding modes of alpha-synuclein revealed by solution NMR spectroscopy. *J Mol Biol* 2009;**390**:775–90.

183. Ferreon AC, Gambin Y, Lemke EA, Deniz AA. Interplay of alpha-synuclein binding and conformational switching probed by single-molecule fluorescence. *Proc Natl Acad Sci USA* 2009;**106**:5645–50.

184. Trexler A, Rhoades E. alpha-Synuclein binds large unilamellar vesicles as an extended helix. *Biochemistry* 2009;**48**:2304–6.

185. Mihajlovic M, Lazaridis T. Membrane-bound structure and energetics of alpha-synuclein. *Proteins* 2008;**70**:761–78.

186. Abedini A, Raleigh DP. A role for helical intermediates in amyloid formation by natively unfolded polypeptides? *Phys Biol* 2009;**6**:15005.

187. Abedini A, Raleigh DP. A critical assessment of the role of helical intermediates in amyloid formation by natively unfolded proteins and polypeptides. *Protein Eng Des Sel* 2009;**22**:453–9.

188. Anderson VL, Ramlall TF, Rospigliosi CC, Webb WW, Eliezer D. Identification of a helical intermediate in trifluoroethanol-induced alpha-synuclein aggregation. *Proc Natl Acad Sci USA* 2010;**107**:18850–5.

189. Aisenbrey C, Borowik T, Bystrom R, Bokvist M, Lindstrom F, Misiak H, et al. How is protein aggregation in amyloidogenic diseases modulated by biological membranes? *Eur Biophys J* 2008;**37**:247–55.

190. Bystrom R, Aisenbrey C, Borowik T, Bokvist M, Lindstrom F, Sani MA, et al. Disordered proteins: biological membranes as two-dimensional aggregation matrices. *Cell Biochem Biophys* 2008;**52**:175–89.

191. Pandey AP, Haque F, Rochet JC, Hovis JS. Clustering of alpha-synuclein on supported lipid bilayers: role of anionic lipid, protein, and divalent ion concentration. *Biophys J* 2009; **96**:540–51.

192. Haque F, Pandey AP, Cambrea LR, Rochet JC, Hovis JS. Adsorption of alpha-synuclein on lipid bilayers: modulating the structure and stability of protein assemblies. *J Phys Chem B* 2010;**114**:4070–81.

193. Pandey AP, Haque F, Rochet JC, Hovis JS. alpha-Synuclein-induced tubule formation in lipid bilayers. *J Phys Chem B* 2011;**115**:5886–93.

194. Bodner CR, Maltsev AS, Dobson CM, Bax A. Differential phospholipid binding of alpha-synuclein variants implicated in Parkinson's disease revealed by solution NMR spectroscopy. *Biochemistry* 2010;**49**:862–71.

195. Bartels T, Ahlstrom LS, Leftin A, Kamp F, Haass C, Brown MF, et al. The N-terminus of the intrinsically disordered protein alpha-synuclein triggers membrane binding and helix folding. *Biophys J* 2010;**99**:2116–24.

196. Vamvaca K, Volles MJ, Lansbury Jr. PT. The first N-terminal amino acids of alpha-synuclein are essential for alpha-helical structure formation in vitro and membrane binding in yeast. *J Mol Biol* 2009;**389**:413–24.

197. Riekkinen P, Rinne UK, Pelliniemi TT, Sonninen V. Interaction between dopamine and phospholipids. Studies of the substantia nigra in Parkinson disease patients. *Arch Neurol* 1975;**32**:25–7.

198. Sevcsik E, Trexler AJ, Dunn JM, Rhoades E. Allostery in a disordered protein: oxidative modifications to alpha-synuclein act distally to regulate membrane binding. *J Am Chem Soc* 2011;**133**:7152–8.

199. Tamamizu-Kato S, Kosaraju MG, Kato H, Raussens V, Ruysschaert JM, Narayanaswami V. Calcium-triggered membrane interaction of the alpha-synuclein acidic tail. *Biochemistry* 2006;**45**:10947–56.

200. Der-Sarkissian A, Jao CC, Chen J, Langen R. Structural organization of alpha-synuclein fibrils studied by site-directed spin labeling. *J Biol Chem* 2003;**278**:37530–5.

201. Chen M, Margittai M, Chen J, Langen R. Investigation of alpha-synuclein fibril structure by site-directed spin labeling. *J Biol Chem* 2007;**282**:24970–9.

202. Del Mar C, Greenbaum EA, Mayne L, Englander SW, Woods Jr. VL. Structure and properties of alpha-synuclein and other amyloids determined at the amino acid level. *Proc Natl Acad Sci USA* 2005;**102**:15477–82.

203. Heise H, Hoyer W, Becker S, Andronesi OC, Riedel D, Baldus M. Molecular-level secondary structure, polymorphism, and dynamics of full-length alpha-synuclein fibrils studied by solid-state NMR. *Proc Natl Acad Sci USA* 2005;**102**:15871–6.

204. Vilar M, Chou HT, Luhrs T, Maji SK, Riek-Loher D, Verel R, et al. The fold of alpha-synuclein fibrils. *Proc Natl Acad Sci USA* 2008;**105**:8637–42.

205. Comellas G, Lemkau LR, Nieuwkoop AJ, Kloepper KD, Ladror DT, Ebisu R, et al. Structured regions of alpha-synuclein fibrils include the early-onset Parkinson's disease mutation sites. *J Mol Biol* 2011;**411**:881–95.

206. Sato T, Kienlen-Campard P, Ahmed M, Liu W, Li H, Elliott JI, et al. Inhibitors of amyloid toxicity based on beta-sheet packing of Abeta40 and Abeta42. *Biochemistry* 2006;**45**:5503–16.

207. Sciarretta KL, Boire A, Gordon DJ, Meredith SC. Spatial separation of beta-sheet domains of beta-amyloid: disruption of each beta-sheet by N-methyl amino acids. *Biochemistry* 2006;**45**:9485–95.

208. Sievers SA, Karanicolas J, Chang HW, Zhao A, Jiang L, Zirafi O, et al. Structure-based design of non-natural amino-acid inhibitors of amyloid fibril formation. *Nature* 2011;**475**:96–100.

209. Karpinar DP, Balija MB, Kugler S, Opazo F, Rezaei-Ghaleh N, Wender N, et al. Pre-fibrillar alpha-synuclein variants with impaired beta-structure increase neurotoxicity in Parkinson's disease models. *EMBO J* 2009;**28**:3256–68.

210. Winner B, Jappelli R, Maji SK, Desplats PA, Boyer L, Aigner S, et al. In vivo demonstration that alpha-synuclein oligomers are toxic. *Proc Natl Acad Sci USA* 2011;**108**:4194–9.

211. Lee SJ, Lim HS, Masliah E, Lee HJ. Protein aggregate spreading in neurodegenerative diseases: problems and perspectives. *Neurosci Res* 2011;**70**:339–48.

212. Lee HJ, Patel S, Lee SJ. Intravesicular localization and exocytosis of alpha-synuclein and its aggregates. *J Neurosci* 2005;**25**:6016–24.

213. Emmanouilidou E, Melachroinou K, Roumeliotis T, Garbis SD, Ntzouni M, Margaritis LH, et al. Cell-produced alpha-synuclein is secreted in a calcium-dependent manner by exosomes and impacts neuronal survival. *J Neurosci* 2010;**30**:6838–51.

214. Jang A, Lee HJ, Suk JE, Jung JW, Kim KP, Lee SJ. Non-classical exocytosis of alpha-synuclein is sensitive to folding states and promoted under stress conditions. *J Neurochem* 2010;**113**:1263–74.

215. Alvarez-Erviti L, Seow Y, Schapira AH, Gardiner C, Sargent IL, Wood MJ, et al. Lysosomal dysfunction increases exosome-mediated alpha-synuclein release and transmission. *Neurobiol Dis* 2011;**42**:360–7.

216. Danzer KM, Ruf WP, Putcha P, Joyner D, Hashimoto T, Glabe C, et al. Heat-shock protein 70 modulates toxic extracellular alpha-synuclein oligomers and rescues trans-synaptic toxicity. *FASEB J* 2011;**25**:326–36.

217. Hansen C, Angot E, Bergstrom AL, Steiner JA, Pieri L, Paul G, et al. alpha-Synuclein propagates from mouse brain to grafted dopaminergic neurons and seeds aggregation in cultured human cells. *J Clin Invest* 2011;**121**:715–25.

218. Lee HJ, Suk JE, Bae EJ, Lee JH, Paik SR, Lee SJ. Assembly-dependent endocytosis and clearance of extracellular alpha-synuclein. *Int J Biochem Cell Biol* 2008;**40**:1835–49.

219. Desplats P, Lee HJ, Bae EJ, Patrick C, Rockenstein E, Crews L, et al. Inclusion formation and neuronal cell death through neuron-to-neuron transmission of alpha-synuclein. *Proc Natl Acad Sci USA* 2009;**106**:13010–5.

220. Braak H, Del Tredici K, Rub U, de Vos RA, Jansen Steur EN, Braak E. Staging of brain pathology related to sporadic Parkinson's disease. *Neurobiol Aging* 2003;**24**:197–211.

221. Zhang W, Wang T, Pei Z, Miller DS, Wu X, Block ML, et al. Aggregated alpha-synuclein activates microglia: a process leading to disease progression in Parkinson's disease. *FASEB J* 2005;**19**:533–42.

222. Zhang W, Dallas S, Zhang D, Guo JP, Pang H, Wilson B, et al. Microglial PHOX and Mac-1 are essential to the enhanced dopaminergic neurodegeneration elicited by A30P and A53T mutant alpha-synuclein. *Glia* 2007;**55**:1178–88.

223. Lee SJ. Origins and effects of extracellular alpha-synuclein: implications in Parkinson's disease. *J Mol Neurosci* 2008;**34**:17–22.

224. Reynolds AD, Glanzer JG, Kadiu I, Ricardo-Dukelow M, Chaudhuri A, Ciborowski P, et al. Nitrated alpha-synuclein-activated microglial profiling for Parkinson's disease. *J Neurochem* 2008;**104**:1504–25.

225. Su X, Federoff HJ, Maguire-Zeiss KA. Mutant alpha-synuclein overexpression mediates early proinflammatory activity. *Neurotox Res* 2009;**16**:238–54.

226. Mattson MP. Neuronal life-and-death signaling, apoptosis, and neurodegenerative disorders. *Antioxid Redox Signal* 2006;**8**:1997–2006.

227. Rochet JC. Novel therapeutic strategies for the treatment of protein-misfolding diseases. *Expert Rev Mol Med* 2007;**9**:1–34.

228. Wassef R, Haenold R, Hansel A, Brot N, Heinemann SH, Hoshi T. Methionine sulfoxide reductase A and a dietary supplement S-methyl-L-cysteine prevent Parkinson's-like symptoms. *J Neurosci* 2007;**27**:12808–16.

229. Liu F, Hindupur J, Nguyen JL, Ruf KJ, Zhu J, Schieler JL, et al. Methionine sulfoxide reductase A protects dopaminergic cells from Parkinson's disease-related insults. *Free Radic Biol Med* 2008;**45**:242–55.

230. Levine RL, Moskovitz J, Stadtman ER. Oxidation of methionine in proteins: roles in antioxidant defense and cellular regulation. *IUBMB Life* 2000;**50**:301–7.

231. Zhang XH, Weissbach H. Origin and evolution of the protein-repairing enzymes methionine sulphoxide reductases. *Biol Rev Camb Philos Soc* 2008;**83**:249–57.

232. Zhu M, Qin ZJ, Hu D, Munishkina LA, Fink AL. Alpha-synuclein can function as an antioxidant preventing oxidation of unsaturated lipid in vesicles. *Biochemistry* 2006;**45**:8135–42.

233. Moskovitz J, Weissbach H, Brot N. Cloning the expression of a mammalian gene involved in the reduction of methionine sulfoxide residues in proteins. *Proc Natl Acad Sci USA* 1996;**93**:2095–9.

234. Petropoulos I, Mary J, Perichon M, Friguet B. Rat peptide methionine sulphoxide reductase: cloning of the cDNA, and down-regulation of gene expression and enzyme activity during aging. *Biochem J* 2001;**355**:819–25.

235. Botella JA, Bayersdorfer F, Schneuwly S. Superoxide dismutase overexpression protects dopaminergic neurons in a Drosophila model of Parkinson's disease. *Neurobiol Dis* 2008;**30**:65–73.

236. Kensler TW, Wakabayashi N, Biswal S. Cell survival responses to environmental stresses via the Keap1-Nrf2-ARE pathway. *Annu Rev Pharmacol Toxicol* 2007;**47**:89–116.

237. Vargas MR, Johnson JA. The Nrf2-ARE cytoprotective pathway in astrocytes. *Expert Rev Mol Med* 2009;**11**:e17.

238. Sykiotis GP, Bohmann D. Stress-activated cap'n'collar transcription factors in aging and human disease. *Sci Signal* 2010;**3**:re3.

239. Barone MC, Sykiotis GP, Bohmann D. Genetic activation of Nrf2 signaling is sufficient to ameliorate neurodegenerative phenotypes in a Drosophila model of Parkinson's disease. *Dis Model Mech* 2011;**4**:701–7.

240. Muchowski PJ, Wacker JL. Modulation of neurodegeneration by molecular chaperones. *Nat Rev Neurosci* 2005;**6**:11–22.

241. Dedmon MM, Christodoulou J, Wilson MR, Dobson CM. Heat shock protein 70 inhibits alpha-synuclein fibril formation via preferential binding to prefibrillar species. *J Biol Chem* 2005;**280**:14733–40.

242. Huang C, Cheng H, Hao S, Zhou H, Zhang X, Gao J, et al. Heat shock protein 70 inhibits alpha-synuclein fibril formation via interactions with diverse intermediates. *J Mol Biol* 2006;**364**:323–36.

243. Luk KC, Mills IP, Trojanowski JQ, Lee VM. Interactions between Hsp70 and the hydrophobic core of alpha-synuclein inhibit fibril assembly. *Biochemistry* 2008;**47**:12614–25.

244. Roodveldt C, Bertoncini CW, Andersson A, van der Goot AT, Hsu ST, Fernandez-Montesinos R, et al. Chaperone proteostasis in Parkinson's disease: stabilization of the Hsp70/alpha-synuclein complex by Hip. *EMBO J* 2009;**28**:3758–70.

245. Scherzer CR, Eklund AC, Morse LJ, Liao Z, Locascio JJ, Fefer D, et al. Molecular markers of early Parkinson's disease based on gene expression in blood. *Proc Natl Acad Sci USA* 2007;**104**:955–60.

246. Hinault MP, Cuendet AF, Mattoo RU, Mensi M, Dietler G, Lashuel HA, et al. Stable alpha-synuclein oligomers strongly inhibit chaperone activity of the Hsp70 system by weak interactions with J-domain co-chaperones. *J Biol Chem* 2010;**285**:38173–82.

247. Auluck PK, Chan HY, Trojanowski JQ, Lee VM, Bonini NM. Chaperone suppression of alpha-synuclein toxicity in a Drosophila model for Parkinson's disease. *Science* 2002;**295**:865–8.

248. Auluck PK, Meulener MC, Bonini NM. Mechanisms of suppression of {alpha}-synuclein neurotoxicity by geldanamycin in Drosophila. *J Biol Chem* 2005;**280**:2873–8.

249. Klucken J, Shin Y, Masliah E, Hyman BT, McLean PJ. Hsp70 reduces alpha-synuclein aggregation and toxicity. *J Biol Chem* 2004;**279**:25497–502.

250. Outeiro TF, Putcha P, Tetzlaff JE, Spoelgen R, Koker M, Carvalho F, et al. Formation of toxic oligomeric alpha-synuclein species in living cells. *PLoS One* 2008;**3**:e1867.

251. Liu F, Nguyen JL, Hulleman JD, Li L, Rochet J-C. Mechanisms of DJ-1 neuroprotection in a cellular model of Parkinson's disease. *J Neurochem* 2008;**105**:2435–53.

252. Shimshek DR, Mueller M, Wiessner C, Schweizer T, van der Putten PH. The HSP70 molecular chaperone is not beneficial in a mouse model of alpha-synucleinopathy. *PLoS One* 2010;**5**:e10014.

253. Rekas A, Adda CG, Andrew Aquilina J, Barnham KJ, Sunde M, Galatis D, et al. Interaction of the molecular chaperone alphaB-crystallin with alpha-synuclein: effects on amyloid fibril formation and chaperone activity. *J Mol Biol* 2004;**340**:1167–83.

254. Rekas A, Jankova L, Thorn DC, Cappai R, Carver JA. Monitoring the prevention of amyloid fibril formation by alpha-crystallin. Temperature dependence and the nature of the aggregating species. *FEBS J* 2007;**274**:6290–304.

255. Waudby CA, Knowles TP, Devlin GL, Skepper JN, Ecroyd H, Carver JA, et al. The interaction of alphaB-crystallin with mature alpha-synuclein amyloid fibrils inhibits their elongation. *Biophys J* 2010;**98**:843–51.

256. Ghosh JG, Houck SA, Clark JI. Interactive sequences in the molecular chaperone, human alphaB crystallin modulate the fibrillation of amyloidogenic proteins. *Int J Biochem Cell Biol* 2008;**40**:954–67.

257. Ahmad MF, Raman B, Ramakrishna T, Rao Ch M. Effect of phosphorylation on alpha B-crystallin: differences in stability, subunit exchange and chaperone activity of homo and mixed oligomers of alpha B-crystallin and its phosphorylation-mimicking mutant. *J Mol Biol* 2008;**375**:1040–51.

258. Zourlidou A, Payne Smith MD, Latchman DS. HSP27 but not HSP70 has a potent protective effect against alpha-synuclein-induced cell death in mammalian neuronal cells. *J Neurochem* 2004;**88**:1439–48.

259. Outeiro TF, Klucken J, Strathearn KE, Liu F, Nguyen P, Rochet JC, et al. Small heat shock proteins protect against alpha-synuclein-induced toxicity and aggregation. *Biochem Biophys Res Commun* 2006;**351**:631–8.

260. Pountney DL, Treweek TM, Chataway T, Huang Y, Chegini F, Blumbergs PC, et al. Alpha B-crystallin is a major component of glial cytoplasmic inclusions in multiple system atrophy. *Neurotox Res* 2005;**7**:77–85.

261. Abou-Sleiman PM, Healy DG, Quinn N, Lees AJ, Wood NW. The role of pathogenic DJ-1 mutations in Parkinson's disease. *Ann Neurol* 2003;**54**:283–6.

262. Bonifati V, Rizzu P, van Baren MJ, Schaap O, Breedveld GJ, Krieger E, et al. Mutations in the DJ-1 gene associated with autosomal recessive early-onset parkinsonism. *Science* 2003;**299**:256–9.

263. Hague S, Rogaeva E, Hernandez D, Gulick C, Singleton A, Hanson M, et al. Early-onset Parkinson's disease caused by a compound heterozygous DJ-1 mutation. *Ann Neurol* 2003;**54**:271–4.

264. Clark LN, Afridi S, Mejia-Santana H, Harris J, Louis ED, Cote LJ, et al. Analysis of an early-onset Parkinson's disease cohort for DJ-1 mutations. *Mov Disord* 2004;**19**:796–800.

265. Hering R, Strauss KM, Tao X, Bauer A, Woitalla D, Mietz EM, et al. Novel homozygous p. E64D mutation in DJ1 in early onset Parkinson disease (PARK7). *Hum Mutat* 2004;**24**:321–9.

266. Nuytemans K, Theuns J, Cruts M, Van Broeckhoven C. Genetic etiology of Parkinson disease associated with mutations in the SNCA, PARK2, PINK1, PARK7, and LRRK2 genes: a mutation update. *Hum Mutat* 2010;**31**:763–80.

267. Honbou K, Suzuki NN, Horiuchi M, Niki T, Taira T, Ariga H, et al. The crystal structure of DJ-1, a protein related to male fertility and Parkinson's disease. *J Biol Chem* 2003;**278**:31380–4 [Epub 32003 Jun 31388].

268. Huai Q, Sun Y, Wang H, Chin LS, Li L, Robinson H, et al. Crystal structure of DJ-1/RS and implication on familial Parkinson's disease. *FEBS Lett* 2003;**549**:171–5.

269. Lee SJ, Kim SJ, Kim IK, Ko J, Jeong CS, Kim GH, et al. Crystal structures of human DJ-1 and Escherichia coli Hsp31, which share an evolutionarily conserved domain. *J Biol Chem* 2003;**278**:44552–9.

270. Tao X, Tong L. Crystal structure of human DJ-1, a protein associated with early onset Parkinson's disease. *J Biol Chem* 2003;**278**:31372–9.

271. Wilson MA, Collins JL, Hod Y, Ringe D, Petsko GA. The 1.1-A resolution crystal structure of DJ-1, the protein mutated in autosomal recessive early onset Parkinson's disease. *Proc Natl Acad Sci USA* 2003;**100**:9256–61.

272. Bandyopadhyay S, Cookson MR. Evolutionary and functional relationships within the DJ1 superfamily. *BMC Evol Biol* 2004;**4**:6.

273. Lucas JI, Marin I. A new evolutionary paradigm for the Parkinson disease gene DJ-1. *Mol Biol Evol* 2007;**24**:551–61.

274. Lev N, Roncevic D, Ickowicz D, Melamed E, Offen D. Role of DJ-1 in Parkinson's disease. *J Mol Neurosci* 2006;**29**:215–25.

275. Kahle PJ, Waak J, Gasser T. DJ-1 and prevention of oxidative stress in Parkinson's disease and other age-related disorders. *Free Radic Biol Med* 2009;**47**:1354–61.

276. Canet-Aviles RM, Wilson MA, Miller DW, Ahmad R, McLendon C, Bandyopadhyay S, et al. The Parkinson's disease protein DJ-1 is neuroprotective due to cysteine-sulfinic acid-driven mitochondrial localization. *Proc Natl Acad Sci USA* 2004;**101**:9103–8.

277. Taira T, Saito Y, Niki T, Iguchi-Ariga SM, Takahashi K, Ariga H. DJ-1 has a role in antioxidative stress to prevent cell death. *EMBO Rep* 2004;**5**:213–8.

278. Shendelman S, Jonason A, Martinat C, Leete T, Abeliovich A. DJ-1 is a redox-dependent molecular chaperone that inhibits α-synuclein aggregate formation. *PLoS Biol* 2004;**2**:e362.

279. Zhou W, Zhu M, Wilson MA, Petsko GA, Fink AL. The oxidation state of DJ-1 regulates its chaperone activity toward alpha-synuclein. *J Mol Biol* 2006;**356**:1036–48.

280. Logan T, Clark L, Ray SS. Engineered disulfide bonds restore chaperone-like function of DJ-1 mutants linked to familial Parkinson's disease. *Biochemistry* 2010;**49**:5624–33.

281. Witt AC, Lakshminarasimhan M, Remington BC, Hasim S, Pozharski E, Wilson MA. Cysteine pKa depression by a protonated glutamic acid in human DJ-1. *Biochemistry* 2008;**47**: 7430–40.

282. Macedo MG, Anar B, Bronner IF, Cannella M, Squitieri F, Bonifati V, et al. The DJ-1L166P mutant protein associated with early onset Parkinson's disease is unstable and forms higher-order protein complexes. *Hum Mol Genet* 2003;**12**:2807–16.

283. Moore DJ, Zhang L, Dawson TM, Dawson VL. A missense mutation (L166P) in DJ-1, linked to familial Parkinson's disease, confers reduced protein stability and impairs homo-oligomerization. *J Neurochem* 2003;**87**:1558–67.

284. Gorner K, Holtorf E, Odoy S, Nuscher B, Yamamoto A, Regula JT, et al. Differential effects of Parkinson's disease-associated mutations on stability and folding of DJ-1. *J Biol Chem* 2004;**279**:6943–51.

285. Olzmann JA, Brown K, Wilkinson KD, Rees HD, Huai Q, Ke H, et al. Familial Parkinson's disease-associated L166P mutation disrupts DJ-1 protein folding and function. *J Biol Chem* 2004;**279**:8506–15.

286. Hulleman JD, Mirzaei H, Guigard E, Taylor KL, Ray SS, Kay CM, et al. Destabilization of DJ-1 by familial substitution and oxidative modifications: implications for Parkinson's disease. *Biochemistry* 2007;**46**:5776–89.

287. Lakshminarasimhan M, Maldonado MT, Zhou W, Fink AL, Wilson MA. Structural impact of three Parkinsonism-associated missense mutations on human DJ-1. *Biochemistry* 2008;**47**:1381–92.

288. Malgieri G, Eliezer D. Structural effects of Parkinson's disease linked DJ-1 mutations. *Protein Sci* 2008;**17**:855–68.

289. Choi J, Sullards MC, Olzmann JA, Rees HD, Weintraub ST, Bostwick DE, et al. Oxidative damage of DJ-1 is linked to sporadic Parkinson and Alzheimer diseases. *J Biol Chem* 2006;**281**:10816–24.

290. Meulener MC, Xu K, Thomson L, Ischiropoulos H, Bonini NM. Mutational analysis of DJ-1 in Drosophila implicates functional inactivation by oxidative damage and aging. *Proc Natl Acad Sci USA* 2006;**103**:12517–22.

291. Zhou W, Freed CR. DJ-1 up-regulates glutathione synthesis during oxidative stress and inhibits A53T alpha-synuclein toxicity. *J Biol Chem* 2005;**280**:43150–8.

292. Batelli S, Albani D, Rametta R, Polito L, Prato F, Pesaresi M, et al. DJ-1 modulates alpha-synuclein aggregation state in a cellular model of oxidative stress: relevance for Parkinson's disease and involvement of HSP70. *PLoS One* 2008;**3**:e1884.

293. Li HM, Niki T, Taira T, Iguchi-Ariga SM, Ariga H. Association of DJ-1 with chaperones and enhanced association and colocalization with mitochondrial Hsp70 by oxidative stress. *Free Radic Res* 2005;**39**:1091–9.

294. Deeg S, Gralle M, Sroka K, Bahr M, Wouters FS, Kermer P. BAG1 restores formation of functional DJ-1 L166P dimers and DJ-1 chaperone activity. *J Cell Biol* 2010;**188**:505–13.

295. Bennett MC, Bishop JF, Leng Y, Chock PB, Chase TN, Mouradian MM. Degradation of alpha-synuclein by proteasome. *J Biol Chem* 1999;**274**:33855–8.

296. Tofaris GK, Layfield R, Spillantini MG. Alpha-synuclein metabolism and aggregation is linked to ubiquitin-independent degradation by the proteasome. *FEBS Lett* 2001;**509**:22–6.

297. McNaught KS, Mytilineou C, Jnobaptiste R, Yabut J, Shashidharan P, Jennert P, et al. Impairment of the ubiquitin-proteasome system causes dopaminergic cell death and inclusion body formation in ventral mesencephalic cultures. *J Neurochem* 2002;**81**:301–6.

298. Webb JL, Ravikumar B, Atkins J, Skepper JN, Rubinsztein DC. Alpha-synuclein is degraded by both autophagy and the proteasome. *J Biol Chem* 2003;**278**:25009–13.

299. Machiya Y, Hara S, Arawaka S, Fukushima S, Sato H, Sakamoto M, et al. Phosphorylated alpha-synuclein at Ser-129 is targeted to the proteasome pathway in a ubiquitin-independent manner. *J Biol Chem* 2010;**285**:40732–44.

300. Ancolio K, Alves da Costa C, Ueda K, Checler F. Alpha-synuclein and the Parkinson's disease-related mutant Ala53Thr-alpha-synuclein do not undergo proteasomal degradation in HEK293 and neuronal cells. *Neurosci Lett* 2000;**285**:79–82.

301. Petrucelli L, O'Farrell C, Lockhart PJ, Baptista M, Kehoe K, Vink L, et al. Parkin protects against the toxicity associated with mutant α-synuclein: proteasome dysfunction selectively affects catecholaminergic neurons. *Neuron* 2002;**36**:1007–19.

302. Vogiatzi T, Xilouri M, Vekrellis K, Stefanis L. Wild type α-synuclein is degraded by chaperone mediated autophagy and macroautophagy in neuronal cells. *J Biol Chem* 2008;**283**:23542–56.

303. Stefanis L, Larsen KE, Rideout HJ, Sulzer D, Greene LA. Expression of A53T mutant but not wild-type alpha-synuclein in PC12 cells induces alterations of the ubiquitin-dependent degradation system, loss of dopamine release, and autophagic cell death. *J Neurosci* 2001;**21**:9549–60.

304. Tanaka Y, Engelender S, Igarashi S, Rao RK, Wanner T, Tanzi RE, et al. Inducible expression of mutant alpha-synuclein decreases proteasome activity and increases sensitivity to mitochondria-dependent apoptosis. *Hum Mol Genet* 2001;**10**:919–26.

305. Snyder H, Mensah K, Theisler C, Lee J, Matouschek A, Wolozin B. Aggregated and monomeric alpha-synuclein bind to the S6' proteasomal protein and inhibit proteasomal function. *J Biol Chem* 2003;**278**:11753–9.

306. Lindersson E, Beedholm R, Hojrup P, Moos T, Gai W, Hendil KB, et al. Proteasomal inhibition by alpha-synuclein filaments and oligomers. *J Biol Chem* 2004;**279**:12924–34.

307. Emmanouilidou E, Stefanis L, Vekrellis K. Cell-produced alpha-synuclein oligomers are targeted to, and impair, the 26S proteasome. *Neurobiol Aging* 2010;**31**:953–68.

308. Zhang NY, Tang Z, Liu CW. Alpha-synuclein protofibrils inhibit 26S proteasome-mediated protein degradation: understanding the cytotoxicity of protein protofibrils in neurodegenerative diseases pathogenesis. *J Biol Chem* 2008;**283**:20288–98.

309. Lee FK, Wong AK, Lee YW, Wan OW, Chan HY, Chung KK. The role of ubiquitin linkages on alpha-synuclein induced-toxicity in a Drosophila model of Parkinson's disease. *J Neurochem* 2009;**110**:208–19.

310. Cuervo AM, Stefanis L, Fredenburg R, Lansbury PT, Sulzer D. Impaired degradation of mutant alpha-synuclein by chaperone-mediated autophagy. *Science* 2004;**305**:1292–5.

311. Mak SK, McCormack AL, Manning Bog AB, Cuervo AM, Di Monte DA. Lysosomal degradation of alpha-synuclein in vivo. *J Biol Chem* 2010;**285**:13621–9.

312. Martinez-Vicente M, Talloczy Z, Kaushik S, Massey AC, Mazzulli J, Mosharov EV, et al. Dopamine-modified alpha-synuclein blocks chaperone-mediated autophagy. *J Clin Invest* 2008;**118**:777–88.

313. Xilouri M, Vogiatzi T, Vekrellis K, Park D, Stefanis L. Abberant alpha-synuclein confers toxicity to neurons in part through inhibition of chaperone-mediated autophagy. *PLoS One* 2009;4:e5515.

314. Ebrahimi-Fakhari D, Cantuti-Castelvetri I, Fan Z, Rockenstein E, Masliah E, Hyman BT, et al. Distinct roles in vivo for the ubiquitin-proteasome system and the autophagy-lysosomal pathway in the degradation of {alpha}-synuclein. *J Neurosci* 2011;**31**:14508–20.

315. Winslow AR, Chen CW, Corrochano S, Acevedo-Arozena A, Gordon DE, Peden AA, et al. alpha-Synuclein impairs macroautophagy: implications for Parkinson's disease. *J Cell Biol* 2010;**190**:1023–37.

316. Yu WH, Dorado B, Figueroa HY, Wang L, Planel E, Cookson MR, et al. Metabolic activity determines efficacy of macroautophagic clearance of pathological oligomeric alpha-synuclein. *Am J Pathol* 2009;**175**:736–47.

317. Sarkar S, Davies JE, Huang Z, Tunnacliffe A, Rubinsztein DC. Trehalose, a novel mTOR-independent autophagy enhancer, accelerates the clearance of mutant huntingtin and {alpha}-synuclein. *J Biol Chem* 2007;**282**:5641–52.

318. Sarkar S, Floto RA, Berger Z, Imarisio S, Cordenier A, Pasco M, et al. Lithium induces autophagy by inhibiting inositol monophosphatase. *J Cell Biol* 2005;**170**:1101–11.

319. Sarkar S, Perlstein EO, Imarisio S, Pineau S, Cordenier A, Maglathlin RL, et al. Small molecules enhance autophagy and reduce toxicity in Huntington's disease models. *Nat Chem Biol* 2007;**3**:331–8.

320. Crews L, Spencer B, Desplats P, Patrick C, Paulino A, Rockenstein E, et al. Selective molecular alterations in the autophagy pathway in patients with Lewy body disease and in models of alpha-synucleinopathy. *PLoS One* 2010;**5**:e9313.

321. Riedel M, Goldbaum O, Schwarz L, Schmitt S, Richter-Landsberg C. 17-AAG induces cytoplasmic alpha-synuclein aggregate clearance by induction of autophagy. *PLoS One* 2010;**5**:e8753.

322. Ishikawa A, Tsuji S. Clinical analysis of 17 patients in 12 Japanese families with autosomal-recessive type juvenile parkinsonism. *Neurology* 1996;**47**:160–6.

323. Kitada T, Asakawa S, Hattori N, Matsumine H, Yamamura Y, Minoshima S, et al. Mutations in the parkin gene cause autosomal recessive juvenile parkinsonism. *Nature* 1998;**392**:605–8.

324. van Nuenen BF, Weiss MM, Bloem BR, Reetz K, van Eimeren T, Lohmann K, et al. Heterozygous carriers of a Parkin or PINK1 mutation share a common functional endophenotype. *Neurology* 2009;**72**:1041–7.

325. Wang C, Ma H, Feng X, Xie S, Chan P. Parkin dosage mutations in patients with early-onset sporadic and familial Parkinson's disease in Chinese: an independent pathogenic role. *Brain Res* 2010;**1358**:30–8.

326. Deshaies RJ, Joazeiro CA. RING domain E3 ubiquitin ligases. *Annu Rev Biochem* 2009;**78**:399–434.

327. Shimura H, Hattori N, Kubo S, Mizuno Y, Asakawa S, Minoshima S, et al. Familial Parkinson disease gene product, parkin, is a ubiquitin-protein ligase. *Nat Genet* 2000;**25**:302–5.

328. Zhang Y, Gao J, Chung KK, Huang H, Dawson VL, Dawson TM. Parkin functions as an E2-dependent ubiquitin-protein ligase and promotes the degradation of the synaptic vesicle-associated protein, CDCrel-1. *Proc Natl Acad Sci USA* 2000;**97**:13354–9.

329. Imai Y, Soda M, Takahashi R. Parkin suppresses unfolded protein stress-induced cell death through its E3 ubiquitin-protein ligase activity. *J Biol Chem* 2000;**275**:35661–4.

330. West AB, Dawson VL, Dawson TM. The role of Parkin in Parkinson's disease. In: Dwson TM, editor. *Parkinson's disease: genetics and pathogenesis.* New York: Informa Healthcare; 2007. pp. 199–218.

331. Dawson TM, Dawson VL. The role of parkin in familial and sporadic Parkinson's disease. *Mov Disord* 2010;**25**(Suppl. 1):S32–9.

332. de la Torre ER, Gomez-Suaga P, Martinez-Salvador M, Hilfiker S. Posttranslational modifications as versatile regulators of parkin function. *Curr Med Chem* 2011;**18**:2477–85.

333. Winklhofer KF, Henn IH, Kay-Jackson PC, Heller U, Tatzelt J. Inactivation of parkin by oxidative stress and C-terminal truncations: a protective role of molecular chaperones. *J Biol Chem* 2003;**278**:47199–208.

334. Chung KK, Thomas B, Li X, Pletnikova O, Troncoso JC, Marsh L, et al. S-nitrosylation of parkin regulates ubiquitination and compromises parkin's protective function. *Science* 2004;**304**:1328–31.

335. LaVoie MJ, Ostaszewski BL, Weihofen A, Schlossmacher MG, Selkoe DJ. Dopamine covalently modifies and functionally inactivates parkin. *Nat Med* 2005;**11**:1214–21.

336. Meng F, Yao D, Shi Y, Kabakoff J, Wu W, Reicher J, et al. Oxidation of the cysteine-rich regions of parkin perturbs its E3 ligase activity and contributes to protein aggregation. *Mol Neurodegener* 2011;**6**:34.

337. Moszczynska A, Yamamoto BK. Methamphetamine oxidatively damages parkin and decreases the activity of 26S proteasome in vivo. *J Neurochem* 2011;**116**:1005–17.

338. Wong ES, Tan JM, Wang C, Zhang Z, Tay SP, Zaiden N, et al. Relative sensitivity of parkin and other cysteine-containing enzymes to stress-induced solubility alterations. *J Biol Chem* 2007;**282**:12310–8.

339. Kawahara K, Hashimoto M, Bar-On P, Ho GJ, Crews L, Mizuno H, et al. alpha-Synuclein aggregates interfere with Parkin solubility and distribution: role in the pathogenesis of Parkinson disease. *J Biol Chem* 2008;**283**:6979–87.

340. Corti O, Hampe C, Koutnikova H, Darios F, Jacquier S, Prigent A, et al. The p38 subunit of the aminoacyl-tRNA synthetase complex is a Parkin substrate: linking protein biosynthesis and neurodegeneration. *Hum Mol Genet* 2003;**12**:1427–37.

341. Ko HS, von Coelln R, Sriram SR, Kim SW, Chung KK, Pletnikova O, et al. Accumulation of the authentic parkin substrate aminoacyl-tRNA synthetase cofactor, p38/JTV-1, leads to catecholaminergic cell death. *J Neurosci* 2005;**25**:7968–78.

342. Ko HS, Kim SW, Sriram SR, Dawson VL, Dawson TM. Identification of far upstream element-binding protein-1 as an authentic Parkin substrate. *J Biol Chem* 2006;**281**:16193–6.

343. Shin JH, Ko HS, Kang H, Lee Y, Lee YI, Pletinkova O, et al. PARIS (ZNF746) repression of PGC-1alpha contributes to neurodegeneration in Parkinson's disease. *Cell* 2011;**144**:689–702.

344. Zheng B, Liao Z, Locascio JJ, Lesniak KA, Roderick SS, Watt ML, et al. PGC-1alpha, a potential therapeutic target for early intervention in Parkinson's disease. *Sci Transl Med* 2010;**2**:52ra73.

345. Pacelli C, De Rasmo D, Signorile A, Grattagliano I, di Tullio G, D'Orazio A, et al. Mitochondrial defect and PGC-1alpha dysfunction in parkin-associated familial Parkinson's disease. *Biochim Biophys Acta* 2011;**1812**:1041–53.

346. Finck BN, Kelly DP. PGC-1 coactivators: inducible regulators of energy metabolism in health and disease. *J Clin Invest* 2006;**116**:615–22.

347. St-Pierre J, Drori S, Uldry M, Silvaggi JM, Rhee J, Jager S, et al. Suppression of reactive oxygen species and neurodegeneration by the PGC-1 transcriptional coactivators. *Cell* 2006;**127**:397–408.
348. Wu Z, Boss O. Targeting PGC-1 alpha to control energy homeostasis. *Expert Opin Ther Targets* 2007;**11**:1329–38.
349. Lin JD. Minireview: the PGC-1 coactivator networks: chromatin-remodeling and mitochondrial energy metabolism. *Mol Endocrinol* 2009;**23**:2–10.
350. Geisler S, Holmstrom KM, Skujat D, Fiesel FC, Rothfuss OC, Kahle PJ, et al. PINK1/Parkin-mediated mitophagy is dependent on VDAC1 and p62/SQSTM1. *Nat Cell Biol* 2010;**12**:119–31.
351. Chew KC, Matsuda N, Saisho K, Lim GG, Chai C, Tan HM, et al. Parkin mediates apparent e2-independent monoubiquitination in vitro and contains an intrinsic activity that catalyzes polyubiquitination. *PLoS One* 2011;**6**:e19720.
352. Mukhopadhyay D, Riezman H. Proteasome-independent functions of ubiquitin in endocytosis and signaling. *Science* 2007;**315**:201–5.
353. Komander D. The emerging complexity of protein ubiquitination. *Biochem Soc Trans* 2009;**37**:937–53.
354. Sakata E, Yamaguchi Y, Kurimoto E, Kikuchi J, Yokoyama S, Yamada S, et al. Parkin binds the Rpn10 subunit of 26S proteasomes through its ubiquitin-like domain. *EMBO Rep* 2003;**4**:301–6.
355. Um JW, Im E, Lee HJ, Min B, Yoo L, Yoo J, et al. Parkin directly modulates 26S proteasome activity. *J Neurosci* 2010;**30**:11805–14.
356. Petrucelli L, O'Farrell C, Lockhart PJ, Baptista M, Kehoe K, Vink L, et al. Parkin protects against the toxicity associated with mutant alpha-synuclein: proteasome dysfunction selectively affects catecholaminergic neurons. *Neuron* 2002;**36**:1007–19.
357. Valente EM, Abou-Sleiman PM, Caputo V, Muqit MM, Harvey K, Gispert S, et al. Hereditary early-onset Parkinson's disease caused by mutations in PINK1. *Science* 2004;**304**:1158–60.
358. Samaranch L, Lorenzo-Betancor O, Arbelo JM, Ferrer I, Lorenzo E, Irigoyen J, et al. PINK1-linked parkinsonism is associated with Lewy body pathology. *Brain* 2010;**133**:1128–42.
359. Farrer M, Chan P, Chen R, Tan L, Lincoln S, Hernandez D, et al. Lewy bodies and parkinsonism in families with parkin mutations. *Ann Neurol* 2001;**50**:293–300.
360. Pramstaller PP, Schlossmacher MG, Jacques TS, Scaravilli F, Eskelson C, Pepivani I, et al. Lewy body Parkinson's disease in a large pedigree with 77 Parkin mutation carriers. *Ann Neurol* 2005;**58**:411–22.
361. Dagda RK, Zhu J, Chu CT. Mitochondrial kinases in Parkinson's disease: converging insights from neurotoxin and genetic models. *Mitochondrion* 2009;**9**:289–98.
362. Guo M. What have we learned from Drosophila models of Parkinson's disease? *Prog Brain Res* 2010;**184**:3–16.
363. Whitworth AJ, Pallanck LJ. The PINK1/Parkin pathway: a mitochondrial quality control system? *J Bioenerg Biomembr* 2009;**41**:499–503.
364. Greene JC, Whitworth AJ, Kuo I, Andrews LA, Feany MB, Pallanck LJ. Mitochondrial pathology and apoptotic muscle degeneration in Drosophila parkin mutants. *Proc Natl Acad Sci USA* 2003;**100**:4078–83.
365. Pesah Y, Pham T, Burgess H, Middlebrooks B, Verstreken P, Zhou Y, et al. Drosophila parkin mutants have decreased mass and cell size and increased sensitivity to oxygen radical stress. *Development* 2004;**131**:2183–94.
366. Whitworth AJ, Theodore DA, Greene JC, Benes H, Wes PD, Pallanck LJ. Increased glutathione S-transferase activity rescues dopaminergic neuron loss in a Drosophila model of Parkinson's disease. *Proc Natl Acad Sci USA* 2005;**102**:8024–9.

367. Palacino JJ, Sagi D, Goldberg MS, Krauss S, Motz C, Wacker M, et al. Mitochondrial dysfunction and oxidative damage in parkin-deficient mice. *J Biol Chem* 2004;**279**:18614–22.

368. Clark IE, Dodson MW, Jiang C, Cao JH, Huh JR, Seol JH, et al. Drosophila pink1 is required for mitochondrial function and interacts genetically with parkin. *Nature* 2006;**441**:1162–6.

369. Park J, Lee SB, Lee S, Kim Y, Song S, Kim S, et al. Mitochondrial dysfunction in Drosophila PINK1 mutants is complemented by parkin. *Nature* 2006;**441**:1157–61.

370. Yang Y, Gehrke S, Imai Y, Huang Z, Ouyang Y, Wang JW, et al. Mitochondrial pathology and muscle and dopaminergic neuron degeneration caused by inactivation of Drosophila Pink1 is rescued by Parkin. *Proc Natl Acad Sci USA* 2006;**103**:10793–8.

371. Yun J, Cao JH, Dodson MW, Clark IE, Kapahi P, Chowdhury RB, et al. Loss-of-function analysis suggests that Omi/HtrA2 is not an essential component of the PINK1/PARKIN pathway in vivo. *J Neurosci* 2008;**28**:14500–10.

372. Ibanez P, Lesage S, Lohmann E, Thobois S, De Michele G, Borg M, et al. Mutational analysis of the PINK1 gene in early-onset parkinsonism in Europe and North Africa. *Brain* 2006;**129**:686–94.

373. Kitada T, Tong Y, Gautier CA, Shen J. Absence of nigral degeneration in aged parkin/DJ-1/PINK1 triple knockout mice. *J Neurochem* 2009;**111**:696–702.

374. Muftuoglu M, Elibol B, Dalmizrak O, Ercan A, Kulaksiz G, Ogus H, et al. Mitochondrial complex I and IV activities in leukocytes from patients with parkin mutations. *Mov Disord* 2004;**19**:544–8.

375. Hoepken HH, Gispert S, Morales B, Wingerter O, Del Turco D, Mulsch A, et al. Mitochondrial dysfunction, peroxidation damage and changes in glutathione metabolism in PARK6. *Neurobiol Dis* 2007;**25**:401–11.

376. Stichel CC, Zhu XR, Bader V, Linnartz B, Schmidt S, Lubbert H. Mono- and double-mutant mouse models of Parkinson's disease display severe mitochondrial damage. *Hum Mol Genet* 2007;**16**:2377–93.

377. Exner N, Treske B, Paquet D, Holmstrom K, Schiesling C, Gispert S, et al. Loss-of-function of human PINK1 results in mitochondrial pathology and can be rescued by parkin. *J Neurosci* 2007;**27**:12413–8.

378. Gautier CA, Kitada T, Shen J. Loss of PINK1 causes mitochondrial functional defects and increased sensitivity to oxidative stress. *Proc Natl Acad Sci USA* 2008;**105**:11364–9.

379. Wood-Kaczmar A, Gandhi S, Yao Z, Abramov AY, Miljan EA, Keen G, et al. PINK1 is necessary for long term survival and mitochondrial function in human dopaminergic neurons. *PLoS One* 2008;**3**:e2455.

380. Poole AC, Thomas RE, Andrews LA, McBride HM, Whitworth AJ, Pallanck LJ. The PINK1/Parkin pathway regulates mitochondrial morphology. *Proc Natl Acad Sci USA* 2008;**105**:1638–43.

381. Gegg ME, Cooper JM, Schapira AH, Taanman JW. Silencing of PINK1 expression affects mitochondrial DNA and oxidative phosphorylation in dopaminergic cells. *PLoS One* 2009;**4**:e4756.

382. Sandebring A, Thomas KJ, Beilina A, van der Brug M, Cleland MM, Ahmad R, et al. Mitochondrial alterations in PINK1 deficient cells are influenced by calcineurin-dependent dephosphorylation of dynamin-related protein 1. *PLoS One* 2009;**4**:e5701.

383. Dagda RK, Cherra 3rd SJ, Kulich SM, Tandon A, Park D, Chu CT. Loss of PINK1 function promotes mitophagy through effects on oxidative stress and mitochondrial fission. *J Biol Chem* 2009;**284**:13843–55.

384. Morais VA, Verstreken P, Roethig A, Smet J, Snellinx A, Vanbrabant M, et al. Parkinson's disease mutations in PINK1 result in decreased Complex I activity and deficient synaptic function. *EMBO Mol Med* 2009;**1**:99–111.

385. Grunewald A, Voges L, Rakovic A, Kasten M, Vandebona H, Hemmelmann C, et al. Mutant Parkin impairs mitochondrial function and morphology in human fibroblasts. *PLoS One* 2010;**5**:e12962.

386. Shim JH, Yoon SH, Kim KH, Han JY, Ha JY, Hyun DH, et al. The antioxidant Trolox helps recovery from the familial Parkinson's disease-specific mitochondrial deficits caused by PINK1- and DJ-1-deficiency in dopaminergic neuronal cells. *Mitochondrion* 2011;**11**:707-15.

387. Billia F, Hauck L, Konecny F, Rao V, Shen J, Mak TW. PTEN-inducible kinase 1 (PINK1)/Park6 is indispensable for normal heart function. *Proc Natl Acad Sci USA* 2011;**108**:9572-7.

388. Schmidt S, Linnartz B, Mendritzki S, Sczepan T, Lubbert M, Stichel CC, et al. Genetic mouse models for Parkinson's disease display severe pathology in glial cell mitochondria. *Hum Mol Genet* 2011;**20**:1197-211.

389. Chen H, Chan DC. Mitochondrial dynamics—fusion, fission, movement, and mitophagy—in neurodegenerative diseases. *Hum Mol Genet* 2009;**18**:R169-76.

390. Gandre-Babbe S, van der Bliek AM. The novel tail-anchored membrane protein Mff controls mitochondrial and peroxisomal fission in mammalian cells. *Mol Biol Cell* 2008;**19**:2402-12.

391. Otera H, Wang C, Cleland MM, Setoguchi K, Yokota S, Youle RJ, et al. Mff is an essential factor for mitochondrial recruitment of Drp1 during mitochondrial fission in mammalian cells. *J Cell Biol* 2011;**191**:1141-58.

392. Fuller MT. Spermatogenesis. In: Martinez-Arias A, Bate M, editors. *The development of Drosophila melanogaster*. Cold Spring Harbor, NY: Cold Spring Harbor Press; 1993. pp. 71-147.

393. Deng H, Dodson MW, Huang H, Guo M. The Parkinson's disease genes pink1 and parkin promote mitochondrial fission and/or inhibit fusion in Drosophila. *Proc Natl Acad Sci USA* 2008;**105**:14503-8.

394. Riparbelli MG, Callaini G. The Drosophila parkin homologue is required for normal mitochondrial dynamics during spermiogenesis. *Dev Biol* 2007;**303**:108-20.

395. Yang Y, Ouyang Y, Yang L, Beal MF, McQuibban A, Vogel H, et al. Pink1 regulates mitochondrial dynamics through interaction with the fission/fusion machinery. *Proc Natl Acad Sci USA* 2008;**105**:7070-5.

396. Park J, Lee G, Chung J. The PINK1-Parkin pathway is involved in the regulation of mitochondrial remodeling process. *Biochem Biophys Res Commun* 2009;**378**:518-23.

397. Mortiboys H, Thomas KJ, Koopman WJ, Klaffke S, Abou-Sleiman P, Olpin S, et al. Mitochondrial function and morphology are impaired in parkin-mutant fibroblasts. *Ann Neurol* 2008;**64**:555-65.

398. Yu W, Sun Y, Guo S, Lu B. The PINK1/Parkin pathway regulates mitochondrial dynamics and function in mammalian hippocampal and dopaminergic neurons. *Hum Mol Genet* 2011;**20**:3227-40.

399. Lutz AK, Exner N, Fett ME, Schlehe JS, Kloos K, Lammermann K, et al. Loss of parkin or PINK1 function increases Drp1-dependent mitochondrial fragmentation. *J Biol Chem* 2009;**284**:22938-51.

400. Dimmer KS, Fritz S, Fuchs F, Messerschmitt M, Weinbach N, Neupert W, et al. Genetic basis of mitochondrial function and morphology in Saccharomyces cerevisiae. *Mol Biol Cell* 2002;**13**:847-53.

401. Altmann K, Westermann B. Role of essential genes in mitochondrial morphogenesis in Saccharomyces cerevisiae. *Mol Biol Cell* 2005;**16**:5410-7.

402. Ichishita R, Tanaka K, Sugiura Y, Sayano T, Mihara K, Oka T. An RNAi screen for mitochondrial proteins required to maintain the morphology of the organelle in Caenorhabditis elegans. *J Biochem* 2008;**143**:449-54.

403. Goldman SJ, Taylor R, Zhang Y, Jin S. Autophagy and the degradation of mitochondria. *Mitochondrion* 2010;**10**:309-15.

404. Twig G, Shirihai OS. The interplay between mitochondrial dynamics and mitophagy. *Antioxid Redox Signal* 2011;**14**:1939–51.

405. Hyde BB, Twig G, Shirihai OS. Organellar vs cellular control of mitochondrial dynamics. *Semin Cell Dev Biol* 2010;**21**:575–81.

406. Twig G, Graf SA, Wikstrom JD, Mohamed H, Haigh SE, Elorza A, et al. Tagging and tracking individual networks within a complex mitochondrial web with photoactivatable GFP. *Am J Physiol Cell Physiol* 2006;**291**:C176–84.

407. Twig G, Elorza A, Molina AJ, Mohamed H, Wikstrom JD, Walzer G, et al. Fission and selective fusion govern mitochondrial segregation and elimination by autophagy. *EMBO J* 2008;**27**:433–46.

408. Acin-Perez R, Hoyos B, Zhao F, Vinogradov V, Fischman DA, Harris RA, et al. Control of oxidative phosphorylation by vitamin A illuminates a fundamental role in mitochondrial energy homoeostasis. *FASEB J* 2010;**24**:627–36.

409. Matsuda N, Sato S, Shiba K, Okatsu K, Saisho K, Gautier CA, et al. PINK1 stabilized by mitochondrial depolarization recruits Parkin to damaged mitochondria and activates latent Parkin for mitophagy. *J Cell Biol* 2010;**189**:211–21.

410. Narendra DP, Jin SM, Tanaka A, Suen DF, Gautier CA, Shen J, et al. PINK1 is selectively stabilized on impaired mitochondria to activate Parkin. *PLoS Biol* 2010;**8**:e1000298.

411. Vives-Bauza C, Zhou C, Huang Y, Cui M, de Vries RL, Kim J, et al. PINK1-dependent recruitment of Parkin to mitochondria in mitophagy. *Proc Natl Acad Sci USA* 2010;**107**:378–83.

412. Rakovic A, Grunewald A, Seibler P, Ramirez A, Kock N, Orolicki S, et al. Effect of endogenous mutant and wild-type PINK1 on Parkin in fibroblasts from Parkinson disease patients. *Hum Mol Genet* 2010;**19**:3124–37.

413. Ziviani E, Tao RN, Whitworth AJ. Drosophila parkin requires PINK1 for mitochondrial translocation and ubiquitinates mitofusin. *Proc Natl Acad Sci USA* 2010;**107**:5018–23.

414. Kawajiri S, Saiki S, Sato S, Sato F, Hatano T, Eguchi H, et al. PINK1 is recruited to mitochondria with parkin and associates with LC3 in mitophagy. *FEBS Lett* 2010;**584**:1073–9.

415. Van Laar VS, Arnold B, Cassady SJ, Chu CT, Burton EA, Berman SB. Bioenergetics of neurons inhibit the translocation response of Parkin following rapid mitochondrial depolarization. *Hum Mol Genet* 2011;**20**:927–40.

416. Seibler P, Graziotto J, Jeong H, Simunovic F, Klein C, Krainc D. Mitochondrial Parkin recruitment is impaired in neurons derived from mutant PINK1 induced pluripotent stem cells. *J Neurosci* 2011;**31**:5970–6.

417. Zhou C, Huang Y, Shao Y, May J, Prou D, Perier C, et al. The kinase domain of mitochondrial PINK1 faces the cytoplasm. *Proc Natl Acad Sci USA* 2008;**105**:12022–7.

418. Whitworth AJ, Lee JR, Ho VM, Flick R, Chowdhury R, McQuibban GA. Rhomboid-7 and HtrA2/Omi act in a common pathway with the Parkinson's disease factors Pink1 and Parkin. *Dis Model Mech* 2008;**1**:168–74 [discussion 173].

419. McQuibban GA, Lee JR, Zheng L, Juusola M, Freeman M. Normal mitochondrial dynamics requires rhomboid-7 and affects Drosophila lifespan and neuronal function. *Curr Biol* 2006;**16**:982–9.

420. Jin SM, Lazarou M, Wang C, Kane LA, Narendra DP, Youle RJ. Mitochondrial membrane potential regulates PINK1 import and proteolytic destabilization by PARL. *J Cell Biol* 2010;**191**:933–42.

421. Deas E, Plun-Favreau H, Gandhi S, Desmond H, Kjaer S, Loh SH, et al. PINK1 cleavage at position A103 by the mitochondrial protease PARL. *Hum Mol Genet* 2011;**20**:867–79.

422. Shi G, Lee JR, Grimes DA, Racacho L, Ye D, Yang H, et al. Functional alteration of PARL contributes to mitochondrial dysregulation in Parkinson's disease. *Hum Mol Genet* 2011;**20**:1966–74.

423. Meissner C, Lorenz H, Weihofen A, Selkoe DJ, Lemberg MK. The mitochondrial intramembrane protease PARL cleaves human Pink1 to regulate Pink1 trafficking. *J Neurochem* 2011;**117**:856–67.
424. Haucke V, Ocana CS, Honlinger A, Tokatlidis K, Pfanner N, Schatz G. Analysis of the sorting signals directing NADH-cytochrome b5 reductase to two locations within yeast mitochondria. *Mol Cell Biol* 1997;**17**:4024–32.
425. Gegg ME, Cooper JM, Chau KY, Rojo M, Schapira AH, Taanman JW. Mitofusin 1 and mitofusin 2 are ubiquitinated in a PINK1/parkin-dependent manner upon induction of mitophagy. *Hum Mol Genet* 2010;**19**:4861–70.
426. Poole AC, Thomas RE, Yu S, Vincow ES, Pallanck L. The mitochondrial fusion-promoting factor mitofusin is a substrate of the PINK1/parkin pathway. *PLoS One* 2010;**5**:e10054.
427. Tanaka A, Cleland MM, Xu S, Narendra DP, Suen DF, Karbowski M, et al. Proteasome and p97 mediate mitophagy and degradation of mitofusins induced by Parkin. *J Cell Biol* 2011;**191**:1367–80.
428. Chan NC, Salazar AM, Pham AH, Sweredoski MJ, Kolawa NJ, Graham RL, et al. Broad activation of the ubiquitin-proteasome system by Parkin is critical for mitophagy. *Hum Mol Genet* 2011;**20**:1726–37.
429. Glauser L, Sonnay S, Stafa K, Moore DJ. Parkin promotes the ubiquitination and degradation of the mitochondrial fusion factor mitofusin 1. *J Neurochem* 2011;**118**:636–45.
430. Rakovic A, Grunewald A, Kottwitz J, Bruggemann N, Pramstaller PP, Lohmann K, et al. Mutations in PINK1 and Parkin impair ubiquitination of Mitofusins in human fibroblasts. *PLoS One* 2011;**6**:e16746.
431. Chen D, Gao F, Li B, Wang H, Xu Y, Zhu C, et al. Parkin mono-ubiquitinates Bcl-2 and regulates autophagy. *J Biol Chem* 2010;**285**:38214–23.
432. Yoshii SR, Kishi C, Ishihara N, Mizushima N. Parkin mediates proteasome-dependent protein degradation and rupture of the outer mitochondrial membrane. *J Biol Chem* 2011;**286**:19630–40.
433. Narendra D, Kane LA, Hauser DN, Fearnley IM, Youle RJ. p62/SQSTM1 is required for Parkin-induced mitochondrial clustering but not mitophagy; VDAC1 is dispensable for both. *Autophagy* 2010;**6**:1090–106.
434. Wang H, Song P, Du L, Tian W, Yue W, Liu M, et al. Parkin ubiquitinates Drp1 for proteasome-dependent degradation: implication of dysregulated mitochondrial dynamics in Parkinson disease. *J Biol Chem* 2011;**286**:11649–58.
435. Head B, Griparic L, Amiri M, Gandre-Babbe S, van der Bliek AM. Inducible proteolytic inactivation of OPA1 mediated by the OMA1 protease in mammalian cells. *J Cell Biol* 2009;**187**:959–66.
436. Raasi S, Wolf DH. Ubiquitin receptors and ERAD: a network of pathways to the proteasome. *Semin Cell Dev Biol* 2007;**18**:780–91.
437. Lee JY, Nagano Y, Taylor JP, Lim KL, Yao TP. Disease-causing mutations in parkin impair mitochondrial ubiquitination, aggregation, and HDAC6-dependent mitophagy. *J Cell Biol* 2010;**189**:671–9.
438. Ding WX, Ni HM, Li M, Liao Y, Chen X, Stolz DB, et al. Nix is critical to two distinct phases of mitophagy, reactive oxygen species-mediated autophagy induction and Parkin-ubiquitin-p62-mediated mitochondrial priming. *J Biol Chem* 2010;**285**:27879–90.
439. Okatsu K, Saisho K, Shimanuki M, Nakada K, Shitara H, Sou YS, et al. p62/SQSTM1 cooperates with Parkin for perinuclear clustering of depolarized mitochondria. *Genes Cells* 2010;**15**:887–900.
440. Gandhi S, Wood-Kaczmar A, Yao Z, Plun-Favreau H, Deas E, Klupsch K, et al. PINK1-associated Parkinson's disease is caused by neuronal vulnerability to calcium-induced cell death. *Mol Cell* 2009;**33**:627–38.

441. Akundi RS, Huang Z, Eason J, Pandya JD, Zhi L, Cass WA, et al. Increased mitochondrial calcium sensitivity and abnormal expression of innate immunity genes precede dopaminergic defects in Pink1-deficient mice. *PLoS One* 2011;**6**:e16038.

442. Heeman B, Van den Haute C, Aelvoet SA, Valsecchi F, Rodenburg RJ, Reumers V, et al. Depletion of PINK1 affects mitochondrial metabolism, calcium homeostasis and energy maintenance. *J Cell Sci* 2011;**124**:1115–25.

443. Pridgeon JW, Olzmann JA, Chin LS, Li L. PINK1 protects against oxidative stress by phosphorylating mitochondrial chaperone TRAP1. *PLoS Biol* 2007;**5**:e172.

444. Beilina A, Van Der Brug M, Ahmad R, Kesavapany S, Miller DW, Petsko GA, et al. Mutations in PTEN-induced putative kinase 1 associated with recessive parkinsonism have differential effects on protein stability. *Proc Natl Acad Sci USA* 2005;**102**:5703–8.

445. Lin W, Kang UJ. Characterization of PINK1 processing, stability, and subcellular localization. *J Neurochem* 2008;**106**:464–74.

446. Takatori S, Ito G, Iwatsubo T. Cytoplasmic localization and proteasomal degradation of N-terminally cleaved form of PINK1. *Neurosci Lett* 2008;**430**:13–7.

447. Haque ME, Thomas KJ, D'Souza C, Callaghan S, Kitada T, Slack RS, et al. Cytoplasmic Pink1 activity protects neurons from dopaminergic neurotoxin MPTP. *Proc Natl Acad Sci USA* 2008;**105**:1716–21.

448. Lin W, Kang UJ. Structural determinants of PINK1 topology and dual subcellular distribution. *BMC Cell Biol* 2010;**11**:90.

449. Murata H, Sakaguchi M, Jin Y, Sakaguchi Y, Futami J, Yamada H, et al. A new cytosolic pathway from a Parkinson disease-associated kinase, BRPK/PINK1: activation of AKT via mTORC2. *J Biol Chem* 2011;**286**:7182–9.

450. Aleyasin H, Rousseaux MW, Marcogliese PC, Hewitt SJ, Irrcher I, Joselin AP, et al. DJ-1 protects the nigrostriatal axis from the neurotoxin MPTP by modulation of the AKT pathway. *Proc Natl Acad Sci USA* 2010;**107**:3186–91.

451. Xiong H, Wang D, Chen L, Choo YS, Ma H, Tang C, et al. Parkin, PINK1, and DJ-1 form a ubiquitin E3 ligase complex promoting unfolded protein degradation. *J Clin Invest* 2009;**119**:650–60.

452. Thomas KJ, McCoy MK, Blackinton J, Beilina A, van der Brug M, Sandebring A, et al. DJ-1 acts in parallel to the PINK1/parkin pathway to control mitochondrial function and autophagy. *Hum Mol Genet* 2011;**20**:40–50.

453. Nagakubo D, Taira T, Kitaura H, Ikeda M, Tamai K, Iguchi-Ariga SM, et al. DJ-1, a novel oncogene which transforms mouse NIH3T3 cells in cooperation with ras. *Biochem Biophys Res Commun* 1997;**231**:509–13.

454. Wilson MA. The role of cysteine oxidation in DJ-1 function and dysfunction. *Antioxid Redox Signal* 2011;**15**:111–22.

455. Blackinton J, Lakshminarasimhan M, Thomas KJ, Ahmad R, Greggio E, Raza AS, et al. Formation of a stabilized cysteine sulfinic acid is critical for the mitochondrial function of the parkinsonism protein DJ-1. *J Biol Chem* 2009;**284**:6476–85.

456. Waak J, Weber SS, Gorner K, Schall C, Ichijo H, Stehle T, et al. Oxidizable residues mediating protein stability and cytoprotective interaction of DJ-1 with apoptosis signal-regulating kinase 1. *J Biol Chem* 2009;**284**:14245–57.

457. Zhang L, Shimoji M, Thomas B, Moore DJ, Yu SW, Marupudi NI, et al. Mitochondrial localization of the Parkinson's disease related protein DJ-1: implications for pathogenesis. *Hum Mol Genet* 2005;**14**:2063–73.

458. Ooe H, Taira T, Iguchi-Ariga SM, Ariga H. Induction of reactive oxygen species by bisphenol A and abrogation of bisphenol A-induced cell injury by DJ-1. *Toxicol Sci* 2005;**88**:114–26.

459. Blackinton J, Ahmad R, Miller DW, van der Brug MP, Canet-Aviles RM, Hague SM, et al. Effects of DJ-1 mutations and polymorphisms on protein stability and subcellular localization. *Brain Res Mol Brain Res* 2005;**134**:76–83.
460. Lev N, Ickowicz D, Melamed E, Offen D. Oxidative insults induce DJ-1 upregulation and redistribution: implications for neuroprotection. *Neurotoxicology* 2008;**29**:397–405.
461. Junn E, Jang WH, Zhao X, Jeong BS, Mouradian MM. Mitochondrial localization of DJ-1 leads to enhanced neuroprotection. *J Neurosci Res* 2009;**87**:123–9.
462. Hao LY, Giasson BI, Bonini NM. DJ-1 is critical for mitochondrial function and rescues PINK1 loss of function. *Proc Natl Acad Sci USA* 2010;**107**:9747–52.
463. Krebiehl G, Ruckerbauer S, Burbulla LF, Kieper N, Maurer B, Waak J, et al. Reduced basal autophagy and impaired mitochondrial dynamics due to loss of Parkinson's disease-associated protein DJ-1. *PLoS One* 2010;**5**:e9367.
464. Irrcher I, Aleyasin H, Seifert EL, Hewitt SJ, Chhabra S, Phillips M, et al. Loss of the Parkinson's disease-linked gene DJ-1 perturbs mitochondrial dynamics. *Hum Mol Genet* 2010;**19**:3734–46.
465. Goldberg MS, Fleming SM, Palacino JJ, Cepeda C, Lam HA, Bhatnagar A, et al. Parkin-deficient mice exhibit nigrostriatal deficits but not loss of dopaminergic neurons. *J Biol Chem* 2003;**278**:43628–35.
466. Von Coelln R, Thomas B, Savitt JM, Lim KL, Sasaki M, Hess EJ, et al. Loss of locus coeruleus neurons and reduced startle in parkin null mice. *Proc Natl Acad Sci USA* 2004;**101**:10744–9.
467. Dauer W, Kholodilov N, Vila M, Trillat AC, Goodchild R, Larsen KE, et al. Resistance of alpha-synuclein null mice to the parkinsonian neurotoxin MPTP. *Proc Natl Acad Sci USA* 2002;**99**:14524–9.
468. Ellis CE, Murphy EJ, Mitchell DC, Golovko MY, Scaglia F, Barcelo-Coblijn GC, et al. Mitochondrial lipid abnormality and electron transport chain impairment in mice lacking alpha-synuclein. *Mol Cell Biol* 2005;**25**:10190–201.
469. Hsu LJ, Sagara Y, Arroyo A, Rockenstein E, Sisk A, Mallory M, et al. Alpha-synuclein promotes mitochondrial deficit and oxidative stress. *Am J Pathol* 2000;**157**:401–10.
470. Orth M, Tabrizi SJ, Schapira AH, Cooper JM. Alpha-synuclein expression in HEK293 cells enhances the mitochondrial sensitivity to rotenone. *Neurosci Lett* 2003;**351**:29–32.
471. Parihar MS, Parihar A, Fujita M, Hashimoto M, Ghafourifar P. Mitochondrial association of alpha-synuclein causes oxidative stress. *Cell Mol Life Sci* 2008;**65**:1272–84.
472. Smith WW, Jiang H, Pei Z, Tanaka Y, Morita H, Sawa A, et al. Endoplasmic reticulum stress and mitochondrial cell death pathways mediate A53T mutant alpha-synuclein-induced toxicity. *Hum Mol Genet* 2005;**14**:3801–11.
473. Shavali S, Brown-Borg HM, Ebadi M, Porter J. Mitochondrial localization of alpha-synuclein protein in alpha-synuclein overexpressing cells. *Neurosci Lett* 2008;**439**:125–8.
474. Nakamura K, Nemani VM, Wallender EK, Kaehlcke K, Ott M, Edwards RH. Optical reporters for the conformation of alpha-synuclein reveal a specific interaction with mitochondria. *J Neurosci* 2008;**28**:12305–17.
475. Nakamura K, Nemani VM, Azarbal F, Skibinski G, Levy JM, Egami K, et al. Direct membrane association drives mitochondrial fission by the Parkinson disease-associated protein {alpha}-synuclein. *J Biol Chem* 2011;**286**:20710–26.
476. Kamp F, Exner N, Lutz AK, Wender N, Hegermann J, Brunner B, et al. Inhibition of mitochondrial fusion by alpha-synuclein is rescued by PINK1, Parkin and DJ-1. *EMBO J* 2011;**29**:3571–89.
477. Devi L, Raghavendran V, Prabhu BM, Avadhani NG, Anandatheerthavarada HK. Mitochondrial import and accumulation of alpha-synuclein impair complex I in human dopaminergic neuronal cultures and Parkinson disease brain. *J Biol Chem* 2008;**283**:9089–100.

478. Kowald A, Kirkwood TB. Evolution of the mitochondrial fusion-fission cycle and its role in aging. *Proc Natl Acad Sci USA* 2011;**108**:10237–42.

479. Chen H, Chan DC. Physiological functions of mitochondrial fusion. *Ann N Y Acad Sci* 2010;**1201**:21–5.

480. Cassidy-Stone A, Chipuk JE, Ingerman E, Song C, Yoo C, Kuwana T, et al. Chemical inhibition of the mitochondrial division dynamin reveals its role in Bax/Bak-dependent mitochondrial outer membrane permeabilization. *Dev Cell* 2008;**14**:193–204.

Huntington Disease and the Huntingtin Protein

Zhiqiang Zheng and
Marc I. Diamond

*Hope Center for Neurological Diseases,
Knight-Alzheimer Disease Research Center,
Department of Neurology, Washington
University in St. Louis, St. Louis, Missouri,
USA*

Huntington disease (HD) is a devastating neurodegenerative disease that derives from CAG repeat expansion in the huntingtin gene. The clinical syndrome consists of progressive personality changes, movement disorder, and dementia and can develop in children and adults. The huntingtin protein is required for human development and normal brain function. It is subject to posttranslational modification, and some events, such as phosphorylation, can play an enormous role in regulating toxicity of the huntingtin protein. The function of huntingtin in the cell is unknown, and it may play a role as a scaffold. Multiple mouse models of HD have now been created with fragments and full-length protein. The models show variable fidelity to the disease in

Progress in Molecular Biology
and Translational Science, Vol. 107
DOI: 10.1016/B978-0-12-385883-2.00010-2

189

terms of genetics, pathology, and rates of progression. Pathogenesis of HD involves cleavage of the protein and is associated with neuronal accumulation of aggregated forms. The potential mechanisms of neurodegeneration are myriad, including primary effects of protein homeostasis, gene expression, and mitochondrial dysfunction. Specific therapeutic approaches are similarly varied and include efforts to reduce huntingtin gene expression, protein accumulation, and protein aggregation.

I. Introduction

Huntington disease (HD) is an autosomal dominant neurodegenerative disorder caused by the expansion of the CAG codon repeat within the huntingtin gene. It affects approximately 7 per 100,000.[1] CAG encodes the amino acid glutamine, so the expansion results in an abnormally long glutamine tract within the N-terminus of the huntingtin protein (Htt). In addition to HD, eight other dominantly inherited neurodegenerative diseases have been described, each caused by CAG expansion. These disorders are collectively termed "polyglutamine" diseases. They are prototypical protein misfolding disorders—in all cases studied, the expanded polyglutamine tracts destabilize the native protein conformation. Penetrance is virtually 100%, with the exception of spinobulbar muscular atrophy, a motor neuron disease and sensory neuronopathy caused by polyglutamine expansion in the androgen receptor.[2-4] This disease afflicts only males because dihydrotestosterone initiates pathogenesis.[5]

II. Genetic and Clinical Features

Normal individuals have fewer than 36 CAG repeats in the huntingtin gene and most commonly about 15–25 repeats. Repeats between 27 and 35 are unstable during meiosis, which can lead to new expansion of CAG repeats into the disease range in subsequent generations. When repeats are present in the pathogenic range, repeat expansion inherited by children leads to "genetic anticipation," or earlier disease onset than was observed in the parent. This is usually seen when fathers pass the mutant gene to their children, which may be due to the fact that repeat instability is observed during spermatogenesis.[6] Virtually all individuals with 38 or more CAG repeats are predicted to develop HD symptoms if they live long enough. However, individuals with fewer than 40 repeats may only manifest symptoms at a late age.[7] The length of the expanded CAG repeats correlates inversely with the age at symptom onset.[8,9] Extremely long CAG repeats (>60) are highly associated with early onset HD, also known as juvenile HD. It is estimated that CAG repeat length accounts for approximately 70% of the variance in age of onset,[8,10] which means that other genetic and environmental factors also play an important role in determining pathogenesis.

HD is particularly devastating because of its multiple cognitive and motor features, which have a relentless course over 10–20 years. Juvenile HD is very rapid, and the affected children seldom survive into adulthood. The general features of each stage of disease are detailed below, but the timing and constellation of features are quite variable among patients. Early HD symptoms relate to frontal/executive dysfunction and include progressive personality changes, for example, irritability, and difficulty with multitasking and prioritization of activities. Psychiatric disturbances, especially depression, as well as slight involuntary movements and poor coordination may also be present.[1] In the middle phase of the disease, in addition to progressive cognitive disturbances, a movement disorder becomes more prominent, with frequent involuntary twitching, and writhing movements of the face, trunk, and extremities termed "chorea." Disturbances of gait manifest, with frequent falls.[11] A thought disorder may also appear, with bizarre ideation that can merge into frank psychosis.[12] In the later stages of the disease, chorea can become quite severe, and a hypermetabolic state produces progressive weight loss.[13] Difficulty in swallowing leads to choking and coughing episodes. At this point, the cognitive disturbance is severe enough to produce disability. Patients are unable to maintain their usual employment, and often require significant caretaking, if not institutionalization.[1] In the final stages, patients become bedbound and usually succumb to wasting and aspiration pneumonia.

The Htt protein is widely expressed and causes pathology throughout the brain. This has been readily observed using serial MRI in presymptomatic and symptomatic patients.[14] Striking atrophy occurs in the striatum, especially the caudate and putamen. Pathology here involves vulnerable GABAergic medium-sized spiny neurons.[15] Striatal atrophy develops long before symptom onset, with ≈50% reduction in striatal volume at the time of diagnosis.[16,17] Yet it is misleading pathologically to look solely to the striatum. HD also affects other brain regions, including cerebral cortex (layers III, V, and VI), hippocampus, thalamus, globus pallidus, subthalamic nucleus, substantia nigra, white matter, and cerebellum.[7,18] The involvement of so many brain structures helps explain the highly variable clinical presentations of the disease. The cloning of the Htt gene made it possible to recognize that Htt fragments accumulate as intracellular inclusions in multiple brain regions,[19] highlighting the role of protein misfolding and aggregation in this disorder.

III. Structural Features of the Huntingtin Protein

Htt and its homologs are evolutionarily conserved among many vertebrates and invertebrates. For example, the coding region of Htt from humans and the Fugu fish is 69% identical at the nucleotide level.[20] The gene-encoding human Htt spans over 180kb, with 67 exons. Human Htt is ubiquitously expressed, with the highest levels in the central nervous system.[21] The human Htt gene consists of two alternatively polyadenylated forms, which differ in size by 3kb. The longer transcript is predominantly expressed in brain, whereas the shorter one is more widely expressed.[22,23] Within the cell, Htt is associated with multiple organelles, including the nucleus, plasma membrane, endocytic and autophagic vesicles, endoplasmic reticulum, endosome, Golgi, and mitochondria.[24] This may reflect a role as a scaffold protein that links multiple factors and is consistent with its large, multidomain structure. Htt is required for normal embryonic development.[25,26]

Full-length Htt is composed of $\approx$3114 amino acids, producing a protein of $\approx$350kDa. Due to its size, it has not yet been possible to generate full-length Htt crystals. However, the structures of several domains have been described in relative detail. To elucidate the role of the N-terminal polyglutamine tract in normal Htt function, an ablation of the seven CAG repeats in the mouse gene was performed.[27] Development was normal (unlike knockout of the Htt gene, which is embryonic lethal), and adult mice exhibited only subtle defects in learning and memory. This suggests that the polyglutamine stretch contributes to the subtle regulation of neural function by Htt but is not required for Htt's essential functions.

Htt also contains HEAT (*Htt*, *e*longation factor 3, the PR65/*A* subunit of protein phosphatase 2A and the lipid kinase *T*or) repeats. HEAT domains consist of $\approx$50 amino acids and comprise two antiparallel α-helices forming a hairpin that is normally involved in protein–protein interactions. A total of 16 HEAT repeats have been identified in Htt, organized into four clusters.[28,29] The biological role of HEAT domains in Htt function is unknown.

Wild-type Htt is normally found in the cytoplasm, but mutant Htt accumulates within the nucleus, especially as N-terminal truncation fragments.[19,21,30–33] A conserved nuclear export signal (NES) in the C-terminus has been identified.[34] Another potential NES exists in the N-terminal 17 amino acid region that has been demonstrated to interact with the nuclear pore protein (TPR), which is involved in the export of proteins from nucleus into cytoplasm.[35]

Multiple proteolytic cleavage sites have been identified in Htt, which likely generate a wide range of N-terminal fragments. Putative Htt proteases include caspases, calpain, aspartic endopeptidases, and matrix metalloproteinases. Htt is cleaved by caspase 3 *in vitro* at amino acids 513 and 552,[36,37] caspase 2 at amino acids 552,[38] and caspase 6 at amino acids 586.[37] Two putative calpain sites are located at amino acids 469 and 536.[39,40] The matrix metalloproteinase, MMP-10, which is active in HD mice, cleaves Htt at amino acid 402.[41] Htt fragments with mutated cleavage sites, which are resistant to the cleavage by some of these enzymes, exhibit less toxicity *in vitro* and *in vivo*, suggesting that proteolytic processing of full-length Htt is probably an important event in the initiation of HD pathogenesis.[36,42,43]

Htt also has many posttranslational modifications, including ubiquitination, phosphorylation, acetylation, sumoylation, and palmitoylation. These could regulate its stability, localization, and function. Ubiquitination targets Htt for degradation.[30,44] Three lysines in its N-terminus, K6, K9 and K15, have been identified as sumoylation sites.[45] While ubiquitination of Htt may serve a protective role by increasing its clearance, Htt sumoylation at these lysines exacerbates polyglutamine-mediated neurodegeneration in a *Drosophila* model of HD.[45] Many phosphorylation sites in the Htt N-terminus also have been identified. Phosphorylation of threonine 3 may affect Htt aggregation and toxicity.[46] It has also been observed that the IkappaB kinase (IKK) complex phosphorylates Htt at serine 13 and 16, which can promote the modification of the adjacent lysines and enhance Htt clearance. Expansion of the polyglutamine tract may reduce Htt phosphorylation, helping promote accumulation of mutant Htt.[47] Additionally, the importance of these residues has been clearly shown in a full-length Htt mouse model based on bacterial artificial chromosome (BAC) expression. In this work, phosphomimic aspartate substitutions at serines 13 and 16 completely prevented polyglutamine-mediated toxicity, indicating that these amino acids may be very important determinants of pathogenesis.[48] Several other phosphorylation sites in Htt have been identified, including S421 by Akt; S434, S1181, and S1201 by Cdk5t.[49–52] Further studies, especially using phosphomimetic mutations in full-length mouse models, will be needed to clarify the role of phosphorylation at these sites in controlling Htt function and toxicity. Additionally, it has been shown that Htt acetylation at lysine 444 is required for degradation by the macroautophagy pathway.[53] Finally, palmitoylation at C214 by HIP14 may alter Htt aggregation. Polyglutamine tract expansion decreases the interaction between HIP14 and Htt, resulting in decreased palmitoylation, increased inclusions, and enhanced neurotoxicity (Fig. 1).[54]

IV. Htt Function(s)

Htt probably plays multiple roles in development and physiology. It is clearly essential for embryonic development, as Htt gene knockout is lethal in mice by embryonic day 8.5.[25,26] Increased cellular death in Htt knockout animals may indicate an anti-apoptotic role for Htt, which is supported by the

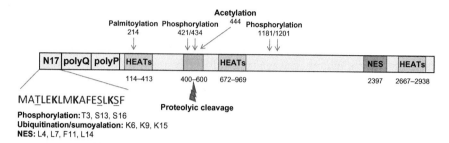

Fig. 1. Diagram of regions of the Htt protein. Full-length htt is composed of ~3144 amino acids, with a molecular weight of 350kDa. The N-terminal 17 amino acid (N17) region is highly conserved and contains multiple posttranslational modifications, such as phosphorylation and ubiquitination, which could strongly affect the pathogenesis. Following the N17 region are the polyglutamine (polyQ) and polyproline (polyP) regions. The function of polyproline is still largely unclear. Studies have suggested a hot spot of proteolyic cleavage exists between amino acids 400 and 600. Putative proteases include caspases 2, 3, 6, calpain, and matrix metalloproteinase 10. Up to 16 HEAT repeats with unknown functions have been identified in Htt. Multiple regions of phosphorylation have been identified. The C-terminus contains a leucine-rich nuclear export signal (NES), which may target full-length Htt to cytoplasm. Another leucine-rich NES could exist in the N17 region. (See Color Insert.)

findings that overexpression of full-length wild-type Htt in cultured striatal neurons protects them from apoptosis, whereas depletion of wild-type Htt makes neurons more sensitive to apoptotic insults.[55–57] The molecular mechanism of the anti-apoptotic function of Htt is unclear. One possible mechanism is that Htt can bind and prevent the formation of the pro-apoptotic Hip1-HIPPI (Hip1 protein interactor) complex.[58] Another possible mechanism is that Htt can directly block the activation caspase 3 and 9.[55,59]

Many studies have implicated Htt in transcription. Htt interacts with various transcription factors and transcriptional regulatory proteins. For example, Htt has been proposed to regulate the production of the brain-derived neurotrophic factor (BDNF).[60,61] BDNF transcription is suppressed by the binding of the repressor element-1 silencing transcription factor (REST)/neuro-restrictive silencer factor (NRSF) to its upstream DNA element neuron-restrictive silencer element.[62,63] Htt acts as positive regulator of BDNF transcription by preventing the recruitment of REST/NRSF to its DNA response element. Evidence suggests Htt may not interact directly with REST/NRSF, but rather forms a complex with HAP1 and RILP, which can directly bind REST/NRSF.[63] It is unclear which, if any, of the HD symptoms might result from the effect on BDNF.

By interacting with a variety of endocytic/trafficking proteins, such as α-adaptin, HIP1, HIP14, HAP1, HAP40, PACSIN1 and SH3GL3, clathrin, and dynamin, Htt also may participate in long- and short-range axonal transport and vesicle trafficking. For example, in addition to the role of transcriptional regulation of BDNF, full-length wild-type Htt also stimulates BDNF vesicular trafficking in neurons,[64] while knockdown of Htt reduces such transport.[65] Htt is enriched at synaptic terminals and interacts with cytoskeletal and synaptic vesicle proteins to regulate synaptic activity in neurons. By interacting with the SH3 domain of PSD95, a key protein in synaptic transmission, Htt might regulate synaptic plasticity.[66]

V. Animal Models of Huntington Disease

With the cloning of the HD gene, multiple model organisms, such as *Caenorhabditis elegans* (worm), *Drosophila* (fruit fly), and *Danio rerio* (zebrafish), have now been used to model polyglutamine toxicity.[67–71] These are advantageous in their rapid development of pathology and low cost. They also exhibit many features of the human disease, such as progressive neuronal cell death, reduced lifespan, aggregation of Htt, and formation of inclusions. These models are especially useful in the identification of pathways involved in HD and the screening of potential suppressors of HD pathology. However, it is still uncertain to what extent these will predict effects of chemical and genetic modifiers of pathology in mammals. Thus they are probably best used to identify candidates for further study in mammalian systems, rather than to vet candidates identified in mammalian systems.

Many mouse HD models now exist, both transgenic and knock-in. A list follows of the commonly used models.

A. R6/1 and R6/2

The transgenic R6/1 and R6/2 HD mouse lines contain ≈1kb of human HD promoter region followed by exon 1 of the Htt gene, carrying 115–156 CAG repeats and 262bp of intron 1.[19,72] R6/2 mice develop progressive neurological phenotypes, such as reduced brain and striatal volume, extensive intraneuronal inclusion formation, and relatively little cell death. Mice exhibit a resting tremor and shuddering movements as early as 5 weeks of age. They have progressive motor deficits and seizures. The phenotype is rapidly progressive, and these animals die between 12 and 18 weeks, depending on the laboratory setting and strain background. Their predictable onset and rapid course has made them a mainstay of preclinical drug trials. However, their distribution of pathology is not anatomically faithful to human disease, and thus they may serve best as a model of "polyglutaminopathy" (which is still very useful), rather than true HD. The R6/1 line has a more slowly progressive phenotype than the R6/2 line.

B. N171-82Q

The N171-82Q transgenic HD mouse line was generated by expressing a cDNA encoding an N-terminal fragment (171 amino acids) of huntingtin with 82 glutamines under the control of the mouse prion promoter.[73] Starting at the age of 10 weeks, these mice develop behavioral abnormalities including coordination loss, tremors, hypokinesis, and abnormal gait. Intranuclear inclusion bodies and neuritic aggregates are also observed in multiple types of neurons.

C. YAC Transgenic HD Models

Yeast artificial chromosome (YAC) transgenic HD mice were generated by expressing the full-length human Htt gene with varying numbers of CAG repeats.[74] YAC46 and YAC72 were first created. YAC72 mice show early electrophysiological abnormalities and selective degeneration of medium spiny neurons by 12 months. Interestingly, neurodegeneration in these mice occurs in the absence of obvious Htt inclusion formation. To create more rapid disease onset and reduce inter-animal variability, YAC128 was generated (128 CAG repeats).[33,75] This line exhibits uniform phenotypes as early as 3 months of age. A progressive motor deficit on the rotarod occurs at the age of 6 months. Severe neurodegeneration occurs, with striatal atrophy at 9 months, and cortical atrophy at 12 months of age.

D. BACHD

Recently, new full-length transgenic HD mice have been generated via the BAC technique. Like the YAC models, the BAC transgenic allows expression of the human Htt gene within the context of the surrounding genomic regulatory elements. The BAC readily enables modification of the Htt gene within this context, which facilitates the study of various mutants. In BACDHD mice, full-length human Htt with 97 CAG repeats is expressed under the control of endogenous Htt regulatory elements. These mice exhibit progressive motor deficits in rotarod performance starting as early as 2 months of age, neuronal synaptic dysfunction, and late-onset selective neurodegeneration in both striatum and cortex.[76] This model has also been used with great success to define the effect of specific N-terminal mutations in the Htt gene.[48]

E. HD Knock-in

In transgenic mice, the gene of interest is randomly inserted into the genome. This may interfere with the activity of other genes. Also, overexpression of the transgene may produce artifactual results. To overcome these drawbacks, "knock-in" HD mice have been generated in which the endogenous mouse Htt gene is replaced by a mutated Htt that carries expanded CAG repeats. The knock-in models are generally considered the most "faithful" to the human disease in that mutated Htt is placed in the appropriate context within the mouse genome and expression is controlled by the endogenous Htt promoter. However, knock-in HD mice generally exhibit subtle phenotypes. Neuropathology is observed at a very old age, even with the presence of a very long polyglutamine tract.[77–79] While these animals may be less useful for preclinical studies, they may be very useful for studies of more subtle HD phenotypes, for example, network dysfunction or early metabolic changes.

F. Other Mammalian HD Models

A transgenic rat HD model has been generated by expressing a truncated Htt fragment carrying 51 CAG repeats under the control of the endogenous rat Htt promoter.[80] This model exhibits adult-onset neurological phenotypes, slowly progressive motor dysfunction, and formation of intracellular inclusions in neurons. *In vivo* imaging has demonstrated significant striatal shrinkage, suggesting this model could be useful for preclinical imaging studies. A transgenic monkey model has been developed by expressing Htt exon 1 with 84 CAG repeats fused to the green fluorescent protein (GFP). This model shows extensive neuronal inclusions and mimics some important clinical features of HD, such as dystonia and chorea (Table I).[81]

VI. Pathogenesis of HD

A. Proteolytic Cleavage

The proteolysis of full-length of Htt, which results in N-terminal fragments containing the polyglutamine tract, appears to be a crucial step for progression of HD. The truncated N-terminal fragments tend to display more aggregation and neuronal toxicity than full-length Htt. The presence of these truncated N-terminal fragments in both HD mice and patients has been shown.[31,40,82] So far, a variety of proteases that cleave Htt have been identified, including caspases 2, 3, 6, 7, calpains, and aspartic endopeptidases. Some protease inhibitors, or mutations that block Htt cleavage, have been demonstrated to slow HD progression in mice.[37,42] Cleavage of Htt could produce a repertoire of fragments, each with a distinct propensity to cause disease. Further work will determine the specific role of these different truncated N-terminal fragments in HD progression, and to what extent a single protease can be targeted to slow pathogenesis.

B. Misfolding and Aggregation of Htt

The role of protein aggregation in the pathogenesis of HD is a confusing and controversial area. Htt *inclusions* are formed within cells. They comprise macromolecular aggregates of Htt, but the distribution of aggregate types and structures is poorly understood. Aggregates of many types have been

TABLE I

MOUSE MODELS OF HUNTINGTON DISEASE

Model	Promoter	Construct	Polyglutamine length	Age of onset	Neuropathology	Behavior phenotype
R6/1	Human Htt	~1.9kb fragment of the 5'-human Htt gene	115 for R6/1	5 months for R6/1	Intranuclear and neuropil aggregates throughout the brain; global brain atrophy; minimal cell death	Tremors and gait abnormalities; rotarod deficit; clasping behavior; learning deficit
R6/2			144 for R6/2	2 months for R6/2		
N171-82Q	Mouse prion	N-terminal 171 amino acids of human Htt	82	5 months	Inclusions in striatum, cortex, hippocampus and amygdala; degeneration in striatum	Tremors and gait abnormalities; rotarod deficit; loss of coordination; hypokinesis
YAC128	Human Htt	YAC expressing full-length human Htt	128	3 months	Inclusions in striatum; neuron loss in striatum.	Rotarod deficit; clasping; gait abnormalities; circling behavior
BACHD	Human Htt	BAC expressing full-length human Htt	97	3 months	Synaptic dysfunction; cortical and striatal atrophy.	Rotarod deficit
HdhQ92–111	Mouse Htt	Replace mouse Htt exon 1 with a human mutant exon 1	92–111	20 months	No brain atrophy and neuronal loss; gliosis in striatum	No obvious behavior deficit
HdhQ140	Mouse Htt	Replace the endogenous polyglutamine with expanded polyglutamine	140	12 months	Nuclear and neuropil inclusions in striatum, cortex, nucleus accumbens, and olfactory tubercle	Increased locomotor activity and rearing at 1 month of age; hypoactivity and gait abnormalities later
HdhQ150	Mouse Htt promoter	Replace the endogenous polyglutamine with expanded polyglutamine	150	4 months	Nuclear inclusions in striatum; striatal gliosis	Clasping behavior; gait abnormalities; rotarod deficit; hypoactivity

described. Fibrils are long, ordered, *para*-crystalline polymers of Htt peptides that are readily sedimented by centrifugation. Oligomers are small, ordered structures that may range in size from two to possibly hundreds of monomers. Other ordered structures also may exist (e.g., annuli). In addition, *amorphous aggregates* form that lack ordered structure. There is also likely to be considerable conformational diversity even among these structures. An emerging consensus holds that oligomers, rather than large fibrils or intracellular inclusions, are the toxic agents in HD. It is clear that Htt forms inclusions in patients, but their prevalence within affected regions is not completely certain.[19,83–85]

Expanded polyglutamine Htt is aggregation-prone *in vitro* and *in vivo*, and a similar polyglutamine length threshold exists for *in vitro* aggregation and for *in vivo* toxicity in patients.[86] Aggregation of Htt fragments depends highly on the length of polyglutamine, fragment size, and intracellular protein interactors.[87–90] Polyglutamine expansion could promote aggregation by facilitating a conformational transition to a novel aggregation-prone form of Htt. Either a monomer or an oligomer then could serve to nucleate (seed) aggregate growth, producing structures rich in β-sheets.[91]

It is very difficult to dissociate Htt cellular toxicity from its ability to aggregate using site-directed mutagenesis—mutations in Htt that block aggregation also block its toxicity.[92] Yet, it is clearly possible to dissociate *inclusion formation* from toxicity.[93] Although *in vivo* studies have shown an association of inclusions with HD progression,[72,94] several lines of evidence also suggest little correlation between inclusions and behavioral dysfunction or neuronal loss.[74–76] In addition, in some HD animal models, little inclusion formation is observed during HD progression, which has caused some to argue that aggregation may not be required for HD pathogenesis. However, it is likely that "microaggregates," or oligomers, are present before the appearance of large protein deposits. Emerging *in vitro* and *in vivo* evidence has shown the existence of such Htt oligomers, which appear to be more toxic than Htt fibrils, and are thus likely to be a principal pathogenic agent.[83,84,95,96]

C. Transcriptional Dysregulation

A number of studies have documented alterations of gene expression that are associated with HD pathogenesis. The challenge has been to sort out *pathogenic* changes, that is, those that, if corrected, would alter the course of disease, from *epiphenomena* which simply reflect metabolic changes in sick cells. Htt with expanded polyglutamine has been reported to bind and change the activity of many transcription factors and cofactors, including CREB-binding protein (CBP), p300/CBP-associated factor (p/CAF), p53, sp1, TAFII130, and PQBO1.[97,98] Htt binding may sequester these transcriptional factors/cofactors away from their normal DNA elements. For example, binding of mutated Htt to CBP leads to relocation of CBP from the nucleus into Htt inclusions.[99] Another well-characterized transcription factor, REST/NRSF, binds the NRSE within the BDNF promoter to suppress its transcription.[62,63] Wild-type Htt binds cytosolic REST/NRSF and prevents its translocation to the nucleus, whereas Htt with expanded polyglutamine interacts with REST/NRSF less effectively, resulting in the accumulation of REST/NRSF in the nucleus. Mutated Htt also occupies the promoter region of PGC-1α by associating with the CREB–TAF4 complex. This is reported to suppress expression of PGC-1α. Reduced PGC-1α mRNA expression has been found in both mouse and human HD brains and may partly account for the impaired energy metabolism observed in HD.[100] So far, it has not been possible to demonstrate a single gene or group of genes whose elimination or overexpression could replicate the HD phenotype, so that the question of whether HD truly derives from a primary genetic disturbance remains uncertain.

D. Mitochondrial Dysfunction

As with transcriptional dysregulation, it is not yet clear whether the mitochondrial dysfunction found in HD is a cause, or simply an epiphenomenon, of neuronal toxicity. Indeed, mitochondrial dysfunction and oxidative stress have been implicated in a variety of neurodegenerative diseases.[101,102] Mitochondria are an important source of reactive oxygen species (ROS) and also are a key target for ROS damage. ROS can cause peroxidation of lipids and oxidative damage of DNA by forming 8-hydroxy-2′-deoxyguanosine (8-OHdG). Increased striatal lipid peroxidation is found to parallel the neurological phenotype in HD transgenic mice.[103] Oxidative damage to DNA also is found in both HD patients and transgenic mice.[104,105] Mutated Htt has been reported to damage mitochondria by multiple mechanisms, including reduction of energy production and membrane potential, alteration in calcium homeostasis, enhancement of oxidative stress, and triggering of the caspase-mediated apoptotic pathways.[106–109] These mitochondrial defects might be caused by an interaction between mutant Htt and the outer membrane of mitochondria.[109–111] Htt could also indirectly impact mitochondrial biogenesis and function through effects on PGC-1α gene expression.

It has been demonstrated that mitochondrial energy metabolism is impaired in both HD patients and HD mouse models.[112,113] Decreased activity of several important enzymes involved in energy metabolism has been reported, including respiratory chain complexes I, II, III, and IV.[114–116] Reduction ATP production and energy failure could make striatal cells more vulnerable to other pathological insults. Mitochondria isolated from cells expressing expanded Htt also have decreased membrane potential. Neuronal cells derived from knock-in HD mice HdhQ111 exhibit high sensitivity to the calcium-induced permeability transition pore opening. It has also been demonstrated that an expanded N-terminal Htt fragment, but not wild-type Htt, significantly decreases the calcium concentration threshold necessary to trigger the mito-chondrial permeability transition pore opening and increases the propensity for calcium-induced cytochrome *c* release.[110] Mitochondria from clonal striatal cells expressing mutated Htt also show a reduced buffering capacity for calci-um uptake.[117]

E. Defects in Htt Clearance: UPS and Autophagy (See Chapter "Protein Quality Control in Neurodegenerative Disease")

The ubiquitin–proteasome system (UPS) mediates attachment of ubiquitin to target proteins and their subsequent degradation by the proteasome. Dys-function of the UPS in HD has been demonstrated.[30,118,119] This may be linked more broadly to general defects in protein quality control that are induced by a single misfolded species.[120] Htt aggregates are modified with ubiquitin and colocalize with proteasome subunits, suggesting that aggregated or misfolded Htt may be targeted for degradation by the UPS. It has been proposed that sequestration of UPS components into Htt inclusions alters their subcellular localization and thus the efficiency of the UPS.[19,30] Alternatively, polyglutamine-containing proteins may not be degraded efficiently within the proteasome, and expanded polyglutamine sequences trapped in the proteasome may block the entry of other substrates into the barrel of its catalytic core.[121] Finally, it has been proposed that inefficient degradation of expanded polyglutamine tracts may lead to the production of aggregation-prone fragments.[119,122] The importance of Htt degradation in the aggregation process is also linked to its effect on protein levels. Proteasome inhibitors increase the formation of Htt inclusions, whereas en-hancement of proteasome degradation by overexpression of heat shock proteins or chaperones decreases Htt aggregation.[123–125]

Multiple studies have now shown that autophagy also plays a role in the clearance of huntingtin aggregates.[126,127] Activation of autophagy, either by chemical treatments (e.g., rapamycin) or by overexpression of autophagy com-ponents, increases the clearance of mutated Htt and reduces its aggrega-tion.[126,128] Posttranslational modification of the mutant Htt by acetylation at lysine residue 444 (K444) also enhances its clearance by increasing Htt delivery to the autophagosome.[53]

VII. Therapeutic Targets for HD

There are no effective therapies for HD, and treatments are limited to controlling psychiatric symptoms and chorea. Various therapeutic strategies are now proposed but are still in their experimental stages.

A. Transcription

Alterations in gene expression in animal models of HD have led some to propose that correcting these deficits will rescue neuronal dysfunction and degeneration. BDNF is perhaps the most compelling target.[60,129,130] Likewise, disruption of PGC1-α transcription by Htt has also been proposed as a mechanism for defective mitochondrial function.[100,131,132] For most transcriptional changes, however, it remains unclear whether they represent a cause or consequence of HD pathology. In certain cases, gene expression alterations might in fact be adaptive. Mutant Htt has been proposed to disrupt the activity of many key transcriptional factors or transcription-regulatory proteins, generally leading to the downregulation of target gene expression. A number of compounds have been proposed to restore normal transcription activity in HD animal models. For example, HDAC inhibitors such as suberoylanilide hydroxamic acid (SAHA), sodium butyrate, and phenylbutyrate have been administered to experimental animals,[133–137] but the enzymatic targets of these compounds remain elusive, and may not be acetyl transferases that are directly related to transcription. Indeed, some studies suggest that HDAC inhibitors may be neuroprotective by increasing acetylation of other nonhistone proteins, such as tubulin,[138] or by increasing levels of heat shock proteins.[139] Other effects of HDAC inhibitors, such as prevention of Htt aggregation, anti-inflammation, and anti-apoptotic properties, may also mediate their neuroprotection in HD.[140,141] Further studies are needed to clarify the contributions of epigenetic modifications and nonhistone effects when applying HDAC inhibitors as HD therapies. HDAC inhibitors are of limited potential because of their high toxicity, and thus identification of more specific targets could improve their utility.

B. Mitochondrial Dysfunction

Mutant Htt, directly or indirectly, is associated with mitochondrial dysfunction, resulting in impaired energy production, increased intracellular calcium, oxidative stress, and apoptosis. Creatine, which stimulates mitochondrial respiration and has antioxidant properties, has alleviated symptoms in mouse HD models.[142,143] Coenzyme Q_{10}, which promotes energy production, is also neuroprotective in R6/2 HD mice.[144,145] However, clinical trials with these compounds did not show significant benefits in HD patients.[146,147] While PGC1-α, a transcriptional regulator of mitochondrial biogenesis and respiration, has been implicated as a potential target of HD therapy,[100,131,132] it is not clear yet how this pathway could be targeted specifically with small molecules.

C. Targeting Mutant Htt Protein

1. PROTEOLYSIS

If Htt proteolysis initiates pathogenesis, then compounds that decrease its cleavage could have therapeutic value. Caspase 6 has been implicated as a potential therapeutic target, since mutation of a cleavage site for this protease reduced toxicity of the full-length protein.[42] No inhibitors of this enzyme are yet in clinical trials. Minocycline had small effects when administered to R6/2 mice, improving lifespan.[148,149] This compound may inhibit caspase activity, although it has been shown to produce benefits only in a mouse model that expresses an Htt fragment, and thus it is unlikely to function via preventing Htt cleavage. Nonetheless, a clinical trial with minocycline is currently being conducted (www.huntington-study-group.com) because this is an FDA-approved drug. If specific proteases are clearly implicated in Htt proteolysis, this may create new opportunities to design more targeted therapies.

2. PROTEIN CLEARANCE

Activation of autophagy may enhance clearance of aggregated Htt and offer protection in HD.[150] A principal challenge in developing therapies will be to activate components of this pathway that are selective for protein aggregation, without inducing more general toxicity. The mTOR inhibitor rapamycin, which induces autophagy, has been shown to be beneficial in cell, *Drosophila*, and mouse HD models.[126,151,152] However, strong side effects, especially immuno-suppression, prevent its chronic use in HD patients. Recently, new compounds were described to selectively stimulate the clearance of mutant Htt; however, the underlying molecular mechanisms remain elusive.[153–155] Given the widespread role of protein aggregation in neurodegenerative diseases, this mode of therapy could be applicable more generally.

3. PROTEIN AGGREGATION

Much effort has been devoted to the discovery of compounds that inhibit Htt aggregation. Many compounds that block Htt aggregation directly or indirectly have been identified in cell-based assays and in *Drosophila* models.[156–159] However, effects of these compounds in HD mouse models are so far modest, at best.[156,157,160,161] This may be due to poor bioavailability, and in the case of compounds that stoichiometrically inhibit Htt aggregation via direct interaction, it may be especially difficult to achieve sufficient brain concentrations without concurrent toxicity. Induction of molecular chaperones, such as Hsp70, may promote protein refolding or degradation of mutant Htt.[162–164] It remains unclear, however, whether global induction of chaperones can be accomplished safely on a chronic basis. A better understanding of protein interactions and regulatory pathways that control Htt misfolding and aggregation could help create more effective therapies using this strategy, as the targets now are somewhat limited.

4. GENE THERAPY

Gene therapy using RNA interference (RNAi) or antisense oligonucleotides is promising for a variety of genetic disorders. Indeed, a phase I clinical trial using antisense oligonucleotides to treat patients suffering from amyotrophic lateral sclerosis due to mutation in SOD1 is now under way (http://www.alsa.org/patient/drug.cfm?id=1592). The use of shRNA- or siRNA-based knockdown of mutant Htt expression has been validated in HD mouse models,[165,166] as has the use of antisense oligonucleotides.[167] Several issues must be addressed before these approaches can be applied to patients. Off-target effects are probably more widespread than generally appreciated.[168,169] Delivery of siRNA or antisense oligonucleotides requires chronic administration into the cerebrospinal fluid (CSF) via catheter and whole brain penetration is not fully established for large animals. Finally, the safety of virus-based gene knockdown in the brain is not yet readily controlled or proven safe. Although there are still significant hurdles, the potential of genetic therapy in HD is nonetheless promising.

5. ARTIFICIAL PEPTIDES OR INTRABODIES

Intrabodies are recombinant antibody fragments that target intracellular Htt.[170–173] The protein specificity created by a peptide has allowed targeting of various domains of the Htt molecule, including the N-terminal region and the polyglutamine tract, to reduce aggregation and promote protein clearance.[174] This therapeutic strategy is still in a very early stage and will need more work to solve technical issues such as stability and specificity of intrabodies and optimal delivery methods, yet it could offer highly selective modulation of intracellular Htt clearance and aggregation.

VIII. Conclusion

Genetic and experimental evidence suggests that most polyglutamine proteins, including Htt, exert toxic effects as a result of an abnormal conformation induced by the polyglutamine expansion. This is likely distinct from their normal function and linked to aggregation and abnormal protein interactions. This contrasts other types of disease-causing mutations that accentuate or inhibit normal enzymatic activities. It is still unknown what Htt's crucial cellular functions are, and indeed, it is likely that the protein has several, given its large size and multiple conserved regions. Despite the fact that Htt's disease-causing activities represent an apparent gain of toxic function, its normal interactions within the cell will hold important clues to how this pathogenic activity is released, for example, via proteolysis, posttranslational modifications, or protein contacts that prevent or promote misfolding or regulate clearance. This has obvious implications for Htt-specific therapy. Genetic therapies will become more feasible as delivery systems evolve, while nonspecific therapies aimed at bolstering general cellular resistance to protein misfolding and metabolic defects may also play a role. The plethora of mouse models will facilitate preclinical testing, especially as our understanding develops about how best to use them. Thus, the treatment of HD is likely to depend on a detailed and evolving knowledge of Htt biology, coupled with interventions that target crucial events in the pathogenic cascade.

REFERENCES

1. Bates G, Harper P, Jones L, editors. *Huntington's disease*. New York: Oxford University Press; 2002.
2. Harding AE. Molecular genetics and clinical aspects of inherited disorders of nerve and muscle. *Curr Opin Neurol Neurosurg* 1992;**5**:600–4.
3. Kennedy WR, Alter M, Sung JH. Progressive proximal spinal and bulbar muscular atrophy of late onset. A sex-linked recessive trait. *Neurology* 1968;**18**:671–80.
4. La Spada AR, Wilson EM, Lubahn DB, Harding AE, Fischbeck KH. Androgen receptor gene mutations in X-linked spinal and bulbar muscular atrophy. *Nature* 1991;**352**:77–9.
5. Katsuno M, Adachi H, Kume A, Li M, Nakagomi Y, Niwa H, et al. Testosterone reduction prevents phenotypic expression in a transgenic mouse model of spinal and bulbar muscular atrophy. *Neuron* 2002;**35**:843–54.
6. Duyao M, Ambrose C, Myers R, Novelletto A, Persichetti F, Frontali M, et al. Trinucleotide repeat length instability and age of onset in Huntington's disease. *Nat Genet* 1993;**4**:387–92.
7. Walker FO. Huntington's disease. *Lancet* 2007;**369**:218–28.
8. Andrew SE, Goldberg YP, Kremer B, Telenius H, Theilmann J, Adam S, et al. The relationship between trinucleotide (CAG) repeat length and clinical features of Huntington's disease. *Nat Genet* 1993;**4**:398–403.
9. Rubinsztein DC, Barton DE, Davison BC, Ferguson-Smith MA. Analysis of the huntingtin gene reveals a trinucleotide-length polymorphism in the region of the gene that contains two CCG-rich stretches and a correlation between decreased age of onset of Huntington's disease and CAG repeat number. *Hum Mol Genet* 1993;**2**:1713–5.
10. Brinkman RR, Mezei MM, Theilmann J, Almqvist E, Hayden MR. The likelihood of being affected with Huntington disease by a particular age, for a specific CAG size. *Am J Hum Genet* 1997;**60**:1202–10.
11. Grimbergen YA, Knol MJ, Bloem BR, Kremer BP, Roos RA, Munneke M. Falls and gait disturbances in Huntington's disease. *Mov Disord* 2008;**23**:970–6.
12. van Duijn E, Kingma EM, van der Mast RC. Psychopathology in verified Huntington's disease gene carriers. *J Neuropsychiatry Clin Neurosci* 2007;**19**:441–8.
13. Politis M, Pavese N, Tai YF, Tabrizi SJ, Barker RA, Piccini P. Hypothalamic involvement in Huntington's disease: an in vivo PET study. *Brain* 2008;**131**:2860–9.

14. Kloppel S, Henley SM, Hobbs NZ, Wolf RC, Kassubek J, Tabrizi SJ, et al. Magnetic resonance imaging of Huntington's disease: preparing for clinical trials. *Neuroscience* 2009;**164**:205–19.
15. Mitchell IJ, Cooper AJ, Griffiths MR. The selective vulnerability of striatopallidal neurons. *Prog Neurobiol* 1999;**59**:691–719.
16. Aylward EH, Brandt J, Codori AM, Mangus RS, Barta PE, Harris GJ. Reduced basal ganglia volume associated with the gene for Huntington's disease in asymptomatic at-risk persons. *Neurology* 1994;**44**:823–8.
17. Aylward EH, Sparks BF, Field KM, Yallapragada V, Shpritz BD, Rosenblatt A, et al. Onset and rate of striatal atrophy in preclinical Huntington disease. *Neurology* 2004;**63**:66–72.
18. Vonsattel JP. Huntington disease models and human neuropathology: similarities and differences. *Acta Neuropathol* 2008;**115**:55–69.
19. Davies SW, Turmaine M, Cozens BA, DiFiglia M, Sharp AH, Ross CA, et al. Formation of neuronal intranuclear inclusions underlies the neurological dysfunction in mice transgenic for the HD mutation. *Cell* 1997;**90**:537–48.
20. Sathasivam K, Baxendale S, Mangiarini L, Bertaux F, Hetherington C, Kanazawa I, et al. Aberrant processing of the Fugu HD (FrHD) mRNA in mouse cells and in transgenic mice. *Hum Mol Genet* 1997;**6**:2141–9.
21. DiFiglia M, Sapp E, Chase K, Schwarz C, Meloni A, Young C, et al. Huntingtin is a cytoplasmic protein associated with vesicles in human and rat brain neurons. *Neuron* 1995;**14**:1075–81.
22. Lin B, Rommens JM, Graham RK, Kalchman M, MacDonald H, Nasir J, et al. Differential 3′ polyadenylation of the Huntington disease gene results in two mRNA species with variable tissue expression. *Hum Mol Genet* 1993;**2**:1541–5.
23. Lin B, Nasir J, MacDonald H, Hutchinson G, Graham RK, Rommens JM, et al. Sequence of the murine Huntington disease gene: evidence for conservation, alternate splicing and polymorphism in a triplet (CCG) repeat [corrected]. *Hum Mol Genet* 1994;**3**:85–92.
24. Trottier Y, Devys D, Imbert G, Saudou F, An I, Lutz Y, et al. Cellular localization of the Huntington's disease protein and discrimination of the normal and mutated form. *Nat Genet* 1995;**10**:104–10.
25. Nasir J, Floresco SB, O'Kusky JR, Diewert VM, Richman JM, Zeisler J, et al. Targeted disruption of the Huntington's disease gene results in embryonic lethality and behavioral and morphological changes in heterozygotes. *Cell* 1995;**81**:811–23.
26. Duyao MP, Auerbach AB, Ryan A, Persichetti F, Barnes GT, McNeil SM, et al. Inactivation of the mouse Huntington's disease gene homolog Hdh. *Science* 1995;**269**:407–10.
27. Clabough EB, Zeitlin SO. Deletion of the triplet repeat encoding polyglutamine within the mouse Huntington's disease gene results in subtle behavioral/motor phenotypes in vivo and elevated levels of ATP with cellular senescence in vitro. *Hum Mol Genet* 2006;**15**:607–23.
28. Andrade MA, Bork P. HEAT repeats in the Huntington's disease protein. *Nat Genet* 1995;**11**:115–6.
29. Tartari M, Gissi C, Lo Sardo V, Zuccato C, Picardi E, Pesole G, et al. Phylogenetic comparison of huntingtin homologues reveals the appearance of a primitive polyQ in sea urchin. *Mol Biol Evol* 2008;**25**:330–8.
30. DiFiglia M, Sapp E, Chase KO, Davies SW, Bates GP, Vonsattel JP, et al. Aggregation of huntingtin in neuronal intranuclear inclusions and dystrophic neurites in brain. *Science* 1997;**277**:1990–3.
31. Landles C, Sathasivam K, Weiss A, Woodman B, Moffitt H, Finkbeiner S, et al. Proteolysis of mutant huntingtin produces an exon 1 fragment that accumulates as an aggregated protein in neuronal nuclei in Huntington disease. *J Biol Chem* 2010;**285**:8808–23.
32. Becher MW, Kotzuk JA, Sharp AH, Davies SW, Bates GP, Price DL, et al. Intranuclear neuronal inclusions in Huntington's disease and dentatorubral and pallidoluysian atrophy:

correlation between the density of inclusions and IT15 CAG triplet repeat length. *Neurobiol Dis* 1998;4:387–97.

33. Van Raamsdonk JM, Murphy Z, Slow EJ, Leavitt BR, Hayden MR. Selective degeneration and nuclear localization of mutant huntingtin in the YAC128 mouse model of Huntington disease. *Hum Mol Genet* 2005;14:3823–35.

34. Xia J, Lee DH, Taylor J, Vandelft M, Truant R. Huntingtin contains a highly conserved nuclear export signal. *Hum Mol Genet* 2003;12:1393–403.

35. Cornett J, Cao F, Wang CE, Ross CA, Bates GP, Li SH, et al. Polyglutamine expansion of huntingtin impairs its nuclear export. *Nat Genet* 2005;37:198–204.

36. Wellington CL, Ellerby LM, Hackam AS, Margolis RL, Trifiro MA, Singaraja R, et al. Caspase cleavage of gene products associated with triplet expansion disorders generates truncated fragments containing the polyglutamine tract. *J Biol Chem* 1998;273:9158–67.

37. Wellington CL, Singaraja R, Ellerby L, Savill J, Roy S, Leavitt B, et al. Inhibiting caspase cleavage of huntingtin reduces toxicity and aggregate formation in neuronal and nonneuronal cells. *J Biol Chem* 2000;275:19831–8.

38. Hermel E, Gafni J, Propp SS, Leavitt BR, Wellington CL, Young JE, et al. Specific caspase interactions and amplification are involved in selective neuronal vulnerability in Huntington's disease. *Cell Death Differ* 2004;11:424–38.

39. Gafni J, Ellerby LM. Calpain activation in Huntington's disease. *J Neurosci* 2002;22:4842–9.

40. Kim YJ, Yi Y, Sapp E, Wang Y, Cuiffo B, Kegel KB, et al. Caspase 3-cleaved N-terminal fragments of wild-type and mutant huntingtin are present in normal and Huntington's disease brains, associate with membranes, and undergo calpain-dependent proteolysis. *Proc Natl Acad Sci USA* 2001;98:12784–9.

41. Miller JP, Holcomb J, Al-Ramahi I, de Haro M, Gafni J, Zhang N, et al. Matrix metalloproteinases are modifiers of huntingtin proteolysis and toxicity in Huntington's disease. *Neuron* 2010;67:199–212.

42. Graham RK, Deng Y, Slow EJ, Haigh B, Bissada N, Lu G, et al. Cleavage at the caspase-6 site is required for neuronal dysfunction and degeneration due to mutant huntingtin. *Cell* 2006;125:1179–91.

43. Gafni J, Hermel E, Young JE, Wellington CL, Hayden MR, Ellerby LM. Inhibition of calpain cleavage of huntingtin reduces toxicity: accumulation of calpain/caspase fragments in the nucleus. *J Biol Chem* 2004;279:20211–20.

44. Kalchman MA, Graham RK, Xia G, Koide HB, Hodgson JG, Graham KC, et al. Huntingtin is ubiquitinated and interacts with a specific ubiquitin-conjugating enzyme. *J Biol Chem* 1996;271:19385–94.

45. Steffan JS, Agrawal N, Pallos J, Rockabrand E, Trotman LC, Slepko N, et al. SUMO modification of Huntingtin and Huntington's disease pathology. *Science* 2004;304:100–4.

46. Aiken CT, Steffan JS, Guerrero CM, Khashwji H, Lukacsovich T, Simmons D, et al. Phosphorylation of threonine 3: implications for huntingtin aggregation and neurotoxicity. *J Biol Chem* 2009;284:29427–36.

47. Thompson LM, Aiken CT, Kaltenbach LS, Agrawal N, Illes K, Khoshnan A, et al. IKK phosphorylates Huntingtin and targets it for degradation by the proteasome and lysosome. *J Cell Biol* 2009;187:1083–99.

48. Gu X, Greiner ER, Mishra R, Kodali R, Osmand A, Finkbeiner S, et al. Serines 13 and 16 are critical determinants of full-length human mutant huntingtin induced disease pathogenesis in HD mice. *Neuron* 2009;64:828–40.

49. Humbert S, Bryson EA, Cordelieres FP, Connors NC, Datta SR, Finkbeiner S, et al. The IGF-1/Akt pathway is neuroprotective in Huntington's disease and involves Huntingtin phosphorylation by Akt. *Dev Cell* 2002;2:831–7.

50. Luo S, Vacher C, Davies JE, Rubinsztein DC. Cdk5 phosphorylation of huntingtin reduces its cleavage by caspases: implications for mutant huntingtin toxicity. *J Cell Biol* 2005;**169**:647–56.

51. Warby SC, Chan EY, Metzler M, Gan L, Singaraja RR, Crocker SF, et al. Huntingtin phosphorylation on serine 421 is significantly reduced in the striatum and by polyglutamine expansion in vivo. *Hum Mol Genet* 2005;**14**:1569–77.

52. Anne SL, Saudou F, Humbert S. Phosphorylation of huntingtin by cyclin-dependent kinase 5 is induced by DNA damage and regulates wild-type and mutant huntingtin toxicity in neurons. *J Neurosci* 2007;**27**:7318–28.

53. Jeong H, Then F, Melia Jr. TJ, Mazzulli JR, Cui L, Savas JN, et al. Acetylation targets mutant huntingtin to autophagosomes for degradation. *Cell* 2009;**137**:60–72.

54. Yanai A, Huang K, Kang R, Singaraja RR, Arstikaitis P, Gan L, et al. Palmitoylation of huntingtin by HIP14 is essential for its trafficking and function. *Nat Neurosci* 2006;**9**:824–31.

55. Rigamonti D, Bauer JH, De-Fraja C, Conti L, Sipione S, Sciorati C, et al. Wild-type huntingtin protects from apoptosis upstream of caspase-3. *J Neurosci* 2000;**20**:3705–13.

56. Zhang Y, Li M, Drozda M, Chen M, Ren S, Mejia Sanchez RO, et al. Depletion of wild-type huntingtin in mouse models of neurologic diseases. *J Neurochem* 2003;**87**:101–6.

57. Leavitt BR, van Raamsdonk JM, Shehadeh J, Fernandes H, Murphy Z, Graham RK, et al. Wild-type huntingtin protects neurons from excitotoxicity. *J Neurochem* 2006;**96**:1121–9.

58. Gervais FG, Singaraja R, Xanthoudakis S, Gutekunst CA, Leavitt BR, Metzler M, et al. Recruitment and activation of caspase-8 by the Huntingtin-interacting protein Hip-1 and a novel partner Hippi. *Nat Cell Biol* 2002;**4**:95–105.

59. Rigamonti D, Sipione S, Goffredo D, Zuccato C, Fossale E, Cattaneo E. Huntingtin's neuroprotective activity occurs via inhibition of procaspase-9 processing. *J Biol Chem* 2001;**276**:14545–8.

60. Zuccato C, Ciammola A, Rigamonti D, Leavitt BR, Goffredo D, Conti L, et al. Loss of huntingtin-mediated BDNF gene transcription in Huntington's disease. *Science* 2001;**293**:493–8.

61. Zuccato C, Cattaneo E. Brain-derived neurotrophic factor in neurodegenerative diseases. *Nat Rev Neurol* 2009;**5**:311–22.

62. Zuccato C, Belyaev N, Conforti P, Ooi L, Tartari M, Papadimou E, et al. Widespread disruption of repressor element-1 silencing transcription factor/neuron-restrictive silencer factor occupancy at its target genes in Huntington's disease. *J Neurosci* 2007;**27**:6972–83.

63. Zuccato C, Tartari M, Crotti A, Goffredo D, Valenza M, Conti L, et al. Huntingtin interacts with REST/NRSF to modulate the transcription of NRSE-controlled neuronal genes. *Nat Genet* 2003;**35**:76–83.

64. Gauthier LR, Charrin BC, Borrell-Pages M, Dompierre JP, Rangone H, Cordelieres FP, et al. Huntingtin controls neurotrophic support and survival of neurons by enhancing BDNF vesicular transport along microtubules. *Cell* 2004;**118**:127–38.

65. Her LS, Goldstein LS. Enhanced sensitivity of striatal neurons to axonal transport defects induced by mutant huntingtin. *J Neurosci* 2008;**28**:13662–72.

66. Smith R, Brundin P, Li JY. Synaptic dysfunction in Huntington's disease: a new perspective. *Cell Mol Life Sci* 2005;**62**:1901–12.

67. Faber PW, Alter JR, MacDonald ME, Hart AC. Polyglutamine-mediated dysfunction and apoptotic death of a Caenorhabditis elegans sensory neuron. *Proc Natl Acad Sci USA* 1999;**96**:179–84.

68. Parker JA, Connolly JB, Wellington C, Hayden M, Dausset J, Neri C. Expanded polyglutamines in Caenorhabditis elegans cause axonal abnormalities and severe dysfunction of PLM mechanosensory neurons without cell death. *Proc Natl Acad Sci USA* 2001;**98**:13318–23.

69. Gunawardena S, Her LS, Brusch RG, Laymon RA, Niesman IR, Gordesky-Gold B, et al. Disruption of axonal transport by loss of huntingtin or expression of pathogenic polyQ proteins in Drosophila. *Neuron* 2003;**40**:25–40.

70. Marsh JL, Thompson LM. Drosophila in the study of neurodegenerative disease. *Neuron* 2006;**52**:169–78.

71. Miller VM, Nelson RF, Gouvion CM, Williams A, Rodriguez Lebron E, Harper SQ, et al. CHIP suppresses polyglutamine aggregation and toxicity in vitro and in vivo. *J Neurosci* 2005;**25**:9152–61.

72. Mangiarini L, Sathasivam K, Seller M, Cozens B, Harper A, Hetherington C, et al. Exon 1 of the HD gene with an expanded CAG repeat is sufficient to cause a progressive neurological phenotype in transgenic mice. *Cell* 1996;**87**:493–506.

73. Schilling G, Becher MW, Sharp AH, Jinnah HA, Duan K, Kotzuk JA, et al. Intranuclear inclusions and neuritic aggregates in transgenic mice expressing a mutant N-terminal fragment of huntingtin. *Hum Mol Genet* 1999;**8**:397–407.

74. Hodgson JG, Agopyan N, Gutekunst CA, Leavitt BR, LePiane F, Singaraja R, et al. A YAC mouse model for Huntington's disease with full-length mutant huntingtin, cytoplasmic toxicity, and selective striatal neurodegeneration. *Neuron* 1999;**23**:181–92.

75. Slow EJ, van Raamsdonk J, Rogers D, Coleman SH, Graham RK, Deng Y, et al. Selective striatal neuronal loss in a YAC128 mouse model of Huntington disease. *Hum Mol Genet* 2003;**12**:1555–67.

76. Gray M, Shirasaki DI, Cepeda C, Andre VM, Wilburn B, Lu XH, et al. Full-length human mutant huntingtin with a stable polyglutamine repeat can elicit progressive and selective neuropathogenesis in BACHD mice. *J Neurosci* 2008;**28**:6182–95.

77. Wheeler VC, White JK, Gutekunst CA, Vrbanac V, Weaver M, Li XJ, et al. Long glutamine tracts cause nuclear localization of a novel form of huntingtin in medium spiny striatal neurons in HdhQ92 and HdhQ111 knock-in mice. *Hum Mol Genet* 2000;**9**:503–13.

78. Lin CH, Tallaksen-Greene S, Chien WM, Cearley JA, Jackson WS, Crouse AB, et al. Neurological abnormalities in a knock-in mouse model of Huntington's disease. *Hum Mol Genet* 2001;**10**:137–44.

79. Menalled LB, Sison JD, Dragatsis I, Zeitlin S, Chesselet MF. Time course of early motor and neuropathological anomalies in a knock-in mouse model of Huntington's disease with 140 CAG repeats. *J Comp Neurol* 2003;**465**:11–26.

80. von Horsten S, Schmitt I, Nguyen HP, Holzmann C, Schmidt T, Walther T, et al. Transgenic rat model of Huntington's disease. *Hum Mol Genet* 2003;**12**:617–24.

81. Yang SH, Cheng PH, Banta H, Piotrowska-Nitsche K, Yang JJ, Cheng EC, et al. Towards a transgenic model of Huntington's disease in a non-human primate. *Nature* 2008;**453**:921–4.

82. Wellington CL, Ellerby LM, Gutekunst CA, Rogers D, Warby S, Graham RK, et al. Caspase cleavage of mutant huntingtin precedes neurodegeneration in Huntington's disease. *J Neurosci* 2002;**22**:7862–72.

83. Sathasivam K, Lane A, Legleiter J, Warley A, Woodman B, Finkbeiner S, et al. Identical oligomeric and fibrillar structures captured from the brains of R6/2 and knock-in mouse models of Huntington's disease. *Hum Mol Genet* 2010;**19**:65–78.

84. Legleiter J, Mitchell E, Lotz GP, Sapp E, Ng C, DiFiglia M, et al. Mutant huntingtin fragments form oligomers in a polyglutamine length-dependent manner in vitro and in vivo. *J Biol Chem* 2010;**285**:14777–90.

85. Sapp E, Schwarz C, Chase K, Bhide PG, Young AB, Penney J, et al. Huntingtin localization in brains of normal and Huntington's disease patients. *Ann Neurol* 1997;**42**:604–12.

86. Wanker EE. Protein aggregation and pathogenesis of Huntington's disease: mechanisms and correlations. *Biol Chem* 2000;**381**:937–42.

87. Martindale D, Hackam A, Wieczorek A, Ellerby L, Wellington C, McCutcheon K, et al. Length of huntingtin and its polyglutamine tract influences localization and frequency of intracellular aggregates. *Nat Genet* 1998;**18**:150–4.

88. Hackam AS, Singaraja R, Wellington CL, Metzler M, McCutcheon K, Zhang TQ, et al. The influence of Huntingtin protein size on nuclear localization and cellular toxicity. *J Cell Biol* 1998;**141**:1097–105.

89. Li SH, Li XJ. Aggregation of N-terminal huntingtin is dependent on the length of its glutamine repeats. *Hum Mol Genet* 1998;**7**:777–82.

90. Chen SM, Ferrone FA, Wetzel R. Huntington's disease age-of-onset linked to polyglutamine aggregation nucleation. *Proc Natl Acad Sci USA* 2002;**99**:11884–9.

91. Zuccato C, Valenza M, Cattaneo E. Molecular mechanisms and potential therapeutical targets in Huntington's disease. *Physiol Rev* 2010;**90**:905–81.

92. Ross CA, Poirier MA. Opinion: what is the role of protein aggregation in neurodegeneration? *Nat Rev Mol Cell Biol* 2005;**6**:891–8.

93. Arrasate M, Mitra S, Schweitzer ES, Segal MR, Finkbeiner S. Inclusion body formation reduces levels of mutant huntingtin and the risk of neuronal death. *Nature* 2004;**431**:805–10.

94. Yamamoto A, Lucas JJ, Hen R. Reversal of neuropathology and motor dysfunction in a conditional model of Huntington's disease. *Cell* 2000;**101**:57–66.

95. Poirier MA, Li H, Macosko J, Cai S, Amzel M, Ross CA. Huntingtin spheroids and protofibrils as precursors in polyglutamine fibrilization. *J Biol Chem* 2002;**277**:41032–7.

96. Wacker JL, Zareie MH, Fong H, Sarikaya M, Muchowski PJ. Hsp70 and Hsp40 attenuate formation of spherical and annular polyglutamine oligomers by partitioning monomer. *Nat Struct Mol Biol* 2004;**11**:1215–22.

97. Dunah AW, Jeong H, Griffin A, Kim YM, Standaert DG, Hersch SM, et al. Sp1 and TAFII130 transcriptional activity disrupted in early Huntington's disease. *Science* 2002;**296**:2238–43.

98. Okazawa H. Polyglutamine diseases: a transcription disorder? *Cell Mol Life Sci* 2003;**60**:1427–39.

99. Nucifora Jr. FC, Sasaki M, Peters MF, Huang H, Cooper JK, Yamada M, et al. Interference by huntingtin and atrophin-1 with cbp-mediated transcription leading to cellular toxicity. *Science* 2001;**291**:2423–8.

100. Cui L, Jeong H, Borovecki F, Parkhurst CN, Tanese N, Krainc D. Transcriptional repression of PGC-1alpha by mutant huntingtin leads to mitochondrial dysfunction and neurodegeneration. *Cell* 2006;**127**:59–69.

101. Trushina E, McMurray CT. Oxidative stress and mitochondrial dysfunction in neurodegenerative diseases. *Neuroscience* 2007;**145**:1233–48.

102. Moreira PI, Zhu X, Wang X, Lee HG, Nunomura A, Petersen RB, et al. Mitochondria: a therapeutic target in neurodegeneration. *Biochim Biophys Acta* 2010;**1802**:212–20.

103. Perez-Severiano F, Rios C, Segovia J. Striatal oxidative damage parallels the expression of a neurological phenotype in mice transgenic for the mutation of Huntington's disease. *Brain Res* 2000;**862**:234–7.

104. Polidori MC, Mecocci P, Browne SE, Senin U, Beal MF. Oxidative damage to mitochondrial DNA in Huntington's disease parietal cortex. *Neurosci Lett* 1999;**272**:53–6.

105. Bogdanov MB, Andreassen OA, Dedeoglu A, Ferrante RJ, Beal MF. Increased oxidative damage to DNA in a transgenic mouse model of Huntington's disease. *J Neurochem* 2001;**79**:1246–9.

106. Browne SE, Bowling AC, MacGarvey U, Baik MJ, Berger SC, Muqit MM, et al. Oxidative damage and metabolic dysfunction in Huntington's disease: selective vulnerability of the basal ganglia. *Ann Neurol* 1997;**41**:646–53.

107. Browne SE, Ferrante RJ, Beal MF. Oxidative stress in Huntington's disease. *Brain Pathol* 1999;**9**:147–63.

108. Brustovetsky N, LaFrance R, Purl KJ, Brustovetsky T, Keene CD, Low WC, et al. Age-dependent changes in the calcium sensitivity of striatal mitochondria in mouse models of Huntington's disease. *J Neurochem* 2005;**93**:1361–70.

109. Panov AV, Gutekunst CA, Leavitt BR, Hayden MR, Burke JR, Strittmatter WJ, et al. Early mitochondrial calcium defects in Huntington's disease are a direct effect of polyglutamines. *Nat Neurosci* 2002;**5**:731–6.

110. Choo YS, Johnson GV, MacDonald M, Detloff PJ, Lesort M. Mutant huntingtin directly increases susceptibility of mitochondria to the calcium-induced permeability transition and cytochrome c release. *Hum Mol Genet* 2004;**13**:1407–20.

111. Orr AL, Li S, Wang CE, Li H, Wang J, Rong J, et al. N-terminal mutant huntingtin associates with mitochondria and impairs mitochondrial trafficking. *J Neurosci* 2008;**28**:2783–92.

112. Seong IS, Ivanova E, Lee JM, Choo YS, Fossale E, Anderson M, et al. HD CAG repeat implicates a dominant property of huntingtin in mitochondrial energy metabolism. *Hum Mol Genet* 2005;**14**:2871–80.

113. Jenkins BG, Koroshetz WJ, Beal MF, Rosen BR. Evidence for impairment of energy metabolism in vivo in Huntington's disease using localized 1H NMR spectroscopy. *Neurology* 1993;**43**:2689–95.

114. Arenas J, Campos Y, Ribacoba R, Martin MA, Rubio JC, Ablanedo P, et al. Complex I defect in muscle from patients with Huntington's disease. *Ann Neurol* 1998;**43**:397–400.

115. Browne SE. Mitochondria and Huntington's disease pathogenesis: insight from genetic and chemical models. *Ann N Y Acad Sci* 2008;**1147**:358–82.

116. Gu M, Gash MT, Mann VM, Javoy-Agid F, Cooper JM, Schapira AH. Mitochondrial defect in Huntington's disease caudate nucleus. *Ann Neurol* 1996;**39**:385–9.

117. Milakovic T, Quintanilla RA, Johnson GV. Mutant huntingtin expression induces mitochondrial calcium handling defects in clonal striatal cells: functional consequences. *J Biol Chem* 2006;**281**:34785–95.

118. Bence NF, Sampat RM, Kopito RR. Impairment of the ubiquitin-proteasome system by protein aggregation. *Science* 2001;**292**:1552–5.

119. Holmberg CI, Staniszewski KE, Mensah KN, Matouschek A, Morimoto RI. Inefficient degradation of truncated polyglutamine proteins by the proteasome. *EMBO J* 2004;**23**:4307–18.

120. Gidalevitz T, Ben-Zvi A, Ho KH, Brignull HR, Morimoto RI. Progressive disruption of cellular protein folding in models of polyglutamine diseases. *Science* 2006;**311**:1471–4.

121. Imarisio S, Carmichael J, Korolchuk V, Chen CW, Saiki S, Rose C, et al. Huntington's disease: from pathology and genetics to potential therapies. *Biochem J* 2008;**412**:191–209.

122. Venkatraman P, Wetzel R, Tanaka M, Nukina N, Goldberg AL. Eukaryotic proteasomes cannot digest polyglutamine sequences and release them during degradation of polyglutamine-containing proteins. *Mol Cell* 2004;**14**:95–104.

123. Carmichael J, Chatellier J, Woolfson A, Milstein C, Fersht AR, Rubinsztein DC. Bacterial and yeast chaperones reduce both aggregate formation and cell death in mammalian cell models of Huntington's disease. *Proc Natl Acad Sci USA* 2000;**97**:9701–5.

124. Warrick JM, Chan HY, Gray-Board GL, Chai Y, Paulson HL, Bonini NM. Suppression of polyglutamine-mediated neurodegeneration in Drosophila by the molecular chaperone HSP70. *Nat Genet* 1999;**23**:425–8.

125. Wyttenbach A, Carmichael J, Swartz J, Furlong RA, Narain Y, Rankin J, et al. Effects of heat shock, heat shock protein 40 (HDJ-2), and proteasome inhibition on protein aggregation in cellular models of Huntington's disease. *Proc Natl Acad Sci USA* 2000;**97**:2898–903.

126. Ravikumar B, Vacher C, Berger Z, Davies JE, Luo S, Oroz LG, et al. Inhibition of mTOR induces autophagy and reduces toxicity of polyglutamine expansions in fly and mouse models of Huntington disease. *Nat Genet* 2004;**36**:585–95.

127. Sarkar S, Rubinsztein DC. Huntington's disease: degradation of mutant huntingtin by autophagy. *FEBS J* 2008;**275**:4263–70.
128. Jia K, Hart AC, Levine B. Autophagy genes protect against disease caused by polyglutamine expansion proteins in Caenorhabditis elegans. *Autophagy* 2007;**3**:21–5.
129. Xie Y, Hayden MR, Xu B. BDNF overexpression in the forebrain rescues Huntington's disease phenotypes in YAC128 mice. *J Neurosci* 2010;**30**:14708–18.
130. Lynch G, Kramar EA, Rex CS, Jia Y, Chappas D, Gall CM, et al. Brain-derived neurotrophic factor restores synaptic plasticity in a knock-in mouse model of Huntington's disease. *J Neurosci* 2007;**27**:4424–34.
131. St-Pierre J, Drori S, Uldry M, Silvaggi JM, Rhee J, Jager S, et al. Suppression of reactive oxygen species and neurodegeneration by the PGC-1 transcriptional coactivators. *Cell* 2006;**127**:397–408.
132. Weydt P, Pineda VV, Torrence AE, Libby RT, Satterfield TF, Lazarowski ER, et al. Thermoregulatory and metabolic defects in Huntington's disease transgenic mice implicate PGC-1alpha in Huntington's disease neurodegeneration. *Cell Metab* 2006;**4**:349–62.
133. Minamiyama M, Katsuno M, Adachi H, Waza M, Sang C, Kobayashi Y, et al. Sodium butyrate ameliorates phenotypic expression in a transgenic mouse model of spinal and bulbar muscular atrophy. *Hum Mol Genet* 2004;**13**:1183–92.
134. Hockly E, Richon VM, Woodman B, Smith DL, Zhou X, Rosa E, et al. Suberoylanilide hydroxamic acid, a histone deacetylase inhibitor, ameliorates motor deficits in a mouse model of Huntington's disease. *Proc Natl Acad Sci USA* 2003;**100**:2041–6.
135. Ferrante RJ, Kubilus JK, Lee J, Ryu H, Beesen A, Zucker B, et al. Histone deacetylase inhibition by sodium butyrate chemotherapy ameliorates the neurodegenerative phenotype in Huntington's disease mice. *J Neurosci* 2003;**23**:9418–27.
136. Gardian G, Browne SE, Choi DK, Klivenyi P, Gregorio J, Kubilus JK, et al. Neuroprotective effects of phenylbutyrate in the N171-82Q transgenic mouse model of Huntington's disease. *J Biol Chem* 2005;**280**:556–63.
137. Steffan JS, Bodai L, Pallos J, Poelman M, McCampbell A, Apostol BL, et al. Histone deacetylase inhibitors arrest polyglutamine-dependent neurodegeneration in Drosophila. *Nature* 2001;**413**:739–43.
138. Hubbert C, Guardiola A, Shao R, Kawaguchi Y, Ito A, Nixon A, et al. HDAC6 is a microtubule-associated deacetylase. *Nature* 2002;**417**:455–8.
139. Kovacs JJ, Murphy PJ, Gaillard S, Zhao X, Wu JT, Nicchitta CV, et al. HDAC6 regulates Hsp90 acetylation and chaperone-dependent activation of glucocorticoid receptor. *Mol Cell* 2005;**18**:601–7.
140. Iwata A, Riley BE, Johnston JA, Kopito RR. HDAC6 and microtubules are required for autophagic degradation of aggregated huntingtin. *J Biol Chem* 2005;**280**:40282–92.
141. Giorgini F, Moller T, Kwan W, Zwilling D, Wacker JL, Hong S, et al. Histone deacetylase inhibition modulates kynurenine pathway activation in yeast, microglia, and mice expressing a mutant huntingtin fragment. *J Biol Chem* 2008;**283**:7390–400.
142. Andreassen OA, Dedeoglu A, Ferrante RJ, Jenkins BG, Ferrante KL, Thomas M, et al. Creatine increase survival and delays motor symptoms in a transgenic animal model of Huntington's disease. *Neurobiol Dis* 2001;**8**:479–91.
143. Ferrante RJ, Andreassen OA, Jenkins BG, Dedeoglu A, Kuemmerle S, Kubilus JK, et al. Neuroprotective effects of creatine in a transgenic mouse model of Huntington's disease. *J Neurosci* 2000;**20**:4389–97.
144. Ferrante RJ, Andreassen OA, Dedeoglu A, Ferrante KL, Jenkins BG, Hersch SM, et al. Therapeutic effects of coenzyme Q10 and remacemide in transgenic mouse models of Huntington's disease. *J Neurosci* 2002;**22**:1592–9.

145. Schilling G, Coonfield ML, Ross CA, Borchelt DR. Coenzyme Q10 and remacemide hydrochloride ameliorate motor deficits in a Huntington's disease transgenic mouse model. *Neurosci Lett* 2001;**315**:149-53.

146. Huntington Study Group. A randomized, placebo-controlled trial of coenzyme Q10 and remacemide in Huntington's disease. *Neurology* 2001;**57**:397-404.

147. Verbessem P, Lemiere J, Eijnde BO, Swinnen S, Vanhees L, Van Leemputte M, et al. Creatine supplementation in Huntington's disease: a placebo-controlled pilot trial. *Neurology* 2003;**61**:925-30.

148. Chen M, Ona VO, Li M, Ferrante RJ, Fink KB, Zhu S, et al. Minocycline inhibits caspase-1 and caspase-3 expression and delays mortality in a transgenic mouse model of Huntington disease. *Nat Med* 2000;**6**:797-801.

149. Wang X, Zhu S, Drozda M, Zhang W, Stavrovskaya IG, Cattaneo E, et al. Minocycline inhibits caspase-independent and -dependent mitochondrial cell death pathways in models of Huntington's disease. *Proc Natl Acad Sci USA* 2003;**100**:10483-7.

150. Rubinsztein DC, Gestwicki JE, Murphy LO, Klionsky DJ. Potential therapeutic applications of autophagy. *Nat Rev Drug Discov* 2007;**6**:304-12.

151. Ravikumar B, Duden R, Rubinsztein DC. Aggregate-prone proteins with polyglutamine and polyalanine expansions are degraded by autophagy. *Hum Mol Genet* 2002;**11**:1107-17.

152. Berger Z, Ravikumar B, Menzies FM, Oroz LG, Underwood BR, Pangalos MN, et al. Rapamycin alleviates toxicity of different aggregate-prone proteins. *Hum Mol Genet* 2006;**15**:433-42.

153. Coufal M, Maxwell MM, Russel DE, Amore AM, Altmann SM, Hollingsworth ZR, et al. Discovery of a novel small-molecule targeting selective clearance of mutant huntingtin fragments. *J Biomol Screen* 2007;**12**:351-60.

154. Sarkar S, Perlstein EO, Imarisio S, Pineau S, Cordenier A, Maglathlin RL, et al. Small molecules enhance autophagy and reduce toxicity in Huntington's disease models. *Nat Chem Biol* 2007;**3**:331-8.

155. Sarkar S, Krishna G, Imarisio S, Saiki S, O'Kane CJ, Rubinsztein DC. A rational mechanism for combination treatment of Huntington's disease using lithium and rapamycin. *Hum Mol Genet* 2008;**17**:170-8.

156. Heiser V, Scherzinger E, Boeddrich A, Nordhoff E, Lurz R, Schugardt N, et al. Inhibition of huntingtin fibrillogenesis by specific antibodies and small molecules: implications for Huntington's disease therapy. *Proc Natl Acad Sci USA* 2000;**97**:6739-44.

157. Zhang X, Smith DL, Meriin AB, Engemann S, Russel DE, Roark M, et al. A potent small molecule inhibits polyglutamine aggregation in Huntington's disease neurons and suppresses neurodegeneration in vivo. *Proc Natl Acad Sci USA* 2005;**102**:892-7.

158. Desai UA, Pallos J, Ma AA, Stockwell BR, Thompson LM, Marsh JL, et al. Biologically active molecules that reduce polyglutamine aggregation and toxicity. *Hum Mol Genet* 2006;**15**:2114-24.

159. Pollitt SK, Pallos J, Shao J, Desai UA, Ma AA, Thompson LM, et al. A rapid cellular FRET assay of polyglutamine aggregation identifies a novel inhibitor. *Neuron* 2003;**40**:685-94.

160. Li M, Huang Y, Ma AA, Lin E, Diamond MI. Y-27632 improves rotarod performance and reduces huntingtin levels in R6/2 mice. *Neurobiol Dis* 2009;**36**:413-20.

161. Tanaka M, Machida Y, Niu S, Ikeda T, Jana NR, Doi H, et al. Trehalose alleviates polyglutamine-mediated pathology in a mouse model of Huntington disease. *Nat Med* 2004;**10**:148-54.

162. Sittler A, Lurz R, Lueder G, Priller J, Lehrach H, Hayer-Hartl MK, et al. Geldanamycin activates a heat shock response and inhibits huntingtin aggregation in a cell culture model of Huntington's disease. *Hum Mol Genet* 2001;**10**:1307-15.

163. Hay DG, Sathasivam K, Tobaben S, Stahl B, Marber M, Mestril R, et al. Progressive decrease in chaperone protein levels in a mouse model of Huntington's disease and induction of stress proteins as a therapeutic approach. *Hum Mol Genet* 2004;**13**:1389–405.

164. Katsuno M, Sang C, Adachi H, Minamiyama M, Waza M, Tanaka F, et al. Pharmacological induction of heat-shock proteins alleviates polyglutamine-mediated motor neuron disease. *Proc Natl Acad Sci USA* 2005;**102**:16801–6.

165. Xia H, Mao Q, Eliason SL, Harper SQ, Martins IH, Orr HT, et al. RNAi suppresses polyglutamine-induced neurodegeneration in a model of spinocerebellar ataxia. *Nat Med* 2004;**10**:816–20.

166. Harper SQ, Staber PD, He X, Eliason SL, Martins IH, Mao Q, et al. RNA interference improves motor and neuropathological abnormalities in a Huntington's disease mouse model. *Proc Natl Acad Sci USA* 2005;**102**:5820–5.

167. Hu J, Matsui M, Gagnon KT, Schwartz JC, Gabillet S, Arar K, et al. Allele-specific silencing of mutant huntingtin and ataxin-3 genes by targeting expanded CAG repeats in mRNAs. *Nat Biotechnol* 2009;**27**:478–84.

168. Denovan-Wright EM, Rodriguez-Lebron E, Lewin AS, Mandel RJ. Unexpected off-targeting effects of anti-huntingtin ribozymes and siRNA in vivo. *Neurobiol Dis* 2008;**29**:446–55.

169. Jackson AL, Bartz SR, Schelter J, Kobayashi SV, Burchard J, Mao M, et al. Expression profiling reveals off-target gene regulation by RNAi. *Nat Biotechnol* 2003;**21**:635–7.

170. Colby DW, Chu Y, Cassady JP, Duennwald M, Zazulak H, Webster JM, et al. Potent inhibition of huntingtin aggregation and cytotoxicity by a disulfide bond-free single-domain intracellular antibody. *Proc Natl Acad Sci USA* 2004;**101**:17616–21.

171. Khoshnan A, Ko J, Patterson PH. Effects of intracellular expression of anti-huntingtin antibodies of various specificities on mutant huntingtin aggregation and toxicity. *Proc Natl Acad Sci USA* 2002;**99**:1002–7.

172. McLear JA, Lebrecht D, Messer A, Wolfgang WJ. Combinational approach of intrabody with enhanced Hsp70 expression addresses multiple pathologies in a fly model of Huntington's disease. *FASEB J* 2008;**22**:2003–11.

173. Wang CE, Zhou H, McGuire JR, Cerullo V, Lee B, Li SH, et al. Suppression of neuropil aggregates and neurological symptoms by an intracellular antibody implicates the cytoplasmic toxicity of mutant huntingtin. *J Cell Biol* 2008;**181**:803–16.

174. Kazantsev A, Walker HA, Slepko N, Bear JE, Preisinger E, Steffan JS, et al. A bivalent Huntingtin binding peptide suppresses polyglutamine aggregation and pathogenesis in Drosophila. *Nat Genet* 2002;**30**:367–76.

The Complex Molecular Biology of Amyotrophic Lateral Sclerosis (ALS)

Rachel L. Redler[*,†,‡] and
Nikolay V. Dokholyan[*,†,‡]

[*]Department of Biochemistry and
Biophysics, University of North Carolina,
Chapel Hill, North Carolina, USA

[†]Curriculum in Bioinformatics and
Computational Biology, University of North
Carolina, Chapel Hill, North Carolina, USA

[‡]Center for Computational and Systems
Biology, University of North Carolina,
Chapel Hill, North Carolina, USA

Amyotrophic lateral sclerosis (ALS) is an adult-onset neurodegenerative disorder that causes selective death of motor neurons followed by paralysis and death. A subset of ALS cases is caused by mutations in the gene for Cu, Zn superoxide dismutase (SOD1), which impart a toxic gain of function to this antioxidant enzyme. This neurotoxic property is widely believed to stem from an increased propensity to misfold and aggregate caused by decreased stability of the native homodimer or a tendency to lose stabilizing posttranslational modifications. Study of the molecular mechanisms of SOD1-related ALS has revealed a complex array of interconnected pathological processes, including glutamate excitotoxicity, dysregulation of neurotrophic factors and axon guidance

Progress in Molecular Biology
and Translational Science, Vol. 107
DOI: 10.1016/B978-0-12-385883-2.00002-3

215

proteins, axonal transport defects, mitochondrial dysfunction, deficient protein quality control, and aberrant RNA processing. Many of these pathologies are directly exacerbated by misfolded and aggregated SOD1 and/or cytosolic calcium overload, suggesting the primacy of these events in disease etiology and their potential as targets for therapeutic intervention.

I. ALS Is a Deadly Neurodegenerative Disorder

Amyotrophic lateral sclerosis (ALS) was first described by the noted French neurologist Jean-Martin Charcot in 1869, who connected the progressive paralytic syndrome with lesions in both white and gray matter of the central nervous system (CNS).[1] Over 140 years later, ALS is the most common adult-onset motor neuron disorder, affecting approximately 1–2 per 100,000 people worldwide. Considering the short course of disease progression (death/ tracheotomy typically within 2–5 years of diagnosis), 1 in every 800 individuals is expected to face ALS in his/her lifetime.[2–4]

As described by Charcot, ALS involves degeneration of the upper motor neurons (UMN) of the motor cortex and of the lower motor neurons (LMN), which extend through the brainstem and spinal cord to innervate skeletal muscle. Though the upper and lower motor systems are known to be interconnected, controlling voluntary muscle movement in concert, the primary site of dysfunction in ALS has long been a source of debate.[5–7] Questions of UMN/LMN primacy aside, ALS is clearly specific for motor neurons and largely spares cognitive ability, sensation, and autonomic nervous functions. Muscles controlling eye movement and the pelvic floor are the only skeletal musculature left unaffected. However, in a minority of cases (5–10%), patients also develop frontotemporal lobar dementia (FTLD). It has been suggested that a greater percentage of patients experience some cognitive change (such as loss in executive function) without crossing the threshold required for a diagnosis of dementia.[8]

Clinical presentation varies but most commonly consists of weakness, fasciculations (twitching muscles), and/or hyperreflexivity of facial muscles (bulbar onset) or limbs (spinal onset). Interestingly, initial symptoms usually appear at a focal site and later spread along contiguous anatomic paths.[9] Diagnosis is achieved by a combination of clinical examination and electromyography (EMG), in which positive, sharp waves and fibrillation potentials provide evidence for active denervation. The El Escorial criteria were developed in 1990 and are still utilized to diagnose and classify ALS cases as "possible," "probable," or "definite"[10] (Fig. 1). Guidelines on implementation of the El Escorial criteria have been revised to place greater emphasis on electrophysiological abnormalities, which can be detected earlier and thus facilitate timely diagnosis.[11]

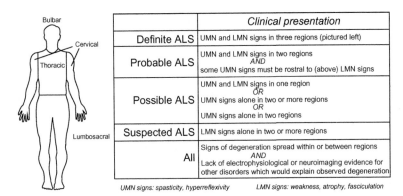

	Clinical presentation
Definite ALS	UMN and LMN signs in three regions (pictured left)
Probable ALS	UMN and LMN signs in two regions *AND* some UMN signs must be rostral to (above) LMN signs
Possible ALS	UMN and LMN signs in one region *OR* UMN signs alone in two or more regions *OR* UMN signs alone in two regions
Suspected ALS	LMN signs alone in two or more regions
All	Signs of degeneration spread within or between regions *AND* Lack of electrophysiological or neuroimaging evidence for other disorders which would explain observed degeneration

UMN signs: spasticity, hyperreflexivity   *LMN signs: weakness, atrophy, fasciculation*

FIG. 1. El Escorial criteria for diagnosis of ALS. UMN = upper motor neuron; LMN = lower motor neuron.

II. Etiology of ALS

The majority of ALS cases (≈82%) are sporadic (SALS), having no apparent heritability.[9] Up to 5% of SALS cases are caused by mutations in the 43-kDa *trans*-activating response region DNA-binding protein (TDP-43). TDP-43 mutations have also been linked to ≈3% of inherited, or "familial" ALS (FALS).[12] The most commonly occurring mutations in patients with FALS are found in the gene for Cu, Zn superoxide dismutase (SOD1) and account for approximately 20% of all FALS.[13,14] Most of these mutations are missense mutations that cause autosomal dominant ALS, except the D90A polymorphism, which can also behave as a recessive mutation.[15] FALS-causative mutations have also been found in genetic loci corresponding to alsin, a guanine exchange factor for Rac1 that plays a role in cytoskeletal dynamics[16,17]; senataxin, a DNA/RNA helicase that may be involved in RNA processing[18,19]; vesicle-associated membrane protein-associated protein B (VAPB), which facilitates intracellular vesicular trafficking[20]; and angiogenin (ANG)[21–23] (Table I). Some polymorphisms found in patients with ALS do not segregate completely with disease and may represent genetic risk factors rather than causative mutations. Mutations in the neurofilament-heavy subunit,[24,25] vascular endothelial growth factor (VEGF),[26] and ciliary neurotrophic factor (CNTF)[27,28] fall under this category. All genetic loci that have been reported as putative modifiers of ALS susceptibility are listed in the ALS Online Genetics Database (http://alsod.iop.kcl.ac.uk).

There is evidence to suggest that specific environmental factors play a prominent role in the etiology of some ALS cases. Geographically limited populations with dramatically increased ALS incidence, such as inhabitants of the Kii peninsula in Japan,[29] the Chamorro people of Guam, Gulf War veterans,[30,31] and Italian soccer players,[32] certainly lead one to suspect the

TABLE I
GENETIC LOCI ASSOCIATED WITH ALS

	Locus	Chromosome	Gene	Characteristics
Classical ALS	ALS1	12q22.1	Superoxide dismutase 1 (SOD1)	AD, adult onset
	ALS2	2q33	Alsin	AR, juvenile onset
	ALS3	18q21	Unknown	AD, adult onset
	ALS4	9q34	Senataxin (SETX)	AD, juvenile onset
	ALS5	15q15.1–21.1	Unknown	AR, juvenile onset
	ALS6	16q12.1–12.2	Fused in sarcoma (FUS)	AD, adult onset
	ALS7	20p13	Unknown	AD, adult onset
	ALS8	20q13.33	Vesicle-associated membrane protein-associated protein B (VAPB)	AD, adult onset
	ALS9	14q11	Angiogenin (ANG)	AD, adult onset
	ALS10	1p36	Tar DNA-binding protein (TARDBP)	AD, adult onset
	ALSX	X	Unknown	XD, adult onset
Atypical ALS	ALS/FTLD	9q21–22 9p13.3–21.3	Unknown	XD, adult onset
	ALS–PDC	17q21.1	Membrane-associated protein tau (MAPT)	AD, adult onset

AD, autosomal dominant; AR, autosomal recessive; XD, X-linked dominant; FTLD, frontotemporal lobar dementia; PDC, Parkinsonism-dementia complex.
Updated references for each locus at the ALS Online Genetics Database (http://alsod.iop.kcl.ac.uk).
Gray-shaded areas are alternated with unshaded rows to improve readability of the table

environment as a potential modifier of disease susceptibility. There also have been reports of ALS in individuals with intense exposure to particular stressors, such as harsh chemicals and heavy metals,[33,34] viral infection,[35] electrical shock,[36] and traumatic nerve injury.[37] Most of these reports, however, involve a very small number of cases and do not permit rigorous evaluation of these stressors as potential risk factors for ALS.

The most convincing instance of a causal link between ALS and toxin exposure is the case of the Chamorro population. A cycad indigenous to Guam produces the neurotoxin β-methylamino-L-alanine (BMAA) in its seeds, which are eaten by flying foxes as well as ground into flour by the Chamorro. While the dosage of BMAA resulting from a reasonable human consumption of cycad flour is far below the threshold necessary to provoke neurodegeneration in primates,[38] this potent neurotoxin is enriched 100-fold in the tissues of the flying fox, a delicacy to the Chamorro.[39,40] Furthermore, BMAA is found in the brain tissue from Chamorros who succumb to ALS, but not those who die of other causes, and the prevalence of ALS among this population dropped after overhunting thinned the flying fox population.[39]

While providing a convincing causal link between BMAA exposure and ALS, it is tempting to dismiss the case of the Chamorro as inapplicable to disease risk in the general population. However, BMAA-producing cyanobacteria are present in many ecosystems, and a recent study of Baltic Sea marine life revealed that BMAA is concentrated in the tissues of organisms at higher trophic levels, such as fish and mollusks, that are consumed by humans.[41] BMAA also was found in cyanobacteria-containing sand from Qatar,[42] raising the possibility that Gulf veterans may have been exposed to this toxin through inhalation. Furthermore, the incidence of ALS diagnosis is elevated 10–25–fold among residents of Enfield, New Hampshire, a town bordering Lake Mascoma, which is subject to frequent "blooms" of cyanobacteria.[43] While no conclusive statements can be made from these few examples, the worldwide prevalence of cyanobacteria seems a compelling reason to investigate ALS risk associated with BMAA exposure.

III. SOD1-Related Pathology as a General Model for ALS

The discovery of SOD1's role in FALS[14] offered the first insight into the molecular mechanisms of ALS, and the study of SOD1-mediated pathology has contributed much to our current understanding of the disease. The majority of *in vivo* work has utilized transgenic mice expressing FALS mutants of human SOD1, which develop a progressive motor neuron syndrome reminiscent of the human ALS phenotype (reviewed in Ref. 44). The sporadic disease differs little clinically from SOD1-related FALS, leading to the widespread supposition that all cases of ALS share some common mechanism(s) of pathology.[2,45,46] In reviewing the proposed molecular bases of ALS, we focus on the contribution of SOD1, a well-studied cause of ALS that may exhibit pathogenic mechanisms common to other forms of the disease.

IV. Misfolding and Aggregation Is the Most Likely Source of SOD1 Toxicity

SOD1 is a ubiquitous cytosolic enzyme whose primary function is the dismutation of the superoxide radical (O_2^-) to a less oxidizing species (H_2O_2) via a bound Cu^{2+} ion. Although this enzyme plays an important role as a cellular antioxidant, the ability of SOD1 mutants to selectively kill motor neurons is not linked to a loss of dismutase function. Not only do many FALS mutants retain enzymatic activity at or near wild-type levels,[47–49] but SOD1 null mice do not exhibit neurodegeneration.[50] Furthermore, the toxicity of SOD1 mutants cannot be reversed by coexpression of wild-type SOD1.[51] This evidence has led to widespread acceptance of the hypothesis that SOD1 mutants acquire a novel toxic property independent of their enzymatic function.

Despite over 15 years of research, the mode(s) by which SOD1 mutants selectively kill motor neurons have not been clearly delineated. However, a large body of evidence implicates a common propensity to misfold and aggregate as the primary toxic gain of function. Destabilization of the native fold is an attractive hypothesis for SOD1 mutant pathogenicity, offering a plausible explanation for the common disease outcome of over 140 mutants spanning the sequence and structure. Early *in silico* studies by our laboratory predicted that a majority of SOD1 mutations would destabilize the native fold or quaternary structure,[52] a trend that since has been verified experimentally.[53–56] Especially severe destabilization caused by certain mutations could account for their inherently higher aggressiveness (short disease duration).[57,58] Indeed, several recent analyses of *in vitro* SOD1 mutant behavior and FALS patient survival showed that protein instability and increased aggregation rate correlated with decreased survival time[59,60] (Fig. 2). Furthermore, the presence of SOD1-immunoreactive proteinaceous aggregates in SALS patient motor neurons[62–65] suggests that aberrant oligomerization of SOD1 could be a common feature of ALS, regardless of genotype. It thus appears that ALS is a protein conformational disorder, akin to other neurodegenerative diseases such as Alzheimer's, Parkinson's, and Huntington's.[2]

Though a primary role for SOD1 aggregation in FALS seems likely, deconstruction of the molecular determinants and mechanisms of this process is incomplete. SOD1 is an extremely stable enzyme in its fully mature, homodimeric form, remaining active in the presence of 6 M guanidinium chloride or 8 M urea.[66,67] SOD1 owes its extraordinary stability largely to the coordination of Zn^{2+}, which constrains the relatively unstructured electrostatic and zinc-binding loops, "tethering" them together and protecting the protein core, an eight-stranded Greek key β-barrel[53,68,69] (Fig. 3). The catalytic copper ion and an intrasubunit disulfide bridge between Cys-57 and Cys-146 appear to contribute relatively little to monomer thermodynamic stability, but the latter modification constrains loop mobility and facilitates dimer formation.[66,68,74] Metal-bound, disulfide-oxidized SOD1 forms an exceptionally stable homodimer, with low nanomolar binding affinity.[75,76] These maturation events are mutually interdependent—metal binding promotes disulfide bond formation, disulfide bond formation and metal binding promote dimerization, and dimeric SOD1 is more resistant to disulfide reduction/metal loss.[68,75,77]

In vitro studies show that dimer dissociation is a necessary initiating step in SOD1 aggregation.[76,78] The resultant monomeric SOD1 is more susceptible to the loss of the stabilizing zinc ion and disulfide bridge,[79,80] leading to freer loop movement[81] and exposure of β-barrel edge strands.[68,82] Dynamical studies of

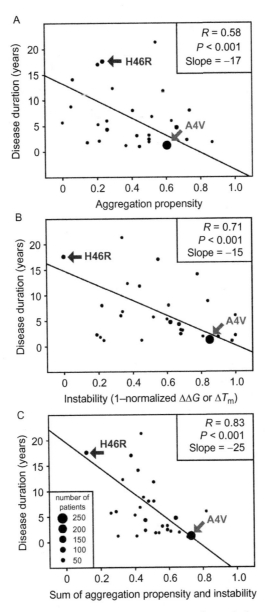

Fɪɢ. 2. Protein instability and aggregation propensity correlate with shorter survival times in SOD1-related FALS. In all panels, data are weighted for linear regression analysis according to the number of patients for which survival data was available. (A) The aggregation propensities of FALS-causative SOD1 mutations are calculated using a rederivation of the Chiti–Dobson equation,[61] which was validated by comparison with available experimental data, and normalized such that the

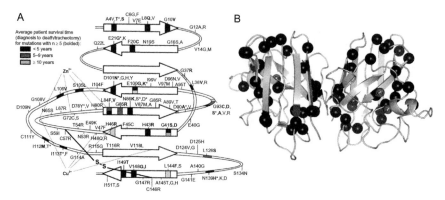

FIG. 3. Location of FALS-causative mutations on the SOD1 structure. (A) Map of SOD1 secondary structure showing locations of FALS missense mutations, residues that coordinate Cu^{2+} (His-46, His-48, His-120, His-63) and Zn^{2+} (His-63, His-71, His-80, and Asp-83) ions, and the intramolecular disulfide bridge. Arrows indicate β-strands. Epidemiological data were taken from Refs. 45,60,70–73; mutations with survival data for at least five patients are bolded, and the residue position is shaded to indicate the corresponding average survival time. For positions with more than one $n \geq 5$ mutation, the upper color corresponds to the first mutation listed. (B) Crystal structure (PDB code 1spd) of fully mature (metal-bound, disulfide-intact) homodimeric SOD1 with positions of aggressive (survival time <5 years) mutations indicated by black spheres.

wild-type and FALS mutant SOD1 revealed a transient "excited state" whose population is enhanced by mutations and zinc loss, but unaffected by disulfide status.[83] Increased surface hydrophobicity of metal-free, disulfide-reduced mutant SOD1 was shown directly by Tiwari *et al.* using 1-anilinonaphthalene-8-sulfonic acid (ANS), a fluorescent dye that binds to hydrophobic surfaces.[84] Munch *et al.* obtained similar results using a different hydrophobic dye, Sypro Orange, and found that increased exposure of hydrophobic regions precedes aggregation.[85] A general model of SOD1 aggregation in ALS has emerged in which dimer dissociation and subsequent metal loss (and/or disulfide reduction) induce structural distortions that favor assembly into non-native oligomers (oligomers other than the native homodimer) (Fig. 4). FALS mutations promote aggregation by increasing the tendency of SOD1 to lose its stabilizing

least and most aggregation-prone proteins have values of 0 and 1, respectively. (B) Protein instability is taken as 1 minus the change in free energy of unfolding or change in melting temperature of the mutant protein compared to the wild type (from published *in vitro* data). Instability values are normalized such that the most and least stable proteins have values of 0 and 1, respectively. (C) Normalized sums of aggregation propensity and protein instability values plotted against patient survival. From Wang *et al.*[60] (For color version of this figure, the reader is referred to the Web version of this chapter.)

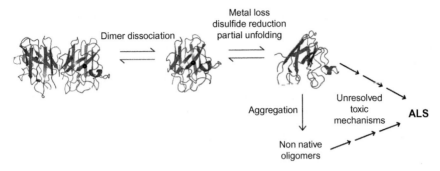

FIG. 4. General mechanism of SOD1 aggregation.

posttranslational modifications and/or by decreasing the intrinsic stability of the apo-monomer.[52,54–56,68,86–88] Substantial gaps remain in our understanding of the relationship between SOD1 aggregation and ALS pathology. These pertain to aggregate structure, mechanism of formation, and toxicity.

A. SOD1 Aggregate Structure

No high-resolution structural information is available for misfolded monomeric SOD1 or nonnative oligomers. The transient nature of many structurally-perturbed SOD1 species makes their isolation for study impractical. However, misfolded dimeric or monomeric SOD1 can be detected using an antibody specific for residues 145–151, which are normally buried within the native dimer interface.[89] SOD1 monomers with a more substantially disrupted fold can be tracked using an antibody recognizing the natively buried residues of β strand 4 (residues 42–48).[90] Chromatographic methods have also been utilized to isolate misfolded SOD1 using their affinity to hydrophobic resins.[91] Continued study using these and similar methods will be useful in tracking the spatial and temporal distribution of misfolded SOD1 in cell culture, transgenic mouse models, and patients with ALS, providing insight into the molecular determinants and cellular consequences of SOD1 destabilization.

Electron microscopic, immunohistological, and biochemical studies have shed some light on the structural properties of SOD1 aggregates. Both insoluble, detergent-resistant aggregates and soluble oligomers have been noted in cell culture, transgenic mice, and *in vitro*.[63,64,92–94] These species contain metal-free SOD1 that is full length and usually lacks the native disulfide bridge.[95] Aggregates formed *in vitro* under near-physiological conditions are often fibrillar and bind thioflavin T (ThT$^+$, suggestive of amyloid character),[86,96–98] while *in vivo* aggregates sometime appear amorphous or pore-shaped and do not always bind amyloid-sensitive dyes.[90,93,99–101] Soluble misfolded SOD1 populates a wide range of oligomeric states and also accumulates as non-native monomers, dimers, or trimers.[62,91,96] The instability of some soluble oligomers may preclude the use of static structural techniques, such as X-ray crystallography, to determine structural details, but solution-state methods such as nuclear magnetic resonance (NMR) or limited proteolysis, especially coupled with computational structural modeling, may yield insights into their conformations.

B. Mechanism of SOD1 Aggregation

The likelihood that misfolded SOD1 samples a multitude of conformational states also complicates detailed mechanistic study of oligomer formation. However, it is clear that posttranslational modifications of the SOD1 polypeptide modulate oligomer formation to some extent. As discussed above, the native intramolecular disulfide bridge and metal binding both impart exceptional stability to SOD1 and, unsurprisingly, loss of these factors drives misfolding and aggregation. However, reduction of the native Cys-57–Cys-146 disulfide has been putatively linked to the initiation, but not elongation, of amyloid-like fibril formation in vitro.[86,97] Disulfide-intact, but metal-free, SOD1 incubated at physiological pH and temperature can be induced to aggregate by disrupting noncovalent interactions with a chaotrope, but treatment with a reducing agent instead results in a 20-fold shorter lag period.[97] Disulfide bond reduction, while apparently dispensable for fibril formation in vitro, may specifically accelerate nucleation. Indeed, the presence of a small amount of disulfide-reduced wild-type or mutant SOD1 appeared to "recruit" disulfide-intact wild-type SOD1 into fibrils without the need for additional reducing agent.[97] The mechanism by which disulfide-reduced SOD1 facilitates fibril nucleation has not yet been demonstrated, although the requirement of Cys-57 and Cys-146 suggests that intermolecular cross-linking between these two residues may play a role.[97] It is also unclear whether in vivo SOD1 aggregation, which is not always amyloid-like, proceeds by elongation of nuclei.

The two free cysteines in SOD1, at positions 6 and 111, also appear to be involved in SOD1 oligomer assembly. In vitro aggregation of metal-free wild-type SOD1 coincides with a loss of free cysteines and oligomer formation is attenuated by mutations at either or both sites,[96,102] leading to the hypothesis that intermolecular disulfide cross-linking mediates oligomerization. However, more recent studies in mutant SOD1 transgenic mice show that aberrant disulfide linkages are present only in large-scale aggregates appearing late in the disease.[103,104] A secondary role for intermolecular disulfide cross-linking in aggregation is unsurprising given the reducing environment of the cytosol and may be due to "trapping" of SOD1 in a misfolded state after an initial destabilizing trigger, such as Zn^{2+} loss or altered conformational dynamics resulting from mutation.[87,88] Cell culture experiments reveal a key role for

Cys-111 in the promotion of SOD1 oligomerization, as mutation of this residue, but not Cys-6, attenuated oligomer formation and protected cells from mutant SOD1-mediated toxicity.[105] It could be that the higher solvent accessibility of Cys-111 (and thus, increased susceptibility to aberrant intermolecular disulfide cross-linking) accounts for its particular importance in SOD1 oligomerization. However, recent investigations by our laboratory offer an alternate interpretation of this phenomenon. We recently confirmed earlier reports that Cys-111 forms a mixed disulfide with glutathione and showed that this modification is abundant in human tissue. Interestingly, Cys-111 glutathionylation triggers dissociation of both wild-type and FALS mutant dimers *in vitro*, thus promoting the first step in SOD1 aggregation.[105a] The characterization of intermolecular disulfide formation as a nonessential late event in oligomerization suggests that Cys-111 may primarily promote aggregation by its ability to be glutathionylated, a modification that destabilizes the native homodimer. Treatments used to prevent Cys-111-mediated SOD1 aggregation in previous cell culture experiments,[105] such as addition of a reducing agent and overexpression of glutaredoxin, would remove the glutathione moiety in addition to reducing intermolecular disulfides. Therefore, further study would be useful to resolve the contributions of Cys-111 glutathionylation and intermolecular disulfide bond formation in oligomer formation.

An emerging question in the study of mutant-mediated SOD1 aggregation is the extent of involvement of the wild-type protein. Since most FALS patients with SOD1 mutations are heterozygous, recent studies have utilized transgenic mice expressing both human wild-type and FALS mutant SOD1 to more accurately recapitulate SOD1 behavior *in vivo*. Coexpression of SOD1^{WT} exacerbates the disease phenotypes of SOD1^{G93A}[106,107] SOD1^{G85R}[108] SOD1^{L126Z}, and SOD1^{A4V} mice,[92] hastening the appearance of cellular pathologies and shortening survival times (Fig. 5). The effect of the wild-type protein on SOD1^{A4V} mice is particularly dramatic; even though FALS patients with this mutation exhibit particularly rapid disease progression, mice expressing only SOD1^{A4V} do not develop motor neuron disease within their lifetimes.[48] The toxic effect of coexpressing wild-type protein may be a simple issue of protein copy number. An earlier study of G85R mice[51] did not find any effect of human wild-type coexpression on survival, but both SOD1^{G85R} and SOD1^{WT} were expressed at lower levels than in the more recent model.[108] The observation that mutant SOD1 toxicity depends heavily on protein abundance, while not surprising, is troubling since nearly all mutant SOD1 transgenic mice substantially overexpress the protein.[44] However, mice overexpressing SOD1^{WT} alone, while exhibiting minor deficits in motor function, do not experience paralysis or die prematurely.[107] Thus, FALS mutants clearly possess intrinsic pathogenicity independent of gene dosage. Mutant-wild-type heterodimers and disulfide-linked aggregates containing both wild-type and mutant SOD1 have been

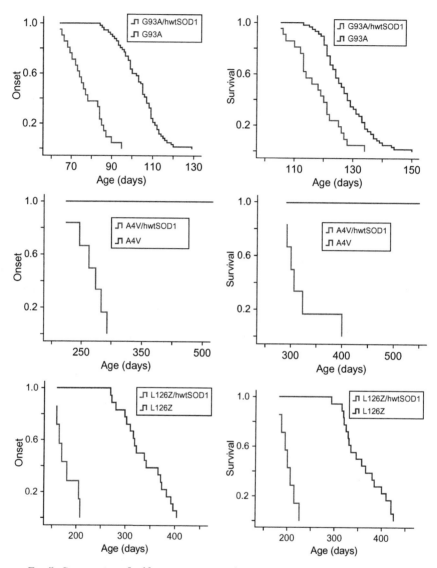

Fig. 5. Coexpression of wild-type SOD1 exacerbates the phenotype of FALS mutant trans-genic mice. Survival and symptom onset are plotted versus age for mice expressing G93A, A4V, and L126Z (truncation) mutants of SOD1 with and without coexpression of the human wild-type enzyme (hwtSOD1). From Deng et al.[92] Copyright 2006 National Academy of Sciences, USA. (See Color Insert.)

observed,[92,108] suggesting that wild-type SOD1 is "recruited" into non-native oligomers by pathogenic mutants, possibly under conditions of oxidative stress. These studies present an incomplete picture of the role of SOD1[WT] in aggregation but highlight the need for further scrutiny of the physiological relevance of commonly used transgenic mouse models.

C. Toxicity of SOD1 Aggregates

While misfolding and aggregation has been convincingly linked to ALS pathogenesis, the species responsible for motor neuron death has not been identified. Insoluble inclusion bodies appear in brain stem and spinal cord coincident with symptom onset and accumulate progressively in the terminal stages,[109–113] leading to an initial belief that large-scale aggregates are themselves toxic. However, the ability to detect soluble misfolded SOD1 led to the discovery that these non-native forms are present from birth[91,114] and selectively enriched in motor neurons[89,91] of FALS transgenic mice. It thus appears that small misfolded SOD1 may be the actual toxic culprit(s), present throughout life but causing symptoms only when cells can no longer keep their deleterious effects in check. In such a scenario, assembly of soluble misfolded SOD1 into relatively inert inclusions is expected to be neuroprotective, a phenomenon that has been demonstrated for aggregation of Aβ and huntingtin in Alzheimer's and Huntington's diseases, respectively.[115–117] However, the relative toxicities of small soluble oligomers and large-scale aggregates of SOD1 remain to be directly proven. Similarly, no consensus has yet been reached on the mode(s) by which non-native SOD1 kills cells. The evidence at present, though sometimes contradictory, identifies a diverse set of targeted organelles, signaling pathways, and other cellular processes. In the remainder of this chapter, we will discuss the various pathological processes occurring in ALS, with special attention to a potential causal role for misfolded and/or aggregated SOD1.

V. Motor Neuron Death in ALS: Apoptotic Versus Necrotic, Cell-Autonomous Versus Non-Cell-Autonomous

Classification of motor neuron death in ALS remains controversial. Spinal cord motor neurons of ALS patients and transgenic mice overexpress the pro-apoptotic BH3-only protein Bax,[118] and knocking out this protein in SOD1[G93A] mice results in delayed disease onset.[119] However, activation of "executioner" caspases (caspase-3, caspase-6, and caspase-7) is not always seen[120–122] and the morphology of dying motor neurons is often uncharacteristic of apoptotic bodies.[123,124] The current model for neuronal death in ALS is the one that has characteristics of both apoptosis and necrosis, with "necrotic-like" and "apoptotic-like" processes dominating in different cell types and/or disease stages that have yet to be delineated.[121,125]

Another question pertaining to classification of cell death in ALS is the autonomy of this process in motor neurons. A cell-autonomous "dying forward" process was long assumed, in which dysfunction within motor neurons, independent of input from surrounding cells, leads to their death and a subsequent denervation of motor endplates. However, several studies using cell-specific expression of mutant SOD1 support a prominent role of non-neuronal cells in promoting cell death. The most striking evidence against cell-autonomous motor neuron death is the reported lack of ALS phenotype of transgenic mice expressing mutant SOD1 under a neuron-specific promoter.[126,127] Mice with supraendogenous neuron-specific expression do experience neurodegeneration but show different pathological changes compared to transgenics ubiquitously expressing FALS mutant SOD1. Symptom onset occurs later, is diffuse rather than focal, and lacks certain morphological hallmarks such as mitochondrial vacuolization.[128] Astrocytes, supporting cells that neighbor motor neurons, have also been proposed to turn deadly in ALS through defects in glutamate processing and other mechanisms (see below). While mutant SOD1-expressing astrocytes exert toxicity on motor neurons in coculture,[129] astrocyte-specific expression failed to cause motor neuron disease in mice.[130] Mutant SOD1 expression limited to microglia (phagocytic cells in the CNS) or Schwann cells, which myelinate motor axons, likewise produced no ALS phenotype.[44,131,132] Although mutant SOD1 in neurons, astrocytes, and microglia appears insufficient to provoke ALS symptoms in isolation, knockdown in these cell types using Cre–Lox systems or siRNA delays disease onset and/or progression in transgenic mice with ubiquitous expression.[133–135] Surprisingly, Schwann cell-specific knockout of mutant SOD1 was reported to accelerate disease progression.[136] Taken together, these studies highlight the importance of non-neuronal cells in ALS pathogenesis and progression and suggest that the primary site of dysfunction may not be the motor neuron itself (Fig. 6).

Interestingly, skeletal muscle-specific expression of mutant, and to a lesser extent, wild-type, SOD1 in mice was recently shown to cause early motor deficits, followed by neuromuscular junction (NMJ) dismantlement and late-onset motor neuron loss.[137] This result is surprising in light of previous studies showing no effect of mutant SOD1 knockdown in muscle[138,139] but is consistent with reports of muscular defects as the primary pathogenic events in ALS.[140,141] The ability of muscle-restricted expression of mutant SOD1 to provoke motor neuron degeneration, as well as the precedence of neuromuscular denervation in the disease course (Fig. 7), suggests a "dying back" model of ALS where loss of the neuronal cell body is not the initiating event. The primacy of the NMJ in ALS pathogenesis is further supported by studies

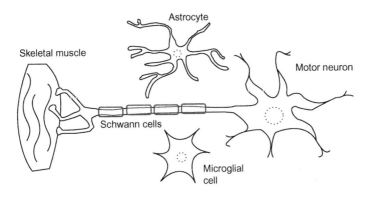

Cell type	Restricted mutSOD1 expression sufficient to cause MND?	Effect of cell type-specific mutSOD1 knockout	Ref.
Motor neuron	Only when overexpressed	Delays onset, early progression	[126–128,133]
Astrocyte	No	Slows progression	[130,135]
Microglia	No	Slows progression	[131,133]
Schwann cell	No	Hastens late progression	[132,136]
Skeletal muscle	Yes	no effect	[137,138]

FIG. 6. Motor neuron death in ALS is not cell autonomous. Table below the figure summarizes findings of studies using transgenic mouse models with tissue-specific mutant SOD1 expression or Cre–Lox/siRNA knockdown of ubiquitously expressed mutant SOD1.

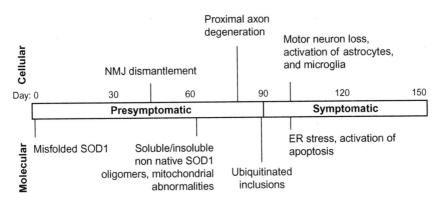

FIG. 7. Timeline of molecular and cellular pathologies in transgenic mice ubiquitously expressing SOD1^{G93A}. The "symptomatic" stage denotes the period following initial onset of muscle weakness and wasting (80–100 days). Dates of pathology appearance taken from.[44,91,114]

showing that inhibition of pro-apoptotic machinery, while completely preventing motor neuron loss in mice, did not prevent denervation and offered little functional improvement or lifespan extension.[119,142,143] The mechanisms by which toxic signals are transmitted from muscle to NMJ to motor axons are unknown, but a recent study of SOD1^{G93A} mice noted increased retrograde axonal transport of proteins related to cellular stress and death.[144] Muscular overexpression of a mitochondrial uncoupling protein, which disrupts ATP synthesis, was also sufficient to induce progressive NMJ dismantlement and motor neuron loss in mice.[145] Loss of compensatory reinnervation may also be involved in NMJ pathology. A skeletal muscle-specific microRNA was recently identified that slows disease progression in SOD1^{G93A} mice by stimulating reformation of neuromuscular synapses with denervated muscles.[146] Growing support for a "dying back" hypothesis of ALS highlights the need for additional investigation of skeletal muscle as a primary site of pathology.

VI. ALS Comprises a Spectrum of Pathologies

On a subcellular level, ALS pathology is staggeringly complex and includes abnormalities in nearly all cellular compartments. Many of these are undoubtedly secondary effects or compensatory mechanisms for an initial dysfunctional "trigger," the identification of which has remained elusive despite nearly 20 years of research on the molecular bases of ALS. We will review some of the more notable and well-studied pathological processes and discuss their relevance to the initial stages of disease, when therapeutic intervention may still be possible.

A. Excitotoxic, Inflammatory, and Oxidative Insults

In 1992, Rothstein *et al.* found defects in glutamate signaling in neuronal tissue from patients who died of ALS but not Alzheimer's and Huntington's diseases,[147] revealing a unique molecular basis for ALS. This phenomenon was later attributed to the selective loss of the astrocytic glutamate transporter EAAT2, which is crucial for prompt clearance of glutamate from the synaptic cleft after firing.[148] Both FALS and SALS patients, and mutant SOD1 mice, have decreased levels of functional EAAT2 protein (also known as GLT1) and increased circulating glutamate in the cerebrospinal fluid (CSF).[149–152] Work in transgenic mice confirms the importance of EAAT2/GLT1-mediated glutamate clearance to motor neuron health. Deletion of this gene is sufficient to induce progressive neurodegeneration,[153] and genetically encoded[154] or exogenously stimulated EAAT2/GLT1 overexpression[155] delays symptom onset in ALS mouse models. The mechanism(s) by which EAAT2/GLT1 is downregulated in ALS are not yet understood, and it is not clear whether decreased mRNA synthesis/stability is a factor. Postmortem spinal cord from ALS patients had normal EAAT2/GLT1 mRNA levels.[156] However, a later analysis of SOD1^{G93A} mice using *in situ* hybridization and qRT-PCR revealed a substantial decrease in EAAT2/GLT1 promoter activity and transcript quantity concomitant with disease onset.[157] EAAT2/GLT1 is directly affected by several deleterious processes that occur in the ALS-affected CNS, suggesting that deficiency of its transport function and subsequent glutamate overload may be a secondary event in ALS pathogenesis. Caspase-3 activation (which is itself a relatively late-occurring phenomenon[158]) results in a truncated, inactive version of EAAT2/GLT1,[159] and oxidative damage to the C-terminus of EAAT2/GLT1 diminishes its ability to transport glutamate.[160–162] EAAT2/GLT1 expression in astrocytes is also subject to modulation by neuronal signaling, via activation of the transcription factors κB motif binding phosphoprotein (KBBP) by presynaptic axons.[157] Synapse loss and denervation in SOD1[G93A] mice results in decreased astrocytic KBBP and diminished expression of EAAT2/GLT1, revealing that the astrocytic glutamate transporter is downregulated in response to synaptic dysfunction.[157] Taken together, these lines of evidence show that deficient glutamate reuptake by astrocytes is induced by preexisting neuronal stress, which is further exacerbated by the resultant excitotoxicity (Fig. 8).

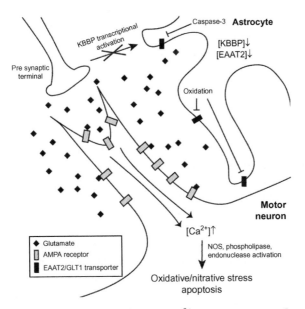

FIG. 8. Glutamate excitotoxicity causes influx of Ca^{2+} to motor neurons and activates apoptosis. ALS patients and mouse models have decreased levels of the astrocytic glutamate transporter EAAT2/GLT1, which clears glutamate from the synapse after firing. Dysfunction in the presynaptic motor axon disrupts activation of the EAAT2/GLT1 transcriptional activator κB motif binding phosphoprotein (KBBP). Deficient EAAT2/GLT1, combined with inactivation of the transporter by oxidative damage and caspase-3-mediated proteolysis, results in persistent glutamate stimulation of the Ca^{2+}-permeable AMPA receptor, which is especially abundant in motor neurons. The resultant calcium influx activates several pro-oxidant and pro-apoptotic factors and prolonged Ca^{2+} excess results in motor neuron death.

Prolonged hyperstimulation by glutamate induces death primarily by allowing persistent calcium influx through the Ca^{2+}-permeable α-amino-3-hydroxy-5-methyl-4-isoxazole propionic acid (AMPA) receptors, which are specifically enriched in motor neurons.[163–166] Excess Ca^{2+} floods into the mitochondria and overwhelms its natural buffering capacity, triggering reactive oxygen species (ROS) production, disrupting protein homeostasis, and eventually activating the apoptotic machinery.[167,168] Cytosolic calcium overload further perpetuates itself by stimulating the opening of ryanodine receptors (RyR) on the endoplasmic reticulum (ER) membrane, allowing Ca^{2+} release from the luminal space into the cytosol.[169] AMPA receptor-mediated calcium influx also increases mutant SOD1 aggregation in cultured motor neurons[170] and mouse models of ALS,[171] which produces additional ER and mitochondrial dysfunction (see below). Glutamate excitotoxicity thus acts synergistically with protein aggregation and mitochondrial/ER dysfunction to stress motor neurons and activate apoptosis in SOD1-related FALS (Fig. 10). Motor neurons are selectively vulnerable to excitotoxic stress due to their abundance of AMPA receptors and low calcium-buffering ability.[172,173] Riluzole, the as-yet sole drug approved for the treatment of ALS, inhibits excitotoxic stress in neurons by slowing glutamate release and blocking AMPA receptors,[174–177] but confers a survival benefit of only few months.[178]

In addition to excessive glutamate, secreted oxidative, nitrative, and inflammatory factors also contribute to motor neuron stress and death in ALS, as illustrated by the cytotoxic effect of ALS patient CSF on healthy rat spinal cord cultures.[179] While it has become clear that neurons, astrocytes, and microglia are all capable of secreting pro-inflammatory cytokines and other inducers of cellular stress and death, the relative contribution of each cell type to the transmittance of stress signals in ALS is unresolved. Astrocytes expressing mutant SOD1 kill wild-type motor neurons in co-culture by secreting an unidentified soluble factor that activates the pro-apoptotic Bax protein.[129] Activated microglia, which are pathologic hallmarks in the CNS of ALS patients and mouse models,[180–184] release a host of inflammatory and pro-apoptotic factors. For example, cyclooxygenase-2 (COX-2) and inducible nitric oxide synthase (iNOS) are activated, resulting in enhanced production of prostaglandins and nitric oxide (NO), respectively,[185] and programmed death signals such as the Fas ligand (FasL) and tumor necrosis factor-alpha (TNF-α) are released.[185,186] A great deal of crosstalk exists between the cellular responses to each individual cytokine. For example, COX-2 and iNOS are activated by TNF-α stimulation of astrocytes,[187] and the transcription of FasL in motor neurons is activated by NO.[186]

In cases of SOD1-related FALS, the mutant protein directly contributes to production of extracellular stressors. First, mutant SOD1 disrupts redox regulation of NADPH oxidase (Nox), a membrane-bound producer of extracellular superoxide, through a direct interaction with Rac1. Oxidizing conditions normally promote the dissociation of the SOD1–Rac1 complex and cessation of Nox activation, but mutant SOD1 remains bound to Rac1 and allows persistent superoxide production under these conditions.[188] Extracellular superoxide may then enhance neuroinflammation by stimulating microglia which are activated by ROS[189] (Fig. 10). Furthermore, mutant SOD1 may itself be a secreted factor that contributes to neurotoxicity. Mutations in SOD1 confer an affinity for the secretory vesicle proteins chromogranins A and B.[190] Localization of the mutant protein to secretory vesicles allows its transport from neurons and astrocytes to the extracellular space, resulting in activation of microglia and motor neuron death.[190] Neuroinflammatory and excitotoxic insults from microglia and astrocytes are thought to primarily affect disease progression rather than representing a primary trigger of disease.[191] However, some biochemical indicators of microglial activation, such their accumulation in the CNS and increased TNF-α production, are present prior to symptom onset in ALS mice.[186,192] Although relief of excitotoxic stress with Riluzole offers limited survival benefit, the combination of this or similar drugs with anti-neuroinflammatory agents may result in more satisfactory functional outcomes.

B. Dysregulation of Neurotrophic Factors and Axon Guidance Proteins

Neural networks are not static entities that remain stable indefinitely after development; rather, they require the continuous input of neurotrophic factors and axon guidance cues secreted by glia and innervated muscle. Loss of survival-promoting neurotrophic signaling has therefore been proposed as a contributing factor to motor neuron demise in ALS. In support of this view, CNTF knockout produces progressive motor neuron death in mice[193] and exacerbates neurodegeneration in the SOD1^{G93A} model.[194] Muscle-specific overexpression of glial cell-derived neurotrophic factor (GDNF) in the G93A mouse preserves NMJs and improves motor neuron survival,[195] suggesting therapeutic potential of neurotrophic factor supplementation. However, deficits in neurotrophic factors are not seen in ALS patients; to the contrary, GDNF and CNTF are upregulated in muscle, CSF, and postmortem spinal cord from ALS patients.[196–199] The overabundance of these factors in symptomatic individuals suggests that their upregulation is part of a defensive response to existing pathology and is ultimately insufficient to halt disease progression. In line with this view, administration of CNTF[200] or brain-derived neurotrophic factor (BDNF)[201] showed no measurable benefit to ALS patients. It is possible that neurotrophic factors hold therapeutic potential if administered early and/or at intact NMJs (recapitulating the beneficial muscle-specific overexpression of GDNF reported by Li et al.[195]), but such a requirement likely precludes their usefulness to ALS patients.

VEGF and ANG, which are involved in maintenance of both neural net-works and vasculature, have also been implicated in ALS.[202,203] VEGF, in particular, has received significant attention as a disease-modifying factor since the discovery that diminished VEGF expression in transgenic mice is sufficient to cause late-onset neurodegeneration.[204] Furthermore, ALS patients have decreased circulating levels of VEGF in CSF compared to healthy controls.[205] Mutant SOD1 directly contributes to VEGF deficiency through binding of the 3'-untranslated region (UTR) of VEGF mRNA, desta-bilizing transcripts and downregulating expression[206,207] (Fig. 10). Loss of VEGF function could also have a genetic basis independent of SOD1 muta-tions. Single-nucleotide polymorphisms in the VEGF promoter region were found to correlate with an increased risk of ALS,[26] although a recent meta-analysis of available genotype data restricts this effect to males.[208] Mutations in ANG were also linked to a small subset of ALS cases.[21–23] ALS-associated ANG mutations occur at functionally important residues involved in catalysis and nuclear localization[22] rather than hampering ANG expression, which is un-changed or even increased in ALS patients and mouse models.[209,210] The common functional consequence of ALS-associated ANG mutations appears to be an inability to promote neural connectivity and survival.[211]

The relative contributions of these proteins' neuroprotective and angiogen-ic properties to CNS health are unknown, but there is evidence that deficien-cies in both functions could promote neurodegeneration. The neuroprotective effect of both VEGF and ANG in cell culture[90,204,211] gives strong evidence for neurotrophic action of these proteins independent of vasculature. However, decreased cerebral blood flow in ALS patients[212,213] and disruptions in the blood–spinal cord barrier of several mouse models[214] indicate that vascular dysfunctions are indeed present in ALS. In mouse models of ALS, overexpres-sion of VEGF or its receptors, or administration of purified VEGF directly to the CNS, resulted in neuroprotection and prolonged survival,[215,216] leading to hope for VEGF administration as a therapeutic strategy. Restoration of ANG activity may also hold therapeutic potential. Further study is needed to explore this possibility and to resolve the molecular details of ANG and VEGF action in the CNS.

One possible mechanism for VEGF-mediated neuroprotection is its antag-onism of the axon guidance protein Sema3A. Sema3A is a member of the semaphorin family of proteins, which guide axons to their targets during development and also play a role in the complex phenomena of neural network refinement and plasticity.[217] Sema3A is a secreted glycoprotein that acts as an axonal chemorepellent through binding of a neuropilin-1/plexin-A coreceptor complex, which triggers downstream cytoskeletal reorganization and axon withdrawal.[218] These receptor components are expressed throughout adult-hood in spinal cord motor axons, a sensitivity which allows them to avoid

Sema3A-producing scar tissue during post-injury regeneration.[219–221] However, postnatal responsiveness to Sema3A may be a liability in individuals with
SOD1 mutations. Terminal Schwann cells of SOD1^{G93A} mice release abnormally high levels of Sema3A into the NMJ before symptom onset,[219] which
would be expected to induce axonal withdrawal from the synapse. Interestingly,
VEGF also binds neuropilin 1, leading some to propose that it prevents denervation in ALS by competing with Sema3A for receptor binding.[219]

In addition to Sema3A, ALS patients show increased expression of other
axonal chemorepellents, including ephrinA1 in motor neurons[196] and Nogo-A
in muscle.[222] Muscle-specific overexpression of the secreted factor Nogo-A
induced axon retraction from the NMJ in mice and higher Nogo-A expression
in a subset of ALS patients correlated with disease severity.[223] Although the
dysregulation of axonal guidance proteins is clearly correlated with ALS pathology, there is no strong evidence for causation. In fact, Nogo-A upregulation
has been reported to occur in response to neuromuscular denervation,[224] not
vice versa. However, overabundance of axonal chemorepellents at the NMJ
following initial retraction would certainly inhibit compensatory reinnervation
and may transform a minor insult into irreparable damage to the neuromuscular synapse (Fig. 9).

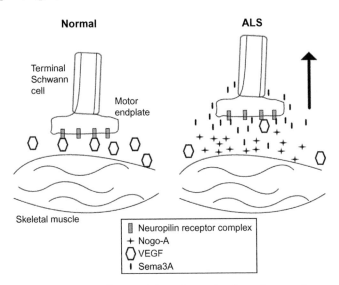

FIG. 9. Aberrant expression of neurotrophic and axonal guidance factors at the neuromuscular
junction of SOD1^{G93A} mice. Expression of the neuroprotective vascular endothelial growth factor
(VEGF) is decreased because of the binding of mutant SOD1 to the 3′-untranslated region of
VEGF mRNA, while expression of the axonal chemorepellents Sema3A and Nogo-A is upregulated
in terminal Schwann cells and muscle, respectively. Loss of trophic support and increased repulsive
cues may induce withdrawal of the motor axon from the neuromuscular synapse.

C. Axonal Structure and Transport Defects

The combination of polarity, high energetic demand, and extreme axon length (up to $1m$)[225] makes axonal integrity paramount to motor neuron viability. Accumulation of neurofilaments, which maintain axonal diameter and structural integrity in motor neurons, is a long-recognized hallmark of ALS pathology in humans and mouse models and is thought to contribute to the selective vulnerability of long, large-caliber motor axons.[2,48,226–228] Neurofilaments consist of light (NF-L), medium (NF-M), and heavy (NF-H) subunits, in equal proportion, and their proper assembly is crucial to the maintenance and extension of vulnerable large-caliber motor axons.[229,230] Misassembly of neurofilaments due to over- or under-expression, mutation, or deficient transport of individual subunits results in their accumulation, further hindrance of axonal transport, and eventual motor neuron death.[231–236] Hyperphosphorylation of neurofilaments also contributes to defective transport by causing their detachment from motor complexes and promoting aberrant self-association.[237–240] In ALS, this phenomenon is attributable to overactivation of p38 MAP kinase and Cdk5, which phosphorylate NF-M and NF-H.[241–243] Neurofilament accumulation appears to be selectively deleterious to axons. Overexpression of NF-H causes sequestration of neurofilaments within the cell body and perikarya and markedly delays disease onset in mouse models of ALS.[244] In addition to relieving the axonal burden of neurofilament aggregates, thus facilitating transport, accumulated neurofilaments in the perikarya are thought to counter glutamate excitotoxicity by chelating excess calcium and/or binding the cytoplasmic domains of glutamate receptors.[245,246]

In addition to neurofilaments, axonal transport of many other cellular components is indispensable for motor neuron health and homeostasis. Transport between the cell body and neuromuscular synapse is mediated by the dynein/dynactin (retrograde) and kinesin (anterograde) motor protein complexes, which carry adaptor-bound cargo along axonal microtubules. Mutant SOD1 mice show presymptomatic defects in both anterograde and retrograde transport,[144,247,248] with a particular retardation in the trafficking of mitochondria[249,250] and cytoskeletal components such as neurofilament and tubulin subunits.[248] Mitochondria are normally enriched near the neuromuscular synapse to meet the high energetic and calcium-buffering needs of the firing axon.[251] Impaired anterograde transport may thus explain the early onset distal axonopathy observed in ALS mouse models. SALS patients show accumulation of mitochondria in proximal axons,[252] which is further evidence for impaired anterograde transport as a fundamentally important mechanism of all ALS pathology. As with several other pathological processes in SOD1-related FALS, misfolding and aggregation directly impair axonal transport though aberrant interactions. Mutant SOD1 acquires the ability to bind motor complexes that are instrumental to both anterograde (kinesin-2[253]) and retrograde (dynein/dynactin[254,255]) axonal transport (Fig. 10). Mutant SOD1 also interacts directly with the 3'-UTR of NF-L mRNA,[256] resulting in decreased expression that is observed in both FALS and SALS patients.[257–259]

Because of its prevalence in human patients and early onset in ALS mouse models, dysregulation of neurofilament transport and metabolism became an early candidate for a common mechanism of ALS pathogenesis. While this hypothesis is attractive due to its apparent specificity for motor axons, evidence for a primary role of axonal neurofilaments in ALS is contradictory. While the aforementioned work by Couillard-Despres et al. indicates that reducing axonal neurofilaments dramatically delays ALS pathology,[244] a second study showed no benefit from sequestration of neurofilaments in the cell body and perikarya.[260] It may be that neurofilament dysfunction modifies neurodegenerative severity in ALS but is not sufficient to cause disease, a possibility that does not preclude defective axonal transport as a primary mechanism of pathogenesis.

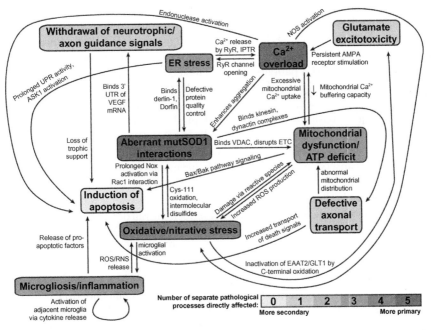

Fig. 10. Diverse pathological processes in SOD1-related FALS are highly interrelated and many stem directly from SOD1 misfolding/aggregation and cytosolic calcium overload. Abbreviations: mutSOD1, mutant SOD1; UTR, untranslated region; VDAC, voltage-dependent anion channel; ETC, electron transport chain; UPR, unfolded protein response; ROS/RNS, reactive oxygen/nitrogen species.

The early retardation of both anterograde and retrograde trafficking in motor axons undoubtedly initiates a spectrum of deleterious effects such as energetic deficiencies at the distal synapse and impaired neuromuscular communication. Further study is crucial to reveal the underlying mechanisms and consequences of axonal transport malfunction in ALS and to identify targets for therapeutic intervention.

D. Mitochondrial Dysfunction

Mitochondrial abnormalities such as swelling and vacuolization are pathological hallmarks in spinal cords of ALS patients and most transgenic mouse models,[49,107,121,261,262] leading to much interest in the mitochondrion's involvement in disease. Perturbed energy homeostasis and ATP deficits are observed in both SOD1[G93A] mice and skeletal muscle biopsies from ALS patients.[263–268] One mechanism by which the FALS mutant G93A impairs cellular respiration is through a novel ability to bind cytosolic malate dehydrogenase, which disrupts the malate–aspartate shuttle.[269] Misfolded and aggregated SOD1 mutants also accumulate on the cytoplasmic face of the outer mitochondrial membrane and bind directly to the voltage-dependent anion channel (VDAC), depolarizing the membrane and disrupting the normal functioning of the electron transport chain (ETC)[266,270–272] (Fig. 10). ETC dysfunction is a notable convergence in the pathologies of sporadic and familial ALS, and ETC inhibition in SALS patients has been linked to mutations in mitochondrial DNA.[267,273,274] Mitochondrial genome instability has been proposed (controversially) to play a central role in the natural aging process,[275–279] which would offer a possible basis for the late onset of disease in SALS. Interestingly, SOD1[G93A] mice and a subset of sporadic ALS patients are hypermetabolic[265,280] and administration of a high-fat diet modestly improved survival in mice.[265] The cause(s) and significance of hypermetabolism are unknown, as are the mechanisms by which aberrant metabolic states mediate toxicity in ALS. The surprising finding that metabolic dysfunction in skeletal muscle can provoke motor neuron death[145] suggests that mitochondrial defects may be central to the retrograde neurodegeneration seen in mouse models of ALS.

Mitochondria are also key players in the buffering of intracellular calcium, which in prolonged excess results in the activation of pro-oxidant and apoptotic factors such as nitric oxide synthase (NOS), phospholipases, and endonucleases.[281,282] ALS mice show a CNS-specific decrease in mitochondrial calcium loading capacity that precedes motor deficits.[283] Likewise, both ALS patients and mouse models have increased intracellular calcium concomitant with mitochondrial damage.[283–285] It is not clear whether decreased mitochondrial buffering capacity precedes cytosolic Ca^{2+} overload or vice versa, since these processes reciprocally enhance each other[167,286] (Fig. 10). Depletion of mitochondrial calcium-buffering ability is particularly deleterious to neurons and skeletal muscle, whose normal functioning involves frequent influxes of calcium to generate action potentials. This, combined with the enrichment of mutant SOD1 in mitochondria of motor neurons,[266,271,272,287,288] muscle,[137] and astrocytes,[289] may account for the sensitivity of these cells to mutant SOD1-mediated toxicity. Disturbance of mitochondrial function may also directly cause cell death by activating the apoptotic cascade. Aberrant localization to the intermembrane space and matrix[288,290,291] disrupts the structural integrity of the organelle, resulting in release of the apoptotic trigger cytochrome c.[292] Misfolded SOD1 monomers and oligomers also provoke apoptosis by associating with the pro-survival factor Bcl-2. The normally anti-apoptotic Bcl-2 exposes a toxic BH3 domain upon mutant SOD1 binding, resulting in cell death and interference of synaptic transmission at the NMJ.[293–295] Given the presymptomatic, cell type-specific recruitment of mutant SOD1 to mitochondria,[271,287] dysfunctional changes in this organelle merit consideration as primary contributors to ALS pathogenesis.

E. Deficient Protein Quality Control

The presence of proteinaceous aggregates in spinal cords of FALS and SALS patients suggests that malfunction or overloading of protein quality control machinery is a common feature of neurodegeneration. The ubiquitin–proteasome system (UPS), in which ubiquitin-tagged proteins are targeted for proteasomal degradation, is one such mechanism of misfolded protein clearance. Degradation of misfolded mutant SOD1 proceeds via the UPS and impedes its functioning by sequestering proteasomal subunits and ubiquitin ligases such as Dorfin,[296–300] while proteasomal inhibition produces a reciprocal enhancement of SOD1 aggregation[93,301,302] (Fig. 10). Ubiquitin- and ubiquitin ligase-positive intraneuronal inclusion bodies are found in FALS mouse models[47,303] and postmortem spinal cord of SALS patients,[124,300,304–306] indicating UPS activity and sequestration in both forms of the disease. Studies of SOD1 mutant transgenic mice reveal proteasomal impairment in the disease-vulnerable spinal cord and brainstem only after the onset of symptoms[307,308]; so UPS dysfunction is unlikely to be an initiator of pathology in SOD1-related FALS. Interestingly, as disease progresses, constitutively active proteasomal components are replaced by inflammatory cytokine-responsive subunits to yield the inducible "immunoproteasome", which degrades proteins into antigenic peptide fragments to be presented by the class 1 major histocompatibility complex.[307,309,310] Inhibition of immunoproteasome formation using a small-molecule anti-inflammatory agent shortens survival in a rat model of ALS,[311] but a more targeted genetic approach involving knockdown of the LMP2 immunoproteasomal subunit yielded no effect on survival in SOD1[G93A] mice.[312] The role of the immunoproteasome in ALS pathology is yet to be precisely determined, but its upregulation may be a response to glia-mediated inflammation in the CNS.[313,314]

Protein quality control by ER-associated degradation (ERAD) is also impaired in ALS, leading to stress signaling that can directly induce motor neuron death via activation of apoptosis. During synthesis and maturation of nascent proteins in the ER, misfolded species are cleared from the luminal space by the ERAD pathway (reviewed in Ref. 315). Dysfunction or overloading of ERAD results in accumulation of unfolded proteins and triggers the unfolded protein response (UPR).[316,317] Mutant SOD1 interferes directly with ERAD by binding to derlin-1, a transmembrane protein responsible for the translocation of misfolded proteins from the ER lumen,[318] as well as the ER-luminal chaperone BiP[319] (Fig. 10). Sustained ER stress in SOD1 mutant mice leads to the activation of ASK1, an apoptotic protein kinase, and survival can be prolonged by ASK1 ablation.[318] Derlin-1 interaction was detected only after symptom onset,[318] but multiple triggers of ER stress are clearly present in ALS, as evidenced by presymptomatic UPR activation in SOD1 mutant mouse models[320] and upregulation of UPR components in SALS patients.[306,321,322] Furthermore, mutations in the UPR protein VAPB have been linked to some ALS cases.[20] Thus, ER stress may not be ruled out as a primary contributor to ALS pathogenesis nor may its involvement be limited to SOD1-related cases.

F. Aberrant RNA Processing

Malfunction and aggregation of two nucleic acid binding proteins was recently shown to be a common causal factor for some cases of both familial and sporadic ALS. Since the surprising observation that the 43-kDa *trans*-activating response region DNA-binding protein (TDP-43) is present in a majority of ubiquitinated proteinaceous inclusions in ALS and frontotemporal lobar degeneration (FTLD),[323,324] over 35 dominant mutations in TDP-43 have been linked to ALS.[309,325–343] TDP-43 is notably excluded from inclusions of patients with SOD1-related FALS,[344] perhaps an indication of divergent pathological mechanisms. However, it was recently shown that the small heat shock protein B8 (HspB8) is involved in clearance of both SOD1 and TDP-43 aggregates,[345] which is evidence that, despite differences in etiology, TDP-43 and SOD1-related ALS may respond to similar therapeutic approaches. Mutations in a second RNA/DNA-binding protein, fused in sarcoma (FUS) (also known as translocation in liposarcoma (TLS)), have also been linked to ALS and FTLD.[346–360] This common genetic basis for ALS and FTLD blurs the distinction between these disorders and may account for their co-occurance in some patients.[8]

The study of TDP-43- and FUS/TLS-related proteinopathies is a burgeoning field, as even the normal functions of these proteins were not well understood prior to the revelation of their roles in neurodegenerative diseases. Both proteins are widely expressed, predominantly nuclear proteins in healthy cells[361] and are involved in RNA processing events such as splicing and transcriptional regulation (reviewed in Ref. 362). Cytoplasmic aggregation and nuclear depletion are early, and perhaps independent, events in TDP-43-related ALS pathology[363–366] and are accompanied by proteolytic cleavage,[324,367,368] hyperphosphorylation,[367,369,370] and ubiquitination.[323,324] The combination of aberrant localization and posttranslational modification of TDP-43 in ALS raises the question of whether TDP-43 pathogenicity is a loss or gain of function. Does toxicity stem from the loss of normal TDP-43 nuclear function, or does cleaved, phosphorylated, or aggregated TDP-43 acquire cytotoxic properties? The TDP-43 C-terminal fragment is produced by caspase-3 cleavage[371–374] and increases in abundance as symptoms progress, suggesting that proteolysis of TDP-43 may be secondary to activation of apoptosis. Cytoplasmic inclusions stain negative for several known binding partners of TDP-43,[375] suggesting that aggregates do not exert toxicity by sequestering these components. However, the possibility that altered TDP-43 disrupts cellular homeostasis through novel aberrant interactions, as is the case with misfolded SOD1, has not been ruled out. FUS/TLS, while also aggregating in the cytoplasm,[354,357,360] appears to retain a more normal pattern of localization in the ALS-affected CNS, and neither phosphorylation nor cleavage is significantly correlated with disease.[354,357,360,370]

Intense study is under way to clarify the cellular functions of TDP-43 and FUS/TLS and the role of mutations in ALS and other neurodegenerative disorders. Interestingly, defective RNA processing has been noted previously in ALS and shown to cause EAAT2 deficiency,[376] a phenomenon that TDP-43 or FUS/TLS dysfunction may explain. Elucidation of the roles of TDP-43 and FUS/TLS in RNA metabolism has the potential to fill gaps in our understanding of numerous pathological deregulatory events in ALS.

VII. Concluding Remarks

The molecular biology of ALS is extraordinarily complex, and identification of the crucial initiating factors has remained elusive. However, a critical need exists for effective therapies to prevent loss of motor function and extend life. This effort should be focused on developing strategies for intervention at primary sites of dysfunction. In the case of SOD1-related FALS, protein misfolding and aggregation and calcium dysregulation drive many of the diverse pathological events in disease progression (Fig. 10) and should thus be considered prime candidates for therapeutic targeting. Mutant SOD1 transgenic mice will continue to be invaluable for mechanistic study of disease and development/evaluation of drug candidates. However, these models should be evaluated critically for relevance based on criteria such as copy numbers of wild-type and mutant SOD1 and presence of posttranslational modifications that affect stability.

REFERENCES

1. Goetz CG. Amyotrophic lateral sclerosis: early contributions of Jean-Martin Charcot. *Muscle Nerve* 2000;**23**:336.
2. Bruijn LI, Miller TM, Cleveland DW. Unraveling the mechanisms involved in motor neuron degeneration in ALS. *Annu Rev Neurosci* 2004;**27**:723.
3. Cleveland DW, Rothstein JD. From Charcot to Lou Gehrig: deciphering selective motor neuron death in ALS 1. *Nat Rev Neurosci* 2001;**2**:806.
4. Rothstein JD. Current hypotheses for the underlying biology of amyotrophic lateral sclerosis. *Ann Neurol* 2009;**65**(Suppl. 1):S3.
5. Chou SM, Norris FH. Amyotrophic lateral sclerosis: lower motor neuron disease spreading to upper motor neurons. *Muscle Nerve* 1993;**16**:864.
6. Eisen A, Weber M. The motor cortex and amyotrophic lateral sclerosis. *Muscle Nerve* 2001;**24**:564.
7. Mochizuki Y, Mizutani T, Takasu T. Amyotrophic lateral sclerosis with marked neurological asymmetry: clinicopathological study. *Acta Neuropathol* 1995;**90**:44.
8. Lomen-Hoerth C, Anderson T, Miller B. The overlap of amyotrophic lateral sclerosis and frontotemporal dementia. *Neurology* 2002;**59**:1077.
9. Ravits JM, La Spada AR. ALS motor phenotype heterogeneity, focality, and spread: deconstructing motor neuron degeneration. *Neurology* 2009;**73**:805.
10. Brooks BR. El Escorial World Federation of Neurology criteria for the diagnosis of amyotrophic lateral sclerosis. Subcommittee on Motor Neuron Diseases/Amyotrophic Lateral Sclerosis of the World Federation of Neurology Research Group on Neuromuscular Diseases and the El Escorial "Clinical limits of amyotrophic lateral sclerosis" workshop contributors. *J Neurol Sci* 1994;**124**(Suppl.):96–107.
11. de Carvalho M, Dengler R, Eisen A, England JD, Kaji R, Kimura J, et al. Electrodiagnostic criteria for diagnosis of ALS. *Clin Neurophysiol* 2008;**119**:497–503.
12. Beleza-Meireles A, Al-Chalabi A. Genetic studies of amyotrophic lateral sclerosis: controversies and perspectives. *Amyotroph Lateral Scler* 2009;**10**:1.
13. Deng HX, Hentati A, Tainer JA, Iqbal Z, Cayabyab A, Hung WY, et al. Amyotrophic lateral sclerosis and structural defects in Cu, Zn superoxide dismutase. *Science* 1993;**261**:1047.
14. Rosen DR, Siddique T, Patterson D, Figlewicz DA, Sapp P, Hentati A, et al. Mutations in Cu/Zn superoxide dismutase gene are associated with familial amyotrophic lateral sclerosis 1. *Nature* 1993;**362**:59.
15. Al-Chalabi A, Andersen PM, Chioza B, Shaw C, Sham PC, Robberecht W, et al. Recessive amyotrophic lateral sclerosis families with the D90A SOD1 mutation share a common founder: evidence for a linked protective factor. *Hum Mol Genet* 1998;**7**:2045.
16. Hadano S, Hand CK, Osuga H, Yanagisawa Y, Otomo A, Devon RS, et al. A gene encoding a putative GTPase regulator is mutated in familial amyotrophic lateral sclerosis 2. *Nat Genet* 2001;**29**:166.
17. Yang Y, Hentati A, Deng HX, Dabbagh O, Sasaki T, Hirano M, et al. The gene encoding alsin, a protein with three guanine-nucleotide exchange factor domains, is mutated in a form of recessive amyotrophic lateral sclerosis. *Nat Genet* 2001;**29**:160.
18. Chance PF, Rabin BA, Ryan SG, Ding Y, Scavina M, Crain B, et al. Linkage of the gene for an autosomal dominant form of juvenile amyotrophic lateral sclerosis to chromosome 9q34. *Am J Hum Genet* 1998;**62**:633.
19. Chen YZ, Bennett CL, Huynh HM, Blair IP, Puls I, Irobi J, et al. DNA/RNA helicase gene mutations in a form of juvenile amyotrophic lateral sclerosis (ALS4). *Am J Hum Genet* 2004;**74**:1128.

20. Nishimura AL, Mitne-Neto M, Silva HC, Richieri-Costa A, Middleton S, Cascio D, et al. A mutation in the vesicle-trafficking protein VAPB causes late-onset spinal muscular atrophy and amyotrophic lateral sclerosis. *Am J Hum Genet* 2004;**75**:822.

21. Greenway MJ, Alexander MD, Ennis S, Traynor BJ, Corr B, Frost E, et al. A novel candidate region for ALS on chromosome 14q11.2. *Neurology* 2004;**63**:1936.

22. Greenway MJ, Andersen PM, Russ C, Ennis S, Cashman S, Donaghy C, et al. ANG mutations segregate with familial and 'sporadic' amyotrophic lateral sclerosis. *Nat Genet* 2006;**38**:411.

23. Wu D, Yu W, Kishikawa H, Folkerth RD, Iafrate AJ, Shen Y, et al. Angiogenin loss-of-function mutations in amyotrophic lateral sclerosis. *Ann Neurol* 2007;**62**:609.

24. Al-Chalabi A, Andersen PM, Nilsson P, Chioza B, Andersson JL, Russ C, et al. Deletions of the heavy neurofilament subunit tail in amyotrophic lateral sclerosis. *Hum Mol Genet* 1999;**8**:157.

25. Figlewicz DA, Krizus A, Martinoli MG, Meininger V, Dib M, Rouleau GA, et al. Variants of the heavy neurofilament subunit are associated with the development of amyotrophic lateral sclerosis. *Hum Mol Genet* 1994;**3**:1757.

26. Lambrechts D, Storkebaum E, Morimoto M, Del-Favero J, Desmet F, Marklund SL, et al. VEGF is a modifier of amyotrophic lateral sclerosis in mice and humans and protects motoneurons against ischemic death. *Nat Genet* 2003;**34**:383.

27. Al-Chalabi A, Scheffler MD, Smith BN, Parton MJ, Cudkowicz ME, Andersen PM, et al. Ciliary neurotrophic factor genotype does not influence clinical phenotype in amyotrophic lateral sclerosis. *Ann Neurol* 2003;**54**:130.

28. Giess R, Goetz R, Schrank B, Ochs G, Sendtner M, Toyka K. Potential implications of a ciliary neurotrophic factor gene mutation in a German population of patients with motor neuron disease. *Muscle Nerve* 1998;**21**:236.

29. Kokubo Y, Kuzuhara S, Narita Y. Geographical distribution of amyotrophic lateral sclerosis with neurofibrillary tangles in the Kii Peninsula of Japan. *J Neurol* 2000;**247**:850–2.

30. Haley RW. Excess incidence of ALS in young Gulf War veterans. *Neurology* 2003;**61**:750.

31. Horner RD, Kamins KG, Feussner JR, Grambow SC, Hoff-Lindquist J, Harati Y, et al. Occurrence of amyotrophic lateral sclerosis among Gulf War veterans. *Neurology* 2003;**61**:742.

32. Chio A, Benzi G, Dossena M, Mutani R, Mora G. Severely increased risk of amyotrophic lateral sclerosis among Italian professional football players. *Brain* 2005;**128**:472.

33. Sutedja NA, Fischer K, Veldink JH, Van Der Heijden GJ, Kromhout H, Heederik D, Huisman MH, Wokke JJ, Van den Berg LH. What we truly know about occupation as a risk factor for ALS: a critical and systematic review. *Amyotroph Lateral Scler* 2008;**10**:295–301.

34. Sutedja NA, Veldink JH, Fischer K, Kromhout H, Heederik D, Huisman MH, Wokke JH, Van den Berg LH. Exposure to chemicals and metals and risk of amyotrophic lateral sclerosis: a systematic review. *Amyotroph Lateral Scler* 2008;**10**:302–9.

35. Mattson MP. Infectious agents and age-related neurodegenerative disorders. *Ageing Res Rev* 2004;**3**:105.

36. Jafari H, Couratier P, Camu W. Motor neuron disease after electric injury. *J Neurol Neurosurg Psychiatry* 2001;**71**:265.

37. Kurtzke JF. Risk factors in amyotrophic lateral sclerosis. *Adv Neurol* 1991;**56**:245.

38. Duncan MW, Steele JC, Kopin IJ, Markey SP. 2-Amino-3-(methylamino)-propanoic acid (BMAA) in cycad flour: an unlikely cause of amyotrophic lateral sclerosis and parkinsonism-dementia of Guam. *Neurology* 1990;**40**:767.

39. Cox PA, Banack SA, Murch SJ. Biomagnification of cyanobacterial neurotoxins and neurodegenerative disease among the Chamorro people of Guam. *Proc Natl Acad Sci USA* 2003;**100**:13380.

40. Cox PA, Banack SA, Murch SJ, Rasmussen U, Tien G, Bidigare RR, et al. Diverse taxa of cyanobacteria produce β-N-methylamino-L-alanine, a neurotoxic amino acid. *Proc Natl Acad Sci USA* 2005;**102**:5074.
41. Jonasson S, Eriksson J, Berntzon L, Spacil Z, Ilag LL, Ronnevi LO, et al. Transfer of a cyanobacterial neurotoxin within a temperate aquatic ecosystem suggests pathways for human exposure. *Proc Natl Acad Sci USA* 2010;**107**:9252.
42. Cox PA, Richer R, Metcalf JS, Banack SA, Codd GA, Bradley WG. Cyanobacteria and BMAA exposure from desert dust: a possible link to sporadic ALS among Gulf War veterans. *Amyotroph Lateral Scler* 2009;**10**(Suppl 2):109.
43. Caller TA, Doolin JW, Haney JF, Murby AJ, West KG, Farrar HE, et al. A cluster of amyotrophic lateral sclerosis in New Hampshire: a possible role for toxic cyanobacteria blooms. *Amyotroph Lateral Scler* 2009;**10**(Suppl. 2):101.
44. Turner BJ, Talbot K. Transgenics, toxicity and therapeutics in rodent models of mutant SOD1-mediated familial ALS. *Prog Neurobiol* 2008;**85**:94.
45. Andersen PM, Nilsson P, Keranen ML, Forsgren L, Hagglund J, Karlsborg M, et al. Phenotypic heterogeneity in motor neuron disease patients with CuZn-superoxide dismutase mutations in Scandinavia. *Brain* 1997;**120**(Pt. 10):1723.
46. Hand CK, Khoris J, Salachas F, Gros-Louis F, Lopes AA, Mayeux-Portas V, et al. A novel locus for familial amyotrophic lateral sclerosis, on chromosome 18q. *Am J Hum Genet* 2002;**70**:251.
47. Bruijn LI, Becher MW, Lee MK, Anderson KL, Jenkins NA, Copeland NG, et al. ALS-linked SOD1 mutant G85R mediates damage to astrocytes and promotes rapidly progressive disease with SOD1-containing inclusions. *Neuron* 1997;**18**:327.
48. Gurney ME, Pu H, Chiu AY, Dal Canto MC, Polchow CY, Alexander DD, et al. Motor neuron degeneration in mice that express a human Cu, Zn superoxide dismutase mutation. *Science* 1994;**264**:1772.
49. Wong PC, Pardo CA, Borchelt DR, Lee MK, Copeland NG, Jenkins NA, et al. An adverse property of a familial ALS-linked SOD1 mutation causes motor neuron disease characterized by vacuolar degeneration of mitochondria. *Neuron* 1995;**14**:1105.
50. Reaume AG, Elliott JL, Hoffman EK, Kowall NW, Ferrante RJ, Siwek DF, et al. Motor neurons in Cu/Zn superoxide dismutase-deficient mice develop normally but exhibit enhanced cell death after axonal injury 1. *Nat Genet* 1996;**13**:43.
51. Bruijn LI, Houseweart MK, Kato S, Anderson KL, Anderson SD, Ohama E, et al. Aggregation and motor neuron toxicity of an ALS-linked SOD1 mutant independent from wild-type SOD1 1. *Science* 1998;**281**:1851.
52. Khare SD, Caplow M, Dokholyan NV. FALS mutations in Cu, Zn superoxide dismutase destabilize the dimer and increase dimer dissociation propensity: a large-scale thermodynamic analysis 1. *Amyloid* 2006;**13**:226.
53. Furukawa Y, O'Halloran TV. Amyotrophic lateral sclerosis mutations have the greatest destabilizing effect on the apo- and reduced form of SOD1, leading to unfolding and oxidative aggregation. *J Biol Chem* 2005;**280**:17266.
54. Hough MA, Grossmann JG, Antonyuk SV, Strange RW, Doucette PA, Rodriguez JA, et al. Dimer destabilization in superoxide dismutase may result in disease-causing properties: structures of motor neuron disease mutants 1. *Proc Natl Acad Sci USA* 2004;**101**:5976.
55. Rodriguez JA, Shaw BF, Durazo A, Sohn SH, Doucette PA, Nersissian AM, et al. Destabilization of apoprotein is insufficient to explain Cu, Zn-superoxide dismutase-linked ALS pathogenesis. *Proc Natl Acad Sci USA* 2005;**102**:10516.
56. Shaw BF, Valentine JS. How do ALS-associated mutations in superoxide dismutase 1 promote aggregation of the protein? 1. *Trends Biochem Sci* 2007;**32**:78.

57. Radunovic A, Leigh PN. Cu/Zn superoxide dismutase gene mutations in amyotrophic lateral sclerosis: correlation between genotype and clinical features. *J Neurol Neurosurg Psychiatry* 1996;**61**:565.

58. Cudkowicz ME, McKenna-Yasek D, Sapp PE, Chin W, Geller B, Hayden DL, et al. Epidemiology of mutations in superoxide dismutase in amyotrophic lateral sclerosis. *Ann Neurol* 1997;**41**:210.

59. Bystrom R, Andersen PM, Grobner G, Oliveberg M. SOD1 mutations targeting surface hydrogen bonds promote amyotrophic lateral sclerosis without reducing apo-state stability. *J Biol Chem* 2010;**285**:19544.

60. Wang Q, Johnson JL, Agar NY, Agar JN. Protein aggregation and protein instability govern familial amyotrophic lateral sclerosis patient survival 1. *PLoS Biol* 2008;**6**:e170. doi:10.1371/journal.pbio.0060170.

61. Chiti F, Stefani M, Taddei N, Ramponi G, Dobson CM. Rationalization of the effects of mutations on peptide and protein aggregation rates. *Nature* 2003;**424**:805–8.

62. Gruzman A, Wood WL, Alpert E, Prasad MD, Miller RG, Rothstein JD, et al. Common molecular signature in SOD1 for both sporadic and familial amyotrophic lateral sclerosis. *Proc Natl Acad Sci USA* 2007;**104**:12524.

63. Shibata N, Asayama K, Hirano A, Kobayashi M. Immunohistochemical study on superoxide dismutases in spinal cords from autopsied patients with amyotrophic lateral sclerosis 2. *Dev Neurosci* 1996;**18**:492.

64. Shibata N, Hirano A, Kobayashi M, Sasaki S, Kato T, Matsumoto S, et al. Cu/Zn superoxide dismutase-like immunoreactivity in Lewy body-like inclusions of sporadic amyotrophic lateral sclerosis 1. *Neurosci Lett* 1994;**179**:149.

65. Matsumoto S, Kusaka H, Ito H, Shibata N, Asayama T, Imai T. Sporadic amyotrophic lateral sclerosis with dementia and Cu/Zn superoxide dismutase-positive Lewy body-like inclusions. *Clin Neuropathol* 1996;**15**:41.

66. Bartnikas TB, Gitlin JD. Mechanisms of biosynthesis of mammalian copper/zinc superoxide dismutase. *J Biol Chem* 2003;**278**:33602.

67. Forman HJ, Fridovich I. On the stability of bovine superoxide dismutase. The effects of metals. *J Biol Chem* 1973;**248**:2645.

68. Ding F, Dokholyan NV. Dynamical roles of metal ions and the disulfide bond in Cu, Zn superoxide dismutase folding and aggregation. *Proc Natl Acad Sci USA* 2008;**105**:19696.

69. Tiwari A, Hayward LJ. Mutant SOD1 instability: implications for toxicity in amyotrophic lateral sclerosis. *Neurodegener Dis* 2005;**2**:115.

70. Andersen PM. Amyotrophic lateral sclerosis associated with mutations in the CuZn superoxide dismutase gene 1. *Curr Neurol Neurosci Rep* 2006;**6**:37.

71. Esteban J, Rosen DR, Bowling AC, Sapp P, Kenna-Yasek D, O'Regan JP, et al. Identification of two novel mutations and a new polymorphism in the gene for Cu/Zn superoxide dismutase in patients with amyotrophic lateral sclerosis 1. *Hum Mol Genet* 1994;**3**:997.

72. Nogales-Gadea G, Garcia-Arumi E, Andreu AL, Cervera C, Gamez J. A novel exon 5 mutation (N139H) in the SOD1 gene in a Spanish family associated with incomplete penetrance. *J Neurol Sci* 2004;**219**:1–6.

73. Prudencio M, Hart PJ, Borchelt DR, Andersen PM. Variation in aggregation propensities among ALS-associated variants of SOD1: correlation to human disease 1. *Hum Mol Genet* 2009;**18**:3217.

74. Banci L, Bertini I, Cantini F, D'Onofrio M, Viezzoli MS. Structure and dynamics of copper-free SOD: the protein before binding copper. *Protein Sci* 2002;**11**:2479.

75. Doucette PA, Whitson LJ, Cao X, Schirf V, Demeler B, Valentine JS, et al. Dissociation of human copper-zinc superoxide dismutase dimers using chaotrope and reductant. Insights into the molecular basis for dimer stability 1. *J Biol Chem* 2004;**279**:54558.

76. Khare SD, Caplow M, Dokholyan NV. The rate and equilibrium constants for a multistep reaction sequence for the aggregation of superoxide dismutase in amyotrophic lateral sclerosis 1. *Proc Natl Acad Sci USA* 2004;**101**:15094.

77. Arnesano F, Banci L, Bertini I, Martinelli M, Furukawa Y, O'Halloran TV. The unusually stable quaternary structure of human Cu, Zn-superoxide dismutase 1 is controlled by both metal occupancy and disulfide status. *J Biol Chem* 2004;**279**:47998.

78. Rakhit R, Crow JP, Lepock JR, Kondejewski LH, Cashman NR, Chakrabartty A. Monomeric Cu, Zn-superoxide dismutase is a common misfolding intermediate in the oxidation models of sporadic and familial amyotrophic lateral sclerosis 1. *J Biol Chem* 2004;**279**:15499.

79. Lindberg MJ, Normark J, Holmgren A, Oliveberg M. Folding of human superoxide dismutase: disulfide reduction prevents dimerization and produces marginally stable monomers. *Proc Natl Acad Sci USA* 2004;**101**:15893.

80. Ray SS, Nowak RJ, Strokovich K, Brown Jr. RH, Walz T, Lansbury Jr. PT. An intersubunit disulfide bond prevents in vitro aggregation of a superoxide dismutase-1 mutant linked to familial amyotrophic lateral sclerosis 1. *Biochemistry* 2004;**43**:4899.

81. Molnar KS, Karabacak NM, Johnson JL, Wang Q, Tiwari A, Hayward LJ, et al. A common property of amyotrophic lateral sclerosis-associated variants: destabilization of the copper/zinc superoxide dismutase electrostatic loop. *J Biol Chem* 2009;**284**:30965.

82. Durazo A, Shaw BF, Chattopadhyay M, Faull KF, Nersissian AM, Valentine JS, et al. Metal-free superoxide dismutase-1 and three different amyotrophic lateral sclerosis variants share a similar partially unfolded β-barrel at physiological temperature. *J Biol Chem* 2009;**284**:34382.

83. Teilum K, Smith MH, Schulz E, Christensen LC, Solomentsev G, Oliveberg M, et al. Transient structural distortion of metal-free Cu/Zn superoxide dismutase triggers aberrant oligomerization 1. *Proc Natl Acad Sci USA* 2009;**106**:18273.

84. Tiwari A, Liba A, Sohn SH, Seetharaman SV, Bilsel O, Matthews CR, et al. Metal deficiency increases aberrant hydrophobicity of mutant superoxide dismutases that cause amyotrophic lateral sclerosis. *J Biol Chem* 2009;**284**:27746.

85. Munch C, Bertolotti A. Exposure of hydrophobic surfaces initiates aggregation of diverse ALS-causing superoxide dismutase-1 mutants. *J Mol Biol* 2010;**399**:512.

86. Furukawa Y, Kaneko K, Yamanaka K, O'Halloran TV, Nukina N. Complete loss of post-translational modifications triggers fibrillar aggregation of SOD1 in the familial form of amyotrophic lateral sclerosis. *J Biol Chem* 2008;**283**:24167.

87. Khare SD, Ding F, Dokholyan NV. Folding of Cu, Zn superoxide dismutase and familial amyotrophic lateral sclerosis. *J Mol Biol* 2003;**334**:515.

88. Khare SD, Dokholyan NV. Common dynamical signatures of familial amyotrophic lateral sclerosis-associated structurally diverse Cu, Zn superoxide dismutase mutants. *Proc Natl Acad Sci USA* 2006;**103**:3147.

89. Rakhit R, Robertson J, Vande VC, Horne P, Ruth DM, Griffin J, et al. An immunological epitope selective for pathological monomer-misfolded SOD1 in ALS. *Nat Med* 2007;**13**:754.

90. Kerman A, Liu HN, Croul S, Bilbao J, Rogaeva E, Zinman L, et al. Amyotrophic lateral sclerosis is a non-amyloid disease in which extensive misfolding of SOD1 is unique to the familial form. *Acta Neuropathol* 2010;**119**:335.

91. Zetterstrom P, Stewart HG, Bergemalm D, Jonsson PA, Graffmo KS, Andersen PM, et al. Soluble misfolded subfractions of mutant superoxide dismutase-1s are enriched in spinal cords throughout life in murine ALS models. *Proc Natl Acad Sci USA* 2007;**104**:14157.

92. Deng HX, Shi Y, Furukawa Y, Zhai H, Fu R, Liu E, et al. Conversion to the amyotrophic lateral sclerosis phenotype is associated with intermolecular linked insoluble aggregates of SOD1 in mitochondria. *Proc Natl Acad Sci USA* 2006;**103**:7142. doi:10.1073/pnas.0602046103.

93. Johnston JA, Dalton MJ, Gurney ME, Kopito RR. Formation of high molecular weight complexes of mutant Cu, Zn-superoxide dismutase in a mouse model for familial amyotrophic lateral sclerosis 1. *Proc Natl Acad Sci USA* 2000;**97**:12571.

94. Shibata N, Hirano A, Kobayashi M, Siddique T, Deng HX, Hung WY, et al. Intense superoxide - dismutase-1 immunoreactivity in intracytoplasmic hyaline inclusions of familial amyotrophic lateral sclerosis with posterior column involvement 1. *J Neuropathol Exp Neurol* 1996;**55**:481.

95. Shaw BF, Lelie HL, Durazo A, Nersissian AM, Xu G, Chan PK, et al. Detergent-insoluble aggregates associated with amyotrophic lateral sclerosis in transgenic mice contain primarily full-length, unmodified superoxide dismutase-1. *J Biol Chem* 2008;**283**:8340.

96. Banci L, Bertini I, Durazo A, Girotto S, Gralla EB, Martinelli M, et al. Metal-free superoxide dismutase forms soluble oligomers under physiological conditions: a possible general mechanism for familial ALS 1. *Proc Natl Acad Sci USA* 2007;**104**:11263.

97. Chattopadhyay M, Durazo A, Sohn SH, Strong CD, Gralla EB, Whitelegge JP, et al. Initiation and elongation in fibrillation of ALS-linked superoxide dismutase. *Proc Natl Acad Sci USA* 2008;**105**:18663.

98. DiDonato M, Craig L, Huff ME, Thayer MM, Cardoso RM, Kassmann CJ, et al. ALS mutants of human superoxide dismutase form fibrous aggregates via framework destabilization 1. *J Mol Biol* 2003;**332**:601.

99. Jonsson PA, Graffmo KS, Andersen PM, Brannstrom T, Lindberg M, Oliveberg M, et al. Disulphide-reduced superoxide dismutase-1 in CNS of transgenic amyotrophic lateral sclerosis models. *Brain* 2006;**129**:451.

100. Matsumoto G, Kim S, Morimoto RI. Huntingtin and mutant SOD1 form aggregate structures with distinct molecular properties in human cells. *J Biol Chem* 2006;**281**:4477.

101. Matsumoto G, Stojanovic A, Holmberg CI, Kim S, Morimoto RI. Structural properties and neuronal toxicity of amyotrophic lateral sclerosis-associated Cu/Zn superoxide dismutase 1 aggregates 1. *J Cell Biol* 2005;**171**:75.

102. Niwa J, Yamada S, Ishigaki S, Sone J, Takahashi M, Katsuno M, et al. Disulfide bond mediates aggregation, toxicity, and ubiquitylation of familial amyotrophic lateral sclerosis-linked mutant SOD1. *J Biol Chem* 2007;**282**:28087.

103. Karch CM, Prudencio M, Winkler DD, Hart PJ, Borchelt DR. Role of mutant SOD1 disulfide oxidation and aggregation in the pathogenesis of familial ALS. *Proc Natl Acad Sci USA* 2009;**106**:7774.

104. Karch CM, Borchelt DR. A limited role for disulfide cross-linking in the aggregation of mutant SOD1 linked to familial amyotrophic lateral sclerosis. *J Biol Chem* 2008;**283**:13528.

105. Cozzolino M, Amori I, Pesaresi MG, Ferri A, Nencini M, Carri MT. Cysteine 111 affects aggregation and cytotoxicity of mutant Cu, Zn-superoxide dismutase associated with familial amyotrophic lateral sclerosis. *J Biol Chem* 2008;**283**:866.

105a. Redler RL, Wilcox KC, Proctor EA, Fee L, Caplow M, Dokholyan NV. Glutathionylation at Cys-111 Induces Dissociation of Wild Type and FALS Mutant SOD1 Dimers. *Biochem* 2011;**50**:7057–66.

106. Fukada K, Nagano S, Satoh M, Tohyama C, Nakanishi T, Shimizu A, et al. Stabilization of mutant Cu/Zn superoxide dismutase (SOD1) protein by coexpressed wild SOD1 protein accelerates the disease progression in familial amyotrophic lateral sclerosis mice. *Eur J Neurosci* 2001;**14**:2032.

107. Jaarsma D, Haasdijk ED, Grashorn JA, Hawkins R, van Duijn W, Verspaget HW, et al. Human Cu/Zn superoxide dismutase (SOD1) overexpression in mice causes mitochondrial vacuolization, axonal degeneration, and premature motoneuron death and accelerates motoneuron disease in mice expressing a familial amyotrophic lateral sclerosis mutant SOD1. *Neurobiol Dis* 2000;**7**:623.

108. Wang L, Deng HX, Grisotti G, Zhai H, Siddique T, Roos RP. Wild-type SOD1 overexpression accelerates disease onset of a G85R SOD1 mouse. *Hum Mol Genet* 2009;**18**:1642.

109. Sasaki S, Warita H, Murakami T, Shibata N, Komori T, Abe K, et al. Ultrastructural study of aggregates in the spinal cord of transgenic mice with a G93A mutant SOD1 gene 1. *Acta Neuropathol* 2005;**109**:247.

110. Turner BJ, Lopes EC, Cheema SS. Neuromuscular accumulation of mutant superoxide dismutase 1 aggregates in a transgenic mouse model of familial amyotrophic lateral sclerosis. *Neurosci Lett* 2003;**350**:132.

111. Wang J, Xu G, Borchelt DR. High molecular weight complexes of mutant superoxide dismutase 1: age-dependent and tissue-specific accumulation. *Neurobiol Dis* 2002;**9**:139.

112. Wang J, Xu G, Gonzales V, Coonfield M, Fromholt D, Copeland NG, et al. Fibrillar inclusions and motor neuron degeneration in transgenic mice expressing superoxide dismutase 1 with a disrupted copper-binding site. *Neurobiol Dis* 2002;**10**:128.

113. Wang J, Xu G, Li H, Gonzales V, Fromholt D, Karch C, et al. Somatodendritic accumulation of misfolded SOD1-L126Z in motor neurons mediates degeneration: alphaB-crystallin modulates aggregation. *Hum Mol Genet* 2005;**14**:2335.

114. Jonsson PA, Ernhill K, Andersen PM, Bergemalm D, Brannstrom T, Gredal O, et al. Minute quantities of misfolded mutant superoxide dismutase-1 cause amyotrophic lateral sclerosis. *Brain* 2004;**127**:73.

115. Arrasate M, Mitra S, Schweitzer ES, Segal MR, Finkbeiner S. Inclusion body formation reduces levels of mutant huntingtin and the risk of neuronal death 1. *Nature* 2004;**431**:805.

116. Caughey B, Lansbury PT. Protofibrils, pores, fibrils, and neurodegeneration: separating the responsible protein aggregates from the innocent bystanders. *Annu Rev Neurosci* 2003;**26**:267.

117. Kirkitadze MD, Bitan G, Teplow DB. Paradigm shifts in Alzheimer's disease and other neurodegenerative disorders: the emerging role of oligomeric assemblies. *J Neurosci Res* 2002;**69**:567.

118. Mattson MP. Apoptosis in neurodegenerative disorders. *Nat Rev Mol Cell Biol* 2000;**1**:120.

119. Gould TW, Buss RR, Vinsant S, Prevette D, Sun W, Knudson CM, et al. Complete dissociation of motor neuron death from motor dysfunction by Bax deletion in a mouse model of ALS. *J Neurosci* 2006;**26**:8774.

120. Li M, Ona VO, Guegan C, Chen M, Jackson-Lewis V, Andrews LJ, et al. Functional role of caspase-1 and caspase-3 in an ALS transgenic mouse model. *Science* 2000;**288**:335.

121. Martin LJ, Liu Z, Chen K, Price AC, Pan Y, Swaby JA, et al. Motor neuron degeneration in amyotrophic lateral sclerosis mutant superoxide dismutase-1 transgenic mice: mechanisms of mitochondriopathy and cell death. *J Comp Neurol* 2007;**500**:20.

122. Vukosavic S, Stefanis L, Jackson-Lewis V, Guegan C, Romero N, Chen C, et al. Delaying caspase activation by Bcl-2: a clue to disease retardation in a transgenic mouse model of amyotrophic lateral sclerosis. *J Neurosci* 2000;**20**:9119.

123. Guegan C, Przedborski S. Programmed cell death in amyotrophic lateral sclerosis. *J Clin Invest* 2003;**111**:153.

124. Migheli A, Atzori C, Piva R, Tortarolo M, Girelli M, Schiffer D, et al. Lack of apoptosis in mice with ALS. *Nat Med* 1999;**5**:966.

125. Martin LJ, Al-Abdulla NA, Brambrink AM, Kirsch JR, Sieber FE, Portera-Cailliau C. Neurodegeneration in excitotoxicity, global cerebral ischemia, and target deprivation: a perspective on the contributions of apoptosis and necrosis. *Brain Res Bull* 1998;**46**:281.

126. Pramatarova A, Laganiere J, Roussel J, Brisebois K, Rouleau GA. Neuron-specific expression of mutant superoxide dismutase 1 in transgenic mice does not lead to motor impairment. *J Neurosci* 2001;**21**:3369.

127. Lino MM, Schneider C, Caroni P. Accumulation of SOD1 mutants in postnatal motoneurons does not cause motoneuron pathology or motoneuron disease. *J Neurosci* 2002;**22**:4825.

128. Jaarsma D, Teuling E, Haasdijk ED, De Zeeuw CI, Hoogenraad CC. Neuron-specific expression of mutant superoxide dismutase is sufficient to induce amyotrophic lateral sclerosis in transgenic mice. *J Neurosci* 2008;**28**:2075.

129. Nagai M, Re DB, Nagata T, Chalazonitis A, Jessell TM, Wichterle H, et al. Astrocytes expressing ALS-linked mutated SOD1 release factors selectively toxic to motor neurons. *Nat Neurosci* 2007;**10**:615.

130. Gong YH, Parsadanian AS, Andreeva A, Snider WD, Elliott JL. Restricted expression of G86R Cu/Zn superoxide dismutase in astrocytes results in astrocytosis but does not cause motoneuron degeneration. *J Neurosci* 2000;**20**:660.

131. Beers DR, Henkel JS, Xiao Q, Zhao W, Wang J, Yen AA, et al. Wild-type microglia extend survival in PU.1 knockout mice with familial amyotrophic lateral sclerosis. *Proc Natl Acad Sci USA* 2006;**103**:16021.

132. Turner BJ, Ackerley S, Davies KE, Talbot K. Dismutase-competent SOD1 mutant accumulation in myelinating Schwann cells is not detrimental to normal or transgenic ALS model mice. *Hum Mol Genet* 2010;**19**:815.

133. Boillee S, Yamanaka K, Lobsiger CS, Copeland NG, Jenkins NA, Kassiotis G, et al. Onset and progression in inherited ALS determined by motor neurons and microglia. *Science* 2006;**312**:1389.

134. Yamanaka K, Boillee S, Roberts EA, Garcia ML, McAlonis-Downes M, Mikse OR, et al. Mutant SOD1 in cell types other than motor neurons and oligodendrocytes accelerates onset of disease in ALS mice. *Proc Natl Acad Sci USA* 2008;**105**:7594.

135. Yamanaka K, Chun SJ, Boillee S, Fujimori-Tonou N, Yamashita H, Gutmann DH, et al. Astrocytes as determinants of disease progression in inherited amyotrophic lateral sclerosis. *Nat Neurosci* 2008;**11**:251.

136. Lobsiger CS, Boillee S, McAlonis-Downes M, Khan AM, Feltri ML, Yamanaka K, et al. Schwann cells expressing dismutase active mutant SOD1 unexpectedly slow disease progression in ALS mice. *Proc Natl Acad Sci USA* 2009;**106**:4465.

137. Wong M, Martin LJ. Skeletal muscle-restricted expression of human SOD1 causes motor neuron degeneration in transgenic mice. *Hum Mol Genet* 2010;**19**:2284.

138. Miller TM, Kim SH, Yamanaka K, Hester M, Umapathi P, Arnson H, et al. Gene transfer demonstrates that muscle is not a primary target for non-cell-autonomous toxicity in familial amyotrophic lateral sclerosis. *Proc Natl Acad Sci USA* 2006;**103**:19546.

139. Towne C, Raoul C, Schneider BL, Aebischer P. Systemic AAV6 delivery mediating RNA interference against SOD1: neuromuscular transduction does not alter disease progression in fALS mice. *Mol Ther* 2008;**16**:1018.

140. Dobrowolny G, Aucello M, Molinaro M, Musaro A. Local expression of mIgf-1 modulates ubiquitin, caspase and CDK5 expression in skeletal muscle of an ALS mouse model. *Neurol Res* 2008;**30**:131.

141. Fischer LR, Culver DG, Tennant P, Davis AA, Wang M, Castellano-Sanchez A, et al. Amyotrophic lateral sclerosis is a distal axonopathy: evidence in mice and man. *Exp Neurol* 2004;**185**:232.

142. Rouaux C, Panteleeva I, Rene F, Gonzalez de Aguilar JL, Echaniz-Laguna A, Dupuis L, et al. Sodium valproate exerts neuroprotective effects in vivo through CREB-binding protein-dependent mechanisms but does not improve survival in an amyotrophic lateral sclerosis mouse model. *J Neurosci* 2007;**27**:5535.

143. Dewil M, dela Cruz VF, Van Den Bosch L, Robberecht W. Inhibition of p38 mitogen activated protein kinase activation and mutant SOD1(G93A)-induced motor neuron death. *Neurobiol Dis* 2007;**26**:332.

144. Perlson E, Jeong GB, Ross JL, Dixit R, Wallace KE, Kalb RG, et al. A switch in retrograde signaling from survival to stress in rapid-onset neurodegeneration. *J Neurosci* 2009;**29**:9903.

145. Dupuis L, Gonzalez de Aguilar JL, Echaniz-Laguna A, Eschbach J, Rene F, Oudart H, et al. Muscle mitochondrial uncoupling dismantles neuromuscular junction and triggers distal degeneration of motor neurons. *PLoS One* 2009;4:e5390.

146. Williams AH, Valdez G, Moresi V, Qi X, McAnally J, Elliott JL, et al. MicroRNA-206 delays ALS progression and promotes regeneration of neuromuscular synapses in mice. *Science* 2009;**326**:1549.

147. Rothstein JD, Martin LJ, Kuncl RW. Decreased glutamate transport by the brain and spinal cord in amyotrophic lateral sclerosis. *N Engl J Med* 1992;**326**:1464.

148. Maragakis NJ, Dykes-Hoberg M, Rothstein JD. Altered expression of the glutamate transporter EAAT2b in neurological disease. *Ann Neurol* 2004;**55**:469.

149. Fray AE, Ince PG, Banner SJ, Milton ID, Usher PA, Cookson MR, et al. The expression of the glial glutamate transporter protein EAAT2 in motor neuron disease: an immunohistochemical study. *Eur J Neurosci* 1998;**10**:2481.

150. Howland DS, Liu J, She Y, Goad B, Maragakis NJ, Kim B, et al. Focal loss of the glutamate transporter EAAT2 in a transgenic rat model of SOD1 mutant-mediated amyotrophic lateral sclerosis (ALS). *Proc Natl Acad Sci USA* 2002;**99**:1604.

151. Rothstein JD, Van KM, Levey AI, Martin LJ, Kuncl RW. Selective loss of glial glutamate transporter GLT-1 in amyotrophic lateral sclerosis. *Ann Neurol* 1995;**38**:73.

152. Sasaki S, Komori T, Iwata M. Excitatory amino acid transporter 1 and 2 immunoreactivity in the spinal cord in amyotrophic lateral sclerosis. *Acta Neuropathol* 2000;**100**:138.

153. Rothstein JD, Dykes-Hoberg M, Pardo CA, Bristol LA, Jin L, Kuncl RW, et al. Knockout of glutamate transporters reveals a major role for astroglial transport in excitotoxicity and clearance of glutamate. *Neuron* 1996;**16**:675.

154. Guo H, Lai L, Butchbach ME, Stockinger MP, Shan X, Bishop GA, et al. Increased expression of the glial glutamate transporter EAAT2 modulates excitotoxicity and delays the onset but not the outcome of ALS in mice. *Hum Mol Genet* 2003;**12**:2519.

155. Rothstein JD, Patel S, Regan MR, Haenggeli C, Huang YH, Bergles DE, et al. β-Lactam antibiotics offer neuroprotection by increasing glutamate transporter expression. *Nature* 2005;**433**:73.

156. Bristol LA, Rothstein JD. Glutamate transporter gene expression in amyotrophic lateral sclerosis motor cortex. *Ann Neurol* 1996;**39**:676.

157. Yang Y, Gozen O, Watkins A, Lorenzini I, Lepore A, Gao Y, et al. Presynaptic regulation of astroglial excitatory neurotransmitter transporter GLT1. *Neuron* 2009;**61**:880.

158. Pasinelli P, Houseweart MK, Brown Jr. RH, Cleveland DW. Caspase-1 and -3 are sequentially activated in motor neuron death in Cu, Zn superoxide dismutase-mediated familial amyotrophic lateral sclerosis. *Proc Natl Acad Sci USA* 2000;**97**:13901.

159. Boston-Howes W, Gibb SL, Williams EO, Pasinelli P, Brown Jr. RH, Trotti D. Caspase-3 cleaves and inactivates the glutamate transporter EAAT2. *J Biol Chem* 2006;**281**:14076.

160. Trotti D, Danbolt NC, Volterra A. Glutamate transporters are oxidant-vulnerable: a molecular link between oxidative and excitotoxic neurodegeneration? *Trends Pharmacol Sci* 1998;**19**:328.

161. Trotti D, Nussberger S, Volterra A, Hediger MA. Differential modulation of the uptake currents by redox interconversion of cysteine residues in the human neuronal glutamate transporter EAAC1. *Eur J Neurosci* 1997;**9**:2207.

162. Volterra A, Trotti D, Tromba C, Floridi S, Racagni G. Glutamate uptake inhibition by oxygen free radicals in rat cortical astrocytes. *J Neurosci* 1994;**14**:2924.

163. Bar-Peled O, O'Brien RJ, Morrison JH, Rothstein JD. Cultured motor neurons possess calcium-permeable AMPA/kainate receptors. *Neuroreport* 1999;**10**:855.

164. Van Den Bosch L, Vandenberghe W, Klaassen H, Van HE, Robberecht W. Ca(2+)-permeable AMPA receptors and selective vulnerability of motor neurons. *J Neurol Sci* 2000;**180**:29.

165. Carriedo SG, Yin HZ, Weiss JH. Motor neurons are selectively vulnerable to AMPA/kainate receptor-mediated injury in vitro. *J Neurosci* 1996;**16**:4069.

166. Lu YM, Yin HZ, Chiang J, Weiss JH. Ca(2+)-permeable AMPA/kainate and NMDA channels: high rate of Ca2+ influx underlies potent induction of injury. *J Neurosci* 1996;**16**:5457.

167. Carriedo SG, Sensi SL, Yin HZ, Weiss JH. AMPA exposures induce mitochondrial Ca(2+) overload and ROS generation in spinal motor neurons in vitro. *J Neurosci* 2000;**20**:240.

168. Grosskreutz J, Van Den Bosch L, Keller BU. Calcium dysregulation in amyotrophic lateral sclerosis. *Cell Calcium* 2010;**47**:165.

169. Jahn K, Grosskreutz J, Haastert K, Ziegler E, Schlesinger F, Grothe C, et al. Temporospatial coupling of networked synaptic activation of AMPA-type glutamate receptor channels and calcium transients in cultured motoneurons. *Neuroscience* 2006;**142**:1019–29.

170. Kim HJ, Im W, Kim S, Kim SH, Sung JJ, Kim M, et al. Calcium-influx increases SOD1 aggregates via nitric oxide in cultured motor neurons 2. *Exp Mol Med* 2007;**39**:574.

171. Tateno M, Sadakata H, Tanaka M, Itohara S, Shin RM, Miura M, et al. Calcium-permeable AMPA receptors promote misfolding of mutant SOD1 protein and development of amyotrophic lateral sclerosis in a transgenic mouse model. *Hum Mol Genet* 2004;**13**:2183.

172. Ince P, Stout N, Shaw P, Slade J, Hunziker W, Heizmann CW, et al. Parvalbumin and calbindin D-28k in the human motor system and in motor neuron disease. *Neuropathol Appl Neurobiol* 1993;**19**:291.

173. Alexianu ME, Ho BK, Mohamed AH, La BV, Smith RG, Appel SH. The role of calcium-binding proteins in selective motoneuron vulnerability in amyotrophic lateral sclerosis. *Ann Neurol* 1994;**36**:846.

174. Lamanauskas N, Nistri A. Riluzole blocks persistent Na+ and Ca2+ currents and modulates release of glutamate via presynaptic NMDA receptors on neonatal rat hypoglossal motoneurons in vitro. *Eur J Neurosci* 2008;**27**:2501.

175. Hubert JP, Delumeau JC, Glowinski J, Premont J, Doble A. Antagonism by riluzole of entry of calcium evoked by NMDA and veratridine in rat cultured granule cells: evidence for a dual mechanism of action. *Br J Pharmacol* 1994;**113**:261.

176. Doble A. The pharmacology and mechanism of action of riluzole. *Neurology* 1996;**47**:S233.

177. Debono MW, Le GJ, Canton T, Doble A, Pradier L. Inhibition by riluzole of electrophysiological responses mediated by rat kainate and NMDA receptors expressed in Xenopus oocytes. *Eur J Pharmacol* 1993;**235**:283.

178. Miller RG, Mitchell JD, Lyon M, Moore DH. Riluzole for amyotrophic lateral sclerosis (ALS)/motor neuron disease (MND). *Cochrane Database Syst Rev* 2007;(1):CD001447.

179. Tikka TM, Vartiainen NE, Goldsteins G, Oja SS, Andersen PM, Marklund SL, et al. Minocycline prevents neurotoxicity induced by cerebrospinal fluid from patients with motor neurone disease. *Brain* 2002;**125**:722.

180. Alexianu ME, Kozovska M, Appel SH. Immune reactivity in a mouse model of familial ALS correlates with disease progression. *Neurology* 2001;**57**:1282.

181. Almer G, Teismann P, Stevic Z, Halaschek-Wiener J, Deecke L, Kostic V, et al. Increased levels of the pro-inflammatory prostaglandin PGE2 in CSF from ALS patients. *Neurology* 2002;**58**:1277.

182. Hall ED, Oostveen JA, Gurney ME. Relationship of microglial and astrocytic activation to disease onset and progression in a transgenic model of familial ALS. *Glia* 1998;**23**:249.

183. Henkel JS, Engelhardt JI, Siklos L, Simpson EP, Kim SH, Pan T, et al. Presence of dendritic cells, MCP-1, and activated microglia/macrophages in amyotrophic lateral sclerosis spinal cord tissue. *Ann Neurol* 2004;**55**:221.

184. Kawamata T, Akiyama H, Yamada T, McGeer PL. Immunologic reactions in amyotrophic lateral sclerosis brain and spinal cord tissue. *Am J Pathol* 1992;**140**:691.

185. Sargsyan SA, Monk PN, Shaw PJ. Microglia as potential contributors to motor neuron injury in amyotrophic lateral sclerosis. *Glia* 2005;**51**:241.

186. Raoul C, Estevez AG, Nishimune H, Cleveland DW, deLapeyriere O, Henderson CE, et al. Motoneuron death triggered by a specific pathway downstream of Fas. potentiation by ALS-linked SOD1 mutations. *Neuron* 2002;**35**:1067.

187. Falsig J, Porzgen P, Lotharius J, Leist M. Specific modulation of astrocyte inflammation by inhibition of mixed lineage kinases with CEP-1347. *J Immunol* 2004;**173**:2762.

188. Harraz MM, Marden JJ, Zhou W, Zhang Y, Williams A, Sharov VS, et al. SOD1 mutations disrupt redox-sensitive Rac regulation of NADPH oxidase in a familial ALS model 1. *J Clin Invest* 2008;**118**:659.

189. Banati RB, Gehrmann J, Schubert P, Kreutzberg GW. Cytotoxicity of microglia. *Glia* 1993;**7**:111.

190. Urushitani M, Sik A, Sakurai T, Nukina N, Takahashi R, Julien JP. Chromogranin-mediated secretion of mutant superoxide dismutase proteins linked to amyotrophic lateral sclerosis. *Nat Neurosci* 2006;**9**:108.

191. Frank-Cannon TC, Alto LT, McAlpine FE, Tansey MG. Does neuroinflammation fan the flame in neurodegenerative diseases? *Mol Neurodegener* 2009;**4**:47.

192. Graber DJ, Hickey WF, Harris BT. Progressive changes in microglia and macrophages in spinal cord and peripheral nerve in the transgenic rat model of amyotrophic lateral sclerosis. *J Neuroinflammation* 2010;**7**:8.

193. Masu Y, Wolf E, Holtmann B, Sendtner M, Brem G, Thoenen H. Disruption of the CNTF gene results in motor neuron degeneration. *Nature* 1993;**365**:27.

194. Giess R, Holtmann B, Braga M, Grimm T, Mnller-Myhsok B, Toyka KV, et al. Early onset of severe familial amyotrophic lateral sclerosis with a SOD-1 mutation: potential impact of CNTF as a candidate modifier gene. *Am J Hum Genet* 2002;**70**:1277.

195. Li W, Brakefield D, Pan Y, Hunter D, Myckatyn TM, Parsadanian A. Muscle-derived but not centrally derived transgene GDNF is neuroprotective in G93A-SOD1 mouse model of ALS. *Exp Neurol* 2007;**203**:457.

196. Jiang YM, Yamamoto M, Kobayashi Y, Yoshihara T, Liang Y, Terao S, et al. Gene expression profile of spinal motor neurons in sporadic amyotrophic lateral sclerosis. *Ann Neurol* 2005;**57**:236.

197. Grundstrom E, Askmark H, Lindeberg J, Nygren I, Ebendal T, Aquilonius SM. Increased expression of glial cell line-derived neurotrophic factor mRNA in muscle biopsies from patients with amyotrophic lateral sclerosis. *J Neurol Sci* 1999;**162**:169.

198. Grundstrom E, Lindholm D, Johansson A, Blennow K, Askmark H. GDNF but not BDNF is increased in cerebrospinal fluid in amyotrophic lateral sclerosis. *Neuroreport* 2000;**11**:1781.

199. Yamamoto M, Mitsuma N, Inukai A, Ito Y, Li M, Mitsuma T, et al. Expression of GDNF and GDNFR-alpha mRNAs in muscles of patients with motor neuron diseases. *Neurochem Res* 1999;**24**:785.

200. A double-blind placebo-controlled clinical trial of subcutaneous recombinant human ciliary neurotrophic factor (rHCNTF) in amyotrophic lateral sclerosis. ALS CNTF Treatment Study Group. *Neurology* 1996;**46**:1244.

201. A controlled trial of recombinant methionyl human BDNF in ALS: The BDNF Study Group (Phase III). *Neurology* 1999;**52**:1427.

202. Sondell M, Lundborg G, Kanje M. Vascular endothelial growth factor has neurotrophic activity and stimulates axonal outgrowth, enhancing cell survival and Schwann cell proliferation in the peripheral nervous system. *J Neurosci* 1999;**19**:5731.

203. Subramanian V, Feng Y. A new role for angiogenin in neurite growth and pathfinding: implications for amyotrophic lateral sclerosis. *Hum Mol Genet* 2007;**16**:1445.

204. Oosthuyse B, Moons L, Storkebaum E, Beck H, Nuyens D, Brusselmans K, et al. Deletion of the hypoxia-response element in the vascular endothelial growth factor promoter causes motor neuron degeneration. *Nat Genet* 2001;**28**:131.

205. Devos D, Moreau C, Lassalle P, Perez T, De SJ, Brunaud-Danel V, et al. Low levels of the vascular endothelial growth factor in CSF from early ALS patients. *Neurology* 2004;**62**:2127.

206. Li X, Lu L, Bush DJ, Zhang X, Zheng L, Suswam EA, et al. Mutant copper zinc superoxide dismutase associated with amyotrophic lateral sclerosis binds to adenine/uridine-rich stability elements in the vascular endothelial growth factor 3'-untranslated region. *J Neurochem* 2009;**108**:1032.

207. Lu L, Zheng L, Viera L, Suswam E, Li Y, Li X, et al. Mutant Cu/Zn-superoxide dismutase associated with amyotrophic lateral sclerosis destabilizes vascular endothelial growth factor mRNA and downregulates its expression. *J Neurosci* 2007;**27**:7929.

208. Lambrechts D, Poesen K, Fernandez-Santiago R, Al-Chalabi A, Del BR, Van Vught PW, et al. Meta-analysis of vascular endothelial growth factor variations in amyotrophic lateral sclerosis: increased susceptibility in male carriers of the -2578AA genotype. *J Med Genet* 2009;**46**:840.

209. Cronin S, Greenway MJ, Ennis S, Kieran D, Green A, Prehn JH, et al. Elevated serum angiogenin levels in ALS. *Neurology* 2006;**67**:1833.

210. Sebastia J, Kieran D, Breen B, King MA, Netteland DF, Joyce D, et al. Angiogenin protects motoneurons against hypoxic injury. *Cell Death Differ* 2009;**16**:1238.

211. Subramanian V, Crabtree B, Acharya KR. Human angiogenin is a neuroprotective factor and amyotrophic lateral sclerosis associated angiogenin variants affect neurite extension/pathfinding and survival of motor neurons. *Hum Mol Genet* 2008;**17**:130.

212. Kobari M, Obara K, Watanabe S, Dembo T, Fukuuchi Y. Local cerebral blood flow in motor neuron disease: correlation with clinical findings. *J Neurol Sci* 1996;**144**:64.

213. Waldemar G, Vorstrup S, Jensen TS, Johnsen A, Boysen G. Focal reductions of cerebral blood flow in amyotrophic lateral sclerosis: a [99mTc]-d,l-HMPAO SPECT study. *J Neurol Sci* 1992;**107**:19.

214. Zhong Z, Deane R, Ali Z, Parisi M, Shapovalov Y, O'Banion MK, et al. ALS-causing SOD1 mutants generate vascular changes prior to motor neuron degeneration. *Nat Neurosci* 2008;**11**:420.

215. Storkebaum E, Lambrechts D, Dewerchin M, Moreno-Murciano MP, Appelmans S, Oh H, et al. Treatment of motoneuron degeneration by intracerebroventricular delivery of VEGF in a rat model of ALS. *Nat Neurosci* 2005;**8**:85.

216. Wang Y, Mao XO, Xie L, Banwait S, Marti HH, Greenberg DA, et al. Vascular endothelial growth factor overexpression delays neurodegeneration and prolongs survival in amyotrophic lateral sclerosis mice. *J Neurosci* 2007;**27**:304.

217. Pasterkamp RJ, Giger RJ. Semaphorin function in neural plasticity and disease. *Curr Opin Neurobiol* 2009;**19**:263.

218. Tannemaat MR, Korecka J, Ehlert EM, Mason MR, van Duinen SG, Boer GJ, et al. Human neuroma contains increased levels of semaphorin 3A, which surrounds nerve fibers and reduces neurite extension in vitro. *J Neurosci* 2007;**27**:14260.

219. De WF, Vo T, Stam FJ, Wisman LA, Bar PR, Niclou SP, et al. The expression of the chemorepellent Semaphorin 3A is selectively induced in terminal Schwann cells of a subset of neuromuscular synapses that display limited anatomical plasticity and enhanced vulnerability in motor neuron disease. *Mol Cell Neurosci* 2006;**32**:102.

220. Giger RJ, Pasterkamp RJ, Heijnen S, Holtmaat AJ, Verhaagen J. Anatomical distribution of the chemorepellent semaphorin III/collapsin-1 in the adult rat and human brain: predominant expression in structures of the olfactory-hippocampal pathway and the motor system. *J Neurosci Res* 1998;**52**:27.

221. Tang XQ, Heron P, Mashburn C, Smith GM. Targeting sensory axon regeneration in adult spinal cord. *J Neurosci* 2007;**27**:6068.
222. Dupuis L, Gonzalez de Aguilar JL, di Scala F, Rene F, de Tapia M, Pradat PF, et al. Nogo provides a molecular marker for diagnosis of amyotrophic lateral sclerosis. *Neurobiol Dis* 2002;**10**:358.
223. Jokic N, Gonzalez de Aguilar JL, Dimou L, Lin S, Fergani A, Ruegg MA, et al. The neurite outgrowth inhibitor Nogo-A promotes denervation in an amyotrophic lateral sclerosis model. *EMBO Rep* 2006;**7**:1162.
224. Magnusson C, Libelius R, Tagerud S. Nogo (Reticulon 4) expression in innervated and denervated mouse skeletal muscle. *Mol Cell Neurosci* 2003;**22**:298.
225. Shaw PJ, Eggett CJ. Molecular factors underlying selective vulnerability of motor neurons to neurodegeneration in amyotrophic lateral sclerosis. *J Neurol* 2000;**247**(Suppl. 1):I17.
226. Carpenter S. Proximal axonal enlargement in motor neuron disease. *Neurology* 1968;**18**:841.
227. Hirano A, Nakano I, Kurland LT, Mulder DW, Holley PW, Saccomanno G. Fine structural study of neurofibrillary changes in a family with amyotrophic lateral sclerosis. *J Neuropathol Exp Neurol* 1984;**43**:471.
228. Tu PH, Gurney ME, Julien JP, Lee VM, Trojanowski JQ. Oxidative stress, mutant SOD1, and neurofilament pathology in transgenic mouse models of human motor neuron disease. *Lab Invest* 1997;**76**:441.
229. Julien JP. Neurofilaments and motor neuron disease. *Trends Cell Biol* 1997;**7**:243.
230. Kawamura Y, Dyck PJ, Shimono M, Okazaki H, Tateishi J, Doi H. Morphometric comparison of the vulnerability of peripheral motor and sensory neurons in amyotrophic lateral sclerosis. *J Neuropathol Exp Neurol* 1981;**40**:667.
231. Beaulieu JM, Robertson J, Julien JP. Interactions between peripherin and neurofilaments in cultured cells: disruption of peripherin assembly by the NF-M and NF-H subunits. *Biochem Cell Biol* 1999;**77**:41.
232. Collard JF, Cote F, Julien JP. Defective axonal transport in a transgenic mouse model of amyotrophic lateral sclerosis. *Nature* 1995;**375**:61.
233. Cote F, Collard JF, Julien JP. Progressive neuronopathy in transgenic mice expressing the human neurofilament heavy gene: a mouse model of amyotrophic lateral sclerosis. *Cell* 1993;**73**:35.
234. Lee MK, Marszalek JR, Cleveland DW. A mutant neurofilament subunit causes massive, selective motor neuron death: implications for the pathogenesis of human motor neuron disease. *Neuron* 1994;**13**:975.
235. Yuan A, Rao MV, Kumar A, Julien JP, Nixon RA. Neurofilament transport in vivo minimally requires hetero-oligomer formation. *J Neurosci* 2003;**23**:9452.
236. Millecamps S, Robertson J, Lariviere R, Mallet J, Julien JP. Defective axonal transport of neurofilament proteins in neurons overexpressing peripherin. *J Neurochem* 2006;**98**:926.
237. Ackerley S, Thornhill P, Grierson AJ, Brownlees J, Anderton BH, Leigh PN, et al. Neurofilament heavy chain side arm phosphorylation regulates axonal transport of neurofilaments. *J Cell Biol* 2003;**161**:489.
238. Jung C, Lee S, Ortiz D, Zhu Q, Julien JP, Shea TB. The high and middle molecular weight neurofilament subunits regulate the association of neurofilaments with kinesin: inhibition by phosphorylation of the high molecular weight subunit. *Brain Res Mol Brain Res* 2005;**141**:151.
239. Shea TB, Zheng YL, Ortiz D, Pant HC. Cyclin-dependent kinase 5 increases perikaryal neurofilament phosphorylation and inhibits neurofilament axonal transport in response to oxidative stress. *J Neurosci Res* 2004;**76**:795.
240. Wagner OI, Ascano J, Tokito M, Leterrier JF, Janmey PA, Holzbaur EL. The interaction of neurofilaments with the microtubule motor cytoplasmic dynein. *Mol Biol Cell* 2004;**15**:5092.

241. Ackerley S, Grierson AJ, Banner S, Perkinton MS, Brownlees J, Byers HL, et al. p38alpha stress-activated protein kinase phosphorylates neurofilaments and is associated with neurofilament pathology in amyotrophic lateral sclerosis. Mol Cell Neurosci 2004;26:354.
242. Guidato S, Tsai LH, Woodgett J, Miller CC. Differential cellular phosphorylation of neurofilament heavy side-arms by glycogen synthase kinase-3 and cyclin-dependent kinase-5. J Neurochem 1996;66.1608.
243. Sun D, Leung CL, Liem RK. Phosphorylation of the high molecular weight neurofilament protein (NF-H) by Cdk5 and p35. J Biol Chem 1996;271:14245.
244. Couillard-Despres S, Zhu Q, Wong PC, Price DL, Cleveland DW, Julien JP. Protective effect of neurofilament heavy gene overexpression in motor neuron disease induced by mutant superoxide dismutase. Proc Natl Acad Sci USA 1998;95:9626.
245. Ehlers MD, Fung ET, O'Brien RJ, Huganir RL. Splice variant-specific interaction of the NMDA receptor subunit NR1 with neuronal intermediate filaments. J Neurosci 1998;18:720.
246. Perrot R, Berges R, Bocquet A, Eyer J. Review of the multiple aspects of neurofilament functions, and their possible contribution to neurodegeneration. Mol Neurobiol 2008;38:27.
247. Warita H, Itoyama Y, Abe K. Selective impairment of fast anterograde axonal transport in the peripheral nerves of asymptomatic transgenic mice with a G93A mutant SOD1 gene. Brain Res 1999;819:120.
248. Williamson TL, Cleveland DW. Slowing of axonal transport is a very early event in the toxicity of ALS-linked SOD1 mutants to motor neurons. Nat Neurosci 1999;2:50.
249. De Vos KJ, Chapman AL, Tennant ME, Manser C, Tudor EL, Lau KF, et al. Familial amyotrophic lateral sclerosis-linked SOD1 mutants perturb fast axonal transport to reduce axonal mitochondria content. Hum Mol Genet 2007;16:2720.
250. Magrane J, Manfredi G. Mitochondrial function, morphology, and axonal transport in amyotrophic lateral sclerosis. Antioxid Redox Signal 2009;11:1615.
251. Rowland KC, Irby NK, Spirou GA. Specialized synapse-associated structures within the calyx of Held. J Neurosci 2000;20:9135.
252. Sasaki S, Iwata M. Impairment of fast axonal transport in the proximal axons of anterior horn neurons in amyotrophic lateral sclerosis. Neurology 1996;47:535.
253. Tateno M, Kato S, Sakurai T, Nukina N, Takahashi R, Araki T. Mutant SOD1 impairs axonal transport of choline acetyltransferase and acetylcholine release by sequestering KAP3. Hum Mol Genet 2009;18:942.
254. Strom AL, Shi P, Zhang F, Gal J, Kilty R, Hayward LJ, et al. Interaction of amyotrophic lateral sclerosis (ALS)-related mutant copper-zinc superoxide dismutase with the dynein-dynactin complex contributes to inclusion formation. J Biol Chem 2008;283:22795.
255. Zhang F, Strom AL, Fukada K, Lee S, Hayward LJ, Zhu H. Interaction between familial amyotrophic lateral sclerosis (ALS)-linked SOD1 mutants and the dynein complex. J Biol Chem 2007;282:16691.
256. Ge WW, Wen W, Strong W, Leystra-Lantz C, Strong MJ. Mutant copper-zinc superoxide dismutase binds to and destabilizes human low molecular weight neurofilament mRNA. J Biol Chem 2005;280:118.
257. Bergeron C, Beric-Maskarel K, Muntasser S, Weyer L, Somerville MJ, Percy ME. Neurofilament light and polyadenylated mRNA levels are decreased in amyotrophic lateral sclerosis motor neurons. J Neuropathol Exp Neurol 1994;53:221.
258. Wong NK, He BP, Strong MJ. Characterization of neuronal intermediate filament protein expression in cervical spinal motor neurons in sporadic amyotrophic lateral sclerosis (ALS). J Neuropathol Exp Neurol 2000;59:972.
259. Menzies FM, Grierson AJ, Cookson MR, Heath PR, Tomkins J, Figlewicz DA, et al. Selective loss of neurofilament expression in Cu/Zn superoxide dismutase (SOD1) linked amyotrophic lateral sclerosis. J Neurochem 2002;82:1118.

260. Eyer J, Cleveland DW, Wong PC, Peterson AC. Pathogenesis of two axonopathies does not require axonal neurofilaments. *Nature* 1998;**391**:584.

261. Bendotti C, Calvaresi N, Chiveri L, Prelle A, Moggio M, Braga M, et al. Early vacuolization and mitochondrial damage in motor neurons of FALS mice are not associated with apoptosis or with changes in cytochrome oxidase histochemical reactivity. *J Neurol Sci* 2001;**191**:25.

262. Martin LJ. Transgenic mice with human mutant genes causing Parkinson's disease and amyotrophic lateral sclerosis provide common insight into mechanisms of motor neuron selective vulnerability to degeneration. *Rev Neurosci* 2007;**18**:115.

263. Bowling AC, Schulz JB, Brown Jr. RH, Beal MF. Superoxide dismutase activity, oxidative damage, and mitochondrial energy metabolism in familial and sporadic amyotrophic lateral sclerosis. *J Neurochem* 1993;**61**:2322.

264. Browne SE, Yang L, DiMauro JP, Fuller SW, Licata SC, Beal MF. Bioenergetic abnormalities in discrete cerebral motor pathways presage spinal cord pathology in the G93A SOD1 mouse model of ALS. *Neurobiol Dis* 2006;**22**:599.

265. Dupuis L, Oudart H, Rene F, Gonzalez de Aguilar JL, Loeffler JP. Evidence for defective energy homeostasis in amyotrophic lateral sclerosis: benefit of a high-energy diet in a transgenic mouse model. *Proc Natl Acad Sci USA* 2004;**101**:11159.

266. Mattiazzi M, D'Aurelio M, Gajewski CD, Martushova K, Kiaei M, Beal MF, et al. Mutated human SOD1 causes dysfunction of oxidative phosphorylation in mitochondria of transgenic mice. *J Biol Chem* 2002;**277**:29626.

267. Vielhaber S, Kunz D, Winkler K, Wiedemann FR, Kirches E, Feistner H, et al. Mitochondrial DNA abnormalities in skeletal muscle of patients with sporadic amyotrophic lateral sclerosis. *Brain* 2000;**123**:1339.

268. Wiedemann FR, Winkler K, Kuznetsov AV, Bartels C, Vielhaber S, Feistner H, et al. Impairment of mitochondrial function in skeletal muscle of patients with amyotrophic lateral sclerosis. *J Neurol Sci* 1998;**156**:65.

269. Mali Y, Zisapels N. Gain of interaction of ALS-linked G93A superoxide dismutase with cytosolic malate dehydrogenase. *Neurobiol Dis* 2008;**32**:133.

270. Jung C, Higgins CM, Xu Z. A quantitative histochemical assay for activities of mitochondrial electron transport chain complexes in mouse spinal cord sections. *J Neurosci Methods* 2002;**114**:165.

271. Liu J, Lillo C, Jonsson PA, Vande VC, Ward CM, Miller TM, et al. Toxicity of familial ALS-linked SOD1 mutants from selective recruitment to spinal mitochondria 2. *Neuron* 2004;**43**:5.

272. Vande VC, Miller TM, Cashman NR, Cleveland DW. Selective association of misfolded ALS-linked mutant SOD1 with the cytoplasmic face of mitochondria. *Proc Natl Acad Sci USA* 2008;**105**:4022.

273. Borthwick GM, Johnson MA, Ince PG, Shaw PJ, Turnbull DM. Mitochondrial enzyme activity in amyotrophic lateral sclerosis: implications for the role of mitochondria in neuronal cell death. *Ann Neurol* 1999;**46**:787.

274. Swerdlow RH, Parks JK, Cassarino DS, Trimmer PA, Miller SW, Maguire DJ, et al. Mitochondria in sporadic amyotrophic lateral sclerosis. *Exp Neurol* 1998;**153**:135.

275. de Grey AD. Mitochondrial mutations in mammalian aging: an over-hasty about-turn? *Rejuvenation Res* 2004;**7**:171.

276. Khrapko K, Vijg J. Mitochondrial DNA mutations and aging: a case closed? *Nat Genet* 2007;**39**:445.

277. Kujoth GC, Hiona A, Pugh TD, Someya S, Panzer K, Wohlgemuth SE, et al. Mitochondrial DNA mutations, oxidative stress, and apoptosis in mammalian aging. *Science* 2005;**309**:481.

278. Trifunovic A, Wredenberg A, Falkenberg M, Spelbrink JN, Rovio AT, Bruder CE, et al. Premature ageing in mice expressing defective mitochondrial DNA polymerase. *Nature* 2004;**429**:417.

279. Vermulst M, Bielas JH, Kujoth GC, Ladiges WC, Rabinovitch PS, Prolla TA, et al. Mitochondrial point mutations do not limit the natural lifespan of mice. *Nat Genet* 2007;**39**:540.
280. Bouteloup C, Desport JC, Clavelou P, Guy N, Derumeaux-Burel H, Ferrier A, et al. Hypermetabolism in ALS patients: an early and persistent phenomenon. *J Neurol* 2009;**256**:1236.
281. Choi DW. Glutamate neurotoxicity and diseases of the nervous system. *Neuron* 1988;**1**:623–34.
282. Dawson VL, Dawson TM, London ED, Bredt DS, Snyder SH. Nitric oxide mediates glutamate neurotoxicity in primary cortical cultures. *Proc Natl Acad Sci USA* 1991;**88**:6368–71.
283. Damiano M, Starkov AA, Petri S, Kipiani K, Kiaei M, Mattiazzi M, et al. Neural mitochondrial Ca2+ capacity impairment precedes the onset of motor symptoms in G93A Cu/Zn-superoxide dismutase mutant mice. *J Neurochem* 2006;**96**:1349.
284. Curti D, Malaspina A, Facchetti G, Camana C, Mazzini L, Tosca P, et al. Amyotrophic lateral sclerosis: oxidative energy metabolism and calcium homeostasis in peripheral blood lymphocytes. *Neurology* 1996;**47**:1060.
285. Siklos L, Engelhardt J, Harati Y, Smith RG, Joo F, Appel SH. Ultrastructural evidence for altered calcium in motor nerve terminals in amyotropic lateral sclerosis. *Ann Neurol* 1996;**39**:203.
286. Dykens JA. Isolated cerebral and cerebellar mitochondria produce free radicals when exposed to elevated CA2+ and Na+: implications for neurodegeneration. *J Neurochem* 1994;**63**:584–91.
287. Higgins CM, Jung C, Ding H, Xu Z. Mutant Cu, Zn superoxide dismutase that causes motoneuron degeneration is present in mitochondria in the CNS. *J Neurosci* 2002;**22**:215.
288. Vijayvergiya C, Beal MF, Buck J, Manfredi G. Mutant superoxide dismutase 1 forms aggregates in the brain mitochondrial matrix of amyotrophic lateral sclerosis mice. *J Neurosci* 2005;**25**:2463.
289. Cassina P, Cassina A, Pehar M, Castellanos R, Gandelman M, de León A, et al. Mitochondrial dysfunction in SOD1G93A-bearing astrocytes promotes motor neuron degeneration: prevention by mitochondrial-targeted antioxidants. *J Neurosci* 2008;**28**:4115.
290. Higgins CM, Jung C, Xu Z. ALS-associated mutant SOD1G93A causes mitochondrial vacuolation by expansion of the intermembrane space and by involvement of SOD1 aggregation and peroxisomes. *BMC Neurosci* 2003;**4**:16.
291. Jaarsma D, Rognoni F, Van DW, Verspaget HW, Haasdijk ED, Holstege JC. CuZn superoxide dismutase (SOD1) accumulates in vacuolated mitochondria in transgenic mice expressing amyotrophic lateral sclerosis-linked SOD1 mutations. *Acta Neuropathol* 2001;**102**:293.
292. Kirkinezos IG, Bacman SR, Hernandez D, Oca-Cossio J, Arias LJ, Perez-Pinzon MA, et al. Cytochrome c association with the inner mitochondrial membrane is impaired in the CNS of G93A-SOD1 mice. *J Neurosci* 2005;**25**:164.
293. Jonas EA. Molecular participants in mitochondrial cell death channel formation during neuronal ischemia. *Exp Neurol* 2009;**218**:203.
294. Pasinelli P, Belford ME, Lennon N, Bacskai BJ, Hyman BT, Trotti D, et al. Amyotrophic lateral sclerosis-associated SOD1 mutant proteins bind and aggregate with Bcl-2 in spinal cord mitochondria. *Neuron* 2004;**43**:19.
295. Pedrini S, Sau D, Guareschi S, Bogush M, Brown Jr. RH, Naniche N, et al. ALS-linked mutant SOD1 damages mitochondria by promoting conformational changes in Bcl-2. *Hum Mol Genet* 2010;**19**:2974.
296. Di NL, Whitson LJ, Cao X, Hart PJ, Levine RL. Proteasomal degradation of mutant superoxide dismutases linked to amyotrophic lateral sclerosis. *J Biol Chem* 2005;**280**:39907.
297. Hoffman EK, Wilcox HM, Scott RW, Siman R. Proteasome inhibition enhances the stability of mouse Cu/Zn superoxide dismutase with mutations linked to familial amyotrophic lateral sclerosis. *J Neurol Sci* 1996;**139**:15.

298. Niwa J, Ishigaki S, Hishikawa N, Yamamoto M, Doyu M, Murata S, et al. Dorfin ubiquitylates mutant SOD1 and prevents mutant SOD1-mediated neurotoxicity. *J Biol Chem* 2002;**277**:36793–8.

299. Urushitani M, Kurisu J, Tsukita K, Takahashi R. Proteasomal inhibition by misfolded mutant superoxide dismutase 1 induces selective motor neuron death in familial amyotrophic lateral sclerosis. *J Neurochem* 2002;**83**:1030.

300. Watanabe M, Dykes-Hoberg M, Culotta VC, Price DL, Wong PC, Rothstein JD. Histological evidence of protein aggregation in mutant SOD1 transgenic mice and in amyotrophic lateral sclerosis neural tissues. *Neurobiol Dis* 2001;**8**:933.

301. Hyun DH, Lee M, Halliwell B, Jenner P. Proteasomal inhibition causes the formation of protein aggregates containing a wide range of proteins, including nitrated proteins. *J Neurochem* 2003;**86**:363.

302. Puttaparthi K, Wojcik C, Rajendran B, DeMartino GN, Elliott JL. Aggregate formation in the spinal cord of mutant SOD1 transgenic mice is reversible and mediated by proteasomes 1. *J Neurochem* 2003;**87**:851.

303. Bendotti C, Atzori C, Piva R, Tortarolo M, Strong MJ, Debiasi S, et al. Activated p38MAPK is a novel component of the intracellular inclusions found in human amyotrophic lateral sclerosis and mutant SOD1 transgenic mice. *J Neuropathol Exp Neurol* 2004;**63**:113.

304. Leigh PN, Whitwell H, Garofalo O, Buller J, Swash M, Martin JE, et al. Ubiquitin-immunoreactive intraneuronal inclusions in amyotrophic lateral sclerosis. Morphology, distribution, and specificity. *Brain* 1991;**114**(Pt. 2):775.

305. Mendonca DM, Chimelli L, Martinez AM. Expression of ubiquitin and proteasome in motorneurons and astrocytes of spinal cords from patients with amyotrophic lateral sclerosis. *Neurosci Lett* 2006;**404**:315.

306. Sasaki S. Endoplasmic reticulum stress in motor neurons of the spinal cord in sporadic amyotrophic lateral sclerosis. *J Neuropathol Exp Neurol* 2010;**69**:346.

307. Cheroni C, Marino M, Tortarolo M, Veglianese P, De BS, Fontana E, et al. Functional alterations of the ubiquitin-proteasome system in motor neurons of a mouse model of familial amyotrophic lateral sclerosis. *Hum Mol Genet* 2009;**18**:82.

308. Sau D, De BS, Vitellaro-Zuccarello L, Riso P, Guarnieri S, Porrini M, et al. Mutation of SOD1 in ALS: a gain of a loss of function. *Hum Mol Genet* 2007;**16**:1604.

309. Kabashi E, Agar JN, Hong Y, Taylor DM, Minotti S, Figlewicz DA, et al. Proteasomes remain intact, but show early focal alteration in their composition in a mouse model of amyotrophic lateral sclerosis. *J Neurochem* 2008;**105**:2353.

310. Puttaparthi K, Elliott JL. Non-neuronal induction of immunoproteasome subunits in an ALS model: possible mediation by cytokines. *Exp Neurol* 2005;**196**:441.

311. Ahtoniemi T, Goldsteins G, Keksa-Goldsteine V, Malm T, Kanninen K, Salminen A, et al. Pyrrolidine dithiocarbamate inhibits induction of immunoproteasome and decreases survival in a rat model of amyotrophic lateral sclerosis. *Mol Pharmacol* 2007;**71**:30.

312. Puttaparthi K, Van KL, Elliott JL. Assessing the role of immuno-proteasomes in a mouse model of familial ALS. *Exp Neurol* 2007;**206**:53.

313. McGeer PL, McGeer EG. Inflammatory processes in amyotrophic lateral sclerosis. *Muscle Nerve* 2002;**26**:459.

314. Papadimitriou D, Le Verche V, Jacquier A, Ikiz B, Przedborski S, Re DB. Inflammation in ALS and SMA: sorting out the good from the evil. *Neurobiol Dis* 2010;**37**:493.

315. Vembar SS, Brodsky JL. One step at a time: endoplasmic reticulum-associated degradation. *Nat Rev Mol Cell Biol* 2008;**9**:944.

316. Kozutsumi Y, Segal M, Normington K, Gething MJ, Sambrook J. The presence of malfolded proteins in the endoplasmic reticulum signals the induction of glucose-regulated proteins. *Nature* 1988;**332**:462.

317. Schroder M, Kaufman RJ. ER stress and the unfolded protein response. *Mutat Res* 2005; **569**:29.
318. Nishitoh H, Kadowaki H, Nagai A, Maruyama T, Yokota T, Fukutomi H, et al. ALS-linked mutant SOD1 induces ER stress- and ASK1-dependent motor neuron death by targeting Derlin-1 1. *Genes Dev* 2008;**22**:1451.
319. Kikuchi H, Almer G, Yamashita S, Guegan C, Nagai M, Xu Z, et al. Spinal cord endoplasmic reticulum stress associated with a microsomal accumulation of mutant superoxide dismutase 1 in an ALS model. *Proc Natl Acad Sci USA* 2006;**103**:6025.
320. Saxena S, Cabuy E, Caroni P. A role for motoneuron subtype-selective ER stress in disease manifestations of FALS mice. *Nat Neurosci* 2009;**12**:627.
321. Atkin JD, Farg MA, Walker AK, McLean C, Tomas D, Horne MK. Endoplasmic reticulum stress and induction of the unfolded protein response in human sporadic amyotrophic lateral sclerosis. *Neurobiol Dis* 2008;**30**:400.
322. Ilieva EV, Ayala V, Jove M, Dalfo E, Cacabelos D, Povedano M, et al. Oxidative and endoplasmic reticulum stress interplay in sporadic amyotrophic lateral sclerosis. *Brain* 2007;**130**:3111.
323. Arai T, Hasegawa M, Akiyama H, Ikeda K, Nonaka T, Mori H, et al. TDP-43 is a component of ubiquitin-positive tau-negative inclusions in frontotemporal lobar degeneration and amyotrophic lateral sclerosis. *Biochem Biophys Res Commun* 2006;**351**:602.
324. Neumann M, Sampathu DM, Kwong LK, Truax AC, Micsenyi MC, Chou TT, et al. Ubiquitinated TDP-43 in frontotemporal lobar degeneration and amyotrophic lateral sclerosis. *Science* 2006;**314**:130.
325. Baumer D, Parkinson N, Talbot K. TARDBP in amyotrophic lateral sclerosis: identification of a novel variant but absence of copy number variation. *J Neurol Neurosurg Psychiatry* 2009;**80**:1283.
326. Corrado L, Ratti A, Gellera C, Buratti E, Castellotti B, Carlomagno Y, et al. High frequency of TARDBP gene mutations in Italian patients with amyotrophic lateral sclerosis. *Hum Mutat* 2009;**30**:688.
327. Daoud H, Valdmanis PN, Kabashi E, Dion P, Dupre N, Camu W, et al. Contribution of TARDBP mutations to sporadic amyotrophic lateral sclerosis. *J Med Genet* 2009;**46**:112.
328. Del BR, Ghezzi S, Corti S, Pandolfo M, Ranieri M, Santoro D, et al. TARDBP (TDP-43) sequence analysis in patients with familial and sporadic ALS: identification of two novel mutations. *Eur J Neurol* 2009;**16**:727.
329. Gitcho MA, Baloh RH, Chakraverty S, Mayo K, Norton JB, Levitch D, et al. TDP-43 A315T mutation in familial motor neuron disease. *Ann Neurol* 2008;**63**:535.
330. Kamada M, Maruyama H, Tanaka E, Morino H, Wate R, Ito H, et al. Screening for TARDBP mutations in Japanese familial amyotrophic lateral sclerosis. *J Neurol Sci* 2009;**284**:69.
331. Kirby J, Goodall EF, Smith W, Highley JR, Masanzu R, Hartley JA, et al. Broad clinical phenotypes associated with TAR-DNA binding protein (TARDBP) mutations in amyotrophic lateral sclerosis. *Neurogenetics* 2010;**11**:217.
332. Kuhnlein P, Sperfeld AD, Vanmassenhove B, Van DV, Lee VM, Trojanowski JQ, et al. Two German kindreds with familial amyotrophic lateral sclerosis due to TARDBP mutations. *Arch Neurol* 2008;**65**:1185.
333. Lemmens R, Race V, Hersmus N, Matthijs G, Van Den Bosch L, Van DP, et al. TDP-43 M311V mutation in familial amyotrophic lateral sclerosis. *J Neurol Neurosurg Psychiatry* 2009;**80**:354.
334. Luquin N, Yu B, Saunderson RB, Trent RJ, Pamphlett R. Genetic variants in the promoter of TARDBP in sporadic amyotrophic lateral sclerosis. *Neuromuscul Disord* 2009;**19**:696.
335. Origone P, Caponnetto C, Bandettini Di PM, Ghiglione E, Bellone E, Ferrandes G, et al. Enlarging clinical spectrum of FALS with TARDBP gene mutations: S393L variant in an

Italian family showing phenotypic variability and relevance for genetic counselling. *Amyotroph Lateral Scler* 2010;**11**:223.

336. Pamphlett R, Luquin N, McLean C, Jew SK, Adams L. TDP-43 neuropathology is similar in sporadic amyotrophic lateral sclerosis with or without TDP-43 mutations. *Neuropathol Appl Neurobiol* 2009;**35**:222.

337. Rutherford NJ, Zhang YJ, Baker M, Gass JM, Finch NA, Xu YF, et al. Novel mutations in TARDBP (TDP-43) in patients with familial amyotrophic lateral sclerosis. *PLoS Genet* 2008;**4**: e1000193.

338. Sreedharan J, Blair IP, Tripathi VB, Hu X, Vance C, Rogelj B, et al. TDP-43 mutations in familial and sporadic amyotrophic lateral sclerosis. *Science* 2008;**319**:1668.

339. Ticozzi N, Leclerc AL, van BM, Keagle P, McKenna-Yasek DM, Sapp PC, et al. Mutational analysis of TARDBP in neurodegenerative diseases. *Neurobiol Aging* 2009;**32**:2096–9.

340. Van Deerlin VM, Leverenz JB, Bekris LM, Bird TD, Yuan W, Elman LB, et al. TARDBP mutations in amyotrophic lateral sclerosis with TDP-43 neuropathology: a genetic and histopathological analysis. *Lancet Neurol* 2008;**7**:409.

341. Williams KL, Durnall JC, Thoeng AD, Warraich ST, Nicholson GA, Blair IP. A novel TARDBP mutation in an Australian amyotrophic lateral sclerosis kindred. *J Neurol Neurosurg Psychiatry* 2009;**80**:1286.

342. Xiong HL, Wang JY, Sun YM, Wu JJ, Chen Y, Qiao K, et al. Association between novel TARDBP mutations and Chinese patients with amyotrophic lateral sclerosis. *BMC Med Genet* 2010;**11**:8.

343. Yokoseki A, Shiga A, Tan CF, Tagawa A, Kaneko H, Koyama A, et al. TDP-43 mutation in familial amyotrophic lateral sclerosis. *Ann Neurol* 2008;**63**:538.

344. Mackenzie IR, Bigio EH, Ince PG, Geser F, Neumann M, Cairns NJ, et al. Pathological TDP-43 distinguishes sporadic amyotrophic lateral sclerosis from amyotrophic lateral sclerosis with SOD1 mutations. *Ann Neurol* 2007;**61**:427.

345. Crippa V, Sau D, Rusmini P, Boncoraglio A, Onesto E, Bolzoni E, et al. The small heat shock protein B8 (HspB8) promotes autophagic removal of misfolded proteins involved in amyotrophic lateral sclerosis (ALS). *Hum Mol Genet* 2010;**19**:3440.

346. Belzil VV, Valdmanis PN, Dion PA, Daoud H, Kabashi E, Noreau A, et al. Mutations in FUS cause FALS and SALS in French and French Canadian populations. *Neurology* 2009;**73**:1176.

347. Blair IP, Williams KL, Warraich ST, Durnall JC, Thoeng AD, Manavis J, et al. FUS mutations in amyotrophic lateral sclerosis: clinical, pathological, neurophysiological and genetic analysis. *J Neurol Neurosurg Psychiatry* 2010;**81**:639.

348. Chio A, Restagno G, Brunetti M, Ossola I, Calvo A, Mora G, et al. Two Italian kindreds with familial amyotrophic lateral sclerosis due to FUS mutation. *Neurobiol Aging* 2009;**30**:1272.

349. Corrado L, Del BR, Castellotti B, Ratti A, Cereda C, Penco S, et al. Mutations of FUS gene in sporadic amyotrophic lateral sclerosis. *J Med Genet* 2010;**47**:190.

350. Damme PV, Goris A, Race V, Hersmus N, Dubois B, Bosch LV, et al. The occurrence of mutations in FUS in a Belgian cohort of patients with familial ALS. *Eur J Neurol* 2010;**17**:754.

351. Dejesus-Hernandez M, Kocerha J, Finch N, Crook R, Baker M, Desaro P, et al. De novo truncating FUS gene mutation as a cause of sporadic amyotrophic lateral sclerosis. *Hum Mutat* 2010;**31**:E1377.

352. Drepper C, Herrmann T, Wessig C, Beck M, Sendtner M. C-terminal FUS/TLS mutations in familial and sporadic ALS in Germany. *Neurobiol Aging* 2011;**32**:548.e1–4.

353. Groen EJ, van Es MA, Van Vught PW, Spliet WG, van Engelen-Lee J, de Visser M, et al. FUS mutations in familial amyotrophic lateral sclerosis in the Netherlands. *Arch Neurol* 2010;**67**:224.

354. Kwiatkowski Jr. TJ, Bosco DA, Leclerc AL, Tamrazian E, Vanderburg CR, Russ C, et al. Mutations in the FUS/TLS gene on chromosome 16 cause familial amyotrophic lateral sclerosis. *Science* 2009;**323**:1205.

355. Lai SL, Abramzon Y, Schymick JC, Stephan DA, Dunckley T, Dillman A, et al. FUS mutations in sporadic amyotrophic lateral sclerosis. *Neurobiol Aging* 2011;**32**:550.e1–4.

356. Suzuki N, Aoki M, Warita H, Kato M, Mizuno H, Shimakura N, et al. FALS with FUS mutation in Japan, with early onset, rapid progress and basophilic inclusion. *J Hum Genet* 2010;**55**:252.

357. Tateishi T, Hokonohara T, Yamasaki R, Miura S, Kikuchi H, Iwaki A, et al. Multiple system degeneration with basophilic inclusions in Japanese ALS patients with FUS mutation. *Acta Neuropathol* 2010;**119**:355.

358. Ticozzi N, Silani V, Leclerc AL, Keagle P, Gellera C, Ratti A, et al. Analysis of FUS gene mutation in familial amyotrophic lateral sclerosis within an Italian cohort. *Neurology* 2009;**73**:1180.

359. Van LT, van der Zee J, Sleegers K, Engelborghs S, Vandenberghe R, Gijselinck I, et al. Genetic contribution of FUS to frontotemporal lobar degeneration. *Neurology* 2010;**74**:366.

360. Vance C, Rogelj B, Hortobagyi T, De Vos KJ, Nishimura AL, Sreedharan J, et al. Mutations in FUS, an RNA processing protein, cause familial amyotrophic lateral sclerosis type 6. *Science* 2009;**323**:1208.

361. Lagier-Tourenne C, Cleveland DW. Rethinking ALS: the FUS about TDP-43. *Cell* 2009;**136**:1001.

362. Lagier-Tourenne C, Polymenidou M, Cleveland DW. TDP-43 and FUS/TLS: emerging roles in RNA processing and neurodegeneration. *Hum Mol Genet* 2010;**19**:R46.

363. Giordana MT, Piccinini M, Grifoni S, De MG, Vercellino M, Magistrello M, et al. TDP-43 redistribution is an early event in sporadic amyotrophic lateral sclerosis. *Brain Pathol* 2010;**20**:351.

364. Tatom JB, Wang DB, Dayton RD, Skalli O, Hutton ML, Dickson DW, et al. Mimicking aspects of frontotemporal lobar degeneration and Lou Gehrig's disease in rats via TDP-43 overexpression. *Mol Ther* 2009;**17**:607.

365. Wegorzewska I, Bell S, Cairns NJ, Miller TM, Baloh RH. TDP-43 mutant transgenic mice develop features of ALS and frontotemporal lobar degeneration. *Proc Natl Acad Sci USA* 2009;**106**:18809.

366. Wils H, Kleinberger G, Janssens J, Pereson S, Joris G, Cuijt I, et al. TDP-43 transgenic mice develop spastic paralysis and neuronal inclusions characteristic of ALS and frontotemporal lobar degeneration. *Proc Natl Acad Sci USA* 2010;**107**:3858.

367. Hasegawa M, Arai T, Nonaka T, Kametani F, Yoshida M, Hashizume Y, et al. Phosphorylated TDP-43 in frontotemporal lobar degeneration and amyotrophic lateral sclerosis. *Ann Neurol* 2008;**64**:60.

368. Igaz LM, Kwong LK, Chen-Plotkin A, Winton MJ, Unger TL, Xu Y, et al. Expression of TDP-43 C-terminal fragments in vitro recapitulates pathological features of TDP-43 proteinopathies. *J Biol Chem* 2009;**284**:8516.

369. Inukai Y, Nonaka T, Arai T, Yoshida M, Hashizume Y, Beach TG, et al. Abnormal phosphorylation of Ser409/410 of TDP-43 in FTLD-U and ALS. *FEBS Lett* 2008;**582**:2899.

370. Neumann M, Kwong LK, Lee EB, Kremmer E, Flatley A, Xu Y, et al. Phosphorylation of S409/410 of TDP-43 is a consistent feature in all sporadic and familial forms of TDP-43 proteinopathies. *Acta Neuropathol* 2009;**117**:137.

371. Dormann D, Capell A, Carlson AM, Shankaran SS, Rodde R, Neumann M, et al. Proteolytic processing of TAR DNA binding protein-43 by caspases produces C-terminal fragments with disease defining properties independent of progranulin. *J Neurochem* 2009;**110**:1082.

372. Nishimoto Y, Ito D, Yagi T, Nihei Y, Tsunoda Y, Suzuki N. Characterization of alternative isoforms and inclusion body of the TAR DNA-binding protein-43. *J Biol Chem* 2010;**285**:608.
373. Zhang YJ, Xu YF, Cook C, Gendron TF, Roettges P, Link CD, et al. Aberrant cleavage of TDP-43 enhances aggregation and cellular toxicity. *Proc Natl Acad Sci USA* 2009;**106**:7607.
374. Zhang YJ, Xu YF, Dickey CA, Buratti E, Baralle F, Bailey R, et al. Progranulin mediates caspase-dependent cleavage of TAR DNA binding protein-43. *J Neurosci* 2007;**27**:10530.
375. Neumann M, Igaz LM, Kwong LK, Nakashima-Yasuda H, Kolb SJ, Dreyfuss G, et al. Absence of heterogeneous nuclear ribonucleoproteins and survival motor neuron protein in TDP-43 positive inclusions in frontotemporal lobar degeneration. *Acta Neuropathol* 2007;**113**:543.
376. Lin CL, Bristol LA, Jin L, Dykes-Hoberg M, Crawford T, Clawson L, et al. Aberrant RNA processing in a neurodegenerative disease: the cause for absent EAAT2, a glutamate transporter, in amyotrophic lateral sclerosis. *Neuron* 1998;**20**:589.

Tau and Tauopathies

GLORIA LEE AND
CHAD J. LEUGERS

*Department of Internal Medicine,
University of Iowa Carver College of
Medicine, Iowa City, Iowa, USA*

Tauopathies are age-related neurodegenerative diseases that are characterized by the presence of aggregates of abnormally phosphorylated tau. As tau was originally discovered as a microtubule-associated protein, it has been hypothesized that neurodegeneration results from a loss of the ability of tau to associate with microtubules. However, tau has been found to have other functions aside from the promotion and stabilization of microtubule assembly. It is conceivable that such functions may be affected by the abnormal phosphorylation of tau and might have consequences for neuronal function or viability. This chapter provides an overview of tau structure, functions, and its involvement in neurodegenerative diseases.

Tau was discovered as a microtubule-associated protein from porcine brain that promoted microtubule assembly *in vitro*.[1] The tau protein sequence determined from murine cDNA was the first to be reported for a microtubule-associated protein,[2] and subsequently antisense treatment of primary neuronal cultures indicated a critical role for tau in axonal development.[3] While studies of two independently generated tau knockout mouse models suggested that mice lacking tau appeared to develop normally, neurons cultured from one such mouse showed detectable slowing in axonal maturation.[4,5] In addition, the defects in neuronal development exhibited by a

Progress in Molecular Biology
and Translational Science, Vol. 107
DOI: 10.1016/B978-0-12-385883-2.00004-7

MAP1B knockout mouse were exacerbated when tau was also deleted.[6] These studies underline the fact that several microtubule-associated proteins exist in the brain and suggest that each might have its own distinct function during development and in the adult. In fact, while tau shares significant sequence homology with microtubule-associated proteins MAP2 and MAP4 in the carboxyl terminus microtubule-binding domain, it shares very little homology in the amino terminal "projection domain."[7,8]

The idea that different microtubule-associated proteins each have distinct functions is clearly illustrated by the fact that only tau has been associated with neurofibrillary tangles in age-related neurodegenerative diseases such as Alzheimer's disease (AD).[9–11] Moreover, mutations in the tau gene, *MAPT*, cause autosomal dominant neurodegenerative diseases such as frontotemporal dementia (FTD) with Parkinsonism linked to chromosome 17,[12–14] and transgenic mouse models expressing mutant tau exhibit neuronal loss (reviewed in Refs. 15–18). A mechanistic understanding of the route by which tau leads to neurodegeneration is still unclear. However, alongside hypotheses based on the loss of tau's ability to stabilize microtubules are other possibilities based on new functions and interactions that have been described for tau. This chapter will summarize recent studies on the interaction of tau with microtubules in addition to investigations indicating that the function of tau extends beyond its actions on microtubules. Much like its microtubule-binding properties, these alternative functions of tau may be regulated by phosphorylation. Therefore, such functions might be altered in the disease state where tau is abnormally phosphorylated and play a role in neuropathological processes.

I. Tau Gene and Isoforms

Tau is encoded by a single gene, *MAPT*, located on chromosome 17q21.[19] *MAPT* is over 50kb in size and comprises two haplotypes, H1 and H2, with multiple variants of each.[20,21] Several tau isoforms are generated by alternative splicing, creating both high- and low-molecular-weight isoforms. The human central nervous system expresses six low-molecular-weight isoforms that range in size from 352 to 441 amino acids (Fig. 1). These isoforms are differentiated by the presence or absence of sequences encoded by *MAPT* exons 2, 3, and 10.[22] Exons 9, 10, 11, and 12 each encode a microtubule-binding motif. The four motifs are imperfect copies of an 18-amino acid sequence termed a "repeat," and each repeat is separated by a 13- to 14-amino acid inter-repeat sequence.[2] Isoforms that include exon 10 are commonly referred to as four-repeat or 4R tau isoforms, while those that exclude exon 10 are referred to as

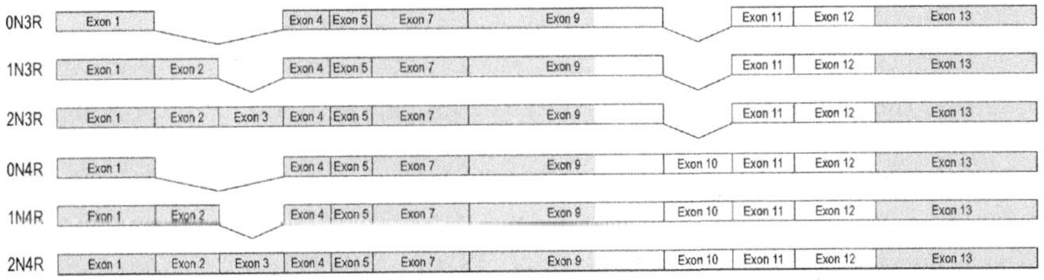

FIG. 1. Tau schematic, drawn to scale, showing the six tau isoforms present in human brain. Exons 2, 3, and 10 are only expressed in the adult. Clear areas each contain a microtubule-binding motif (e.g., exon 10-containing isoforms contain four microtubule-binding motifs).

three-repeat or 3R tau isoforms. Alternative splicing of tau is developmentally regulated, with exons 2, 3, and 10 being expressed only postnatally.[22] Human adult tau has approximately equal representation of 3R and 4R tau isoforms, with the 1N3R and 1N4R being the most abundant forms.[23,24] Alternative splicing of human tau differs from that of rodent tau, as adult rodent tau is predominantly 4R tau.[25] Comparison of the tau sequence from mouse, rat, cow, monkey, goat, and chicken shows high conservation of the microtubule-binding repeats across species.[2,25–28] Tau-like sequences have also been found in frog, nematode, and zebrafish.[29–31]

Because 4R tau isoforms contain a fourth microtubule-binding repeat, adult tau interacts with microtubules more strongly.[32–34] Tau alternative splicing can also affect its phosphorylation, which influences the interaction between tau and microtubules.[35] Phosphorylation is generally higher in fetal tau.[36] When a single tau cDNA is expressed by transfection in cells, several differentially phosphorylated species can be generated.

While mice with a disrupted tau gene are viable, microarray analysis performed on the brains of such mice showed alterations in gene expression relative to wild-type mice.[37] The genes with the highest levels of change did not involve the cytoskeleton, suggesting that the most critical function of tau may not be related to microtubule binding. For example, adult tau knockout mice had increased muscle weakness[38] and were protected against experimentally induced seizures.[39] The idea that tau might play a role in processes other than axonal development is supported by the fact that tau is expressed in nonneuronal cells. Tau expression has been reported in muscle, liver, kidney, and other tissues.[40,41] It has also been found in human breast, prostate, gastric, and pancreatic cancer cell lines and tissues,[42–46] as well as in the muscle cells of individuals with inclusion body myositis.[47] The function of tau in nonneuronal cells remains to be elucidated, and functions outside of the cytoskeleton may have significance for neurodegenerative disease.

II. Tau in Neurodegenerative Disease

While the discovery of tau predated its connection to AD, its importance in neurodegenerative disease has attracted a large community of investigators. AD is characterized by two neuropathological features, namely, senile plaques and neurofibrillary tangles, and tau is the primary component of the neurofibrillary tangles (NFT, reviewed in Refs. 48,49). Senile plaques are made of amyloid β-protein (Aβ), and the gene encoding Aβ has been connected to AD (reviewed in Ref. 50). However, *MAPT* has not been genetically linked to AD. Nevertheless, cultured neurons exposed to Aβ do not undergo cell death in the absence of tau.[51] Likewise, genetically removing tau from animal models that exhibited amyloid plaques lessened the deficits induced by the amyloid.[39] These findings underline a critical role for tau in the neurodegenerative process. Moreover, tau pathology is found in several other age-related neurodegenerative diseases such as progressive supranuclear palsy (PSP), corticobasal degeneration (CBD), Pick's disease, argyrophilic grain disease (AGD), and FTD. Tau pathology involving 4R tau is found in PSP, CBD, and AGD, with tau aggregates being found in the medial temporal lobe, cortex, basal ganglia, subthalamic nucleus, and substantia nigra. Besides neurons, oligodendrocytes and astrocytes can also display tau pathology. Pick's disease has 3R tau in the Pick bodies that are found in the hippocampus and dentate fascia. The clinical presentation of these diseases includes dementia, Parkinsonism, and focal cortical syndrome (reviewed in Refs. 52–55).

The importance of tau in neurodegeneration has been verified by the discovery of *MAPT* mutations in families with FTD with Parkinsonism linked to chromosome 17 (FTDP-17; reviewed in Refs. 56–58). Mutations in the tau gene are responsible for nearly 30% of inherited FTD. These mutations are autosomal dominant and can be located in either coding or noncoding regions. More than 90% of the mutations in coding regions are located in the carboxyl terminal end of tau, with P301L being the most prevalent (Table I). Mutations in noncoding regions are mainly within the intron separating exons 10 and 11. These mutations modulate alternative splicing of the *MAPT* mRNA, resulting in higher than normal levels of 4R tau relative to 3R tau.[59] In addition, the tau haplotype H1c has been linked to PSP.[21,60]

A. Tauopathy Models

Transgenic animals expressing mutant tau cDNAs exhibit tau pathology which increases with age. While these models have shown a variety of traits, perhaps owing to the variety of mutations and gene promoters employed, a striking feature has been neuronal loss and behavioral deficits (reviewed in Refs. 16,17,61). Mouse models expressing only mutant genes involved in amyloid production showed amyloid plaques, but did not acquire neurofibrillary tangles

TABLE I

MISSENSE MUTATIONS IN TAU CAUSING FRONTOTEMPORAL DEMENTIA (REVIEWED IN REFS. 56–58)

Exon 1	Exon 9	Exon 10	Exon 11	Exon 12	Exon 13
R5H, R5L	K257T	N279K	L315R	G335V, G335S	G389R
	I260V	ΔK280	K317M	Q336R	R406W
	L266V	L284L	S320F, S320Y	V337M	T427M
	G272V	N296H, N296N, ΔN296		E342V	
	G273R	P301L, P301S, P301T		S352L	
		G303V		V363I	
		S305N, S305S		K369I	

or suffer neuronal loss.[62] Therefore, inclusion of a mutant tau cDNA in addition to mutant genes involved in amyloid production was integral in creating a triple transgenic mouse model that exhibited both the plaques and tangles characteristic of AD.[63]

Cell culture models that reproduce tau filament formation have been reported.[64,65] However, the polymerization of tau synthesized from *Escherichia coli in vitro*, induced by either arachidonic acid or heparin, has enabled a more extensive analysis of the structural features of tau that are involved in polymerization (reviewed in Refs. 66–68). These studies have led to the identification of specific motifs in the tau repeat region that facilitate the formation of tau filaments.[69] In such assays, FTDP-17-associated missense mutations or tau truncation at Asp421 increased filament formation.[70–72] These results suggest that filament formation in the human neuroblastoma cell culture model may have involved the cleavage of tau at Asp421, as conversion of the neighboring Ser422 to Ala or Glu led to a loss of filament formation in cells.[64] Lastly, the *in vitro* assays have indicated that several phosphomimicking mutations slowed filament formation,[73–75] although some facilitated their formation.[76]

Despite the evidence that tau filaments are a hallmark feature of classic AD pathology, some models hint at the possibility that tau filaments may not be a prerequisite for neurodegeneration. Studies of a mouse model with inducible tau expression have shown that downregulation of tau expression, without a decrease in tangle burden, was sufficient to ameliorate behavioral deficits.[77] Moreover, tauopathy mouse models exist in which behavioral deficits were exhibited without the presence of tangles or neuronal loss.[78,79] In addition, in *Drosophila* and nematode tauopathy models, neuronal loss and behavioral deficits occurred without the formation of tau filaments.[80,81] Taken together, these studies have suggested that tau, in its soluble form, may have unidentified roles in the mechanisms underlying both neuronal cell function and disease. Tau oligomers that occur in advance of tau filaments have been proposed as a critical entity in the neurodegenerative process (reviewed in Ref. 82). These observations, in combination with other recent studies, have raised the possibility that tau filament formation may be a protective mechanism initiated by cells to sequester abnormal tau (reviewed in Refs. 83–85).

In neurodegenerative disease, the phosphorylation state of tau in NFTs and other tau lesions is abnormal, meaning that there is an increase in both the overall number of sites phosphorylated and the level of phosphorylation at particular sites relative to normal adult brain tau. Tau phosphorylation is complex. The 441-residue tau protein has 45 Ser, 35 Thr, and 5 Tyr residues, presenting a multitude of phosphorylation sites. In addition, phosphorylation at some sites facilitates the subsequent phosphorylation of other sites. The effect of phosphorylation depends on the location of the site modified (see Section IV below). Both kinases and phosphatases have been implicated in the appearance of abnormally phosphorylated tau and, despite differences in the morphology of tau lesions among tauopathies, abnormal tau phosphorylation is a common denominator. Interestingly, many sites that are phosphorylated in disease correspond to sites that are phosphorylated during normal brain development.[86–88] The regulation of tau phosphorylation and tau kinases in both development and disease, as well as the functional significance of tau phosphorylation, has been the subject of much investigation. Because of the prevalence of abnormal tau phosphorylation during neurodegeneration, reducing tau phosphorylation as a therapeutic strategy has been investigated (reviewed in Refs. 89,90). A significant challenge has been limiting the action of kinase inhibitors to tau phosphorylation.

The abnormal phosphorylation of tau also occurred in tau transgenic mouse models that express either FTDP-17 mutant or wild-type tau (reviewed in Ref. 91), and reducing tau phosphorylation in one such model lessened aggregated tau and axonal degeneration.[92] Experiments in *Drosophila* tauopathy models have demonstrated that tau phosphorylation is required for neuronal loss.[93] This same study also showed that cell-cycle genes were required for tau-induced neurodegeneration, supporting the hypothesis that neurons die because they are receiving signals to divide (reviewed in Refs. 94,95). The presence of "mitotic" phospho-epitopes in "disease tau," characteristic of the tau expressed during development, had first led to this hypothesis[96–99] and the presence of tetraploid neurons in AD brain has strengthened the hypothesis.[100]

III. Interactions with the Cytoskeleton

A. Microtubule Binding and Assembly

Tau was originally discovered through its ability to promote microtubule assembly, which stems from its ability to modulate the dynamic instability of microtubules.[33,101] The interaction between tau and microtubules depends on the tau microtubule-binding repeats, as well as on the flanking regions upstream and downsteam of the repeats. Defining the role of the flanking regions was largely accomplished by investigating the ability of truncated or point-mutated tau proteins to either associate with microtubules[102–105] or decrease dynamic instability.[33]

Most recently, the physical interaction between tau and microtubules has been delineated using nuclear magnetic resonance (NMR) spectroscopy. Comparison of the NMR spectra of tau in the presence and absence of microtubules indicated that, while all repeats contacted the microtubules, there were specific sequences that were strongly involved in the interaction.[34,106] These sequences included ^{240}KSRLQTAPV^{248}, ^{275}VQIINKKLDLS^{285}, and ^{297}IKHV^{300}. In addition, residues in the flanking regions as far upstream as Ser214 and as far downstream as Lys375 were also involved,[107] with ^{225}KVAVVRT^{231} and ^{370}KIETHK^{375} having especially strong interactions.[34] These data indicated that, in the presence of microtubules, the molecular environment around the tau regions flanking the microtubule repeats changed. Although the simplest explanation is that a direct interaction occurred between these tau areas and the microtubule, one cannot rule out the possibility that flanking regions were involved in intramolecular interactions and that in the presence of microtubules, such interactions were altered.[108]

^{275}VQIINKKLDLS^{285} and ^{297}IKHV^{300} are both in exon 10, a fact that may explain why 4R tau isoforms interact with microtubules more strongly than 3R tau isoforms. Interestingly, in comparing 4R and 3R tau, it has been found that 4R tau could decrease microtubule shortening during dynamic fluctuations in microtubule length whereas 3R tau had no effect in this regard.[109] In addition, 4R tau and 3R tau showed qualitative differences with respect to their actions on the microtubule growth rate and on the behavior of growing microtubule populations.[110] These differences have significance towards both the function of microtubules over the course of development and the consequences of the change in the 4R:3R tau ratio brought on by intronic tau mutations. Lastly, the binding of tau to microtubules has been visualized by cryo-EM, showing that tau bound along individual protofilaments.[111] Synchrotron X-ray scattering data has suggested that tau altered the shape of the protofilament, resulting in changes in the curvature of microtubules and a shift from 13 to 14 microtubules per protofilament.[112]

Several studies have demonstrated that some FTDP-17 missense mutations reduced the ability of tau to promote microtubule assembly.[24,113–115] Additional insights have been gained through NMR data,[116] analysis of microtubule dynamics in cells,[117] and experiments utilizing *Xenopus* oocytes to assess microtubule function.[118] While these investigations have shown that some mutations attenuated the ability of tau to bind to microtubules and to regulate the dynamic instability of microtubules, missense mutations outside of the microtubule repeat region did not replicate these effects, suggesting that different FTDP-linked tau mutations affect tau function in different ways.

The interaction between tau and microtubules is greatly decreased by tau phosphorylation at Ser262 and Ser356, and phosphomimicking replacements at these positions effectively reduce microtubule association *in vitro* and in cells.[119] Other phosphorylation sites shown to have some effects on microtubule association are Ser205, Ser212, Ser214, Thr231, Ser235, Ser396, and Ser404.[33,75,98,120–122] On tubulin, the tau-interacting site is located at the carboxyl terminal end, which is highly acidic. Therefore, the interaction between the basic tau repeat regions and microtubules is thought to be primarily electrostatic in nature. This is consistent with the ability of salt to affect the binding between tau and microtubules. Thus, it is not surprising that the addition of an acidic phosphate group to tau would attenuate its association with microtubules, provided the location of the phosphate was appropriate. A reduction in the microtubule-binding or assembly-promoting ability of tau has been a recurrent theme in hypotheses regarding the role of hyperphosphorylation in mechanisms of neurodegeneration (reviewed in Ref. 123). As a result, development of therapeutics aimed at preserving microtubules has been undertaken (for instance, see Ref. 124).

B. Axonal Transport

Fast axonal transport (FAT) is significantly impaired in a number of neurodegenerative diseases, including tauopathies, and these defects have been linked to alterations in the normal function of tau (reviewed in Ref. 125). For example, in cellular models in which tau was overexpressed, a disruption in the trafficking of membranous vesicles and mitochondria was found.[126–130] Similarly, various mouse models of tauopathy in which wild-type or mutant tau was expressed also demonstrated impaired axonal transport.[131,132] The ability of tau to interfere with axonal transport may arise through direct interactions between tau and transport motor complexes. In fact, tau was able to associate with kinesin, as demonstrated by immunofluorescence and co-immunoprecipitation experiments.[133,134] *In vitro* experiments have further identified a direct interaction between tau and kinesin,[135,136] and between tau and the p150 protein in the dynein–dynactin motor complex.[137] Moreover, these direct interactions between tau and motors decreased the ability of kinesin to attach to microtubules[138] and increased the rate of motor detachment from microtubules,[139,140] although the above studies also reported conflicting data concerning whether tau altered overall cargo transport rates. As these results have shown, the effects of tau on axonal transport may be more complex than simply blocking motor access to the microtubules. For instance, an interaction between tau and c-Jun N-terminal kinase-interacting protein 1 (JIP1) has been proposed to affect the kinesin-I motor complex, causing a relocalization of JIP1 and impaired axonal transport.[141] Also, in *Aplysia*, tau overexpression was capable of causing both a complete stop in transport and a reorganization of microtubule polarity within the axon.[142] The conformation of tau may influence FAT as well since studies performed using squid axoplasm showed that soluble monomeric tau did not affect transport[143] whereas tau filaments or N-terminal fragments significantly reduced transport using a mechanism involving PP1, GSK-3β, and the light chain of the kinesin motor.[144]

In spite of the abundance of data that has implicated tau in the inhibition of FAT, a study showing similar rates of axonal transport in wild-type, tau transgenic, and tau depleted mice has argued against the ability of tau to significantly affect this process.[145] These discrepancies may originate from differences in tau phosphorylation between the experimental systems, differences in the experimental methods used to measure FAT, or differences in the cellular sources used. Further studies are needed to clarify the role of tau in the inhibition of FAT in neurodegenerative diseases.

C. Interactions with Actin

Shortly after it was reported as a microtubule-associated protein, tau was also found to associate with actin *in vitro*.[146,147] The interaction site was subsequently mapped to the microtubule-binding domain, and then specifically to the repeats.[148,149] The functional implications of this interaction have remained obscure, although tau depletion in cultured neurons altered actin morphology in growth cones[150] and inactivation of tau in the growth cones caused collapse of lamellipodia.[151] More recently, the C-terminus of tau was found to co-localize with actin in the growing neurite tips of nerve growth factor (NGF)-differentiated PC12 cells.[152] Nevertheless, tau from *E. coli* failed to bind to actin *in vitro*,[153] suggesting that these previously reported associations might require specific tau phosphorylation or involve intermediates. An association with actin may explain why tau phosphorylated on Ser262, a modification that decreases the affinity of tau for microtubules, had a role in neurite outgrowth.[119,154] Other reports have suggested that tau might affect actin remodeling indirectly. Tau expression antagonized the action of Gem GTPase, a negative regulator of Rho.[37] Rho activation is critical for neurite outgrowth; therefore tau could affect neurite outgrowth by lessening the inhibition of Rho by Gem GTPase. In addition, in fibroblasts treated with platelet–derived growth factor, tau expression delayed actin stress fiber recovery, wherein tau-expressing cells maintained the "high Rac, low Rho" actin morphology characteristic of activated cells.[155] Given the many interactions that have been described for tau (see Section V), as well as its known localization in the axonal growth cone,[156,157] it is probable that tau has a role in orchestrating actin remodeling in response to signaling during neurite outgrowth.

An interaction between tau and actin has also been described in animal models of tauopathy. Actin aggregates analogous to Hirano bodies were found when human mutant tau was expressed in *Drosophila*, and actin aggregates were also identified in mouse tauopathy models.[158] Moreover, actin and actin-related proteins appear in the neuropathology of AD.[159]

IV. Phosphorylation and Other Post-translational Modifications

The phosphorylation of tau on serines and threonines is developmentally regulated. Table II lists the phosphorylated residues that have been identified in fetal and adult rat tau either by mass spectrometry[36,168] or by phospho-specific antibody probes.[98,160,161] These experiments demonstrate that fetal tau is more highly phosphorylated than adult tau. The phosphorylation of tau in AD includes all of the sites shown in Table II.[160,168] Table III lists some commonly used tau antibodies, several of which detect specific phosphorylated sites. In AD, tau is phosphorylated either at sites that are not normally phosphorylated in adult tau or at a higher level at sites that are normally phosphorylated in adult tau.

The phosphorylation of tau causes conformational changes that result in a slowing of its electrophoretic migration, and early structural studies found that tau became more elongated upon phosphorylation.[179] Fluorescence resonance energy transfer studies have suggested that tau normally exists in a conformation in which both the amino and carboxyl termini fold inward. This allowed the carboxyl terminus to simultaneously interact with both the microtubule repeat domain and the amino terminus.[108] Phosphomimicking mutations altered this conformation and generated reactivity to a conformation-specific tau antibody.[180] Such conformational changes may underlie the effects of tau phosphorylation on its interactions with other proteins. Phosphorylation of tau could also affect its proteolytic cleavage.[181]

Several Ser/Thr kinases act on tau (recently reviewed in Ref. 49). Among the best studied are the proline-directed kinases GSK3β, cdk5, MAPK (ERK), JNK (SAPK), and p38. Non-proline-directed kinases MARK, casein kinase I (CKI), PKA, CaMKII, and PKC also phosphorylate tau. In some cases, tau phosphorylation at one site facilitated phosphorylation at another site, known as "priming." This has been demonstrated by the requirement of cdk5- or GSK3β-mediated phosphorylation of residue Ser235 prior to phosphorylation of Thr231 by GSK3β.[122,182] FTDP-17 mutations also promoted phosphorylation *in vitro*.[183] Many studies have been performed either *in vitro* with purified kinases or brain extracts, or in transfected cells where both tau and kinases were overexpressed. A significant challenge, though not unique to tau, has been to identify the kinases that are responsible for the phosphorylation of the endogenous protein in neuronal cells.

TABLE II

PHOSPHORYLATED SER AND THR SITES IN FETAL AND ADULT TAU, NUMBERED ACCORDING TO THE 441-AMINO ACID HUMAN ISOFORM (2N4R) OF TAU

Fetal site	Adult site	Possible kinases													
		CamKII	CKI	Cdc2	Cdk2	Cdk5	Dyrk1a	GSK3β	JNK	MAPK	MARK	p38	PKA	PKB/Akt	TTKI
Thr181	*Thr181**						●		●	●		●			
Ser198															●
Ser199	Ser199*							●					●		●
Ser202	Ser202*						●	●	●	●		●			●
Thr205	Thr205*		●	●	●	●	●	●	●	●			●		
Thr212			●	●	●	●		●	●	●			●		
Ser214	Ser214*			●		●	●	●	●	●				●	
Thr217	Thr217*			●				●	●	●					
Thr231	Thr231					●		●	●	●					
Ser235								●							
Ser262		●									●	●	●		
Ser356	Ser356	●									●	●	●		
Ser396	Ser396		●	●	●	●	●	●	●	●					
Ser400							●								
Ser404	Ser404*		●	●		●	●	●	●	●		●	●		
Ser409	Ser409*	●							●	●			●		
Ser413															
Ser416		●													
Ser422		●					●		●			●			●

Residues in italic indicate that mass spectral data have been obtained for the residue. Asterisk indicates that fetal tau phosphorylation of the residue occurs at a higher level (Refs. 160–167).

TABLE III
MONOCLONAL TAU ANTIBODIES (OTHERS ARE ALSO AVAILABLE)

Name	Epitope	Notes
Alz50	Involves amino terminus and MTBR	Conformation specific[169]
Tau12	Amino acids 9–18	Human specific, total tau[170]
Tau1	Amino acids 189–207	Dephosphorylation specific[171]
Tau5	Amino acids 210–230	Total tau, rodent>human[169]
AT8	Phospho-Ser202/Ser205	[172]
CP13	Phospho-Ser202	[173]
AT100	Phospho-Thr212/Ser214	
CP3	Phospho-Ser214	[173]
AT180	Phospho-Thr231	[174]
CP17	Phospho-Thr231	[175]
TG3	Phospho-Thr231	Conformation specific[176]
CP9	Phospho-Thr231	[173]
12E8	Phospho-Ser262/Ser356	[161]
PG5	Phospho-Ser409	[173]
PHF-1	Phospho-Ser396/Ser404	[177]
9G3	Phospho-Tyr18	[178]

Polyclonal antibodies are also commercially available.

The regulation of tau phosphorylation during development is of significant interest, as several of the sites phosphorylated in disease appear normally during early development. In neuroblastoma cells, Ser214 and Ser262, in addition to several proline-directed sites such as Ser202, Thr205, Thr231, Ser235, Ser396, and Ser404, become highly phosphorylated during mitosis.[96–98] These findings suggest that in developing neurons, tau phosphorylation can be regulated by cell-cycle mechanisms. Changes in tau phosphorylation occurring when neuronal cells were treated with Aβ have also been investigated.[184] These findings have implicated both GSK3β and MAPK[185,186] as kinases involved in the abnormal phosphorylation of tau during AD pathogenesis. The phosphorylation of tau has also been investigated in mouse models of tauopathy as well as in mouse models where kinases or phosphatases were expressed (reviewed in Ref. 91).

Phosphatases also act on tau, and phosphatase inhibition has been suggested as one mechanism by which tau acquires its hyperphosphorylated state during the neurodegenerative process.[187] Both PP1 and PP2A associate with and dephosphorylate tau,[188–191] with PP2A accounting for 70% of the tau phosphatase activity in brain.[192] FTDP-17 mutations reduced the interaction between PP2A and tau, suggesting another route by which these mutations would result in hyperphosphorylation and disease.[193]

Tau contains five potential sites for tyrosine phosphorylation, and Src family tyrosine kinases (Fyn, Src, and Lck), Syk, Abl, and tau-tubulin kinase phosphorylate tau. Direct interactions between tau and the Src homology

3 (SH3) domains of Fyn, Src, and Lck have been demonstrated,[194] and tau also interacts with Abl and Syk.[195,196] Fyn, Src, and Syk phosphorylate tau at Tyr18,[178,196] while Abl phosphorylates Tyr394.[195] Phosphorylated Tyr18 and Tyr394 have been found in early development, in tauopathy mouse models, and in AD brain, reproducing the behavior seen with disease-related Ser/Thr phosphorylated sites.[178,105,107,108] Phospho-Tyr197, a modification that can be generated by tau-tubulin kinase,[162] has been identified along with phospho-Tyr394 in the tau filaments isolated from a tauopathy mouse model. It also occurs in AD brain.[197] The phosphorylation of tyr29 by Lck has also been reported.[199] Functional implications for the tyrosine phosphorylation of tau have yet to be elucidated. However, the presence of these modifications in tau pathology, as well as data implicating Fyn in AD,[200–202] suggests that activated tyrosine kinases will also be a part of the neuropathogenic process.

In addition to being phosphorylated, tau can be O-GlcNAcylated, nitrated, and ubiquitinated. Because O-linked GlcNAcylation of tau occurs on Ser and Thr residues,[203] it has the potential to indirectly regulate tau phosphorylation.[204] Tau nitration on Tyr29 has been found in AD and other tauopathies,[205] and the presence of nitration, which is catalyzed by reactive nitrogen species, is consistent with an elevation of oxidative stress during neurodegeneration. Ubiquitination of tau is readily seen following cotransfection of tau and the E3 ubiquitin ligase CHIP (carboxyl terminus of Hsp70-interacting protein) into nonneuronal cells.[206–208] The presence of ubiquitinated tau is well established in AD,[209] and the specific lysines modified in abnormal tau from AD brain have been identified as Lys254, Lys311, and Lys353.[210] Ubiquitination of tau has been shown to increase soluble tau levels, and to target tau for proteasomal degradation.[206,207,211] Tau can also undergo sumoylation, a ubiquitin-like modification, at Lys340.[212]

V. Other Interactions

A. Phospho-Serine/Threonine-Based Interactions

Tau interacts with Pin1, a prolyl-isomerase that changes the conformation of phospho-Ser/Thr-pro bonds from cis to trans conformation. This interaction was initially shown to involve the WW domain of pin1 that recognizes phospho-Ser/Thr residues and the phospho-Thr231 residue of tau.[213] Subsequently, an interaction between pin1 and the phospho-Thr212 residue of tau was also uncovered by NMR.[214] The presence of pin1 restored the microtubule-polymerizing properties to tau lost following cdc2-mediated phosphorylation[213] and reduced the levels of tau phosphorylation in tau transgenic mice.[215,216] These observations have been attributed to an increase in the susceptibility of tau to phosphatases in the presence of pin1.[217,218]

The opposing effects of pin1 on wild-type versus FTDP-17 mutant tau has presented more evidence for the potential importance of the interaction.[215,216] Also contributing to the interest in the pin1–tau interaction is the finding that the pin1 knockout mice exhibited age-dependent neurodegeneration, increased tau phosphorylation, and neuropathology.[218]

The protein 14-3-3 is a signal transduction protein that exists in several isoforms, of which the 14-3-3ζ isoform has been identified as a tau interactor.[219] 14-3-3ζ has been shown to increase PKA-mediated tau phosphorylation,[219,220] while its effect on GSK3β-mediated tau phosphorylation is less clear.[221–223] The phospho-Ser214 residue on tau is thought to be the primary binding site for 14-3-3ζ,[224] with the added presence of phospho-Ser235 strengthening the interaction.[225] While the impact of 14-3-3ζ on tau phosphorylation in neuronal cells remains to be investigated, evidence that the interaction is influenced by both the phosphorylation state and the isoform of tau[226] suggests that the interaction could potentially have implications for both development and disease.

Tau is able to enhance growth factor-induced MAPK signaling through a mechanism that requires the phosphorylation of tau at Thr231.[227] Tau was phosphorylated on Thr231 in response to NGF, and tau depletion attenuated MAPK activation as well as AP-1 activation.[227] The effect of tau on signaling was independent of an interaction between tau and microtubules. Interestingly, these findings may have significant implications for the role of tau in neurodegenerative disease, as the early appearance of phospho-Thr231-tau,[228,229] as well as an abnormal activation of MAPK, occurs in AD.[230–232] One could speculate that abnormal MAPK signaling induced by various upstream triggers such as Aβ accumulation, oxidative stress, and aberrant growth factor activity would be potentiated by hyperphosphorylated tau, leading to a positive feedback loop where MAPK would phosphorylate tau further. Faulty MAPK signaling might also drive the cell cycle and culminate in neuronal cell death.

B. SH3 Domain Interactions

Within tau exons 7 and 9, upstream of the first microtubule-binding repeat, lies a proline-rich domain containing >20% proline. This region contains seven PXXP motifs that can potentially interact with the SH3 domains commonly found in tyrosine kinases and adapter proteins. In vitro binding assays have demonstrated that a PXXP in tau interacted with the SH3 domain of Src family kinases,[194] and tau also interacted with the SH3 domains of phosphatidylinositol-3 kinase (PI3K), grb2, and phospholipase Cγ (PLCγ).[233] Co-immunoprecipitation experiments have confirmed that tau interacted with Fyn, PI3K, and PLCγ in cells.[43,194,234] A possible functional significance of the tau–Fyn interaction is the upregulation of Fyn kinase activity by tau,[155] a known consequence of SH3 domain interactions for Src family tyrosine kinases. The presence of tau also increased PLCγ activity, though the involvement of the SH3 domain interaction has not been demonstrated.[235] The involvement of the tau–Fyn SH3 interaction in directing the tyrosine phosphorylation of tau has been shown.[236]

Tau phosphorylation affects SH3 domain interactions,[233,236,237] leading to the speculation that these interactions are regulated during development and may have a role in disease. The finding that FTDP-17 tau mutations increased the tau–Fyn SH3 interaction[236] also supports a role for the interaction during neuropathogenesis.

C. Molecular Chaperone Interactions

Tau interacts with both the stress-induced heat shock protein hsp70 and the constitutively expressed heat shock cognate protein hsc70.[207,238,239] Both interactions promote the ubiquitination of tau by CHIP and the proteasome-mediated degradation of tau.[206–208,211,239] The interaction with heat shock proteins may also have a role in the degradation of tau via the autophagy–lysosomal pathway.[240,241] Previous studies have demonstrated that phosphorylation of tau increased both its ubiquitination[207] and its degradation.[242] In addition, hsp27 and hsp90 recognize phosphorylated tau and facilitate its proteasome-mediated degradation.[207,243] These results raise the possibility that, in response to disease-related phosphorylation modifications, the cell attempts to eliminate tau by a mechanism involving interactions with heat shock proteins.

The tau motifs that bind to hsc70 and hsp70 have been identified and correspond to the VQI(I/V) sequences in exons 9 and 10,[238] the motifs that have been found to mediate the β-sheet conformation involved in tau filament formation.[69] Therefore, in addition to potentially directing the degradation of disease tau, the association between hsc70/hsp70 and tau could also be neuroprotective by preventing tau filament formation. The ability of hsp70 to inhibit tau filament formation *in vitro* supports this possibility.[244]

D. Nonmicrotubule Localizations for Tau

Given that certain tau modifications result in a reduced affinity for microtubules, one would predict that nonmicrotubule localizations for tau exist. In fact, the association of tau with two nonmicrotubule structures within cells, the nucleoli and polysomes,[245–247] has long been known. Phosphorylation has been proposed to regulate the localization of tau to the nucleus,[248] and the direct association of tau with nucleic acid *in vitro* has also been reported.[249] It has recently been suggested that tau may contribute to chromosome instability.[250]

Tau that is phosphorylated at Thr231 associated with the microtubule-organizing center[122] and tau that is dephosphorylated at Ser199/Ser202, Ser396/Ser404, or Thr231 associated with the plasma membrane.[156,251,252] Tau dephosphorylated at Ser199/Ser202 was also enriched in the growth cone of primary cultured neurons.[156,253] Moreover, tau associated with lipid rafts, membrane microdomains implicated in signal transduction and growth cone function.[254–257] Evidence indicating that tau–Fyn complexes existed in lipid rafts has also been reported.[254]

The existence of extracellular tau has been recently reported, with the amino terminus being critical for the extracellular localization.[258] Also, an interaction between extracellular tau and muscarinic M1 and M3 receptors has been reported, suggesting that extracellular tau is capable of inducing changes in intracellular calcium.[259] These findings raise the possibility that interneuronal propagation of neurodegenerative disease may involve extracellular tau. The ability of extracellular tau to induce pathology has been explored in mouse and cell culture models.[260,261]

E. Amino Terminus of Tau

Investigations into the properties of truncated tau, terminating in exon 9, have revealed that the amino terminus of tau was capable of associating with the plasma membrane and affecting NGF-mediated neurite outgrowth.[156] Similarly, the amino terminus of tau negatively affected neurite outgrowth in oligodendrocytes.[262] Interestingly, the amino terminus of tau participated in Aβ-oligomer-activated signal transduction pathways where microtubules were disrupted.[263] Because the truncated tau used in these studies contained the proline-rich region, it is conceivable that interactions with proteins such as Src family tyrosine kinases or PI3K may underlie these reported effects. Moreover, by expressing the amino terminus of tau in mice, it has been demonstrated that the localization of Fyn was shifted from the postsynaptic area to the cell soma, because of its interaction with the amino terminus of tau in the soma.[264] As a result, the association of the NMDA receptor with the postsynaptic density was reduced and susceptibility to seizure was also reduced. These data strongly argue for the amino terminus having critical functions in the neuron. Moreover, the expression of the amino terminus of tau was able to lessen the deficits of an APP transgenic mouse model, similar to that achieved when a tau$^{-/-}$ trait was introduced.[39,264]

As a separate consideration, alternative splicing of tau is capable of generating an amino terminal fragment of tau, owing to alternative splice sites in exon 6 that create frameshifts and stop codons.[265] Such tau fragments inhibited FAT.[144] In addition, a toxic amino terminal fragment of tau generated by calpain cleavage has been described in neurons treated with Aβ.[266] The production of this fragment increased in aged primary neuronal cultures and decreased if membrane cholesterol was lowered.[267] Taken together, these observations provide more evidence that nonmicrotubule-associated tau plays an important role in both normal and diseased cells.

VI. Reflections

The idea that tau is more than a microtubule-associated protein is borne out by the fact that tau exists in forms that do not associate with microtubules and interact with many other proteins besides microtubules. And, although tau knockout mice do not exhibit gross defects in brain development, their blunted response to excitotoxic stimuli suggest that tau is important for neuronal function in ways that are not yet understood. In fact, the two genes whose expression was increased the most in tau knockout mouse neurons were c-fos and fosB (Data supplement[37]), which are transcription factors critical for regulating transcription of a diverse range of genes. Such data strongly suggests that tau has a critical role in basic cell growth. It is not possible at present to determine whether this role stems from a function in neuronal or nonneuronal cells.

Three-dimensional structural information for tau would greatly contribute to the understanding of phosphorylation and protein conformation in tau function and the effect of FTDP-17 missense mutations and alternative splicing on tau structure. Because tau has unusual physical properties, obtaining structural data has been challenging. Recent analyses of tau using NMR indicate its potential to provide more information about tau structure.[268]

The ability of tau to interact with a number of signal transduction proteins suggests a possible role for tau in signaling. Tau may participate in the mTor and JNK pathways,[93,269] and we have obtained evidence that tau potentiates NGF-induced MAPK activation.[227] However, despite available co-immunoprecipitation data for some interactions, the specific molecular complexes that engage tau as a signal transduction protein remain to be identified. Also, the functional significance of both the tyrosine phosphorylation of tau and its increased phosphorylation during development remain unclear.

While the microtubule-associated functions of tau are important, its function in signaling may be equally important; but it is unclear which functions are most critical during the neurodegenerative process. Establishing new functions for tau would lead to new hypotheses regarding the connection between tau and neurodegenerative disease. If a nonfilamentous form of hyperphosphorylated tau is responsible for early behavioral deficits, understanding the role of phosphorylated tau during development and in signal transduction may provide clues to pathways that are misregulated during the disease process.

REFERENCES

1. Weingarten MD, Lockwood AH, Hwo SY, Kirschner MW. A protein factor essential for microtubule assembly. *Proc Natl Acad Sci USA* 1975;**72**:1858–62.
2. Lee G, Cowan N, Kirschner M. The primary structure and heterogeneity of tau protein from mouse brain. *Science* 1988;**239**:285–8.

3. Caceres A, Potrebic S, Kosik KS. The effect of tau antisence oligonucleotides on neurite formation of cultured cerebellar macroneurons. *J Neurosci* 1991;**11**:1515–23.

4. Harada A, Oguchi K, Okabe S, Kuno J, Terada S, Ohshima T, et al. Altered microtubule organization in small-calibre axons of mice lacking *tau* protein. *Nature* 1994;**369**:488–91.

5. Dawson HN, Ferreira A, Eyster MV, Ghoshal N, Binder LI, Vitek MP. Inhibition of neuronal maturation in primary hippocampal neurons from tau deficient mice. *J Cell Sci* 2001;**114**:1179–87.

6. Takei Y, Teng J, Harada A, Hirokawa N. Defects in axonal elongation and neuronal migration in mice with disrupted tau and map1b genes. *J Cell Biol* 2000;**150**:989–1000.

7. Lewis SA, Wang DH, Cowan NJ. Microtubule-associated protein MAP2 shares a microtubule binding motif with tau protein. *Science* 1988;**242**:936–9.

8. Chapin SJ, Bulinski JC. Non-neuronal 210×10(3) Mr microtubule-associated protein (MAP4) contains a domain homologous to the microtubule-binding domains of neuronal MAP2 and tau. *J Cell Sci* 1991;**98**:27–36.

9. Wood JG, Mirra SS, Pollock NJ, Binder LI. Neurofibrillary tangles of Alzheimer disease share antigenic determinants with the axonal microtubule-associated protein tau (tau). *Proc Natl Acad Sci USA* 1986;**83**:4040–3 [Published erratum appears in *Proc Natl Acad Sci U S A* 1986 Dec;**83**(24):9773].

10. Grundke II, Iqbal K, Tung YC, Quinlan M, Wisniewski HM, Binder LI. Abnormal phosphorylation of the microtubule-associated protein tau (tau) in Alzheimer cytoskeletal pathology. *Proc Natl Acad Sci USA* 1986;**83**:4913–7.

11. Kosik KS, Joachim CL, Selkoe DJ. Microtubule-associated protein tau (tau) is a major antigenic component of paired helical filaments in Alzheimer disease. *Proc Natl Acad Sci USA* 1986;**83**:4044–8.

12. Hutton M, Lendon CL, Rizzu P, Baker M, Froelich S, Houlden H, et al. Association of missense and 5′-splice-site mutations in tau with the inherited dementia FTDP-17. *Nature* 1998;**393**:702–5.

13. Spillantini MG, Murrell JR, Goedert M, Farlow MR, Klug A, Ghetti B. Mutation in the tau gene in familial multiple system tauopathy with presenile dementia. *Proc Natl Acad Sci USA* 1998;**95**:7737–41.

14. Poorkaj P, Bird TD, Wijsman E, Nemens E, Garruto RM, Anderson L, et al. Tau is a candidate gene for chromosome 17 frontotemporal dementia. *Ann Neurol* 1998;**43**:815–25.

15. Lee VM, Kenyon TK, Trojanowski JQ. Transgenic animal models of tauopathies. *Biochim Biophys Acta* 2005;**1739**:251–9.

16. Gotz J, Ittner LM. Animal models of Alzheimer's disease and frontotemporal dementia. *Nat Rev Neurosci* 2008;**9**:532–44.

17. Denk F, Wade-Martins R. Knock-out and transgenic mouse models of tauopathies. *Neurobiol Aging* 2009;**30**:1–13.

18. Frank S, Clavaguera F, Tolnay M. Tauopathy models and human neuropathology: similarities and differences. *Acta Neuropathol* 2008;**115**:39–53.

19. Neve RL, Harris P, Kosik KS, Kurnit DM, Donlon TA. Identification of cDNA clones for the human microtubule-associated protein tau and chromosomal localization of the genes for tau and microtubule-associated protein 2. *Brain Res* 1986;**387**:271–80.

20. Andreadis A, Brown WM, Kosik KS. Structure and novel exons of the human tau gene. *Biochemistry* 1992;**31**:10626–33.

21. Baker M, Litvan I, Houlden H, Adamson J, Dickson D, Perez-Tur J, et al. Association of an extended haplotype in the tau gene with progressive supranuclear palsy. *Hum Mol Genet* 1999;**8**:711–5.

22. Goedert M, Spillantini MG, Jakes R, Rutherford D, Crowther RA. Multiple isoforms of human microtubule-associated protein tau: sequences and localization in neurofibrillary tangles of Alzheimer's disease. *Neuron* 1989;**3**:519–26.

23. Goedert M, Jakes R. Expression of separate isoforms of human tau protein: correlation with the tau pattern in brain and effects on tubulin polymerization. *EMBO J* 1990;**9**:4225–30.

24. Hong M, Zhukareva V, Vogelsberg-Ragaglia V, Wszolek Z, Reed L, Miller BI, et al. Mutation-specific functional impairments in distinct tau isoforms of hereditary FTDP-17. *Science* 1998;**282**:1914–7.

25. Kosik KS, Orecchio LD, Bakalis S, Neve RL. Developmentally regulated expression of specific tau sequences. *Neuron* 1989;**2**:1389–97.

26. Himmler A, Drechsel D, Kirschner MW, Martin DJ. Tau consists of a set of proteins with repeated C-terminal microtubule-binding domains and variable N-terminal domains. *Mol Cell Biol* 1989;**9**:1381–8.

27. Nelson PT, Stefansson K, Gulcher J, Saper CB. Molecular evolution of tau protein: implications for Alzheimer's disease. *J Neurochem* 1996;**67**:1622–32.

28. Yoshida H, Goedert M. Molecular cloning and functional characterization of chicken brain tau: isoforms with up to five tandem repeats. *Biochemistry* 2002;**41**:15203–11.

29. Goedert M, Baur CP, Ahringer J, Jakes R, Hasegawa M, Spillantini MG, et al. PTL-1, a microtubule-associated protein with tau-like repeats from the nematode Caenorhabditis elegans. *J Cell Sci* 1996;**109**(Pt 11):2661–72.

30. Chen M, Martins RN, Lardelli M. Complex splicing and neural expression of duplicated tau genes in zebrafish embryos. *J Alzheimers Dis* 2009;**18**:305–17.

31. Olesen OF, Kawabata-Fukui H, Yoshizato K, Noro N. Molecular cloning of XTP, a tau-like microtubule-associated protein from Xenopus laevis tadpoles. *Gene* 2002;**283**:299–309.

32. Goode BL, Feinstein SC. Identification of a novel microtubule binding and assembly domain in the developmentally regulated inter-repeat region of tau. *J Cell Biol* 1994;**124**:769–82.

33. Trinczek B, Biernat J, Baumann K, Mandelkow EM, Mandelkow E. Domains of tau protein, differential phosphorylation, and dynamic instability of microtubules. *Mol Biol Cell* 1995;**6**:1887–902.

34. Mukrasch MD, von Bergen M, Biernat J, Fischer D, Griesinger C, Mandelkow E, et al. The "jaws" of the tau-microtubule interaction. *J Biol Chem* 2007;**282**:12230–9.

35. Lindwall G, Cole RD. Phosphorylation affects the ability of tau protein to promote microtubule assembly. *J Biol Chem* 1984;**259**:5301–5.

36. Watanabe A, Hasegawa M, Suzuki M, Takio K, Morishima-Kawashima M, Titani K, et al. In vivo phosphorylation sites in fetal and adult rat tau. *J Biol Chem* 1993;**268**:25712–7.

37. Oyama F, Kotliarova S, Harada A, Ito M, Miyazaki H, Ueyama Y, et al. Gem GTPase and tau: morphological changes induced by gem GTPase in cho cells are antagonized by tau. *J Biol Chem* 2004;**279**:27272–7.

38. Ikegami S, Harada A, Hirokawa N. Muscle weakness, hyperactivity, and impairment in fear conditioning in tau-deficient mice. *Neurosci Lett* 2000;**279**:129–32.

39. Roberson ED, Scearce-Levie K, Palop JJ, Yan F, Cheng IH, Wu T, et al. Reducing endogenous tau ameliorates amyloid beta-induced deficits in an Alzheimer's disease mouse model. *Science* 2007;**316**:750–4.

40. Gu Y, Oyama F, Ihara Y. τ is widely expressed in rat tissues. *J Neurochem* 1996;**76**:1235–44.

41. Kenner L, El-Shabrawi Y, Hutter H, Forstner M, Zatloukal K, Hoefler G, et al. Expression of three- and four-repeat tau isoforms in mouse liver. *Hepatology* 1994;**20**:1086–9.

42. Sangrajrang S, Denoulet P, Millot G, Tatoud R, Podgorniak MP, Tew KD, et al. Estramustine resistance correlates with tau over-expression in human prostatic carcinoma cells. *Int J Cancer* 1998;**77**:626–31.

43. Souter S, Lee G. Microtubule-associated protein tau in human prostate cancer cells: Isoforms, phosphorylation, and interactions. *J Cell Biochem* 2009;**108**:555–64.
44. Rouzier R, Rajan R, Wagner P, Hess KR, Gold DL, Stec J, et al. Microtubule-associated protein tau: a marker of paclitaxel sensitivity in breast cancer. *Proc Natl Acad Sci USA* 2005;**102**:8315–20.
45. Mimori K, Sadanaga N, Yoshikawa Y, Ishikawa K, Hashimoto M, Tanaka F, et al. Reduced tau expression in gastric cancer can identify candidates for successful Paclitaxel treatment. *Br J Cancer* 2006;**94**:1894–7.
46. Jimeno A, Hallur G, Chan A, Zhang X, Cusatis G, Chan F, et al. Development of two novel benzoylphenylurea sulfur analogues and evidence that the microtubule-associated protein tau is predictive of their activity in pancreatic cancer. *Mol Cancer Ther* 2007;**6**:1509–16.
47. Askanas V, Engel WK, Bilak M, Alvarez RB, Selkoe DJ. Twisted tubulofilaments of inclusion body myositis muscle resemble paired helical filaments of Alzheimer brain and contain hyperphosphorylated tau. *Am J Pathol* 1994;**144**:177–87.
48. Ballatore C, Lee VM, Trojanowski JQ. Tau-mediated neurodegeneration in Alzheimer's disease and related disorders. *Nat Rev Neurosci* 2007;**8**:663–72.
49. Chun W, Johnson GV. The role of tau phosphorylation and cleavage in neuronal cell death. *Front Biosci* 2007;**12**:733–56.
50. Hardy J, Selkoe DJ. The amyloid hypothesis of Alzheimer's disease: progress and problems on the road to therapeutics. *Science* 2002;**297**:353–6.
51. Rapoport M, Dawson HN, Binder LI, Vitek MP, Ferreira A. Tau is essential to beta -amyloid-induced neurotoxicity. *Proc Natl Acad Sci USA* 2002;**99**:6364–9.
52. Goedert M. Tau protein and neurodegeneration. *Semin Cell Dev Biol* 2004;**15**:45–9.
53. Yancopoulou D, Spillantini MG. Tau protein in familial and sporadic diseases. *Neuromolecular Med* 2003;**4**:37–48.
54. Dickson DW. Neuropathology of non-Alzheimer degenerative disorders. *Int J Clin Exp Pathol* 2009;**3**:1–23.
55. Ludolph AC, Kassubek J, Landwehrmeyer BG, Mandelkow E, Mandelkow EM, Burn DJ, et al. Tauopathies with parkinsonism: clinical spectrum, neuropathologic basis, biological markers, and treatment options. *Eur J Neurol* 2009;**16**:297–309.
56. Wolfe MS. Tau mutations in neurodegenerative diseases. *J Biol Chem* 2009;**284**:6021–5.
57. van Swieten J, Spillantini MG. Hereditary frontotemporal dementia caused by Tau gene mutations. *Brain Pathol* 2007;**17**:63–73.
58. Lee VM, Goedert M, Trojanowski JQ. Neurodegenerative tauopathies. *Annu Rev Neurosci* 2001;**24**:1121–59.
59. Grover A, Houlden H, Baker M, Adamson J, Lewis J, Prihar G, et al. 5′ splice site mutations in tau associated with the inherited dementia FTDP-17 affect a stem-loop structure that regulates alternative splicing of exon 10. *J Biol Chem* 1999;**274**:15134–43.
60. Myers AJ, Pittman AM, Zhao AS, Rohrer K, Kaleem M, Marlowe L, et al. The MAPT H1c risk haplotype is associated with increased expression of tau and especially of 4 repeat containing transcripts. *Neurobiol Dis* 2007;**25**:561–70.
61. Gotz J, Deters N, Doldissen A, Bokhari L, Ke Y, Wiesner A, et al. A decade of tau transgenic animal models and beyond. *Brain Pathol* 2007;**17**:91–103.
62. McGowan E, Eriksen J, Hutton M. A decade of modeling Alzheimer's disease in transgenic mice. *Trends Genet* 2006;**22**:281–9.
63. Oddo S, Caccamo A, Shepherd JD, Murphy MP, Golde TE, Kayed R, et al. Triple-transgenic model of Alzheimer's disease with plaques and tangles: intracellular Abeta and synaptic dysfunction. *Neuron* 2003;**39**:409–21.
64. Ferrari A, Hoerndli F, Baechi T, Nitsch RM, Gotz J. beta-Amyloid induces paired helical filament-like tau filaments in tissue culture. *J Biol Chem* 2003;**278**:40162–8.

65. DeTure M, Ko LW, Easson C, Yen SH. Tau assembly in inducible transfectants expressing wild-type or FTDP-17 tau. *Am J Pathol* 2002;**161**:1711–22.

66. von Bergen M, Barghorn S, Biernat J, Mandelkow EM, Mandelkow E. Tau aggregation is driven by a transition from random coil to beta sheet structure. *Biochim Biophys Acta* 2005;**1739**:158–66.

67. Binder LI, Guillozet-Bongaarts AL, Garcia-Sierra F, Berry RW. Tau, tangles, and Alzheimer's disease. *Biochim Biophys Acta* 2005;**1739**:216–23.

68. Gamblin TC, Berry RW, Binder LI. Modeling tau polymerization in vitro: a review and synthesis. *Biochemistry* 2003;**42**:15009–17.

69. von Bergen M, Friedhoff P, Biernat J, Heberle J, Mandelkow EM, Mandelkow E. Assembly of τ protein into Alzheimer paired helical filaments depends on a local sequence motif ([306]VQI-VYK[311]) forming β structure. *Proc Natl Acad Sci USA* 2000;**97**:5129–34.

70. Gamblin TC, Chen F, Zambrano A, Abraha A, Lagalwar S, Guillozet AL, et al. Caspase cleavage of tau: linking amyloid and neurofibrillary tangles in Alzheimer's disease. *Proc Natl Acad Sci USA* 2003;**100**:10032–7.

71. Gamblin TC, King ME, Dawson H, Vitek MP, Kuret J, Berry RW, et al. In vitro polymerization of tau protein monitored by laser light scattering: method and application to the study of FTDP-17 mutants. *Biochemistry* 2000;**39**:6136–44.

72. von Bergen M, Barghorn S, Li L, Marx A, Biernat J, Mandelkow EM, et al. Mutations of tau protein in frontotemporal dementia promote aggregation of paired helical filaments by enhancing local beta-structure. *J Biol Chem* 2001;**276**:48165–74.

73. Eidenmuller J, Fath T, Maas T, Pool M, Sontag E, Brandt R. Phosphorylation-mimicking glutamate clusters in the proline-rich region are sufficient to simulate the functional deficiencies of hyperphosphorylated tau protein. *Biochem J* 2001;**357**:759–67.

74. Schneider A, Biernat J, von Bergen M, Mandelkow E, Mandelkow EM. Phosphorylation that detaches tau protein from microtubules (Ser262, Ser214) also protects it against aggregation into Alzheimer paired helical filaments. *Biochemistry* 1999;**38**:3549–58.

75. Sun Q, Gamblin TC. Pseudohyperphosphorylation causing AD-like changes in tau has significant effects on its polymerization. *Biochemistry* 2009;**48**:6002–11.

76. Necula M, Kuret J. Pseudophosphorylation and glycation of tau protein enhance but do not trigger fibrillization in vitro. *J Biol Chem* 2004;**279**:49694–703.

77. SantaCruz K, Lewis J, Spires T, Paulson J, Kotilinek L, Ingelsson M, et al. Tau suppression in a neurodegenerative mouse model improves memory function. *Science* 2005;**309**:476–81.

78. Kimura T, Yamashita S, Fukuda T, Park JM, Murayama M, Mizoroki T, et al. Hyperphosphorylated tau in parahippocampal cortex impairs place learning in aged mice expressing wild-type human tau. *EMBO J* 2007;**26**:5143–52.

79. Taniguchi T, Doe N, Matsuyama S, Kitamura Y, Mori H, Saito N, et al. Transgenic mice expressing mutant (N279K) human tau show mutation dependent cognitive deficits without neurofibrillary tangle formation. *FEBS Lett* 2005;**579**:5704–12.

80. Wittmann CW, Wszolek MF, Shulman JM, Salvaterra PM, Lewis J, Hutton M, et al. Tauopathy in Drosophila: neurodegeneration without neurofibrillary tangles. *Science* 2001;**293**:711–4.

81. Kraemer BC, Zhang B, Leverenz JB, Thomas JH, Trojanowski JQ, Schellenberg GD. Neurodegeneration and defective neurotransmission in a *Caenorhabditis elegans* model of tauopathy. *Proc Natl Acad Sci USA* 2003;**100**:9980–5.

82. Sahara N, Maeda S, Takashima A. Tau oligomerization: a role for tau aggregation intermediates linked to neurodegeneration. *Curr Alzheimer Res* 2008;**5**:591–8.

83. Iqbal K, Liu F, Gong CX, Alonso Adel C, Grundke-Iqbal I. Mechanisms of tau-induced neurodegeneration. *Acta Neuropathol* 2009;**118**:53–69.

84. Bretteville A, Planel E. Tau aggregates: toxic, inert, or protective species? *J Alzheimers Dis* 2008;**14**:431–6.

85. Lee HG, Perry G, Moreira PI, Garrett MR, Liu Q, Zhu X, et al. Tau phosphorylation in Alzheimer's disease: pathogen or protector? *Trends Mol Med* 2005;**11**:164–9.

86. Bramblett GT, Goedert M, Jakes R, Merrick SE, Trojanowski JQ, Lee VM-Y. Abnormal tau phosphorylation at Ser[396] in Alzheimer's disease recapitulates development and contributes to reduced microtubule binding. *Neuron* 1993;**10**:1089–99.

87. Kanemaru K, Takio K, Miura R, Titani K, Ihara Y. Fetal-type phosphorylation of the tau in paired helical filaments. *J Neurochem* 1992;**58**:1667–75.

88. Brion JP, Smith C, Couck AM, Gallo JM, Anderton BH. Developmental changes in tau phosphorylation: fetal tau is transiently phosphorylated in a manner similar to paired helical filament-tau characteristic of Alzheimer's disease. *J Neurochem* 1993;**61**:2071–80.

89. Brunden KR, Trojanowski JQ, Lee VM. Advances in tau-focused drug discovery for Alzheimer's disease and related tauopathies. *Nat Rev Drug Discov* 2009;**8**:783–93.

90. Gong CX, Iqbal K. Hyperphosphorylation of microtubule-associated protein tau: a promising therapeutic target for Alzheimer disease. *Curr Med Chem* 2008;**15**:2321–8.

91. Gotz J, Gladbach A, Pennanen L, van Eersel J, Schild A, David D, et al. Animal models reveal role for tau phosphorylation in human disease. *Biochim Biophys Acta* 2010;**1802**:860–71.

92. Noble W, Planel E, Zehr C, Olm V, Meyerson J, Suleman F, et al. Inhibition of glycogen synthase kinase-3 by lithium correlates with reduced tauopathy and degeneration in vivo. *Proc Natl Acad Sci USA* 2005;**102**:6990–5.

93. Khurana V, Lu Y, Steinhilb ML, Oldham S, Shulman JM, Feany MB. TOR-mediated cell-cycle activation causes neurodegeneration in a Drosophila tauopathy model. *Curr Biol* 2006;**16**:230–41.

94. Yang Y, Herrup K. Cell division in the CNS: protective response or lethal event in post-mitotic neurons? *Biochim Biophys Acta* 2007;**1772**:457–66.

95. Lee HG, Casadesus G, Zhu X, Castellani RJ, McShea A, Perry G, et al. Cell cycle re-entry mediated neurodegeneration and its treatment role in the pathogenesis of Alzheimer's disease. *Neurochem Int* 2009;**54**:84–8.

96. Vincent I, Rosado M, Davies P. Mitotic mechanisms in Alzheimer's disease? *J Cell Biol* 1996;**132**:413–25.

97. Preuss U, Mandelkow EM. Mitotic phosphorylation of tau protein in neuronal cell lines resembles phosphorylation in Alzheimer's disease. *Eur J Cell Biol* 1998;**76**:176–84.

98. Illenberger S, Zheng-Fischhofer Q, Preuss U, Stamer K, Baumann K, Trinczek B, et al. The endogenous and cell cycle-dependent phosphorylation of tau protein in living cells: implications for Alzheimer's disease. *Mol Biol Cell* 1998;**9**:1495–512.

99. Delobel P, Flament S, Hamdane M, Mailliot C, Sambo AV, Begard S, et al. Abnormal Tau phosphorylation of the Alzheimer-type also occurs during mitosis. *J Neurochem* 2002;**83**:412–20.

100. Yang Y, Geldmacher DS, Herrup K. DNA replication precedes neuronal cell death in Alzheimer's disease. *J Neurosci* 2001;**21**:2661–8.

101. Drechsel DN, Hyman AA, Cobb MH, Kirschner MW. Modulation of the dynamic instability of tubulin assembly by the microtubule-associated protein tau. *Mol Biol Cell* 1992;**3**:1141–54.

102. Lee G, Rook SL. Expression of tau protein in non-neuronal cells: microtubule binding and stabilization. *J Cell Sci* 1992;**102**(Pt 2):227–37.

103. Brandt R, Lee G. Functional organization of microtubule-associated protein tau. Identification of regions which affect microtubule growth, nucleation, and bundle formation in vitro. *J Biol Chem* 1993;**268**:3414–9.

104. Gustke N, Trinczek B, Biernat J, Mandelkow EM, Mandelkow E. Domains of tau protein and interactions with microtubules. *Biochemistry* 1994;**33**:9511–22.

105. Goode BL, Denis PE, Panda D, Radeke MJ, Miller HP, Wilson L, et al. Functional interactions between the proline-rich and repeat regions of tau enhance microtubule binding and assembly. *Mol Biol Cell* 1997;**8**:353–65.

106. Mukrasch MD, Biernat J, von Bergen M, Griesinger C, Mandelkow E, Zweckstetter M. Sites of tau important for aggregation populate (beta)-structure and bind to microtubules and polyanions. *J Biol Chem* 2005;**280**:24978–86.

107. Sillen A, Barbier P, Landrieu I, Lefebvre S, Wieruszeski JM, Leroy A, et al. NMR investigation of the interaction between the neuronal protein tau and the microtubules. *Biochemistry* 2007;**46**:3055–64.

108. Jeganathan S, von Bergen M, Brutlach H, Steinhoff HJ, Mandelkow E. Global hairpin folding of tau in solution. *Biochemistry* 2006;**45**:2283–93.

109. Panda D, Samuel JC, Massie M, Feinstein SC, Wilson L. Differential regulation of microtubule dynamics by three- and four-repeat tau: implications for the onset of neurodegenerative disease. *Proc Natl Acad Sci USA* 2003;**100**:9548–53.

110. Levy SF, Leboeuf AC, Massie MR, Jordan MA, Wilson L, Feinstein SC. Three- and four-repeat tau regulate the dynamic instability of two distinct microtubule subpopulations in qualitatively different manners. Implications for neurodegeneration. *J Biol Chem* 2005;**280**:13520–8.

111. Al-Bassam J, Ozer RS, Safer D, Halpain S, Milligan RA. MAP2 and tau bind longitudinally along the outer ridges of microtubule protofilaments. *J Cell Biol* 2002;**157**:1187–96.

112. Choi MC, Raviv U, Miller HP, Gaylord MR, Kiris E, Ventimiglia D, et al. Human microtubule-associated-protein tau regulates the number of protofilaments in microtubules: a synchrotron x-ray scattering study. *Biophys J* 2009;**97**:519–27.

113. Dayanandan R, Van Slegtenhorst M, Mack TG, Ko L, Yen SH, Leroy K, et al. Mutations in tau reduce its microtubule binding properties in intact cells and affect its phosphorylation. *FEBS Lett* 1999;**446**:228–32.

114. Hasegawa M, Smith MJ, Goedert M. Tau proteins with FTDP-17 mutations have a reduced ability to promote microtubule assembly. *FEBS Lett* 1998;**437**:207–10.

115. Barghorn S, Zheng-Fischhofer Q, Ackmann M, Biernat J, von Bergen M, Mandelkow EM, et al. Structure, microtubule interactions, and paired helical filament aggregation by tau mutants of frontotemporal dementias. *Biochemistry* 2000;**39**:11714–21.

116. Fischer D, Mukrasch MD, von Bergen M, Klos-Witkowska A, Biernat J, Griesinger C, et al. Structural and microtubule binding properties of tau mutants of frontotemporal dementias. *Biochemistry* 2007;**46**:2574–82.

117. Bunker JM, Kamath K, Wilson L, Jordan MA, Feinstein SC. FTDP-17 mutations compromise the ability of tau to regulate microtubule dynamics in cells. *J Biol Chem* 2006;**281**:11856–63.

118. Delobel P, Flament S, Hamdane M, Jakes R, Rousseau A, Delacourte A, et al. Functional characterization of FTDP-17 tau gene mutations through their effects on Xenopus oocyte maturation. *J Biol Chem* 2002;**277**:9199–205.

119. Biernat J, Gustke N, Drewes G, Mandelkow EM, Mandelkow E. Phosphorylation of Ser262 strongly reduces binding of tau to microtubules: distinction between PHF-like immunoreactivity and microtubule binding. *Neuron* 1993;**11**:153–63.

120. Leger J, Kempf M, Lee G, Brandt R. Conversion of serine to aspartate imitates phosphorylation-induced changes in the structure and function of microtubule-associated protein tau. *J Biol Chem* 1997;**272**:8441–6.

121. Haase C, Stieler JT, Arendt T, Holzer M. Pseudophosphorylation of tau protein alters its ability for self-aggregation. *J Neurochem* 2004;**88**:1509–20.

122. Cho JH, Johnson GV. Primed phosphorylation of tau at Thr231 by glycogen synthase kinase 3beta (GSK3beta) plays a critical role in regulating tau's ability to bind and stabilize microtubules. *J Neurochem* 2004;**88**:349–58.

123. Feinstein SC, Wilson L. Inability of tau to properly regulate neuronal microtubule dynamics: a loss-of-function mechanism by which tau might mediate neuronal cell death. *Biochim Biophys Acta* 2005;**1739**:268–79.

124. Brunden KR, Zhang B, Carroll J, Yao Y, Potuzak JS, Hogan AM, et al. Epothilone D improves microtubule density, axonal integrity, and cognition in a transgenic mouse model of tauopathy. *J Neurosci* 2010;**30**:13861–6.

125. Morfini GA, Burns M, Binder LI, Kanaan NM, LaPointe N, Bosco DA, et al. Axonal transport defects in neurodegenerative diseases. *J Neurosci* 2009;**29**:12776–86.

126. Ebneth A, Godemann R, Stamer K, Illenberger S, Trinczek B, Mandelkow E. Overexpression of tau protein inhibits kinesin-dependent trafficking of vesicles, mitochondria, and endoplasmic reticulum: implications for Alzheimer's disease. *J Cell Biol* 1998;**143**:777–94.

127. Trinczek B, Ebneth A, Mandelkow EM, Mandelkow E. Tau regulates the attachment/detachment but not the speed of motors in microtubule-dependent transport of single vesicles and organelles. *J Cell Sci* 1999;**112**(Pt 14):2355–67.

128. Hall GF, Lee VM, Lee G, Yao J. Staging of neurofibrillary degeneration caused by human tau overexpression in a unique cellular model of human tauopathy. *Am J Pathol* 2001;**158**:235–46.

129. Stamer K, Vogel R, Thies E, Mandelkow E, Mandelkow EM. Tau blocks traffic of organelles, neurofilaments, and APP vesicles in neurons and enhances oxidative stress. *J Cell Biol* 2002;**156**:1051–63.

130. Mandelkow EM, Stamer K, Vogel R, Thies E, Mandelkow E. Clogging of axons by tau, inhibition of axonal traffic and starvation of synapses. *Neurobiol Aging* 2003;**24**:1079–85.

131. Ishihara T, Hong M, Zhang B, Nakagawa Y, Lee MK, Trojanowski JQ, et al. Age-dependent emergence and progression of a tauopathy in transgenic mice overexpressing the shortest human tau isoform. *Neuron* 1999;**24**:751–62.

132. Ittner LM, Fath T, Ke YD, Bi M, van Eersel J, Li KM, et al. Parkinsonism and impaired axonal transport in a mouse model of frontotemporal dementia. *Proc Natl Acad Sci USA* 2008;**105**:15997–6002.

133. Utton MA, Noble WJ, Hill JE, Anderton BH, Hanger DP. Molecular motors implicated in the axonal transport of tau and alpha-synuclein. *J Cell Sci* 2005;**118**:4645–54.

134. Dubey M, Chaudhury P, Kabiru H, Shea TB. Tau inhibits anterograde axonal transport and perturbs stability in growing axonal neurites in part by displacing kinesin cargo: neurofilaments attenuate tau-mediated neurite instability. *Cell Motil Cytoskeleton* 2008;**65**:89–99.

135. Jancsik V, Filliol D, Rendon A. Tau proteins bind to kinesin and modulate its activation by microtubules. *Neurobiology (Bp)* 1996;**4**:417–29.

136. Cuchillo-Ibanez I, Seereeram A, Byers HL, Leung KY, Ward MA, Anderton BH, et al. Phosphorylation of tau regulates its axonal transport by controlling its binding to kinesin. *FASEB J* 2008;**22**:3186–95.

137. Magnani E, Fan J, Gasparini L, Golding M, Williams M, Schiavo G, et al. Interaction of tau protein with the dynactin complex. *EMBO J* 2007;**26**:4546–54.

138. Seitz A, Kojima H, Oiwa K, Mandelkow EM, Song YH, Mandelkow E. Single-molecule investigation of the interference between kinesin, tau and MAP2c. *EMBO J* 2002;**21**:4896–905.

139. Dixit R, Ross JL, Goldman YE, Holzbaur EL. Differential regulation of dynein and kinesin motor proteins by tau. *Science* 2008;**319**:1086–9.

140. Vershinin M, Carter BC, Razafsky DS, King SJ, Gross SP. Multiple-motor based transport and its regulation by Tau. *Proc Natl Acad Sci USA* 2007;**104**:87–92.

141. Ittner LM, Ke YD, Gotz J. Phosphorylated Tau interacts with c-Jun N-terminal kinase-interacting protein 1 (JIP1) in Alzheimer disease. *J Biol Chem* 2009;**284**:20909–16.

142. Shemesh OA, Erez H, Ginzburg I, Spira ME. Tau-induced traffic jams reflect organelles accumulation at points of microtubule polar mismatching. *Traffic* 2008;**9**:458–71.

143. Morfini G, Pigino G, Mizuno N, Kikkawa M, Brady ST. Tau binding to microtubules does not directly affect microtubule-based vesicle motility. *J Neurosci Res* 2007;**85**:2620–30.

144. LaPointe NE, Morfini G, Pigino G, Gaisina IN, Kozikowski AP, Binder LI, et al. The amino terminus of tau inhibits kinesin-dependent axonal transport: implications for filament toxicity. *J Neurosci Res* 2009;**87**:440–51.
145. Yuan A, Kumar A, Peterhoff C, Duff K, Nixon RA. Axonal transport rates in vivo are unaffected by tau deletion or overexpression in mice. *J Neurosci* 2008;**28**:1682–7.
146. Griffith LM, Pollard TD. Evidence for actin filament-microtubule interaction mediated by microtubule-associated proteins. *J Cell Biol* 1978;**78**:958–65.
147. Griffith LM, Pollard TD. The interaction of actin filaments with microtubules and microtubule-associated proteins. *J Biol Chem* 1982;**257**:9143–51.
148. Correas I, Padilla R, Avila J. The tubulin-binding sequence of brain microtubule-associated proteins, tau and MAP-2, is also involved in actin binding. *Biochem J* 1990;**269**:61–4.
149. Moraga DM, Nunez P, Garrido J, Maccioni RB. A tau fragment containing a repetitive sequence induces bundling of actin filaments. *J Neurochem* 1993;**61**:979–86.
150. DiTella M, Feiguin F, Morfini G, Caceres A. Microfilament-associated growth cone component depends upon Tau for its intracellular localization. *Cell Motil Cytoskeleton* 1994;**29**:117–30.
151. Liu CW, Lee G, Jay DG. Tau is required for neurite outgrowth and growth cone motility of chick sensory neurons. *Cell Motil Cytoskeleton* 1999;**43**:232–42.
152. Yu JZ, Rasenick MM. Tau associates with actin in differentiating PC12 cells. *FASEB J* 2006;**20**:1452–61.
153. Roger B, Al-Bassam J, Dehmelt L, Milligan RA, Halpain S. MAP2c, but not tau, binds and bundles F-actin via its microtubule binding domain. *Curr Biol* 2004;**14**:363–71.
154. Biernat J, Wu YZ, Timm T, Zheng-Fischhofer Q, Mandelkow E, Meijer L, et al. Protein kinase MARK/PAR-1 is required for neurite outgrowth and establishment of neuronal polarity. *Mol Biol Cell* 2002;**13**:4013–28.
155. Sharma VM, Litersky JM, Bhaskar K, Lee G. Tau impacts on growth-factor-stimulated actin remodeling. *J Cell Sci* 2007;**120**:748–57.
156. Brandt R, Leger J, Lee G. Interaction of tau with the neural plasma membrane mediated by tau's amino-terminal projection domain. *J Cell Biol* 1995;**131**:1327–40.
157. Black MM, Slaughter T, Moshiach S, Obrocka M, Fischer I. Tau is enriched on dynamic microtubules in the distal region of growing axons. *J Neurosci* 1996;**16**:3601–19.
158. Fulga TA, Elson-Schwab I, Khurana V, Steinhilb ML, Spires TL, Hyman BT, et al. Abnormal bundling and accumulation of F-actin mediates tau-induced neuronal degeneration in vivo. *Nat Cell Biol* 2007;**9**:139–48.
159. Minamide LS, Striegl AM, Boyle JA, Meberg PJ, Bamburg JR. Neurodegenerative stimuli induce persistent ADF/cofilin-actin rods that disrupt distal neurite function. *Nat Cell Biol* 2000;**2**:628–36.
160. Yu Y, Run X, Liang Z, Li Y, Liu F, Liu Y, et al. Developmental regulation of tau phosphorylation, tau kinases, and tau phosphatases. *J Neurochem* 2009;**108**:1480–94.
161. Seubert P, Mawal-Dewan M, Barbour R, Jakes R, Goedert M, Johnson GV, et al. Detection of phosphorylated Ser262 in fetal tau, adult tau, and paired helical filament tau. *J Biol Chem* 1995;**270**:18917–22.
162. Sato S, Cerny RL, Buescher JL, Ikezu T. Tau-tubulin kinase 1 (TTBK1), a neuron-specific tau kinase candidate, is involved in tau phosphorylation and aggregation. *J Neurochem* 2006;**98**:1573–84.
163. Goedert M, Hasegawa M, Jakes R, Lawler S, Cuenda A, Cohen P. Phosphorylation of microtubule-associated protein tau by stress-activated protein kinases. *FEBS Lett* 1997;**409**:57–62.
164. Liu F, Liang Z, Wegiel J, Hwang YW, Iqbal K, Grundke-Iqbal I, et al. Overexpression of DyrklA contributes to neurofibrillary degeneration in Down syndrome. *FASEB J* 2008;**22**:3224–33.

165. Reynolds CH, Betts JC, Blackstock WP, Nebreda AR, Anderton BH. Phosphorylation sites on tau identified by nanoelectrospray mass spectrometry: differences in vitro between the mitogen-activated protein kinases ERK2, c-Jun N-terminal kinase and P38, and glycogen synthase kinase-3beta. *J Neurochem* 2000;**74**:1587–95.

166. Ryoo SR, Jeong HK, Radnaabazar C, Yoo JJ, Cho HJ, Lee HW, et al. DYRK1A-mediated hyperphosphorylation of Tau. A functional link between Down syndrome and Alzheimer disease. *J Biol Chem* 2007;**282**:34850–7.

167. Shahani N, Brandt R. Functions and malfunctions of the tau proteins. *Cell Mol Life Sci* 2002;**59**:1668–80.

168. Morishima-Kawashima M, Hasegawa M, Takio K, Suzuki M, Yoshida H, Titani K, et al. Proline-directed and non-proline-directed phosphorylation of PHF-tau. *J Biol Chem* 1995;**270**:823–9.

169. Carmel G, Mager EM, Binder LI, Kuret J. The structural basis of monoclonal antibody Alz50's selectivity for Alzheimer's disease pathology. *J Biol Chem* 1996;**271**:32789–95.

170. Horowitz PM, Patterson KR, Guillozet-Bongaarts AL, Reynolds MR, Carroll CA, Weintraub ST, et al. Early N-terminal changes and caspase-6 cleavage of tau in Alzheimer's disease. *J Neurosci* 2004;**24**:7895–902.

171. Kosik KS, Orecchio LD, Binder L, Trojanowski JQ, Lee VM, Lee G. Epitopes that span the tau molecule are shared with paired helical filaments. *Neuron* 1988;**1**:817–25.

172. Mercken M, Vandermeeren M, Lubke U, Six J, Boons J, Van de Voorde A, et al. Monoclonal antibodies with selective specificity for Alzheimer Tau are directed against phosphatase-sensitive epitopes. *Acta Neuropathol* 1992;**84**:265–72.

173. Jicha GA, Weaver C, Lane E, Vianna C, Kress Y, Rockwood J, et al. cAMP-dependent protein kinase phosphorylations on tau in Alzheimer's disease. *J Neurosci* 1999;**19**:7486–94.

174. Goedert M, Jakes R, Crowther RA, Cohen P, Vanmechelen E, Vandermeeren M, et al. Epitope mapping of monoclonal antibodies to the paired helical filaments of Alzheimer's disease: identification of phosphorylation sites in tau protein. *Biochem J* 1994;**301**(Pt 3):871–7.

175. Weaver CL, Espinoza M, Kress Y, Davies P. Conformational change as one of the earliest alterations of tau in Alzheimer's disease. *Neurobiol Aging* 2000;**21**:719–27.

176. Jicha GA, Lane E, Vincent I, Otvos Jr. L, Hoffmann R, Davies P. A conformation- and phosphorylation-dependent antibody recognizing the paired helical filaments of Alzheimer's disease. *J Neurochem* 1997;**69**:2087–95.

177. Otvos Jr. L, Feiner L, Lang E, Szendrei GI, Goedert M, Lee VM. Monoclonal antibody PHF-1 recognizes tau protein phosphorylated at serine residues 396 and 404. *J Neurosci Res* 1994;**39**:669–73.

178. Lee G, Thangavel R, Sharma VM, Litersky JM, Bhaskar K, Fang SM, et al. Phosphorylation of tau by fyn: implications for Alzheimer's disease. *J Neurosci* 2004;**24**:2304–12.

179. Hagestedt T, Lichtenberg B, Wille H, Mandelkow EM, Mandelkow E. Tau protein becomes long and stiff upon phosphorylation: correlation between paracrystalline structure and degree of phosphorylation. *J Cell Biol* 1989;**109**:1643–51.

180. Jeganathan S, Hascher A, Chinnathambi S, Biernat J, Mandelkow EM, Mandelkow E. Proline-directed pseudo-phosphorylation at AT8 and PHF1 epitopes induces a compaction of the paperclip folding of Tau and generates a pathological (MC-1) conformation. *J Biol Chem* 2008;**283**:32066–76.

181. Guillozet-Bongaarts AL, Cahill ME, Cryns VL, Reynolds MR, Berry RW, Binder LI. Pseudophosphorylation of tau at serine 422 inhibits caspase cleavage: in vitro evidence and implications for tangle formation in vivo. *J Neurochem* 2006;**97**:1005–14.

182. Sengupta A, Wu Q, Grundke-Iqbal I, Iqbal K, Singh TJ. Potentiation of GSK-3-catalyzed Alzheimer-like phosphorylation of human tau by cdk5. *Mol Cell Biochem* 1997;**167**:99–105.

183. Alonso Adel C, Mederlyova A, Novak M, Grundke-Iqbal I, Iqbal K. Promotion of hyperphosphorylation by frontotemporal dementia tau mutations. *J Biol Chem* 2004;**279**:34873–81.

184. Busciglio J, Lorenzo A, Yeh J, Yankner BA. β-amyloid fibrils induce tau phosphorylation and loss of microtubule binding. *Neuron* 1995;**14**:879–88.

185. Takashima A, Noguchi K, Michel G, Mercken M, Hoshi M, Ishiguro K, et al. Exposure of rat hippocampal neurons to amyloid beta peptide (25-35) induces the inactivation of phosphatidyl inositol 3 kinase and the activation of tau protein kinase I/glycogen synthase kinase-3 beta. *Neurosci Lett* 1996;**203**:33–6.

186. Rapoport M, Ferreira A. PD98059 prevents neurite degeneration induced by fibrillar beta-amyloid in mature hippocampal neurons. *J Neurochem* 2000;**74**:125–33.

187. Matsuo ES, Shin RW, Billingsley ML, Van deVoorde A, O'Connor M, Trojanowski JQ, et al. Biopsy-derived adult human brain tau is phosphorylated at many of the same sites as Alzheimer's disease paired helical filament tau. *Neuron* 1994;**13**:989–1002.

188. Sontag E, Nunbhakdi-Craig V, Lee G, Bloom GS, Mumby MC. Regulation of the phosphorylation state and microtubule-binding activity of Tau by protein phosphatase 2A. *Neuron* 1996;**17**:1201–7.

189. Sontag E, Nunbhakdi-Craig V, Lee G, Brandt R, Kamibayashi C, Kuret J, et al. Molecular interactions among protein phosphatase 2A, tau, and microtubules. Implications for the regulation of tau phosphorylation and the development of tauopathies. *J Biol Chem* 1999;**274**:25490–8.

190. Liao H, Li Y, Brautigan DL, Gundersen GG. Protein phosphatase 1 is targeted to microtubules by the microtubule-associated protein Tau. *J Biol Chem* 1998;**273**:21901–8.

191. Rahman A, Grundke-Iqbal I, Iqbal K. Phosphothreonine-212 of Alzheimer abnormally hyperphosphorylated tau is a preferred substrate of protein phosphatase-1. *Neurochem Res* 2005;**30**:277–87.

192. Liu F, Grundke-Iqbal I, Iqbal K, Gong CX. Contributions of protein phosphatases PP1, PP2A, PP2B and PP5 to the regulation of tau phosphorylation. *Eur J Neurosci* 2005;**22**:1942–50.

193. Goedert M, Satumtira S, Jakes R, Smith MJ, Kamibayashi C, White 3rd CL, et al. Reduced binding of protein phosphatase 2A to tau protein with frontotemporal dementia and parkinsonism linked to chromosome 17 mutations. *J Neurochem* 2000;**75**:2155–62.

194. Lee G, Newman ST, Gard DL, Band H, Panchamoorthy G. Tau interacts with src-family non-receptor tyrosine kinases. *J Cell Sci* 1998;**111**(Pt 21):3167–77.

195. Derkinderen P, Scales TM, Hanger DP, Leung KY, Byers HL, Ward MA, et al. Tyrosine 394 is phosphorylated in Alzheimer's paired helical filament tau and in fetal tau with c-Abl as the candidate tyrosine kinase. *J Neurosci* 2005;**25**:6584–93.

196. Lebouvier T, Scales TM, Hanger DP, Geahlen RL, Lardeux B, Reynolds CH, et al. The microtubule-associated protein tau is phosphorylated by Syk. *Biochim Biophys Acta* 2008;**1783**:188–92.

197. Vega IE, Cui L, Propst JA, Hutton ML, Lee G, Yen SH. Increase in tau tyrosine phosphorylation correlates with the formation of tau aggregates. *Brain Res Mol Brain Res* 2005;**138**:135–44.

198. Bhaskar K, Hobbs GA, Yen SH, Lee G. Tyrosine phosphorylation of tau accompanies disease progression in transgenic mouse models of tauopathy. *Neuropathol Appl Neurobiol* 2010;**36**:462–77.

199. Williamson R, Scales T, Clark BR, Gibb G, Reynolds CH, Kellie S, et al. Rapid tyrosine phosphorylation of neuronal proteins including tau and focal adhesion kinase in response to amyloid-beta peptide exposure: involvement of Src family protein kinases. *J Neurosci* 2002;**22**:10–20.

200. Chin J, Palop JJ, Yu G-Q, Kojima N, Masliah E, Mucke L. Fyn kinase modulates synaptotoxicity, but not aberrant sprouting, in human amyloid precursor protein transgenic mice. *J Neurosci* 2004;**24**:4692–7.

201. Lambert MP, Barlow AK, Chromy BA, Edwards C, Freed R, Liosatos M, et al. Diffusible, nonfibrillar ligands derived from Abeta1-42 are potent central nervous system neurotoxins. *Proc Natl Acad Sci USA* 1998;**95**:6448–53.

202. Ho GJ, Hashimoto M, Adame A, Izu M, Alford MF, Thal LJ, et al. Altered p59Fyn kinase expression accompanies disease progression in Alzheimer's disease: implications for its functional role. *Neurobiol Aging* 2005;**26**:625–35.

203. Arnold CS, Johnson GV, Cole RN, Dong DL, Lee M, Hart GW. The microtubule-associated protein tau is extensively modified with O-linked N-acetylglucosamine. *J Biol Chem* 1996;**271**:28741–4.

204. Liu F, Iqbal K, Grundke-Iqbal I, Gong CX. Involvement of aberrant glycosylation in phosphorylation of tau by cdk5 and GSK-3beta. *FEBS Lett* 2002;**530**:209–14.

205. Reynolds MR, Reyes JF, Fu Y, Bigio EH, Guillozet-Bongaarts AL, Berry RW, et al. Tau nitration occurs at tyrosine 29 in the fibrillar lesions of Alzheimer's disease and other tauopathies. *J Neurosci* 2006;**26**:10636–45.

206. Petrucelli L, Dickson D, Kehoe K, Taylor J, Snyder H, Grover A, et al. CHIP and Hsp70 regulate tau ubiquitination, degradation and aggregation. *Hum Mol Genet* 2004;**13**:703–14.

207. Shimura H, Schwartz D, Gygi SP, Kosik KS. CHIP-Hsc70 complex ubiquitinates phosphorylated tau and enhances cell survival. *J Biol Chem* 2004;**279**:4869–76.

208. Hatakeyama S, Matsumoto M, Kamura T, Murayama M, Chui DH, Planel E, et al. U-box protein carboxyl terminus of Hsc70-interacting protein (CHIP) mediates poly-ubiquitylation preferentially on four-repeat Tau and is involved in neurodegeneration of tauopathy. *J Neurochem* 2004;**91**:299–307.

209. Morishima-Kawashima M, Hasegawa M, Takio K, Suzuki M, Titani K, Ihara Y. Ubiquitin is conjugated with amino-terminally processed tau in paired helical filaments. *Neuron* 1993;**10**:1151–60.

210. Cripps D, Thomas SN, Jeng Y, Yang F, Davies P, Yang AJ. Alzheimer disease-specific conformation of hyperphosphorylated paired helical filament-Tau is polyubiquitinated through Lys-48, Lys-11, and Lys-6 ubiquitin conjugation. *J Biol Chem* 2006;**281**:10825–38.

211. Dickey CA, Yue M, Lin WL, Dickson DW, Dunmore JH, Lee WC, et al. Deletion of the ubiquitin ligase CHIP leads to the accumulation, but not the aggregation, of both endogenous phospho- and caspase-3-cleaved tau species. *J Neurosci* 2006;**26**:6985–96.

212. Dorval V, Fraser PE. Small ubiquitin-like modifier (SUMO) modification of natively unfolded proteins tau and alpha-synuclein. *J Biol Chem* 2006;**281**:9919–24.

213. Lu PJ, Wulf G, Zhou XZ, Davies P, Lu KP. The prolyl isomerase Pin1 restores the function of Alzheimer-associated phosphorylated tau protein. *Nature* 1999;**399**:784–8.

214. Smet C, Sambo AV, Wieruszeski JM, Leroy A, Landrieu I, Buee L, et al. The peptidyl prolyl cis/trans-isomerase Pin1 recognizes the phospho-Thr212-Pro213 site on Tau. *Biochemistry* 2004;**43**:2032–40.

215. Lim J, Balastik M, Lee TH, Nakamura K, Liou YC, Sun A, et al. Pin1 has opposite effects on wild-type and P301L tau stability and tauopathy. *J Clin Invest* 2008;**118**:1877–89.

216. Yotsumoto K, Saito T, Asada A, Oikawa T, Kimura T, Uchida C, et al. Effect of Pin1 or Microtubule Binding on Dephosphorylation of FTDP-17 Mutant Tau. *J Biol Chem* 2009;**284**:16840–7.

217. Zhou XZ, Kops O, Werner A, Lu PJ, Shen M, Stoller G, et al. Pin1-dependent prolyl isomerization regulates dephosphorylation of Cdc25C and tau proteins. *Mol Cell* 2000;**6**:873–83.

218. Liou YC, Sun A, Ryo A, Zhou XZ, Yu ZX, Huang HK, et al. Role of the prolyl isomerase Pin1 in protecting against age-dependent neurodegeneration. *Nature* 2003;**424**:556–61.

219. Hashiguchi M, Sobue K, Paudel HK. 14-3-3zeta is an effector of tau protein phosphorylation. *J Biol Chem* 2000;**275**:25247–54.

220. Hernandez F, Cuadros R, Avila J. Zeta 14-3-3 protein favours the formation of human tau fibrillar polymers. *Neurosci Lett* 2004;**357**:143–6.

221. Agarwal-Mawal A, Qureshi HY, Cafferty PW, Yuan Z, Han D, Lin R, et al. 14-3-3 connects glycogen synthase kinase-3 beta to tau within a brain microtubule-associated tau phosphorylation complex. *J Biol Chem* 2003;**278**:12722–8.

222. Matthews TA, Johnson GV. 14-3-3Zeta does not increase GSK3beta-mediated tau phosphorylation in cell culture models. *Neurosci Lett* 2005;**384**:211–6.

223. Li T, Paudel HK. 14-3-3zeta facilitates GSK3beta-catalyzed tau phosphorylation in HEK-293 cells by a mechanism that requires phosphorylation of GSK3beta on Ser9. *Neurosci Lett* 2007;**414**:203–8.

224. Sadik G, Tanaka T, Kato K, Yamamori H, Nessa BN, Morihara T, et al. Phosphorylation of tau at Ser214 mediates its interaction with 14-3-3 protein: implications for the mechanism of tau aggregation. *J Neurochem* 2009;**108**:33–43.

225. Sluchanko NN, Seit-Nebi AS, Gusev NB. Phosphorylation of more than one site is required for tight interaction of human tau protein with 14-3-3zeta. *FEBS Lett* 2009;**583**:2739–42.

226. Sadik G, Tanaka T, Kato K, Yanagi K, Kudo T, Takeda M. Differential interaction and aggregation of 3-repeat and 4-repeat tau isoforms with 14-3-3zeta protein. *Biochem Biophys Res Commun* 2009;**383**:37–41.

227. Leugers CJ, Lee G. Tau potentiates nerve growth factor-induced mitogen-activated protein kinase signaling and neurite initiation without a requirement for microtubule binding. *J Biol Chem* 2010;**285**:19125–34.

228. Hasegawa M, Morishima-Kawashima M, Takio K, Suzuki M, Titani K, Ihara Y. Protein sequence and mass spectrometric analyses of tau in the Alzheimer's disease brain. *J Biol Chem* 1992;**267**:17047–54.

229. Vincent I, Zheng JH, Dickson DW, Kress Y, Davies P. Mitotic phosphoepitopes precede paired helical filaments in Alzheimer's disease. *Neurobiol Aging* 1998;**19**:287–96.

230. Pei JJ, Braak H, An WL, Winblad B, Cowburn RF, Iqbal K, et al. Up-regulation of mitogen-activated protein kinases ERK1/2 and MEK1/2 is associated with the progression of neurofibrillary degeneration in Alzheimer's disease. *Brain Res Mol Brain Res* 2002;**109**:45–55.

231. Ferrer I, Blanco R, Carmona M, Ribera R, Goutan E, Puig B, et al. Phosphorylated map kinase (ERK1, ERK2) expression is associated with early tau deposition in neurones and glial cells, but not with increased nuclear DNA vulnerability and cell death, in Alzheimer disease, Pick's disease, progressive supranuclear palsy and corticobasal degeneration. *Brain Pathol* 2001;**11**:144–58.

232. Perry G, Roder H, Nunomura A, Takeda A, Friedlich AL, Zhu X, et al. Activation of neuronal extracellular receptor kinase (ERK) in Alzheimer disease links oxidative stress to abnormal phosphorylation. *Neuroreport* 1999;**10**:2411–5.

233. Reynolds CH, Garwood CJ, Wray S, Price C, Kellie S, Perera T, et al. Phosphorylation regulates tau interactions with Src homology 3 domains of phosphatidylinositol 3-kinase, phospholipase Cgamma1, Grb2, and Src family kinases. *J Biol Chem* 2008;**283**:18177–86.

234. Jenkins SM, Johnson GV. Tau complexes with phospholipase C-gamma in situ. *Neuroreport* 1998;**9**:67–71.

235. Hwang SC, Jhon DY, Bae YS, Kim JH, Rhee SG. Activation of phospholipase C-gamma by the concerted action of tau proteins and arachidonic acid. *J Biol Chem* 1996;**271**:18342–9.

236. Bhaskar K, Yen SH, Lee G. Disease-related modifications in tau affect the interaction between Fyn and Tau. *J Biol Chem* 2005;**280**:35119–25.

237. Zamora-Leon SP, Lee G, Davies P, Shafit-Zagardo B. Binding of Fyn to MAP-2c through an SH3 binding domain. Regulation of the interaction by ERK2. *J Biol Chem* 2001;**276**:39950–8.

238. Sarkar M, Kuret J, Lee G. Two motifs within the tau microtubule-binding domain mediate its association with the hsc70 molecular chaperone. *J Neurosci Res* 2008;**86**:2763–73.

239. Elliott E, Tsvetkov P, Ginzburg I. BAG-1 associates with Hsc70.Tau complex and regulates the proteasomal degradation of Tau protein. *J Biol Chem* 2007;**282**:37276–84.

240. Wang Y, Martinez-Vicente M, Kruger U, Kaushik S, Wong E, Mandelkow EM, et al. Tau fragmentation, aggregation and clearance: the dual role of lysosomal processing. *Hum Mol Genet* 2009;**18**:4153–70.

241. Dolan PJ, Johnson GV. A caspase cleaved form of tau is preferentially degraded through the autophagy pathway. *J Biol Chem* 2010;**285**:21978–87.

242. Dickey CA, Dunmore J, Lu B, Wang JW, Lee WC, Kamal A, et al. HSP induction mediates selective clearance of tau phosphorylated at proline-directed Ser/Thr sites but not KXGS (MARK) sites. *FASEB J* 2006;**20**:753–5.

243. Dickey CA, Kamal A, Lundgren K, Klosak N, Bailey RM, Dunmore J, et al. The high-affinity HSP90-CHIP complex recognizes and selectively degrades phosphorylated tau client proteins. *J Clin Invest* 2007;**117**:648–58.

244. Sahara N, Maeda S, Yoshiike Y, Mizoroki T, Yamashita S, Murayama M, et al. Molecular chaperone-mediated tau protein metabolism counteracts the formation of granular tau oligomers in human brain. *J Neurosci Res* 2007;**85**:3098–108.

245. Loomis PA, Howard TH, Castleberry RP, Binder LI. Identification of nuclear tau isoforms in human neuroblastoma cells. *Proc Natl Acad Sci USA* 1990;**87**:8422–6.

246. Thurston VC, Zinkowski RP, Binder LI. Tau as a nucleolar protein in human nonneural cells in vitro and in vivo. *Chromosoma* 1996;**105**:20–30.

247. Papasozomenos SC, Binder LI. Phosphorylation determines two distinct species of Tau in the central nervous system. *Cell Motil Cytoskeleton* 1987;**8**:210–26.

248. Lefebvre T, Ferreira S, Dupont-Wallois L, Bussiere T, Dupire MJ, Delacourte A, et al. Evidence of a balance between phosphorylation and O-GlcNAc glycosylation of Tau proteins—a role in nuclear localization. *Biochim Biophys Acta* 2003;**1619**:167–76.

249. Krylova SM, Musheev M, Nutiu R, Li Y, Lee G, Krylov SN. Tau protein binds single-stranded DNA sequence specifically—the proof obtained in vitro with non-equilibrium capillary electrophoresis of equilibrium mixtures. *FEBS Lett* 2005;**579**:1371–5.

250. Rossi G, Dalpra L, Crosti F, Lissoni S, Sciacca FL, Catania M, et al. A new function of microtubule-associated protein tau: involvement in chromosome stability. *Cell Cycle* 2008;**7**:1788–94.

251. Maas T, Eidenmuller J, Brandt R. Interaction of tau with the neural membrane cortex is regulated by phosphorylation at sites that are modified in paired helical filaments. *J Biol Chem* 2000;**275**:15733–40.

252. Arrasate M, Perez M, Avila J. Tau dephosphorylation at tau-1 site correlates with its association to cell membrane. *Neurochem Res* 2000;**25**:43–50.

253. Mandell JW, Banker GA. A spatial gradient of tau protein phosphorylation in nascent axons. *J Neurosci* 1996;**16**:5727–40.

254. Klein C, Kramer EM, Cardine AM, Schraven B, Brandt R, Trotter J. Process outgrowth of oligodendrocytes is promoted by interaction of fyn kinase with the cytoskeletal protein tau. *J Neurosci* 2002;**22**:698–707.

255. Kawarabayashi T, Shoji M, Younkin LH, Wen-Lang L, Dickson DW, Murakami T, et al. Dimeric amyloid beta protein rapidly accumulates in lipid rafts followed by apolipoprotein E and phosphorylated tau accumulation in the Tg2576 mouse model of Alzheimer's disease. *J Neurosci* 2004;**24**:3801–9.

256. Williamson R, Usardi A, Hanger DP, Anderton BH. Membrane-bound beta-amyloid oligomers are recruited into lipid rafts by a fyn-dependent mechanism. *FASEB J* 2008;**22**:1552–9.

257. Hernandez P, Lee G, Sjoberg M, Maccioni RB. Tau phosphorylation by cdk5 and Fyn in response to amyloid peptide Abeta (25-35): involvement of lipid rafts. *J Alzheimers Dis* 2009;**16**:149–56.

258. Kim WH, Lee S, Jung C, Ahmed A, Lee G, Hall GF. Interneuronal transfer of human tau between lamprey central neurons *in situ*. *J Alzheimers Dis* 2010;**19**:647–64 in press.

259. Gomez-Ramos A, Diaz-Hernandez M, Rubio A, Miras-Portugal MT, Avila J. Extracellular tau promotes intracellular calcium increase through M1 and M3 muscarinic receptors in neuronal cells. *Mol Cell Neurosci* 2008;**37**:673–81.

260. Clavaguera F, Bolmont T, Crowther RA, Abramowski D, Frank S, Probst A, et al. Transmission and spreading of tauopathy in transgenic mouse brain. *Nat Cell Biol* 2009;**11**:000 13.

261. Frost B, Jacks RL, Diamond MI. Propagation of tau misfolding from the outside to the inside of a cell. *J Biol Chem* 2009;**284**:12845–52.

262. Belkadi A, LoPresti P. Truncated Tau with the Fyn-binding domain and without the microtubule-binding domain hinders the myelinating capacity of an oligodendrocyte cell line. *J Neurochem* 2008;**107**:351–60.

263. King ME, Kan HM, Baas PW, Erisir A, Glabe CG, Bloom GS. Tau-dependent microtubule disassembly initiated by prefibrillar beta-amyloid. *J Cell Biol* 2006;**175**:541–6.

264. Ittner LM, Ke YD, Delerue F, Bi M, Gladbach A, van Eersel J, et al. Dendritic function of tau mediates amyloid-beta toxicity in Alzheimer's disease mouse models. *Cell* 2010;**142**:387–97.

265. Luo MH, Leski ML, Andreadis A. Tau isoforms which contain the domain encoded by exon 6 and their role in neurite elongation. *J Cell Biochem* 2004;**91**:880–95.

266. Park SY, Ferreira A. The generation of a 17 kDa neurotoxic fragment: an alternative mechanism by which tau mediates beta-amyloid-induced neurodegeneration. *J Neurosci* 2005;**25**: 5365–75.

267. Nicholson AM, Ferreira A. Increased membrane cholesterol might render mature hippocampal neurons more susceptible to beta-amyloid-induced calpain activation and tau toxicity. *J Neurosci* 2009;**29**:4640–51.

268. Mukrasch MD, Bibow S, Korukottu J, Jeganathan S, Biernat J, Griesinger C, et al. Structural polymorphism of 441-residue tau at single residue resolution. *PLoS Biol* 2009;**7**:e34.

269. Pei JJ, Hugon J. mTOR-dependent signalling in Alzheimer's disease. *J Cell Mol Med* 2008;**12**:2525–32.

Membrane Pores in the Pathogenesis of Neurodegenerative Disease

Bruce L. Kagan

Department of Psychiatry & Biobehavioral Sciences, David Geffen School of Medicine at UCLA, Semel Institute for Neuroscience and Human Behavior, Los Angeles, California, USA

The neurodegenerative diseases described in this volume, as well as many nonneurodegenerative diseases, are characterized by deposits known as amyloid. Amyloid has long been associated with these various diseases as a pathological marker and has been implicated directly in the molecular pathogenesis of disease. However, increasing evidence suggests that these proteinaceous Congo red staining deposits may not be toxic or destructive of tissue. Recent studies strongly implicate smaller aggregates of amyloid proteins as the toxic species underlying these neurodegenerative diseases. Despite the outward obvious differences among these clinical syndromes, there are some striking similarities in their molecular pathologies. These include dysregulation of intracellular calcium levels, impairment of mitochondrial function, and the ability of virtually all amyloid peptides to form ion-permeable pores in lipid membranes. Pore formation is enhanced by environmental factors that promote protein aggregation and is inhibited by agents, such as Congo red, which prevent aggregation. Remarkably, the pores formed by a variety of amyloid peptides from neurodegenerative and other diseases share a common set of physiologic properties. These include irreversible insertion of the pores in lipid membranes, formation of heterodisperse pore sizes, inhibition by Congo red of pore formation, blockade of pores by zinc, and a relative lack of ion selectivity and voltage dependence. Although there exists some information about the physical structure of these pores, molecular modeling suggests that 4–6-mer amyloid subunits may assemble into 24-mer pore-forming aggregates. The molecular structure of these pores may resemble the β-barrel structure of the toxics pore formed by bacterial toxins, such as staphylococcal α-hemolysin, anthrax toxin, and *Clostridium perfringolysin*.

I. Introduction

Rudolph Virchow observed amorphous starch-like deposits that stained with iodine by light microscopy. He named these "amyloid" in the belief that they were primarily composed of starch. Although later research did find that glycosaminoglycans were always present in amyloid deposits, most of these deposits consist of a single amyloid protein or peptide. When dyes such as Congo red are applied to these deposits, they exhibit green birefringence when seen under cross-polarized light. More detailed studies employing electron microscopy demonstrate that amyloid fibrils have a width of ≈10 nm and a very extended, often indeterminant, length (see Ref. 1 for review of amyloid fibrils). The pathological term "amyloid" has now been applied to dozens of distinct clinical syndromes in spite of the fact that a wide variety of apparently unrelated proteins comprise the amyloid deposits in these different diseases (for a partial list, see Table I). With the exception of the ABri and ADan proteins, there are no sequence homologies among these amyloid-forming proteins and they do not have common biochemical functions. They do, however, all share a common physical chemical property, namely, the ability to form β-sheet secondary structure under the appropriate conditions. Various environmental factors have been demonstrated to convert native proteins into β-sheet amyloid structures. These factors include proteolysis, high concentration, acidic pH, metal binding, amino acid mutation, and interaction with lipid membranes. In some cases, more than one of these factors is at work. Of the neurodegenerative diseases covered in this volume, Alzheimer's disease has been the most intensively studied. The pore-forming abilities of amyloid peptides related to Alzheimer's disease have similarly been the subject of numerous investigations. Although substantial evidence supports a role for β-amyloid in the pathogenesis of Alzheimer's disease, efforts to elucidate the molecular mechanism of pathogenesis have been difficult. However, three biochemical abnormalities have

TABLE I
NEURODEGENERATIVE PROTEIN-FOLDING DISEASES

Disease	Protein	Abbreviation
Alzheimer's disease	Amyloid β-protein	AβPP or APP
Down's syndrome (Trisomy 21)	precursor (Aβ 1–42)	(Aβ 1–42)
Heredity cerebral angiopathy (Dutch)		
Kuru	Prion protein	PrP^c/PrP^{sc}
Gerstmann–Straussler syndrome (GSS)		
Creutzfeld–Jakob disease		
Scrapie (sheep)		
Bovine spongiform encephalopathy ("mad cow")		
Familial amyloid polyneuropathy	Transthyretin	TTR
Familial amyloid polyneuropathy (Finnish)	Gelsolin	Agel
Familial polyneuropathy— Iowa (Irish)	Apolipoprotein A1	ApoA1
Familial British dementia	FBDP	ABri
Familial Danish dementia	FDDP	ADan
Diffuse Lewy body disease	α-Synuclein	AS
Parkinson's disease		
Frontotemporal dementia	Tau	Tau
Amyotrophic lateral sclerosis	Superoxide dismutase-1	SOD1
Triplet repeat diseases (Huntington's, spinocerebellar, ataxias, etc.)	Polyglutamine	PG

been directly linked to the β-amyloid peptide, as well as to other amyloid peptides. These include membrane pore formation, dysregulation of intracellular calcium levels, and mitochondrial depolarization. The latter two biochemical observations result logically and immediately from the observation that amyloid peptides can form pores in lipid membranes. The insertion of irreversible, relatively large, nonselective ion-conducting pores in mitochondrial membranes would immediately lead to mitochondrial membrane depolarization, efflux of calcium, neuronal dysfunction, and eventually cell death. In the rest of this chapter, we will review the evidence for membrane pore formation in the various neurodegenerative diseases and describe in detail why this evidence seems compelling. We will also describe further research directions that are critical for proving or disproving the amyloid pore mechanism.

II. Aggregation and the β-Sheet Conformation

The scientific definition of amyloid is based in part on the ability of these proteins to bind dyes, such as Congo red, that exhibit the characteristic staining pattern under light microscopy. This binding ability and staining pattern reflects the tendency of these peptides to adopt a β-sheet conformation. These β-sheets are then able to stack in an extended manner to form elongated fibrils that precipitate out of solution.[1] Although much effort has been expended in clarifying the detailed structure, biophysical properties, and aggregation pathways of amyloid fibrils, current evidence suggests that the fibrils themselves may have little toxicity to cells or tissues and thus might not play a pathogenic role in disease. Substantial evidence exists that small aggregates or oligomers of amyloid peptides possess toxicity to cells and are likely to cause neuronal dysfunction.[2–4]

The discovery of familial forms of amyloid disease that are generated by specific mutations in these proteins became one of the pillars of the amyloid theory.[5] Specifically, this theory postulated that amyloid proteins, rather than being simply a pathogenetic marker, played an actual etiologic role in cell and organ destruction. This theory was strengthened when it was demonstrated that the Aβ 1–42 peptide exhibited cytotoxicity to neurons *in vitro*.[3] This was soon followed by the demonstration of neurotoxicity of other amyloid peptides, such as PrP (106–126) and α-synuclein.[6,7] Further studies have shown that monomers are nontoxic, but that intermediate aggregation states, including oligomers and protofibrils, possess the toxic function. A pioneering study in this regard was that of Pike *et al.*,[2] which demonstrated that the aggregation state of Aβ affected its cytotoxicity.

Amyloid peptide aggregation is an extremely complicated, nucleation-dependent polymerization process.[8] The nucleation phase of peptide aggregation requires the energetically unfavorable association of multiple peptide monomers to form a nucleus. Aggregate growth from this nucleus, by monomer addition, then occurs rapidly. In such a system, addition of preformed nuclei (seeds) eliminates the nucleation phase, leading to rapid fibril formation. While aggregation was seen to be a necessary prerequisite for toxicity, it was also demonstrated that extended aggregation could actually lead to a decline in toxicity, presumably because aggregates actually became too big to exert their toxic functions.[9] Thus, it was hypothesized that fibril formation might serve a protective function in amyloid diseases. The insoluble amyloid fibers would sequester amyloid protein, preventing the small aggregates from exerting their toxic function by removing them from solution. The sequestration of these fibrils in intracellular inclusion bodies, such as Lewy bodies, lends further credence to the idea that this was a self-protective cell function. A protective function was demonstrated for the inclusion bodies found in Huntington's disease.[10]

Indeed, it was also demonstrated that, in transgenic mouse models of Alzheimer's disease, the learning and memory deficits resulting from altered neuronal function could occur far earlier than the observation of amyloid protein deposits.[11] It was also shown that these learning deficits correlated much more strongly with the presence of Aβ oligomers rather than Aβ fibrils.[12]

Aβ oligomers have also been shown to inhibit synaptic growth and long-term potentiation. This early phase of inducing neuronal dysfunction could be critical in the early phases of diseases such as in Alzheimer's, where subtle deficits in clinical function signal the onset of the illness.

Although the nexus of channel formation, cytotoxicity, β-sheet conformation, and monomer aggregation is now well established, it is less clear how such a diverse variety of primary structures can lead to this common final pathway. The varying proteins and peptides (see Table I) that are involved in amyloid disease show no sequence homologies. Similarly, there is no concordance of native protein function or tertiary or quaternary structure. The only characteristic shared by these proteins appears to be the ability, or perhaps more correctly the liability, to form β-sheet rich aggregates. The various triggers for unfolding vary from disease to disease, but there is usually a critical catalyst, such as a metal ion, a change in pH, exposure to membranes, a change in concentration, or an amino acid mutation, leading to protein destabilization, β-sheet formation, and protein aggregation.

One of the latest steps in the aggregation pathway now appears to be binding and insertion into lipid membranes. It is now well established that the presence of lipid membranes catalyzes β-sheet formation.[13] β-Sheet peptides can have an affinity for membranes, as well as an ability to aggregate and orient themselves so that hydrophobic and hydrophilic surfaces find homologous pairings. With a native protein, new possibilities for hydrogen bonding and β-sheet formation are created.[14] These new bonding opportunities can drive aggregation, which is also aided by the hydrophobic effect. It is energetically unfavorable for hydrophobic residues exposed by the protein's destabilization to be exposed to the aqueous environment. This may drive self-aggregation, as well as interaction with the hydrophobic portions of membranes.

Experimentally, it has been noted that proteins with defects in their hydrogen bonding tend to bind strongly to lipid bilayer membranes.[14] A new consensus appears to be emerging that links the physical chemistry of the β-sheet to the tendency of amyloid proteins to aggregate and interact with lipid membranes. This underlying physical chemistry is strongly reflected in the pore-forming ability of this diverse collection of peptides and proteins. Experimental demonstrations of pore formation are not a mere coincidence, but rather a reflection of the thermodynamic advantage of minimizing free energy, particularly by shielding hydrophobic residues from exposure to the aqueous environment.

III. Aβ

In the early 1990s, the amyloid cascade hypothesis was proposed by Hardy and Higgins.[5] This theory suggested that the Aβ amyloid itself played a pathogenic role in cellular and tissue destruction in amyloid disease. This hypothesis provoked widespread interest in the potential cytotoxicity of Aβ, and in 1993 Arispe et al.[15] reported that the Aβ 1–40 peptide could form ion-permeable channels in planar lipid bilayer membranes. The channels described in this report were voltage independent and relatively cation selective but not discriminative among cations. Calcium, as well as sodium and potassium, were found to permeate these channels, and the authors proposed that disruption of ionic homeostasis could result from the formation of Aβ channels in cellular membranes. The channels described showed several different single-channel conductances, a result suggestive of multiple molecular species forming the channels. Further work also showed that very large conductance channels (up to 5 nanosiemens) (nS) could be observed.[16] Channels this large would cause severe ionic leakages and membrane depolarization in typical neuronal membranes within seconds. The channels had relatively long lifetimes (minutes to hours) and would likely be associated with significant neuronal dysfunction.

This intriguing and provocative hypothesis led several other laboratories to work on Aβ channel activity. However, initial results from different laboratories were conflicting. This probably was the result of rapid and irreproducible aggregation of Aβ into oligomers, protofibrils, and fibrils. This process was poorly understood at the time and poorly controlled in most laboratories. Indeed, some laboratories could not even reproduce their own neurotoxicity or channel-forming effects. Pike et al.[2] demonstrated that cytotoxicity correlated directly with the aggregation state of the peptide, and over time the role of smaller aggregates, including oligomers or protofibrils, was appreciated.[17] Since the pioneering work of Arispe et al.,[15] several other laboratories have demonstrated that Aβ amyloid peptides are capable of forming channels in planar lipid bilayers, liposomes, neurons, oocytes, and fibroblasts.[18] Studies that have correlated toxicity with channel formation have demonstrated a tight linkage. Aβ is also capable of inhibiting long-term potentiation at nanomolar concentrations.[19,20] Channel-forming variants of the Aβ peptide can inhibit LTP, whereas non-channel-forming versions do not.[19] The oligomeric and protofibrillar species of Aβ are solely responsible for the LTP inhibition.[20,21]

Further work on aggregation of Aβ peptides demonstrated that factors such as concentration, solvent exposure, pH, and the presence or absence of nucleation agents could affect both toxicity and channel-forming ability.[2,8,22] Chemicals that inhibit amyloid aggregation, such as Congo red, were reported to block cytotoxicity and inhibit channel formation.[22,23] Hirakura et al. also reported that the single-channel conductance of Aβ and other amyloid peptides depended on the aggregation state. For example, exposing Aβ to organic solvents that tended to dissociate aggregates into monomers lowered the median single-channel conductance observed in lipid membrane experiments.[22] Treatment of Aβ with acidic pH, which tends to increase peptide aggregation, shifted the single-channel conductance distribution to larger conductances. Taken together, these results implied that the heterodisperse Aβ aggregates could form ion channels of varying size and single-channel conductance.

While most studies of Aβ amyloid peptides have looked at Aβ 1–40 and Aβ 1–42, which are the primary species found in amyloid deposits *in vivo*, other Aβ peptides also can form channels in lipid membranes. Mirzabekov *et al.*[24] reported on ion channels formed by Aβ 25–35. These channels were voltage dependent and also relatively poorly selective among common physiologic ions. The concentration dependence of channel formation on Aβ concentration implied that at least three peptide monomers came together to form a channel. Aβ 25–35 not only formed channels but was also able to kill cells. Lin[25] demonstrated that the channel-forming ability of Aβ 25–35 was a necessary condition for cytotoxicity. Peptide variants that could not form channels could not kill cells. However, at least two channel-forming variants did not kill cells, thus channel formation of this shortened Aβ peptide was not sufficient for cytotoxicity.

Lin and Kagan[26] also demonstrated that the composition of lipid membranes affected channel-forming activity. Negatively charged lipids promoted channel formation, while the presence of cholesterol inhibited channel formation. The voltage-dependent opening and closing of the Aβ 25–35 peptide channel was markedly different from the voltage-independent behavior of Aβ 1–40 and 1–42.[26] Aβ peptide variants that possessed fewer than 10 residues were not able to form channels, suggesting that a minimum membrane-spanning length was required for a channel-forming peptide. A β-sheet peptide of 10 residues extends for a length of approximately 30Å, which is just enough to span the hydrophobic portion of the lipid bilayer. This length is also consistent with the length of β-sheets in known channel structures such as staphylococcal α-toxin and anthrax toxin.[27,28] A dramatic exception to this rule was the report that a five-residue Aβ variant, Aβ 31–35, could form ion channels.[29]

After these initial *in vitro* reports of Aβ channel formation, the question immediately arose as to whether Aβ could form channels *in vivo*. Early work demonstrated that Aβ 1–40 and 1–42 could induce currents selective for cations in a preparation of cortical neurons derived from rat.[30,31] These peptides were also observed to induce cation-selective currents in HT cells[32] and ion channels in patches from gonadotropin-releasing hormone-secreting neurons ([33]). No significant physiological differences could be observed between Aβ 1–40 and Aβ 1–42 channel properties.[15,16,22] Lin *et al.*[34] demonstrated that Aβ peptides could transport calcium into liposomes and fibroblasts. Rhee *et al.*[35] and Zhu *et al.*[36] demonstrated that Aβ antibodies, tromethamine, and zinc were capable of inhibiting this cytotoxicity. *In vivo*, Arispe and Doh[37] showed that the cholesterol content of the plasma membrane regulated the ability of Aβ 1–40 and 1–42 to kill cells, and that blockers of the Aβ channel could rescue cells from Aβ cytotoxicity, even at a relatively late stage.[38] They went on to create highly specific blockers, such as the Na 1–7 peptide, which was designed to bind strongly to a hypothesized β-sheet lining the pore region of Aβ 1–42 channels. Remarkably, this peptide is capable of blocking Aβ channels at nanomolar concentrations and preventing cytotoxicity. In contrast, other peptides with similar composition and length, but designed to bind to other regions of Aβ, do not block channel formation or cytotoxicity.

In addition to cytotoxicity, it was also reported that Aβ 1–42 could cause mitochondrial depolarization and the release of cytochrome c.[39] Both these effects are likely related directly to its channel-forming properties. Since channel formation in cell membranes would likely lead to cellular depolarization and calcium dysregulation, it was hypothesized that preventing these effects might ameliorate Aβ-induced toxicity. Liu et al.[40] reported that diazoxide, a K-ATP channel opener, improved memory and decreased Aβ and tau levels in a transgenic mouse model of Alzheimer's disease. This potassium-channel opener would hyperpolarize membranes and thus directly oppose the depolarizing effect of Aβ. One would hypothesize that this might lead to improvements in calcium regulation and calcium homeostasis, and block the inhibition of LTP induced by Aβ peptides, thus resulting in improved memory. Anekonda et al.[41] showed that voltage-dependent calcium channel blockers could protect neurons in culture from the toxicity of an Aβ peptide, APP-CTF. This C-terminal fragment (105kD) of the amyloid precursor protein has also been shown to have channel-forming and cytotoxic properties.[42] Amelioration of cytotoxicity by voltage-dependent calcium channel blockers again suggests that depolarization by Aβ peptides of the cell membrane leading to increased calcium influx and calcium dysregulation is a key mechanism of cytotoxicity.

Further work in lipid bilayers by Kourie et al. has led to the discovery of at least four distinct Aβ 1–40 channel phenotypes.[43] The channels described include channels ranging from 0 to 589 pS $((1pS) = 10^{12}\Omega)$ with properties of bursting, spike generation, and inactivation. These four channel types have reproducible kinetic behavior and current–voltage relationships. They also showed distinctive ion selectivity and blockage sensitivity. This suggested they are indeed distinct molecular species, possibly formed by different forms of aggregates of the Aβ 1–40 peptide. This is consistent with the multiple single-channel conductances reported by previous investigators. Recent structural modeling[44] suggests that Aβ peptides can form highly mobile subunits that aggregate to form relatively large channel structures. These structures, however, are fluid and subject to relatively rapid rearrangement within the membrane. These models are consistent with both the observed electrophysiologic data showing multiple conductance states and the general sizes shown of Aβ channels observed in atomic-force microscopy and electron microscopy, showing channels with an outer diameter of approximately 40Å and an inner diameter of approximately 20Å.[45]

IV. Prion Channels

Prion diseases are sporadic, genetic, or infectious neurodegenerative diseases of animals and humans that are caused by a proteinaceous agent. The transmissible forms of prion diseases, including Creutzfeldt–Jakob disease and bovine spongiform encephalopathy, have brought widespread popular attention to prion disease. Familial forms of the disease are characterized by mutations in the prion protein. A remarkable feature of prion diseases is that they require the conversion of the normal cellular prion protein (PrP^C) to a pathologic form (PrP^{Sc}).[46] This conformational transition involves the conversion of α-helical regions to β-sheets.[47] Prion diseases are also characterized by spongiform degeneration in the brain, which is sometimes accompanied by the deposition of amyloid fibrils, large aggregates of PrP^{Sc}. PrP^C is protease sensitive, whereas PrP^{Sc} has a protease resistant core, PrP 27–30. Different strains of prions exist that are distinguished by their unique conformations.[46] There is also evidence to suggest that neurodegeneration may spread within the brain through a prion-like propagation process.[48]

As with other amyloid diseases, the molecular mechanism of prion pathogenesis has remained elusive, but pore formation has emerged as an attractive hypothesis. Lin et al.[49] first demonstrated that the neurotoxic peptide PrP (106–126) readily formed ion-conducting channels when added to the aqueous phase bounding a planar lipid bilayer membrane. Once inserted in the membrane, channels did not disassociate from the membrane. The open channels exhibited long lifetimes, in the range of minutes, and a heterodisperse set of single-channel conductances, similar to other amyloid channel formers. For example, in 0.1 M sodium chloride, a range of channel conductances from 10 up to 400 pS could be observed. Environmental factors that enhance prion protein aggregation also enhance channel formation, including aging of the peptide and exposure to acidic pH. The latter treatment has also been demonstrated to promote the conversion of α-helical structure in PrP (106–126) into β-sheet.[50] Once inserted, prion protein channels were voltage independent and relatively nonselective, admitting a variety of physiological ions, including sodium, potassium, chloride, and calcium. Exposure to acidic pH (4.5) was also notable, in that it changed the distribution of the single-channel conductances observed, shifting them to greater conductance levels.[49] This would be consistent with the known effect of acid pH on accelerating prion protein aggregation.

In 2000, Manunta et al.[51] reported that PrP (106–126) did not form channels in lipid bilayers, although the aggregation state of their peptide was unclear. However, Kourie and Culverson[52] confirmed that PrP (106–126) did form ion channels. These investigators described a variety of cation channel subtypes formed by PrP (106–126). The biophysical properties of these channel subtypes were reproducible and distinctive and ranged from conductances as low as 40 pS to as high as 1500 pS. This group further went on to demonstrate that PrP channels could be blocked by the antimalarial chemical quinacrine. The mechanism of the blockage was a reduction of mean current through the open channel.[53] Copper ion was also observed to modulate the conductance of PrP (106–126) channels, and these investigators hypothesized a "fast channel block" at the mouth of the pore by copper binding to residues M109 and H111 at the channel mouth.[54]

Further results from this lab also indicated that a much larger PrP fragment, PrP (82–146), could form ion channels that were very similar to the PrP (106–126) channels. The PrP (82–146) fragment is identical to that which is found in the brains of patients with Gerstmann–Straussler–Scheinker disease. By scrambling the sequence of the 127–146 region and the 106–126 region, these investigators showed that it was the 106–126 region that was critical for channel formation.[55]

Other lines of evidence suggest that channel formation is critical in prion disease pathogenesis. In one mutant prion disease in which no amyloid deposits are observed, the mutant prion protein actually becomes a transmembrane protein rather than a GPI (glycero-phosphatdylinositol)-anchored protein.[56] One possibility is that the mutant prion protein becomes a leakage pathway when it becomes transmembrane. Solomon et al.[57] have shown that expression in transfected cells of neurotoxic PrP deletion mutants can induce large, spontaneous ionic currents readily detected by patch-clamp techniques. The most toxic of these mutants is the deletion of 105–125, precisely the region implicated in pore formation in planar lipid bilayers. Intriguingly, the currents observed are similar to those observed in the PrP (106–126) channel, in that they are large and relatively nonselective cation-permeable pores. It is also noteworthy that over-expression of the wild-type cellular prion protein can silence these channels. The channels can also be silenced by treatment with a sulfated glycosaminoglycan. When the authors expressed prion protein molecules with point mutations in the central region corresponding to familial prion disease in humans, they observed similar ionic currents in the cells. These authors hypothesized that wild-type PrP possesses a channel-silencing activity, and that this is disrupted in mutants involved in familial prion disease. Alternatively, the Prp 106–126 deletion may induce wild-type PrP to form a channel similar to PrP 106–126.

Alier et al.[58] used patch-clamp methods to observe that PrP (106–126) could reduce whole cell outward currents and that this effect was attenuated in calcium-free external media. The PrP (106–126) peptide was also observed to depress the potassium-delayed rectifier and transient outward potassium currents.

Henriques et al.[59] observed that PrP (106–126) did not interact with membranes under physiological conditions, but that membrane insertion and leakage could occur when a strong electrostatic interaction was present. They interpret these results as supporting the hypothesis that the normal cellular prion protein is required to mediate the toxicity of PrP (106–126).[46]

Finally, at least one other channel-forming segment of the prion protein has been reported. This is PrP (170–175), which bears a mutation N171S, resulting in schizoaffective disorder.[60] The native fragment 170–175 is not capable of forming pores and there is no evidence that 170–175 (N171S) exists in these patients. Thus, this single amino acid mutation, which confers pore-forming ability on the peptide, also is responsible for a severe neuropsychiatric disease, although not an amyloid disease.

V. α-Synuclein

Parkinson's disease is a neurodegenerative illness primarily affecting dopaminergic neurons in the brain. It is characterized by motor, cognitive, and emotional abnormalities. Although ameliorative treatments are available for Parkinson's disease, none of them has been demonstrated clearly to slow the progressive degenerative course of the illness. Parkinson's disease is also characterized by amyloid deposits. These occur primarily as Lewy bodies in neurons in the brain. The major protein of Lewy bodies is α-synuclein, a 140-residue-long synaptic protein. The biological function of α-synuclein remains unclear, although it is thought to be involved in synaptic vesicle physiology at the presynaptic nerve terminal. The full-length peptide is typically not found in deposits, but rather a fragment referred to as NAC, originally found in Aβ deposits. This acronym stands for the misnomer "nonamyloid component" and consists of residues 66–95 of the full-length α-synuclein. Some familial cases of Parkinson's disease are characterized by mutations in α-synuclein.[61] α-Synuclein was first demonstrated to induce membrane permeability in synthetic lipid vesicles.[62] NAC was subsequently shown to form ion-permeable channels in planar lipid bilayers.[63] Further experiments employing electron microscopy demonstrated annular oligomeric structures resembling pores. Pathogenic Parkinson's mutations were found to accelerate the formation of these structures.[64]

Although amyloid proteins are typically characterized by their tendency to form β-sheet structures, α-synuclein is a natively disordered protein that may adopt many different conformations. When interacting with membranes, monomeric α-synuclein has been shown to adopt a highly α-helical form.[65] Pore formation can be observed from these α-helical monomers. Requirements for pore formation include an anionic lipid in high concentration, phosphatidylethanolamine, and a membrane potential resembling that of a normal cell (inside negative). The E46K and A53T forms of α-synuclein, which cause familial Parkinson's disease, also form pores. However, a third familial Parkinson's disease mutation, A30P, demonstrated a lower affinity for membranes and was not able to form pores. Addition of calcium to the membrane prior to pore formation could inhibit pore formation, and the addition of calcium after pore formation decreased channel conductance. When these authors tested α-synuclein in the membranes, they observed that permeability increased but were not able to observe discrete ion channels. The requirement for phosphatidylethanolamine and a negative membrane potential was not present for the oligomeric α-synuclein, suggesting that a different molecular species was at work through a different molecular membrane mechanism.

The biophysical properties of the NAC channels were found to be quite similar to those of other amyloid channel-forming peptides. For example, NAC channel formation could be inhibited by Congo red and blocked by zinc ion. A wide variety of single-channel conductances of NAC were observed which may have reflected a variety of oligomeric species in the preparation. NAC channels exhibited relatively long lifetimes, irreversible association with the membrane, and poor selectivity among physiologic ions.

Kostka et al.[66] also observed pore formation by α-synuclein oligomers which had been induced by exposure to low concentrations (μm) of iron (Fe^{+3}). Organic solvents were used to trigger aggregation into small oligomers, and iron could subsequently induce further aggregation into larger oligomers. The larger oligomers formed pores in planar lipid bilayers and could be recognized and inhibited by the A11 antibody, which had previously been reported to recognize oligomeric forms of amyloid peptides.

Di Pasquale et al.[67] studied the role of gangliosides in α-synuclein channel formation. They identified a ganglioside-binding domain on α-synuclein, residues 34–50, that was structurally related to the glycosphingolipid-binding domain of microbial protein and other amyloid proteins. They also reported that the disease-linked mutation E46K protein exhibited a stronger affinity for GM3, a minor brain ganglioside that increases in concentration in the brain with aging. E46K mutant α-synuclein formed channels approximately five times less conductive than those of wild types. These mutant channels also had a higher cation selectivity and an asymmetric response to voltage. The presence of GM3, however, in the membrane could correct these altered channel properties and was specific for GM3. They suggested that GM3 might play a protective role in the brain against this mutation.

Kim et al.[68] recently demonstrated pore formation by a variety of oligomers of α-synuclein. They were also able to determine the β structure of these oligomers and show that it was distinct from the β structure exhibited by α-synuclein in the fibril form. These findings are consistent with other findings that oligomers, rather than fibrils, appear to play the key role in disease pathogenesis.[69]

Furukawa et al.[70] reported that changes in membrane permeability linked to mutant α-synuclein molecules play an important role in neural cell degeneration. Specifically, they expressed the α-synuclein mutants A30P and A53T in

dopamine-producing neurons in culture. These cells exhibited higher membrane permeabilities, most likely due to nonselective pore formation by the
mutant α-synuclein molecules. Calcium levels, both basal and in response
to membrane depolarization, were greater in the cells expressing mutant
α-synuclein. Cells could be protected against oxidative stress induced by
calcium by the membrane-permeable calcium chelator BAPTA-AM. Calcium
channel blockers for L-type or N-type channels could not protect the cells,
suggesting that the altered ion and calcium permeability was mediated by
α-synuclein channels.

Taken together, all these results strongly point to a channel-forming role for
α-synuclein in Parkinson's disease pathology. α-Synuclein channels are permeable to calcium and would likely disrupt calcium regulation quite rapidly,
leading to neuronal dysfunction and, ultimately, cell death.

VI. AG Triplet Repeat Diseases/Huntington's

Huntington's disease, a progressive neurodegenerative illness marked by
neuropsychiatric symptoms and choreiform movements, is a member of the
large and expanding group of triplet repeat diseases. These primarily neurological illnesses share an expansion of the CAG codon which specifies glutamine in protein synthesis. Because of this expansion, a large polyglutamine
tract appears in the pathological protein for each disease. Each of these illnesses is characterized by a critical repeat length. If the triplet repeat expansion
is greater than this critical length, clinical illnesses occur, but if it is less than the
critical length, no symptoms are observed. In Huntington's disease, the critical
repeat length is about 37 amino acids.[71] The affected protein in Huntington's
disease, huntingtin, is a protein with uncertain function. However, it is clear
that in the illness, the protein acquires a toxic function, leading to cellular
dysfunction and death. As polyglutamine expansion lengths become longer, the
age of onset drops.[72] This correlation suggests that polyglutamine itself may be
a critical player in the etiology of the illness. On gross pathology, advanced
Huntington's disease exhibits profound neuronal loss in cortex and striatum. In
the early stages of the disease, clinical features such as memory loss, mood
fluctuations, and impaired long-term potentiation can be observed. Triplet
repeat diseases are characterized by the formation of toxic aggregates of the
polyQ-containing proteins. Because these aggregates do not stain with Congo
red and similar dyes, these diseases cannot be considered true amyloidoses.
However, the parallels between triplet repeat diseases and amyloid diseases are
compelling. Transgenic models of Huntington's disease suggest that oligomer
formation is closely tied to the progression of the illness. Experiments in PC12

cells in culture have shown that cells expressing *huntingtin* with a poly-Q repeat of 150 residues become more vulnerable to apoptosis even before aggregates begin to form.

In 2000, Hirakura *et al.*[73] first demonstrated channel formation in planar lipid bilayers by polyglutamine. The channels they observed were similar to those of amyloid protein channels and exhibited a long lifetime, poor ionic selectivity, and heterodisperse single-channel conductances. Their preparation of polyglutamine contained a mixture of polyglutamine lengths. This may have partially accounted for the heterogeneity in single-channel conductances that they observed. They also observed a significant enhancement of channel formation when the polyglutamine was subjected to acidic pH. Consistent with the known properties of polyglutamine, Congo red could not inhibit channel formation. The channels were also not blocked by zinc ions. Thus, the polyglutamine channels exhibited some biophysical properties distinct from those of classical amyloid peptide channels. Monoi *et al.*[74] used a homogeneous preparation of polyglutamine of repeat length 40 to demonstrate a uniform group of channels with a single-channel conductance of 17pS. They also used a control polyglutamine preparation consisting of 29 residues, shorter than the critical repeat length. This preparation failed to form channels. Their model for channel formation, the μ-helix, showed that at least 37 residues in this helical structure would be required to span a membrane. This was consistent with the critical repeat length as observed in clinical illness.

Panov *et al.*[75] showed that mitochondria exhibit a set of physiologic deficits in patients with Huntington's disease. These included a decrease in membrane potential and vulnerability to depolarization. Transgenic mice expressing huntingtin with an elongated polyglutamine tract showed similar physiological alterations in their brain mitochondria. Electron microscopic studies revealed huntingtin in significant quantities on the mitochondrial membranes. Similarly, a nonhuntingtin fusion protein containing a large polyglutamine expansion also was observed to cause similar mitochondrial defects. Taken together, these data suggest that polyglutamine itself is damaging the mitochondria, most likely by a channel or pore mechanism.

VII. Amyotrophic Lateral Sclerosis/Superoxide Dismutase

Amyotrophic lateral sclerosis (ALS) is a progressive neurodegenerative disease characterized by loss of upper motor neuron function. Familial cases of ALS can be caused by mutations in superoxide dismutase 1 (SOD1). The mechanism of SOD toxicity in ALS remains unclear. It has been demonstrated that SOD, when misfolded, can aggregate in a manner analogous to that of amyloid proteins.[76] It has also been shown that these aggregates can form into pore-like structures, as assessed by atomic force microscopy.[77] These investigators used wild-type SOD and three pathogenic mutants to demonstrate that copper-induced oxidation of SOD could trigger aggregation into pore-like assemblies. Although a physiologic ion channel function has not been demonstrated *in vitro* for SOD aggregates, several physiologic studies *in vivo* suggest that pores may play a role in ALS pathogenesis. Vucic *et al.*[78] reported that muscle membrane potential was significantly reduced in familial ALS and spontaneous ALS. Their study suggested that persistent upregulated sodium conductances were associated with axonal degeneration in ALS. Pore formation by a nonselective cationic pore would likely cause both membrane depolarization and sodium conductance upregulation. Meehan *et al.*[79] also reported increased excitability of neurons that were depolarized in an SOD1 mutant mouse model of ALS. These studies showed that the motor neurons, when depolarized, sustained higher frequency firing and persistent inward currents which were activated at lower firing frequencies. Once again, these features would be consistent with the presence of a leakage current caused by a nonspecific pore tending to depolarize the membrane. Pieri *et al.*[80] reported an intrinsic hyperexcitability in the G93A mutant mouse cortical neurons. This was due to higher current density of a persistent sodium current.

Mitochondrial malfunction is also strongly suspected in ALS. Israelson *et al.*[81] demonstrated that misfolded mutant SOD1 could directly bind to and inhibit the voltage-dependent anion channel (VDAC1) in mitochondrial membranes. Working on a transgenic mouse model of familial ALS, these investigators showed that the mutant SOD1 could inhibit the anionic conductance of VDAC, suggesting that SOD1 through this direct binding and inhibition of a mitochondrial ion channel might be contributing directly to mitochondrial dysfunction. While this does not directly favor a channel function for mutant SOD, it does suggest that mutant SOD misfolds in a way that it can directly enter into lipid membranes in the mitochondria. While SOD1 studies in ALS directly demonstrating *in vivo* and *in vitro* channel function are still lacking, the evidence to date suggests that this disease may be analogous to the other neurodegenerative amyloid diseases mentioned above in that the SOD protein may be capable of forming nonspecific leakage pores in plasma and mitochondrial membranes.

VIII. How Channels Damage Neurons

A. Plasma Membranes

Cytotoxicity mediated by the breaching of cell membranes has been well established in the microbial world, including both bacteria and eukaryotic single-celled organisms.[82] It is less well understood in the multicellular kingdom, and particularly in neurons. The bacterial inner cell membrane is not only a demarcator of cells from noncells but also actively used for respiration, active transport, and energy generation. For this reason, the cell membranes of bacteria are substantially less leaky than those of mammalian cells. Leaks in the bacterial inner cell membrane interfere with respiration, active transport, and the energy generation of the bacterium. In addition, the relatively small volume of bacterial cells renders them vulnerable to loss of vital intracellular constituents, such as potassium and magnesium ions, and to the influx of toxic extracellular factors, such as calcium. The fact that channel-forming colicins, defensins, and protegrins kill bacterial cells by inserting into the inner cell membrane, depolarizing the bacteria, and causing the loss of potassium and magnesium is now well established.[45,100,101]

Eukaryotic cells enjoy a measure of protection from channel-forming toxins due to their larger cell volume and the presence of cholesterol and ergosterol in the cell membrane. These agents tend to stiffen cell membranes and prevent insertion of extracellular toxins. The antifungal channel-forming drugs nystatin and amphotericin B are significant exceptions to this rule. They bind specifically to sterols, and this accounts for their ability to kill fungi and mammalian host cells while not affecting bacterial cells. This property also explains their considerable toxicity in humans.[83] These antibiotics also clearly demonstrate that channel formation can be a toxic mechanism for mammalian cells. Cell types differ in their vulnerability to these toxins based on membrane composition, membrane fluidity, membrane potential, ionic pumping capability, and similar factors.[83]

A group of cholesterol-requiring cytolysins constitute a family of channel-forming proteins produced by bacteria. Perfringolysin O of Clostridium is a prime example of these toxins which form giant channels in mammalian cell membranes. Perfringolysin O plays a key role in clostridial pathogenesis. It is curious to note that a key structural motif in channel formation by these cytolysins is the transformation of α-helical regions to β-sheet conformation.[84] Each individual toxin protein contributes two β-sheet hairpins to the ultimate pore structure. Completed pores can consist of up to 50 monomers, with the pore size ranging up to 300Å in diameter. These are among the largest pores described and have been demonstrated to rapidly kill cells as a result of their large size and ionic conductance. On a somewhat smaller scale, staphylococcal α-hemolysin has been demonstrated to form an oligomeric pore of 14 strands in total. This haptomeric oligomer requires two β strands from each monomer.[27] A β-barrel structure also has been described for the pores formed by the outer membrane porins of Gram-negative bacteria.[85] A common theme in all of these channel-forming proteins is the β-barrel structure characterized by a cylindrical array of β strands.

Among mammalian cells, nerve cells find themselves uniquely sensitive to breeches in membrane integrity. The plasma membrane must remain relatively impermeable to ions in order to maintain plasma membrane potential and allow signaling to occur along the membrane of the axon. Small permeability changes are responsible for membrane potential changes, and extraneous leakage pathways affect this signaling quite rapidly. Furthermore, leakage pathways also require increased action of endogenous ion pumps, such as the Na,K-ATPase, which increases metabolic cost. The more leakage pathways that exist, the harder the cell must work and, ultimately, this can lead to depletion of energy stores, cellular dysfunction, and cell death. Poorly selective ion channels can also give rise to disruption of calcium homeostasis. Calcium levels are typically maintained in the micromolar range intracellularly, so brief spikes in calcium concentration can induce intracellular signaling. This is also a metabolically expensive housekeeping function. Increasing intracellular calcium can induce apoptosis and subsequent cell death. Cell membrane depolarization can lead to calcium dysregulation through influx of calcium through the cell membrane. Thus, the insertion of relatively nonselective channels, such as those formed by amyloid proteins, in neuronal membranes would lead rapidly to impaired signaling functions and, in the long run, to more widespread cellular dysfunction and death.

B. Mitochondrial Membranes

As described above, channel formation in plasma membranes of nerve cells is likely to cause immediate impairments in neuronal functioning and ultimately lead to metabolic stress and cell death. Significant evidence, however, points to mitochondrial membranes as a potential target of amyloid peptides. Evolutionarily, mitochondria evolved as symbiotic bacteria which were engulfed by larger cells. Their bacterial origin also is suggested by the similarity of their bimembrane structure (inner and outer membranes) to that of Gram-negative bacteria.[86] The outer membranes of both Gram-negative bacteria and mitochondria contain aqueous pores (known as porins; the mitochondrial porin is called VDAC for voltage-dependent anion channel) that possess significant permeability to physiological ions and small nonelectrolytes. The inner membrane of mitochondria and bacteria is relatively impermeable to ions, which is critical for its role in coupling respiration to oxidative phosphorylation via a proton gradient. Leakage pathways in the mitochondrial or bacterial inner membrane can disrupt energy generation. Significantly, this appears to occur in mitochondria with the appearance of the permeability transition pore (PTP). When this occurs, mitochondrial inner membranes become depolarized, allow cytochrome c and calcium to escape, and signal programmed cell death (apoptosis). It is of interest that a number of apoptosis-related proteins, such as BAX and Bcl-2, can form channels in lipid membranes, although their major function appears to be elsewhere in the apoptoses pathway. The structures of these channel-forming proteins turn out to have significant homologies to the structures of the channel-forming bacterial colicins.[87] Apoptosis is a process known to be elevated in amyloid diseases, and it has been suggested that amyloid proteins may act directly on mitochondria to increase PTP formation and subsequent apoptosis. As an example, Aβ was reported to have direct effects on mitochondria, including loss of membrane potential, PTP opening, and cytochrome c release.[39,88] A number of different amyloid peptides have been reported to cause an increase in oxidative stress. Levels of reactive oxygen species, such as H_2O_2 and superoxide, are linked directly to mitochondrial effects. Another example is the direct effects of polyglutamine on mitochondria. Similar to Aβ, polyglutamine can cause an immediate loss of membrane potential and subsequent increases in PTP formation and apoptosis. Depolarized mitochondria in Huntington's disease appear to occur even prior to clinical features of the disease or to other cellular pathology.[75] This is consistent with the idea that amyloid peptide damage to mitochondria is an early step in the disease process. Since mitochondria also play a key role in regulating calcium stores, the observed loss of calcium homeostasis in many amyloid diseases may also be directly traceable to amyloid peptide effects on mitochondria. Abramov et al.[89] reported that Aβ can directly cause calcium-dependent depolarizations in mitochondria which lead gradually to an overall loss of membrane potential. This slower collapse of membrane potential can be inhibited, to some degree, by the presence of antioxidants. These authors conclude that pore formation by Aβ leads directly to a significant increase in the production of oxygen free radicals. Taken together, these data suggest that amyloid peptides might act, in the very early stages of amyloid disease, to injure mitochondria and impair cellular energy generation and calcium homeostasis.

How amyloid peptides would reach intracellular targets, such as mitochondria, is a complex issue. In some diseases, it is likely that amyloid peptides would reach mitochondria through an endoplasmic reticulum-associated pathway.[90] In other diseases, it is possible that amyloid peptides first bind and insert in plasma membranes and then either cross the membranes completely or are transported through an endocytotic vesicle.[90] Further research should focus on elucidating these pathways, which could be targets for anti-amyloid drug therapy.

IX. Molecular Pore Models

After the first reports of channel formation by Aβ, Durell et al.[91] published several simple models for ion channels induced by Aβ in lipid membranes. These models were based on secondary structure predictions. They inferred that Aβ was predicted to form a β-sheet, which then was followed by an α-helix, β-turn, α-helix, with the C-terminus being particularly hydrophobic. They developed three classes of models. In the first class, the pore was formed by the β-hairpins, and Arispe et al.[38] have recently developed blockers based on this prediction, which bind to the β-hairpin and block the pore. The two other classes were based on the assumption that either the middle helices or the C-terminus helices formed the pore. Because Aβ exhibited multiple single channel conductances which sometimes interconverted, it was presumed possible that all of these structures might be present in the membrane. However, so far there has been experimental evidence supporting only the first model, in which the β-hairpins form the channel. The fact that peptides designed to bind to these β-hairpins do seem to block channel activity both in lipid bilayers and in cell culture suggests that channel blocking may be a viable therapeutic target to pursue in drug development.[38]

More recent models have come from computational methods using molecular dynamic simulations (Fig. 1).[92,93] These models are based on more recent experimental data which show that the amyloid Aβ forms a U-shaped β-strand–β-turn–β strand structure in a lipid membrane environment (Fig. 1). Using this new data, Jang et al. arrived at a structure containing 24 peptides in a circular pore, with both β-strands spanning the membrane. The outer diameter of this pore was 64 Å, rising to 80 Å at the end of the simulation. The inner pore diameter was 20 Å, rising to 27 Å at the end of the simulation. These diameters were in close correspondence with experimental data from atomic force microscopy. In those

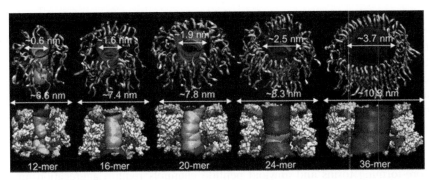

FIG. 1. Averaged pore structures, calculated by the HOLE program, embedded in the averaged channel conformations during the simulations for the 12-, 16-, 20-, 24-, and 36-mer Aβ$_{9-42}$ channels. The 12-, 16-, 20-, and 36-mer structures were obtained from the simulations in the zwitterionic DOPC bilayer. The 24-mer structure was obtained from the anionic bilayer containing POPC and POPG with a molar ratio of 4:1. In the angle views of the pore structure (upper row), whole channel structures are shown with the ribbon representation. In the lateral views of the pore structure (lower row), cross-sectioned channels are given in the surface representation. For the pore structures in the surface representation, the degree of the pore diameter is indicated by the color codes in the order of red<green<blue, but the scale of these colors is relative to each channel. In the channel structures, hydrophobic residues are shown in white, polar and Gly residues are shown in green, positively charged residues are shown in blue, and negatively charged residues are shown in red.[4] (See Color Insert.)

studies, the outer diameter of the Aβ pore was observed to be 80–100Å, and the inner diameter 10–20Å.[94] Atomic force microscopy data also suggested similar pore sizes for a number of different amyloid peptides, including serum amyloid A, β2-microglobulin, α-synuclein, amylin, ABri, and ADan, and prion peptides (Fig. 2). While molecular dynamic simulations cannot be considered conclusive, they do strongly suggest that an oligomeric pore formed from these peptides is indeed possible. An interesting outcome of these simulations was that the 24-mer pore seemed to be formed of subunits that were mobile in the membrane. These subunits appeared to be hexamers and might help account for the variability of single channel conductances observed in experiments. Other limitations on the molecular dynamic studies include the fact that only the 17–42 region of the Aβ was modeled, because there was structural data available for only that portion of the peptide. The omitted 16 N-terminus residues might alter the shape, size, and structure of the pore. The model pore did appear to exhibit cation selectivity. Zinc did not exhibit significant mobility through the pore and lowered the mobility of calcium in the pore when it was bound to glutamate 22. Magnesium and potassium exhibited a high mobility in the pore, consistent with experimental data.

X. β-Sheet Conformation and Amyloid Peptide Channels

A critical step in amyloid channel formation is the folding of the protein into a β-sheet conformation. As described above, this can happen because of the effect of elevated protein concentrations, low pH, amino acid mutations, binding of metal ions, or interaction with lipid membranes. Although the classical channels of muscle and nerve whose structures have been elucidated are primarily found in α-helical structures in the membrane, some bacterial outer membrane porin channels and bacterial toxin channels, including staphylococcal α-hemolysin, anthrax toxin, and clostridial perfringolysin, are β-barrel channels. Numerous similarities between amyloid peptide channels and bacterial pore-forming toxins have been previously described.[17] Both bacterial pore-forming toxins and amyloid peptide channels require aggregation and activation at the membrane surface leading to the process of channel formation. Both classes of proteins exist initially in the soluble state and then undergo a conversion to a membrane-inserted state. A significant conformational change takes place during this transformation as native helical and random coil structures convert to random coil and/or β-sheet. Oligomerization and alignment of β-sheets or strands appears to be critical in the case of the bacterial pore forming toxins but has yet to be established for the amyloid peptides. The β-barrels of the pore-forming toxins are composed of anywhere from 8 to 22 β-strands with an average length of 10–13 residues. The pore diameters can span a range of 15–35Å. It is notable that these pore diameters are in the

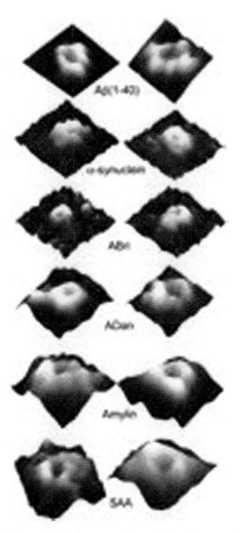

FIG. 2. Individual channel-like structures at high resolution. Two examples are shown for each molecule, in which a central pore can be observed. The number of subunits observed protruding from the surface varies from four to eight subunits. Resolution of AFM images is not enough to resolve individual subunit structures. (Image sizes are 25 nm for Aβ(1–40), 25 nm for α-synuclein, 35 nm for ABri, 20 nm for ADan, 25 nm for amylin, and 20 nm for SAA.[94]) (See Color Insert.)

same range as those observed through atomic force microscopy and predicted by molecular dynamic simulations. It should also be noted, however, that certain classes of bacterial toxins, including streptolysin and perfringolysin O, form extremely large pores with a diameter of 150–450 Å.

Both amyloid peptide channels and pore-forming toxins show heterodispersity in single-channel conductance, which may reflect conformational shifts or changes in the state of aggregation. Both classes of channels are sensitive to membrane composition and fluidity. One aspect in which they differ is that some of the pore-forming toxins require cholesterol for membrane insertion, whereas with the amyloid peptides studied so far, cholesterol is a potent inhibitor of channel formation by amyloid peptides. Although amyloid peptides are not designed to be cytotoxins, it is interesting that they adopt a conformation which mimics that of highly evolved bacterial toxins that have as their goal the killing of mammalian cells. In this view, one can look at amyloid peptide channels as a form of self-directed toxin, although some have argued for an antimicrobial role of beta-amyloid.[95]

Thundimadathil *et al.*[96] demonstrated that model β-sheet peptides $(\times S \times G)^6$ could form large voltage-dependent channels in planar lipid bilayers with physiologic properties resembling those of porins. The presence of subconductance states and complex kinetic behavior also was observed with these short peptides. The relatively high conductance and low selectivity of these channels were similar to those of the amyloid peptide channels and suggested that channel formation might be a generic property of β-sheet peptides. A calculation of diameter based on the single-channel conductance gave approximately 11 Å, similar to that of porins and amyloid peptides. Similar to amyloid peptides, Congo red incubation could inhibit channel formation by inhibiting aggregation. The peptide could also be induced to form fibrillar structures similar to amyloid fibrils in aqueous solution. However, the presence of lipid membranes shifted the peptides into oligomeric structures rather than extended fibrils. The authors also observed the formation of channels with relatively small conductances prior to larger conductance channels appearing in the membrane. This may have been a result of further aggregation of the model β-sheet peptide into larger and larger channel units. Again, this behavior is reminiscent of that of amyloid peptide channels in lipid bilayer membranes. These studies demonstrate that the β-sheet conformation is highly suited for channel formation and tightly links the β-sheet nature of amyloid peptides to their ability to interact strongly with lipid membranes and to form channels.

XI. Reflections

The pore hypothesis of amyloid peptide action was proposed nearly two decades ago[15] and, despite significant evidence in its favor, has yet to be fully confirmed or refuted. This difficulty arises from the nonspecific nature of the channel. The channels formed by amyloid peptides do not possess the characteristic ion selectivity or voltage dependence of the well-known channels of nerve and muscle membranes. They also do not possess the unique single-channel conductance fingerprint for identifying ion channels using electrophysiologic techniques. Furthermore, because they are not proteins in their native conformations, antibodies against the native proteins from which they are derived usually will not recognize them. Thus, the identification of amyloid proteins and peptides *in vivo* has been quite difficult.

An antibody with specificity for amyloid protein oligomers, "A11," was developed by Kayed *et al.*[97] several years ago. However, the amyloid structures recognized by this antibody were not found to form channels, but rather to induce a nonchannel permeability through the membrane. While the reasons for this are not entirely clear, some studies suggest that the oligomers recognized by this antibody are substantially larger than the oligomers that form channels. Furthermore, a recent study has suggested that, in fact, the oligomers recognized by A11 do not permeabilize membranes, but rather a contaminating solvent used in the preparation of these oligomers may be responsible for the nonchannel ion conductance.[98] Thus, the relevance of this nonchannel permeability mechanism has been called into serious question, as well as the relevance of the A11 antibody.

While amyloid pores have been well characterized in *in vitro* systems, such as planar lipid bilayers and liposomes, and also characterized in cell culture systems using patch-clamp techniques, it has been extremely difficult to characterize these pores in an animal model of amyloid disease. The reasons for this include the lack of identifying features, as described above, such as ion selectivity, voltage dependence, antibody recognition, or single-channel conductance. At least one report has identified Aβ pore structures using electron microscopy to study Alzheimer's disease brains.[99] While the identified structures do not appear in tissue from control brain, and they do resemble the pore structures observed *in vitro*, it is impossible to identify them with certainty as Aβ-induced pores. Of course, even if this were possible, it would be critical to identify their functionality as pores, in terms of neuronal currents, ionic homeostasis, etc. Thus, one way to identify these pores would be the development of specific channel blockers that work uniquely on amyloid pores. There is one model for this kind of blocker, which is the NA 1–7 peptide developed by Arispe *et al.*[38] Using a theoretical model of the Aβ pore structure for guidance, they developed a peptide predicted to bond to the β-sheet lining the pore. This peptide, at nanomolar concentrations, does, *in vitro* and in cell culture, strongly block Aβ channels. Thus, this peptide could provide a way to block selectively Aβ channels in an animal model. Technical problems to be solved include delivery of this peptide into the central nervous system, as well as delivery of this peptide to the appropriate cellular compartment. It is one thing to have a peptide in the extracellular fluid, where it could bind and block Aβ channels in the plasma membrane, but quite another thing to have the peptide delivered to mitochondria or other intracellular membranes where Aβ may also be at work.

Despite the difficulties, there are many reasons for viewing the pore hypothesis as compelling. First of all, pore formation is nearly universal among the amyloid peptides that have been tested (see Table II). Second, the pores that are formed by these widely varying sequences of amyloid peptide and proteins are remarkably similar. Specifically, they nearly all show heterodisperse single-channel conductances. They are nearly all voltage independent. They are nearly all irreversibly associated with the membrane once inserted. They generally exhibit very long lifetimes, compared to the millisecond lifetimes of ionic channels in nerve and muscle membranes, and they are usually weakly selective for cations, but not selective in particular for sodium, potassium, or calcium. Congo red, by inhibiting aggregation, can inhibit the formation of amyloid channels and zinc ion at micromolar concentrations and is capable

TABLE II

CHANNEL PROPERTIES OF AMYLOID PEPTIDES FROM NEURODEGENERATIVE DISEASES

Peptide	Single-channel conductance (pS)	Ion selectiviity (permeability ratio)	Blockade by zinc	Inhibition by Congo red	Reference
Aβ 25–35	10–400	Cation ($P_K/P_{Cl}=1.6$)	+	+	24,26
Aβ 1–40	10–2000	Cation ($P_K/P_{Cl}=1.8$)	+	N.D.	15,16
Aβ 1–40	50–4000	Cation ($P_K/P_{Cl}=11.1$)	+	N.D.	15,16
Aβ 1–42	10–2000	Cation ($P_K/P_{Cl}=1.8$)	+	+	15,16,22
CT105 (C-terminus fragment of amyloid precursor protein (APP))	120	Cation	+	+	18
PrP 106–126	10–400	Cation ($P_K/P_{Cl}=2.5$)	+	+	49
PrP 106–126	140,900, 1444	Cation ($P_K/P_{Cl}>10$)	N.D.	N.D.	52
PrP 82–146		Cation (variable)	N.D.	N.D.	55
Polyglutamine	19–220	Nonselective	−	−	73
Polyglutamine	17	Cation	N.D.	N.D.	74
NAC (α-synuclein 65–95)	10–300	Variable	+	+	63
α-Synuclein	50–1300	Cation	N.D.	N.D.	65
α-Synuclein	50–250	Cation	N.D.	N.D.	66

N.D.: not determined.

of blockade of almost all the amyloid channels. Amyloid pores also share responses to zinc blockade, which is voltage independent and partially reflects the heterodisperse nature of the amyloid conductance. The block is one-sided and reversible. The sharing of these features among a wide range of peptides with different primary sequences, different native conformations, different physiologic functions, and different organismal sources strongly suggests that formation of pores by amyloid peptide is a final common pathway, reflecting the specific physical chemistry and the β-sheet nature of these peptides.

In all studies done to date, blockade of amyloid peptide channels is protective. These studies have been limited to cell culture so far, but they are further evidence that the channel mechanism is a mechanism of toxicity of amyloid peptides. While the observation of protection of cells from cytotoxicity by specific blockers of amyloid peptides is highly suggestive, the final proof will require testing in animal models and, ultimately, in humans.

The failure of several anti-amyloid drugs in human clinical trials recently has disappointed many and caused some to question the viability of the amyloid hypothesis. Because inhibiting amyloid aggregation and removing amyloid from the brain do not seem to improve patient outcomes in Alzheimer's disease, some feel that the entire amyloid hypothesis is weak and may not be worth pursuing. However, the channel hypothesis suggests a number of reasons why attempts to remove amyloid from the brain might not succeed. For one thing, experiments with amyloid channels suggest that amyloid inserted into membranes is very hard to remove. Thus, standard techniques using binding or antibodies may not substantially remove amyloid channels inserted in membranes, even while they remove soluble amyloid. Thus, techniques that remove amyloid may not remove the amyloid channels which are responsible for the pathology. Second, inhibitors of aggregation may also fail to inhibit or disaggregate amyloid channels that are already inserted in the membrane. The inhibitors may not be able to penetrate the hydrophobic environment of the membrane, or they may simply not be able to disrupt the very strong association between amyloid monomers, which occurs in the membrane environment. Thus, the amyloid pore hypothesis provides some possible explanation of the recent failures of anti-amyloid drugs in clinical trials.

Blockers with high specificity for amyloid pores and methods to deliver these blockers *in vivo* to animals and humans must be developed if the pore hypothesis is to be tested rigorously.

ACKNOWLEDGMENTS

B. L. K. was supported by a grant from the Alzheimer's Association. We are grateful to Ms. Doris N. Finck for editorial assistance.

REFERENCES

1. Sipe JD, Cohen AS. Review: history of the amyloid fibril. *J Struct Biol* 2000;**130**:88–98.
2. Pike CJ, Burdick D, Walencewicz AJ, Glabe CG, Cotman CW. Neurodegeneration induced by beta-amyloid peptides in vitro: the role of peptide assembly state. *J Neurosci* 1993;**13**:1676–87.
3. Yankner BA, Dawes LR, Fisher S, Villa-Komaroff L, Oster-Granite ML, Neve RL. Neurotoxicity of a fragment of the amyloid precursor associated with Alzheimer's disease. *Science* 1989;**245**(4916):417–20.
4. Reed MN, Hofmeister JJ, Jungbauer L, Welzel AT, Yu C, Sherman MA, et al. Cognitive effects of cell-derived and synthetically derived Aβ oligomers. *Neurobiol Aging* 2011;**32**(10):1784–94 [Epub 2010 Jan 19].
5. Hardy JA, Higgins GA. Alzheimer's disease: the amyloid cascade hypothesis. *Science* 1992;**256**:184–5.
6. Forloni G, Angeretti N, Chiesa R, Monzani E, Salmona M, Bugiani O, et al. Neurotoxicity of a prion protein fragment. *Nature* 1993;**362**(6420):543–6.
7. Xu J, Kao SY, Lee FJ, Song W, Jin LW, Yankner BA. Dopamine-dependent neurotoxicity of alpha-synuclein: a mechanism for selective neurodegeneration in Parkinson disease. *Nat Med* 2002;**8**(6):600–6.
8. Harper JD, Lansbury Jr. PT. Models of amyloid seeding in Alzheimer's disease and scrapie: mechanistic truths and physiological consequences of the time-dependent solubility of amyloid proteins. *Annu Rev Biochem* 1997;**66**:385–407.
9. Hirakura Y, Satoh Y, Hirashima N, Suzuki T, Kagan BL, Kirino Y. Membrane perturbation by the neurotoxic Alzheimer amyloid fragment beta 25-35 requires aggregation and beta-sheet formation. *Biochem Mol Biol Int* 1998;**46**(4):787–94.
10. Arrasate M, Mitra S, Schweitzer ES, Segal MR, Finkbeiner S. Inclusion body formation reduces levels of mutant huntingtin and the risk of neuronal death. *Nature* 2004;**431** (7010):805–10.
11. King DL, Arendash GW, Crawford F, Sterk T, Menendez J, Mullan MJ. Progressive and gender-dependent cognitive impairment in the APP(SW) transgenic mouse model for Alzheimer's disease. *Behav Brain Res* 1999;**103**(2):145–62.
12. Hillen H, Barghorn S, Striebinger A, Labkovsky B, Müller R, Nimmrich V, et al. Generation and therapeutic efficacy of highly oligomer-specific beta-amyloid antibodies. *J Neurosci* 2010;**30**(31):10369–79.
13. Kagan BL, Thundimadathil J. Amyloid peptide pores and the beta sheet conformation. *Adv Exp Med Biol* 2010;**677**:150–67.
14. Fernandez A, Berry RS. Proteins with H-bond packing defects are highly interactive with lipid bilayers: implications for amyloidogenesis. *Proc Natl Acad Sci USA* 2003;**100**:2391–6.
15. Arispe N, Rojas E, Pollard HB. Alzheimer disease amyloid beta protein forms calcium channels in bilayer membranes: blockade by tromethamine and aluminum. *Proc Natl Acad Sci USA* 1993;**90**(2):567–71.
16. Arispe N, Pollard HB, Rojas E. Giant multilevel cation channels formed by Alzheimer disease amyloid beta-protein [A beta P-(1-40)] in bilayer membranes. *Proc Natl Acad Sci USA* 1993;**90**(22):10573–7.
17. Caughey B, Lansbury Jr. PT. Protofibrils, pores, fibrils, and neurodegeneration: separating the responsible protein aggregates from the innocent bystanders. *Annu Rev Neurosci* 2003;**26**: 267–98.
18. Fraser SP, Suh YH, Djamgoz MB. Ionic effects of the Alzheimer's disease beta-amyloid precursor protein and its metabolic fragments. *Trends J Neurosci* 1997;**20**:67–72.

19. Chen QS, Kagan BL, Hirakura Y, Xie CW. Impairment of hippocampal long-term potentiation by Alzheimer amyloid beta-peptides. *J Neurosci Res* 2000;**60**:65–72.

20. Walsh DM, Klyubin I, Fadeeva JV, Cullen WK, Anwyl R, Wolfe MS, et al. Naturally secreted oligomers of amyloid beta protein potently inhibit hippocampal long-term potentiation in vivo. *Nature* 2002;**416**:535–9.

21. O'Nuallain B, Freir DB, Nicoll AJ, Risse E, Ferguson N, Herron CE, et al. Amyloid beta-protein dimers rapidly form stable synaptotoxic protofibrils. *J Neurosci* 2010,**30**(43).14411–9.

22. Hirakura Y, Lin MC, Kagan BL. Alzheimer amyloid abeta1-42 channels: effects of solvent, pH, and Congo red. *J Neurosci Res* 1999;**57**:458–66.

23. Ingrosso L, Ladogana A, Pocchiar IM. Congo red prolongs the incubation period in scrapie-infected hamsters. *J Virol* 1995;**69**:506–8.

24. Mirzabekov T, Lin MC, Yuan WL, Marshall PJ, Carman M, Tomaselli K, et al. Channel formation in planar lipid bilayers by a neurotoxic fragment of the beta-amyloid peptide. *Biochem Biophys Res Commun* 1994;**202**:1142–8.

25. Lin MC. *Channel formation by amyloidogenic neurotoxic and neurodegenerative disease related peptides.* 1996 Ph.D. dissertation, Division of Neuroscience, UCLA.

26. Lin MC, Kagan BL. Electrophysiologic properties of channels induced by Abeta25-35 in planar lipid bilayers. *Peptides* 2002;**23**(7):1215–28.

27. Song L, Hobaugh MR, Shustak C, Cheley S, Bayley H, Gouaux JE. Structure of staphylococcal alpha-hemolysin, a heptameric transmembrane pore. *Science* 1996;**274**:1859–66.

28. Petosa C, Collier RJ, Klimpel KR, Leppla SH, Liddington RC. Crystal structure of the anthrax toxin protective antigen. *Nature* 1997;**385**:833–8.

29. Qi JS, Qiao JT. Suppression of large conductance Ca2+-activated K+ channels by amyloid beta-protein fragment 31-35 in membrane patches excised from hippocampal neurons. *Sheng Li Xue Bao* 2001;**53**:198–204.

30. Furukawa K, Abe Y, Akaike N. Amyloid beta protein-induced irreversible current in rat cortical neurones. *Neuroreport* 1994;**5**(16):2016–8.

31. Weiss JH, Pike CJ, Cotman CW. Ca2+ channel blockers attenuate beta-amyloid peptide toxicity to cortical neurons in culture. *J Neurochem* 1994;**62**:372–5.

32. Sanderson KL, Butler L, Ingram VM. Aggregates of a beta-amyloid peptide are required to induce calcium currents in neuron-like human teratocarcinoma cells: relation to Alzheimer's disease. *Brain Res* 1997;**744**(1):7–14.

33. Kawahara M, Arispe N, Kuroda Y, Rojas E. Alzheimer's disease amyloid beta-protein forms Zn (2+)-sensitive, cation-selective channels across excised membrane patches from hypothalamic neurons. *Biophys J* 1997;**73**:67–75.

34. Lin H, Zhu YJ, Lal R. Amyloid beta protein (1-40) forms calcium-permeable, Zn2+-sensitive channel in reconstituted lipid vesicles. *Biochemistry* 1999;**38**:11189–96.

35. Rhee SK, Quist AP, Lal R. Amyloid beta protein-(1-42) forms calcium-permeable, Zn2+-sensitive channel. *J Biol Chem* 1998;**273**:13379–82.

36. Zhu YJ, Lin H, Lal R. Fresh and nonfibrillar amyloid beta protein(1-40) induces rapid cellular degeneration in aged human fibroblasts: evidence for AbetaP-channel-mediated cellular toxicity. *FASEB J* 2000;**14**(9):1244–54.

37. Arispe N, Doh M. Plasma membrane cholesterol controls the cytotoxicity of Alzheimer's disease AbetaP (1-40) and (1-42) peptides. *FASEB J* 2002;**16**:1526–36.

38. Diaz JC, Simakova O, Jacobson KA, Arispe N, Pollard HB. Small molecule blockers of the Alzheimer Abeta calcium channel potently protect neurons from Abeta cytotoxicity. *Proc Natl Acad Sci USA* 2009;**106**(9):3348–53.

39. Kim HS, Lee JH, Lee JP, Kim EM, Chang KA, Park CH, et al. Amyloid beta peptide induces cytochrome C release from isolated mitochondria. *Neuroreport* 2002;**13**:1989–93.

40. Liu D, Pitta M, Lee JH, Ray B, Lahiri D, Furukawa K, et al. The KATP channel activator diazoxide ameliorates amyloid-β and tau pathologies and improves memory in the 3xTgAD mouse model of Alzheimer's disease. *J Alzheimers Dis* 2010;**22**:443–57 [Epub ahead of print].

41. Anekonda TS, Quinn JF, Harris C, Frahler K, Wadsworth TL, Woltier RL. L-type voltage-gated calcium channel blockade with isradipine as a therapeutic strategy for Alzheimer's disease. *Neurobiol Dis* 2010;**41**:62–70 [Epub ahead of print].

42. Fraser SP, Suh YH, Chong YH, Djamgoz MB. Membrane currents induced in Xenopus oocytes by the C-terminal fragment of the amyloid protein (APP). *J Neurochem* 1996;**66** (5):2034–40.

43. Kourie JI, Henry CL, Farrelly P. Diversity of amyloid beta protein fragment [1-40]-formed channels. *Cell Mol Neurobiol* 2001;**21**(3):255–84.

44. Jang H, Arce FT, Capone R, Ramachandran S, Lal R, Nussinov R. Misfolded amyloid ion channels present mobile beta-sheet subunits in contrast to conventional ion channels. *Biophys J* 2009;**97**(11):3029–37.

45. Jang H, Teran AF, Ramachandran S, Capone R, Lal R, Nussinov R. Structural convergence among diverse, toxic beta-sheet ion channels. *J Phys Chem B* 2010;**114**(29):9445–51.

46. Cobb NJ, Surewicz WK. Prion diseases and their biochemical mechanisms. *Biochemistry* 2009;**48**(12):2574–85.

47. Pan KM, Baldwin M, Nguyen J, Gasset M, Serban A, Groth D, et al. Conversion of alpha-helices into beta-sheets features in the formation of the scrapie prion proteins. *Proc Natl Acad Sci USA* 1993;**90**:10962–6.

48. Brundin P, Melki R, Kopito R. Prionlike transmission of protein aggregates in neurodegenerative diseases. *Nat Rev Mol Cell Biol* 2010;**11**(4):301–7.

49. Lin MC, Mirzabekov T, Kagan BL. Channel formation by a neurotoxic prion protein fragment. *J Biol Chem* 1997;**272**:44–7.

50. De Gioia L, Selvaggin IC, Ghibaudi E, Diomede L, Bugiani O, Forloni G, et al. Conformational polymorphism of the amyloidogenic and neurotoxic peptide homologous to residues 106-126 of the prion protein. *J Biol Chem* 1994;**269**(11):7859–62.

51. Manunta M, Kunz B, Sandmeier E, Christen P, Schindler H. Reported channel formation by prion protein fragment 106-126 in planar lipid bilayers cannot be reproduced [letter]. *FEBS Lett* 2000;**474**(2–3):255–6.

52. Kourie JI, Culverson A. Prion peptide fragment PrP[106-126] forms distinct cation channel types. *J Neurosci Res* 2000;**62**:120–33.

53. Farrelly PV, Kenna BL, Laohachai KL, Bahadi R, Salmona M, Forloni G, et al. Quinacrine blocks PrP (106–126)-formed channels. *J Neurosci Res* 2003;**74**(6):934–41.

54. Kourie JI, Kenna BL, Tew D, Jobling MF, Curtain CC, Masters CL, et al. Copper modulation of ion channels of PrP[106-126] mutant prion peptide fragments. *J Membr Biol* 2003;**193** (1):35–45.

55. Bahadi R, Farrelly PV, Kenna BL, Kourie JI, Tagliavini F, Forloni G, et al. Channels formed with a mutant prion protein PrP(82-146) homologous to a 7-kDa fragment in diseased brain of GSS patients. *Am J Physiol Cell Physiol* 2003;**285**(4):C862–72.

56. Hegde RS, Mastrianni JA, Scott MR, DeFea KA, Tremblay P, Torchia M, et al. A transmembrane form of the prion protein in neurodegenerative disease. *Science* 1998;**279**:827–34.

57. Solomon IH, Huettner JE, Harris DA. Neurotoxic mutants of the prion protein induce spontaneous ionic currents in cultured cells. *J Biol Chem* 2010;**285**(34):26719–26.

58. Alier K, Li Z, Mactavish D, Westaway D, Jhamandas JH. Ionic mechanisms of action of prion protein fragment PrP(106-126) in rat basal forebrain neurons. *J Neurosci Res* 2010;**88** (10):2217–27.

59. Henriques ST, Pattenden LK, Aguilar MI, Castanho MA. PrP(106-126) does not interact with membranes under physiological conditions. *Biophys J* 2008;**95**:1877–89.

60. Berest V, Rutkowski M, Rolka K, Łegowska A, Debska G, Stepkowski D, et al. The prion peptide forms ion channels in planar lipid bilayers. *Cell Mol Biol Lett* 2003;**8**(2):353–62.
61. Bekris LM, Mata IF, Zabetian CP. The genetics of Parkinson disease. *J Geriatr Psychiatry Neurol* 2010;**23**(4):228–42 Epub 2010 Oct 11.
62. Volles MJ, Lansbury Jr. PT. Vesicle permeabilization by protofibrillar alpha-synuclein is sensitive to Parkinson's disease-linked mutations and occurs by a pore-like mechanism. *Biochemistry* 2002;**11**(11):4595 602.
63. Azimova RK, Kagan BL. Ion channels formed by a fragment of alpha-synuclein (NAC) in lipid membranes. *Biophys J* 2003;**84**(2):53a.
64. Lashuel HA, Petre BM, Wall J, Simon M, Nowak RJ, Walz T, et al. Alpha-synuclein, especially the Parkinson's disease-associated mutants, forms pore-like annular and tubular protofibrils. *J Mol Biol* 2002;**322**(5):1089–102.
65. Zakharov SD, Hulleman JD, Dutseva EA, Antonenko YN, Rochet JC, Cramer WA. Helical alpha-synuclein forms highly conductive ion channels. *Biochemistry* 2007;**46**:14369–79.
66. Kostka M, Högen T, Danzer KM, Levin J, Habeck M, Wirth A, et al. Single particle characterization of iron-induced pore-forming alpha-synuclein oligomers. *J Biol Chem* 2008;**283**(16):10992–1003.
67. Di Pasquale E, Fantini J, Chahinian H, Maresca M, Taïeb N, Yahi N. Altered ion channel formation by the Parkinson's-disease-linked E46K mutant of alpha-synuclein is corrected by GM3 but not by GM1 gangliosides. *J Mol Biol* 2010;**397**(1):202–18.
68. Kim HY, Cho MK, Kumar A, Maier E, Siebenhaar C, Becker S, et al. Structural properties of pore-forming oligomers of alpha-synuclein. *J Am Chem Soc* 2009;**131**(47):17482–9.
69. Winner B, Jappelli R, Maji SK, Desplats PA, Boyer L, Aigner S, et al. *In vivo* demonstration that alpha-synuclein oligomers are toxic. *Proc Natl Acad Sci USA* 2011;**108**(10):4194–9 [Epub 2011 Feb 15].
70. Furukawa K, Matsuzaki-Kobayashi M, Hasegawa T, Kikuchi A, Sugeno N, Itoyama Y, et al. Plasma membrane ion permeability induced by mutant alpha-synuclein contributes to the degeneration of neural cells. *J Neurochem* 2006;**97**:1071–7.
71. Walker FO. Huntington's disease. *Lancet* 2007;**369**(9557):218–28.
72. Li SH, Li XJ. Huntingtin and its role in neuronal degeneration. *Neuroscientist* 2004;**10**:467–75.
73. Hirakura Y, Azimov R, Azimova R, Kagan BL. Polyglutamine-induced ion channels: a possible mechanism for the neurotoxicity of Huntington and other CAG repeat diseases. *J Neurosci Res* 2000;**60**:490–4.
74. Monoi H, Futaki S, Kugimiya S, Minakata H, Yoshihara K. Poly-L-glutamine forms cation channels: relevance to the pathogenesis of the polyglutamine diseases. *Biophys J* 2000;**78**(6):2892–9.
75. Panov AV, Gutekunst CA, Leavitt BR, Hayden MR, Burke JR, Strittmatter WJ, et al. Early mitochondrial calcium defects in Huntington's disease are a direct effect of polyglutamines. *Nat Neurosci* 2002;**5**:731–6.
76. Chattopadhyay M, Valentine JS. Aggregation of copper-zinc superoxide dismutase in familial and sporadic ALS. *Antioxid Redox Signal* 2009;**11**(7):1603–14.
77. Chung J, Yang H, de Beus MD, Ryu CY, Cho K, Colón W. Cu/Zn superoxide dismutase can form pore-like structures. *Biochem Biophys Res Commun* 2003;**312**(4):873–6.
78. Vucic S, Nicholson GA, Kiernan MC. Cortical excitability in hereditary motor neuronopathy with pyramidal signs: comparison with ALS. *J Neurol Neurosurg Psychiatry* 2010;**81**(1):97–100.
79. Meehan CF, Moldovan M, Marklund SL, Graffmo KS, Nielsen JB, Hultborn H. Intrinsic properties of lumbar motoneurones in the adult G127insTGGG superoxide dismutase-1

mutant mouse in vivo: evidence for increased persistent inward currents. *Acta Physiol (Oxf)* 2010;**200**:361–76. doi:10.1111/j.1748-1716.2010.02188.x [Epub ahead of print].

80. Pieri M, Carunchio I, Curcio L, Mercuri NB, Zona C. Increased persistent sodium current determines cortical hyperexcitability in a genetic model of amyotrophic lateral sclerosis. *Exp Neurol* 2009;**215**(2):368–79.

81. Israelson A, Arbel N, Da Cruz S, Ilieva H, Yamanaka K, Shoshan-Barmatz V, et al. Misfolded mutant SOD1 directly inhibits VDAC1 conductance in a mouse model of inherited ALS. *Neuron* 2010;**67**(4):575–87.

82. Schein SJ, Kagan BL, Finkelstein A. Colicin K acts by forming voltage-dependent channels in phospholipid bilayer membranes. *Nature* 1978;**276**:159–63.

83. Ng AW, Wasan KM, Lopez-Berestein G. Development of liposomal polyene antibiotics: an historical perspective. *J Pharm Pharm Sci* 2003;**6**(1):67–83.

84. Ramachandran R, Tweten RK, Johnson AE. Membrane-dependent conformational changes initiate cholesterol-dependent cytolysin oligomerization and intersubunit beta-strand alignment. *Nat Struct Mol Biol* 2004;**11**(8):697–705.

85. Cowan SW, Schirmer T, Rummel G, Steiert M, Ghosh R, Pauptit RA, et al. Crystal structures explain functional properties of two E coli porins. *Nature* 1992;**358**(6389):727–33.

86. Zeth K. Structure and evolution of mitochondrial outer membrane proteins of beta-barrel topology. *Biochim Biophys Acta* 2010;**1797**(6–7):1292–9 [Epub 2010 May 5].

87. Lazebnik Y. Why do regulators of apoptosis look like bacterial toxins? *Curr Biol* 2001;**11**(19): R767–8.

88. Parks JK, Smith TS, Trimmer PA, Bennett Jr. JP, Parker Jr. WD. Neurotoxic Abeta peptides increase oxidative stress in vivo through NMDA-receptor and nitric-oxide-synthase mechanisms, and inhibit complex IV activity and induce a mitochondrial permeability transition in vitro. *J Neurochem* 2001;**76**:1050–6.

89. Abramov AY, Canevari L, Duchen MR. Beta-amyloid peptides induce mitochondrial dysfunction and oxidative stress in astrocytes and death of neurons through activation of NADPH oxidase. *J Neurosci* 2004;**24**(2):565–75.

90. Vetrivel KS, Thinakaran G. Amyloidogenic processing of beta-amyloid precursor protein in intracellular compartments. *Neurology* 2006;**66**(2 Suppl. 1):S69–73.

91. Durell SR, Guy HR, Arispe N, Rojas E, Pollard HB. Theoretical models of the ion channel structure of amyloid beta-protein. *Biophys J* 1994;**67**(6):2137–45.

92. Jang H, Zheng J, Lal R, Nussinov R. New structures help the modeling of toxic amyloidbeta ion channels. *Trends Biochem Sci* 2008;**33**:91–100.

93. Shafrir Y, Durell S, Arispe N, Guy HR. Models of membrane-bound Alzheimer's Abeta peptide assemblies. *Proteins* 2010;**78**:3473–87 [Epub ahead of print].

94. Quist A, Doudevski I, Lin H, Azimova R, Ng D, Frangione B, et al. Amyloid ion channels: a common structural link for protein-misfolding disease. *Proc Natl Acad Sci USA* 2005; **102**(30):10427–32.

95. Soscia SJ, Kirby JE, Washicosky KJ, Tucker SM, Ingelsson M, Hyman B, et al. The Alzheimer's disease-associated amyloid beta-protein is an antimicrobial peptide. *PLoS One* 2010;**5** (3):e9505.

96. Thundimadathil J, Roeske RW, Guo L. Effect of membrane mimicking environment on the conformation of a pore-forming (xSxG)6 peptide. *Biopolymers* 2006;**84**:317–28.

97. Kayed R, Sokolov Y, Edmonds B, McIntire TM, Milton SC, Hall JE, et al. Permeabilization of lipid bilayers is a common conformation-dependent activity of soluble amyloid oligomers in protein misfolding diseases. *J Biol Chem* 2004;**279**(45):46363–6.

98. Capone R, Quiroz FG, Prangkio P, Saluja I, Sauer AM, Bautista MR, et al. Amyloid-beta-induced ion flux in artificial lipid bilayers and neuronal cells: resolving a controversy. *Neurotox Res* 2009;**16**(1):1–13.

99. Inoue S. *In situ* Abeta pores in AD brain are cylindrical assembly of Abeta protofilaments. *Amyloid* 2008;**15**(4):223–33.

100. Kagan BL, Selsted ME, Ganz T, Lehrer RI. Antimicrobial defensin peptides form voltage-dependent ion-permeable channels in planar lipid bilayer membranes. *Proc Natl Acad Sci USA* 1990;**87**:210–4.

101. Sokolov Y, Mirzabekov T, Martin DW, Lehrer RI, Kagan BL. Membrane channel formation by antimicrobial protegrins. *Biochim Biophys Acta* 1999,**1420**(1–2):23–9.

Protein Quality Control in Neurodegenerative Disease

Jason E. Gestwicki[*] and
Dan Garza[†]

[*]Department of Pathology and the Life
Sciences Institute, University of Michigan,
Ann Arbor, Michigan, USA

[†]Proteostasis Therapeutics Inc., Cambridge,
Massachusetts, USA

The accumulation of misfolded proteins is a common feature of many neuro-degenerative diseases. These observations suggest a potential link between these disorders and protein quality control, a collection of cellular pathways that sense damage to proteins and facilitate their turnover. Consistent with this idea, activation of quality control components, such as molecular chaperones, has been shown to be protective in multiple neurodegenerative disease models. In addition, key studies have suggested that quality control deteriorates with age, further supporting a relationship between these processes. In this chapter, we discuss the evidence linking neurodegeneration to quality control and present the emerging models. We also speculate on why proper quality control is so difficult for certain proteins.

Copyright 2012, Elsevier Inc.
All rights reserved.
1877-1173/12 $35.00

A common feature of many neurodegenerative diseases is the accumulation of misfolded proteins. As discussed in the previous chapters, the predominant offending protein is distinct in each disease (e.g., polyglutamine-expanded huntingtin in Huntington's disease). These disease-associated proteins are prone to misfolding, accumulation, and aggregation. More specifically, oligo-mers of these misfolded proteins are thought to cause gain-of-function proteo-toxicity through a number of mechanisms.

These observations suggest that protein quality control might play a role in neurodegenerative disease.[1–4] Briefly, quality control is a term used to describe a collection of pathways responsible for monitoring protein integrity.[5–7] This system comprises multiple "arms," including the molecular chaperones, the unfolded protein response (UPR) system, the autophagy–lysozome pathway, and the ubiquitin–proteasome system. Together, these components assist in most stages of a protein's life cycle, including its synthesis, folding, assembly/disassembly into complexes, and its degradation and turnover. Through this process, the quality control system continuously monitors the integrity of protein folding, and if misfolding is detected, then triage decisions are made to limit potential proteotoxic damage. In this way, quality control helps main-tain proper protein homeostasis (proteostasis).

Based on these observations, an emerging model is that the cellular capacity for quality control is one important variable in determining the onset, severity, and progression of many neurodegenerative diseases. Why do specific proteins chronically misfold in these patients? Why are they not removed or detoxified? Does the capacity of quality control influence neurodegeneration?

I. Overview of Protein Quality Control

To better understand the links between neurodegenerative diseases, pro-tein folding, and quality control, we begin with an overview of the interrelated "arms" of the quality control system. Each of these topics is associated with an expansive and informative literature. We include here key references and recent reviews to provide the foundation for discussing how these systems impact neurodegenerative disease.

A. Molecular Chaperones

The molecular chaperones are a large family of highly conserved proteins that assist in quality control through their roles in protein folding and turnover. One major, ubiquitous class of molecular chaperones is the heat shock pro-teins,[8,9] so named because the expression of these proteins is sensitive to stress, such as elevated temperature and oxidative damage. However, most heat shock proteins are also expressed under normal conditions and their activities are critical for normal quality control. This property is highlighted by the fact that some of the abundant heat shock proteins make up 1–2% of the total cellular protein under resting conditions.

The heat shock proteins are named by their approximate molecular weights, with the common classes being Hsp100, Hsp90, Hsp70, Hsp60, Hsp40, and the small heat shock proteins. These classes share no apparent sequence or structural homology. Instead, they seem to occupy specific niches in protein quality control. For example, the Hsp100-type chaperones are AAA$^+$ ATPases that are thought to be primarily involved in protein disaggregation and the primary folding of select substrates.[10,11] Hsp90 has been shown to bind largely folded substrates, such as nuclear hormone receptors and some kinases, to protect them from degradation and assist their remodeling in the presence of ligands.[12] Hsp70 is a family of 70-kDa chaperones that are believed to be involved in primary folding and triage decisions, linking molecular chaperones to the autophagy and ubiquitin–proteasome systems.[13,14] Small heat shock proteins, such as Hsp27 and Hsp33, are ATP-independent chaperones that appear to have specific roles in blocking protein aggregation during stressful conditions.[15] Additionally, these individual classes of heat shock proteins often operate together. For example, Hsp70 and Hsp90 work together to regulate the activity and stability of hormone receptors[12] and they can "hand" substrates to each other.[16] Similarly, Hsp70 and the small heat shock protein Hsp27 bind a model misfolded protein, cystathionine synthase, and their relative availability dictates whether it will be folded or degraded.[17]

The various activities of the chaperones arise from their direct physical contacts with substrates. Consequently, the heat shock proteins are thought to have only loose substrate selectivity, binding rather promiscuously to exposed hydrophobic regions.[18,19] Consistent with this idea, many of the heat shock proteins have hydrophobic "patches" that act as cognate surfaces for binding to aberrantly misfolded proteins. Hsp70 family members contain a hydrophobic pocket capable of binding 8–10 consecutive, nonpolar amino acids in an extended conformation.[20] Other chaperones, such as members of the GroEL or Hsp100 class, assemble into large, homo-oligomeric structures with spacious central cavities. Substrates encapsulated in these cavities are exposed to dynamic, hydrophobic surfaces that facilitate folding or degradation.[11] These chaperones, including the ClpP and ClpX AAA$^+$ ATPases, have remarkable substrate promiscuity, with only a few peptide-like features required for efficient recognition.[21] Interestingly, other chaperones, such as trigger factor, engage their substrates via polar contact surfaces.[22] Trigger factor also binds its substrates using a relatively large surface, perhaps allowing recognition of many different nonconserved substrates by facilitating multipoint binding, thereby distributing the binding free energy. Consistent with this idea, other chaperones are enriched for intrinsically disordered regions, perhaps providing them with the flexibility to interact with large areas of variable cognate surfaces that possess different topologies.[23]

Anfinsen suggested that protein folding information is coded in the amino acid sequence.[24] In large part, folding proceeds through internalization of hydrophobic residues and formation of other favorable contacts, which make the folded state energetically favorable. This idea is commonly represented by a folding trajectory energy diagram, in which an ensemble of initial elongated structures proceeds through key intermediate states (or "saddle points") to eventually arrive at the low-energy folded structure (Fig. 1). Failure to proceed on this route, either through a thermodynamic or kinetic trap, results in stalled folding and accumulation of an intermediate. Because folding is driven, in part, by internalization of hydrophobic residues, these folding intermediates are often expected to have exposed hydrophobic regions. Thus, they pose a potential danger to the cell because they can be prone to energetically favorable self-association, especially in the crowded cytosol or ER lumen. To combat this possibility, multiple systems exist for assisting in protein folding in cells. For example, a chaperone apparatus facilitates cotranslational folding[25] and the surface topology of the ribosome seems to play a productive role in this process.[26] In addition, other members of the molecular chaperone system have specific roles during conditions of stress, such as thermal or osmotic shock. Environmental stressors have the potential to "reset" the folding landscape and create conditions that melt sensitive proteins. Thus, stress conditions require additional chaperone activities in the cytosol, ER lumen, and other

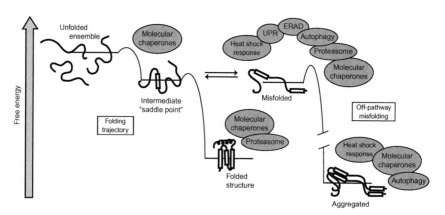

FIG. 1. Oversight of protein folding, misfolding, and turnover by cellular quality control. A simplified schematic of the energetically favorable folding trajectory is shown, starting with an ensemble of unfolded states, proceeding through a folding intermediate (saddle point), and continuing to the low-energy folded state. After folding, turnover is used to recover amino acids. Failure to proceed on the folding trajectory can lead to accumulation of off-pathway misfolded proteins, which can subsequently aggregate. As shown, the various components of the quality control network monitor each aspect of this process.

compartments. A family of stress-inducible chaperones appears to be particularly important in blocking denaturation, inhibiting aggregation, and favoring protein refolding.

In all these activities, molecular chaperones typically bind to hydrophobic regions, shielding exposed hydrophobic patches from bulk solvent and nonproductive interactions. Of course, protein:chaperone contacts also create a natural block to native protein folding because the chaperone-bound hydrophobic regions typically need to be buried in the final folded protein structure. Thus, it is important that protein–chaperone interactions be reversible and, for most chaperones, reversibility is driven by ATP binding and hydrolysis.[9] For example, electron microscopy studies have uncovered the dynamic, nucleotide-dependent cycling of Hsp90 structure.[27] Briefly, apoHsp90 exists in an open dimeric form, capable of loosely binding to substrates. Binding to ATP closes the dimer and increases substrate affinity, while hydrolysis of nucleotide triggers conformational changes that eventually lead to ADP and substrate release. This reversibility is critical because it allows the released substrates another chance to proceed down the folding trajectory. The classic example of this mechanism is the GroEL–GroES chaperone system, which traps misfolded proteins in its interior cavity and uses ATP hydrolysis to drive conformational changes that iteratively expose the substrate to relatively hydrophobic and hydrophilic environments, facilitating translocation, denaturation, and attempts at refolding.[11]

It should be emphasized that, although this discussion is largely focused on initial folding events and chaperone activities associated with stress, chaperones also have important activities on "post-folding" substrates and advanced-folding intermediates. For example, they assist in assembly/disassembly of protein complexes (e.g., clathrin), the transport of polypeptides across membranes (e.g., into the mitochondria), and in the remodeling of active sites during ligand engagement (e.g., nuclear hormone receptors). In addition, many chaperones, especially of the AAA^+ ATPase class, have disaggregation activity, permitting recovery of substrates from off-trajectory fates or storage forms.[10] In this case, ATP turnover can be used to actively "pull apart" aggregated proteins, facilitating their mobilization, recovery, or triage.[11] Thus, chaperone activity is required during many stages of the lifecycle of a protein, whenever hydrophobic regions could be inappropriately exposed.

The folded state of a protein can be partially described by a thermal melting curve, with an associated K_m value which indicates the stability of its protein fold. Each native wild-type protein will be associated with a specific K_m value and a ΔG of folding (Fig. 2). However, errors in translation produce unintended amino acid substitutions on the order of one per 100 amino acids.[164] In addition, this value does not include cumulative errors in mis-aminoacylation or DNA and RNA synthesis. Thus, most (or all) proteins will accumulate errors and a majority of these mutants are likely to adversely impact folding.

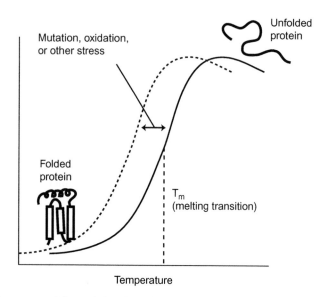

FIG. 2. Protein stability and the effects of mutation and stress. Mis-incorporation of amino acids, natural mutations, environmental stresses, and disease-associated conditions might create scenarios that destabilize proteins and favor their denaturation. As shown in Fig. 1, this process can favor accumulation of misfolded and aggregation-prone intermediates.

Chaperones protect against this problem by facilitating conditions that allow for robust folding across a range of K_m values. This situation could explain why chaperones are so abundant and why there are so few "free" chaperones. Essentially, chaperones are thought to buffer against intrinsic errors in protein synthesis, as well as changes in temperature or environment. Finally, this system creates a clever solution for living organisms because these same "errors" in protein folding are also essential for evolution. Accordingly, chaperones have been found to play an important role in molecular evolution. In a series of illuminating experiments, Rutherford and Lindquist found that mutations in Hsp90 allowed previously silent phenotypic variations to emerge in flies.[28] Interestingly, when these phenotypes were enriched, they eventually became independent of Hsp90, suggesting one possible mechanism by which chaperones produce conditions favorable for evolution of new traits.

B. Heat Shock Response

Although quality control is important for normal proteostasis, cells must also be responsive to acutely stressful situations, such as temperature variations, periods of rapid growth, high demands on secretion of extracellular factors, or oxidative injury. Under these conditions, the total folding and triage capacity of the cell might not match the needs under that particular set of environmental conditions, necessitating the upregulation of the heat shock proteins and other quality control components. This process is typically called the heat shock response (HSR) and it is controlled, in large part, by the transcription factor, heat shock factor 1 (HSF1). HSF1 is normally bound to Hsp90 in an inactive and monomeric state in resting cells.[29] Accumulation of misfolded proteins, which might be associated with a sudden change in the folding environment, leads to preferential binding to HSP90 and subsequent release of HSF1. HSF1 then forms a trimer and gets localized to the nucleus, where it activates transcription of heat shock proteins and many other pro-survival factors.[30,31] Thus, HSF1 is a critical component of quality control, acting both as a sensor and as a master transcriptional regulator. In addition, its activity appears to help coordinate stress responses in multicellular organisms.[32]

Although heat shock proteins are abundant in resting cells, HSF1 activation significantly elevates their numbers. For example, Hsp90 can comprise 1–2% of total cellular protein under resting conditions and as much as 10–20% after heat shock. These observations suggest that the HSF1-responsive transcriptional programs are highly tuned to deal with stress. In addition to its roles in stress, its activity appears to be developmentally regulated, with the quality control capacity of the cell/organism matched with the required mitotic or metabolic rate. Moreover, this capacity is likely adjusted to respond to specialized functions, such as high secretion rates in active pancreatic beta cells. Thus, the HSR, although important for response to stress, serves many purposes in normal physiology.

HSF1 activity appears to play a critical role in cancer and has emerged as a possible therapeutic target.[33] Many HSF1 targets, such as molecular chaperones and other quality control components, play important roles in apoptosis.[34,35] Typically, these roles are associated with pro-survival responsibilities that support cellular viability. For example, under pathophysiological conditions of unrestricted growth (e.g., cancer), high HSR activity is likely conducive to a strongly pro-folding and pro-survival environment. In turn, these conditions would be expected to support high rates of protein synthesis and net stabilization of oncogene products. However, it is also clear that quality control pathways can promote apoptosis, especially in the apoptosis "arm."[36] Thus, any discussion of the role of quality control in neurodegenerative disease should be tempered by a realization that high folding capacity is not a universally favorable goal. The key is proper balance among protein synthesis, folding, function and turnover.

C. Unfolded Protein Response

While the HSR occurs in answer to the accumulation of misfolded proteins in the cytosol, the UPR serves an analogous function in the ER.[37,38] This process involves at least three distinct signaling pathways, each regulated through activation of a membrane-bound receptor, namely, IRE1, PERK, and ATF-6, which monitor protein folding in the ER lumen. In resting cells, these receptors are bound to the ER-resident Hsp70 isoform, BiP or Grp78. In a mechanism conceptually similar to that discussed for HSF1, accumulation of misfolded proteins in the ER is thought to release the receptors from BiP, activating them and triggering a series of signaling cascades. The outcome of these pathways is that protein synthesis is slowed and the expression of quality control components, such as BiP and calnexin, is elevated. Thus, this system monitors folding capacity in the ER and adjusts it in response to flux through the organelle. In addition, the UPR increases the levels or activity of proteins needed for removing misfolded substrates by ER-associated degradation (ERAD)[7] and it initiates an apoptosis program if prolonged stress is detected.

Key targets of the UPR, and essential components of protein quality control in the ER lumen, are the protein disulfide isomerases (PDIs). These chaperones play specific roles in protein folding by facilitating the formation of proper disulfide linkages. They lower the barrier to disulfide interchange by forming covalent adducts with substrate proteins and, thus, assist in the search for the correct combination.[39,40] This step can often be rate-limiting for protein folding, creating a special requirement for this activity along the folding trajectory. Similarly, a family of peptidyl prolyl isomerases (PPIases) acts on another common "saddle point" event in protein folding, namely, isomerization of proline residues.[41] These activities are essential components of the UPR pathway, as they greatly speed the folding process and reduce the dwell time for potentially proteotoxic folding intermediates.

The lumen of the ER provides a number of specific challenges to protein folding, including its high protein content, strict requirements for cargo sorting, and the need for proteins to be first unfolded before they enter the space. Similar challenges are imposed in other organelles. However, less is known about quality control in some compartments. Extensive work has been performed on mitochondrial quality control, which provides an illustration of how folding and triage may be handled.[42,43] The major components of mitochondrial quality control include (a) proteases of the AAA$^+$ ATPase family (such as Lon protease), which degrade damaged and misfolded proteins; (b) ubiquitination and reverse transport of substrates to the cytosolic proteasome system; (c) organelle-specific chaperones; and (d) mitophagy or degradation of entire mitochondria. In addition, reactive oxygen species (ROS) pose a constant danger to the proteins, lipids, and nucleic acids of the mitochondrium, requiring a number of quality control components that focus on antioxidative activities (such as methionine sulfoxide reductases).[44] For example, the Lon protease seems to specifically recognize and degrade aconitase that has been oxidatively damaged.[45] Even with these measures, damaged proteins tend to accumulate in the aged mitochondria of some tissues, perhaps because of aging-related decreases in Lon activity.[46]

D. Autophagy–Lysosome

If a protein is targeted for triage by the quality control systems, cellular degradation pathways are used to recycle amino acids and avoid accumulation of potentially toxic aggregates. These pathways include macroautophagy, a process of enclosing cytoplasmic cargo in a double-membrane vesicle (the autophagosome) and delivering the contents to the lysosome.[47–49] Autophagy is commonly induced under nutrient-starvation conditions, allowing the cell to recover amino acids and other metabolites. It is initiated by phosphoinositide 3-kinase activity and an initiating factor, beclin (Atg6 in yeast). Through the activity of the ubiquitin-like protein, LC3, and other autophagy-related proteins, the autophagosome membrane assembles and is fused with the lysosome. This system can degrade entire organelles (e.g., damaged mitochondria) and substantial volumes of cytosol, including protein aggregates that are resistant to other degradation pathways. The other major autophagic pathway is called chaperone-mediated autophagy (CMA). This process presumably involves more discrete selection of misfolded substrates by chaperones. Using signals that are beginning to be understood, Hsp70 family members seem to recruit misfolded or damaged proteins to LAMP2 in the lysosomal membrane and facilitate their degradation. Recent reviews provide additional molecular details into these two autophagy pathways.[50–52]

Regardless of the entry pathway, cargo proteins are delivered to the lysosome for degradation. The lysosome is a membrane-bound organelle that sequesters a number of hydrolase activities (especially proteases, lipases, amylases, and nucleases) and allows local generation of a highly acidic compartment ($\sim$pH 4.5). Of particular relevance to neurodegenerative disease, proteotoxic aggregates are delivered to the lysosome by autophagy pathways.[53,54] This might be one of the few ways to remove large, preformed aggregates from the cell. Moreover, lysosome dysfunction is often critical in neurodegenerative disease etiology, suggesting that this compartment might be either overutilized or otherwise compromised.

As mentioned above, there must be a balance between pro-survival activities and unfettered cell division and growth. As concrete evidence of this idea, flux through the autophagy pathway is often elevated in cancer cells,[55] while failure to correctly induce autophagy appears to be linked with Huntington's disease.[56] This balance appears to be critical in disease. For example, autophagy in neurons initially appears to play a pro-survival role, while aging converts this process to a pro-death role.[57] Similarly, the increased autophagic flux in cancer cells allows rapid turnover and recycling in a system that is likely to produce more "mutant" proteins, but activation of autophagy is also pro-apoptotic beyond an ill-defined threshold. This balance appears to involve communication with other protein quality control pathways, such as the UPR.[57]

E. Ubiquitin–Proteasome

The proteasome is a large, cytoplasmic, cylindrical complex containing three distinct proteolytic activities. The most common form of the proteasome is the 26S particle, which is composed of a 20S core particle and two 19S "cap" particles. The 20S subunit contains a channel with chymotrypsin-like, trypsin-like, and peptidyl–glutamyl hydrolyzing activities. Together, these activities degrade substrate proteins that enter the 20S cavity. The 19S particles interact with the two ends of the 20S particle, acting as gatekeepers by specifically recognizing ubiquitin-modified substrates and other proteins tagged for destruction. In addition, these 19S particles also have ATP-dependent unfolding activity. This activity is important because proteins must be partially denatured to gain access to the relatively narrow ($\approx 50\,\text{Å}$) channel of the 20S particle. It has been hypothesized that larger aggregates, such as those commonly observed in neurodegenerative disease, might be resistant to the proteasome activity, in part because of their relative resistance to unfolding.[54] Although the proteasome is primarily located in the cytosol, potential substrates are also trafficked to this system from other compartments, including the ER, mitochondria, and plasma membrane.[58–60]

Substrates are tagged for degradation by either polyubiquitination or through ubiquitin-independent mechanisms.[61] It is not yet entirely clear how all substrates are selected for degradation by these pathways. However the N-end rule pathway provides some clues. The N-end rule describes an observation that the identity of the extreme N-terminal residue of a protein can dictate its relative stability in the cytosol.[62] In this pathway, proteins with the appropriate residues, such as Phe, Trp, and Leu, at their N-termini are relatively rapidly turned over via ubiquitin-dependent proteasomal clearance. However, other mechanisms also seem to contribute to determining stability, and a unifying theorem of turnover has not yet emerged. For example, ubiquitin-dependent mechanisms appear to also operate in autophagosome/lysosomal targeting. Further, the exact roles of the molecular chaperones in choosing substrates for degradation are not yet known. These questions will likely be of critical importance for understanding how quality control impacts neurodegenerative disease.

F. Summary of Protein Quality Control

Together, the protein quality control components (e.g., chaperones, UPR, autophagy, ERAD, proteasome) provide critical oversight by facilitating folding, reducing accumulation of proteotoxic substrates, and helping to maintain proteostasis (see Fig. 1). In neurodegenerative disease, misfolded substrates accumulate, suggesting a deficit in the capacity of quality control or some failure to properly triage toxic substrates. These observations raise intriguing questions about the roles of quality control pathways in neurodegeneration, proteostasis, longevity, and aging. How does chronic dysfunction of quality control impact disease progression? How do changes in one "arm" of quality control impact the entire system? Which components of quality control might serve as the best therapeutic targets for neurodegenerative diseases? Can these targets have influence over multiple types of neurodegenerative disease? In the next sections, we examine representative proteotoxic proteins whose accumulation is associated with specific neurodegenerative diseases. The intent is to provide a brief overview, while recent reviews (*vide infra*) can be consulted for more in-depth analysis.

II. Proteotoxic Proteins and Quality Control

A number of proteins are associated with specific neurodegenerative diseases. These proteins share common traits, such as proteotoxicity and the propensity to form aggregates *in vitro* and *in vivo*. Many of these proteins form amyloids, a specific type of protein aggregate that has features of high β-sheet content and reactivity with histological dyes, such as thioflavin T.[63] A few of these substrates are relatively newly characterized (e.g., TDP-43) and less is known about their relationships with protein quality control. Here, we focus on some of best characterized proteotoxic systems in which clear roles for protein quality control have been elucidated. Through these examples, we will outline some general themes that might be important for future consideration.

A. Polyglutamines

Some proteins, including huntingtin, androgen receptor, and ataxin-2, contain stretches of contiguous glutamine residues.[64] These polyglutamine regions are typically less than 35 residues in length. However, expansion of these regions to 40 or more residues makes them prone to aggregation.[65] Interestingly, polyQ-expanded proteins have been found to be relatively resistant to proteolysis and they are poor substrates for the proteasome, suggesting that they might avoid proper turnover,[54] although this conclusion is controversial.[66] In key studies, the Morimoto group has shown that accumulation of polyQ aggregates can affect cellular proteostasis, allowing misfolding of proteins that, in the absence of the polyQ protein, would be folded and stable.[67,68] Some of these aberrant interaction partners seem to include important transcription factors and chaperones.[69] In fact, a network of polyQ-interacting proteins appears to be required for toxicity in a yeast model.[70] Together, these studies suggest that aggregation of polyQ, its avoidance of proper degradation, and its ability to disrupt global proteostasis might contribute to disease pathogenesis.

In support of this idea, numerous genetic and biochemical studies have suggested that increasing the capacity of the protein quality control pathways can block polyQ-related pathology. For example, in yeast, worm, and fly models, polyQ-associated phenotypes are suppressed by concurrent overexpression of molecular chaperones, including Hsp70 and Hsp40.[71–77] The interactions between the polyQ and chaperones appear to be dependent on molecular features of the polyQ. Early studies established that the interactions with Hsp70 and Hsp40 were dependent on polyQ length.[78] Moreover, Hsp27 was found to specifically bind to the regions flanking the polyQ expansion to impact self-assembly and toxicity.[79,80] In one recent and interesting example, the Hsp70–Hsp40 complex was found to selectively associate with small aggregates of huntingtin polyQ, suggesting roles of these chaperones during the critical oligomeric stages of aggregation.[81] Consistent with these general ideas, knockdown of Hsp70 exacerbates polyQ phenotypes in flies[82] and mice.[83] In addition to the direct effects of chaperones on the ability of polyQ proteins to aggregate, these factors also can facilitate their degradation. For example, some Hsp40s, such as HSJ1a, increase ubiquitylation of polyQ in neuron models.[84] Together, these studies suggest that the levels of chaperones and the overall cellular folding capacity might influence the propensity for polyQ aggregation and its cellular impact. However, in most cases, the exact molecular mechanisms of protection are not yet clear. How do chaperones guide polyQ substrates to degradation pathways? What are the exact molecular features being recognized? Why does this system fail in diseased patients?

In addition to molecular chaperones, other components of the quality control pathways appear to impact the assembly and toxicity of polyQ fragments. For example, large polyQ aggregates are effectively cleared from the cytosol (but not in the nucleus) by autophagy.[85] Further, induction of an HSR, through either genetic or pharmacologic intervention, also mitigates polyQ assembly and phenotypes.[86–90] Bien and Bonini conducted genome-wide screens in flies to reveal both chaperone and ubiquitin pathways as major modifiers of toxicity of polyQ-expanded ataxin-3,[71] and in support of this idea, Zoghbi et al. performed yeast two-hybrid studies to implicate these pathways and others in ataxia.[91] Together, these studies have strongly suggested a link between polyQ-associated pathology and the capacity of the quality control network.

B. Microtubule-Binding Protein Tau

Accumulation of hyperphosphorylated tau is associated with a number of neurodegenerative disorders, including Alzheimer's disease (AD), Pick's disease, and frontotemporal dementia with Parkinsonism linked to chromosome

17 (FTDP17).[92,93] Moreover, tau has recently been implicated in mediating the proteotoxic effects of amyloid β-protein (Aβ) in models of AD.[94,95] Thus, there is increasing interest in understanding how tau is processed by cellular quality control systems.

As discussed in Chapter 8, tau is abundantly expressed in the brain, where it is composed of 11 exons that are alternatively spliced into at least six major isoforms. In addition, tau is extensively altered by posttranslational modifications, including phosphorylation and proteolytic processing.[96] Together, these modifications create a structurally diverse population of tau proteins in neurons, but the individual cellular activities and relevance of these isoforms are not yet clear. One common theme, however, is that the hyperphosphorylated and proteolyzed versions accumulate in AD and other neurodegenerative diseases.[97] These forms of tau have a tendency to aggregate, forming insoluble inclusions, termed neurofibrillary tangles (NFTs), and soluble oligomers that have been linked to gain-of-function proteotoxicity in disease.[98] Some of these posttranslational modifications appear to reduce tau's affinity for microtubules, which might further contribute to their tendency to self-associate.

Tau is an intrinsically disordered protein (IDP) and thus it presents many potential opportunities for interactions with chaperones.[23,99,100] Consistent with this idea, the stability and activity of tau appear to be regulated by both Hsp90 and Hsp70. Hsp70 is thought to bind directly to tau and the binding sites have been mapped in vitro.[101] Chemical inhibition of either of these chaperones leads to tau's rapid degradation by the proteosome.[102–104] Interestingly, it was recently reported that inhibition of Hsp70 in transgenic mice reduces soluble tau levels and improves cognition, suggesting that this chaperone plays an important role in both controlling tau stability and determining proteotoxicity.[105] However, neither Hsp70 nor Hsp90 works alone in this tau triage process, and many of their associated co-chaperones may also be involved.[106,107] Hsp27 also binds tau and regulates its stability.[108] Proteasomal degradation of tau appears to depend on the E3 ligase CHIP, which forms a complex with both Hsp70 and Hsp90.[102,109] The co-chaperones BAG-1, BAG-2, FKBP52, and PP5 also have been implicated in helping chaperones identify hyperphosphorylated tau.[110–113] PP5 is a particularly intriguing chaperone partner, as it is a protein phosphatase that has been shown to directly act on hyperphosphorylated tau in an Hsp90-directed fashion. In addition to chaperone-mediated clearance by the proteosome, tau also appears to be removed by macroautophagy and CMA, suggesting that multiple cellular pathways can act on this substrate.[114,115] Together, these observations suggest a model in which the cellular levels of tau and its isoforms are tightly regulated by a network of quality control systems. In this model, aging-related decreases in quality control capacity or mutational damage to tau might tip the balance toward accumulation of aggregation-prone and proteotoxic structures.

C. Amyloid β-Protein

Aβ is produced from the proteolytic cleavage of the amyloid precursor protein (APP), and as discussed in chapter Alzheimer's Disease and the Amyloid β-Protein, its deposition into fibrils is characteristic of AD. Like other neurodegenerative disease-related proteins, Aβ self-assembles into aggregates. *In vitro* and in cells, addition of molecular chaperones, including cytoplasmic and ER-resident Hsp70s, Hsp90, various Hsp40s, and small heat shock proteins, can block Aβ aggregation.[116] In worm models, Hsp70 and Hsp16.2 mitigate Aβ-related defects, further supporting a role for quality control.[117,118] Interestingly, the Good group found that multiple small heat shock proteins could block Aβ aggregation, but only Hsp20 was effective at suppressing toxicity.[119] In one recent report, treatment with heat shock appeared to suppress Aβ toxicity, suggesting that multiple quality control pathways might be involved.[120] However, most chaperones and quality control pathways operate inside the cell, whereas Aβ deposition is typically observed in extracellular regions of the brain. Thus, it is not yet clear how chaperones might impact AD. One possibility is that they impact APP processing in the secretory pathway[121] or that induction of a stress response provides general protection against apoptosis.[122]

D. α-Synuclein

As discussed in chapter Molecular Insights into Parkinson's Disease, Parkinson's disease (PD) is a complex set of disorders that involve defects in energy metabolism and are associated with aggregation of α-synuclein into Lewy bodies. Molecular chaperones, including Hsp70, Hsp27, and crystallin, have been found to be associated with α-synuclein in Lewy bodies.[123–125] Further, quantitative proteomics have revealed a number of other quality control components in these aggregates.[126] Finally, mutant forms of α-synuclein have shown impaired susceptibility to degradation by autophagy, suggesting that they might evade normal turnover.[127]

Similar to what has been observed with polyQ substrates, Hsp70 and Hsp90 seem to interact with α-synuclein during relatively early stages of self-assembly.[128,129] Moreover, overexpression of Hsp70 in mouse models also suppresses α-synuclein aggregation and toxicity.[130] The interaction between α-synuclein and Hsp70 occurs through the hydrophobic domain of α-synuclein, consistent with other chaperone substrates.[131] However, recent evidence from a mouse model of α-synucleinopathy suggests that overexpression of Hsp70 is not protective.[124,132] Even though the direct link to Hsp70 is found to be insufficient, other quality control and metabolic pathways appear to impact α-synuclein toxicity, as elucidated by genetic screens[133] and heat shock studies in a yeast model.[134] In addition, the small heat shock protein, Hsp27, seems to be a particularly effective anti-α-synuclein agent[135] and pharmacological activation of a stress response in flies and mice suppresses α-synuclein-associated pathologic effects.[136–138] Similar to what has been observed in other disease models, it seems likely that the effects of protein quality control on α-synuclein toxicity will involve a combination of direct effects on folding, turnover, and subcellular trafficking. Which of these pathways are most important in the pathogenesis of the disease is an important question for future studies.

E. Prions

Prions are a family of proteins found in mammals (chapter The Complex Molecular Biology of Amyotrophic Lateral Sclerosis (ALS)), yeast (chapter Tau and Tauopathies), and prokaryotes that exhibit at least two structural forms, one of which is able to propagate conversion of the other.[139] The conversion of prions to the infectious state proceeds with significant refolding and conformational changes in the prion fold, suggesting that chaperones and components of the quality control system may be associated with this process. Consistent with this idea, extensive work on prions has been done in yeast systems and a clear role for multiple molecular chaperones and quality control systems has emerged.[140] Briefly, *Saccharomyces cerevisiae* harbor a number of natural prions, including Sup35 and Pin1. Studies on these prions have converged on a model in which Hsp70, Hsp40, and Hsp26 are required for the initial "loosening" of prion fibrils, followed by the action of Hsp104 to fully disassemble them into shorter fragments.[141–143] This process appears to be required for propagation of the prion, as knockouts of Hsp104 and Hsp70s, including Ssa and Ssb family members, have altered transmission to daughter cells.[144] However, the relationships between chaperones and prion processing are complex and still being elucidated. For example, chaperones actively avoid formation of toxic intermediates during prion assembly and disassembly.[144,145] Certain Hsp40 proteins can detoxify Rnq1 by prompting formation of aggregates and limiting accumulation of toxic prefibrillar structures,[145] a result that is consistent with experiments performed in polyglutamine disease models.[80] At the same time, supraphysiological overexpression of Hsp70 and small heat shock proteins can suppress prion propagation,[146] presumably through enhanced degradation and altered balance between fibril assembly and disassembly.

Compared to the work performed in yeast, fewer studies have explored quality control in human or mammalian prion systems. Prions cause a number of neurodegenerative diseases in humans, including Creutzfeldt–Jakob disease (CJD). One interesting study has shown that mouse knockouts of HSF1 are more susceptible to toxicity by prions, suggesting broad roles for the HSR.[147] Because prions operate through a folding transition and their infectivity is dependent on a distinct folding state, it seems likely that chaperones will be found to have additional important activities in determining susceptibility to infection. For example, the chaperones GroEL and Hsp104 were found to promote conversion of a mammalian prion to its protease-resistant infectious state when the reactions were seeded with some of the disease-associated protein.[148] In addition, Hsc70 bind to prion protein (PrP), suggesting a potential role for this chaperone.[149] However, it should be noted that the effects of protein quality control components on directly balancing the folding, conformational transitions, turnover, and propagation of prions are likely supplemented by more global effects on antiapoptotic signaling pathways.

III. Reflections and Prospectus

From these observations and many others, it is clear that protein quality control plays a critical role in determining the fate of substrates whose accumulation is associated with neurodegeneration. Not surprisingly, these realizations have led to a second wave of interesting and provocative questions and, in many ways, the studies outlined in this chapter are only the beginning.

(1) Why are older individuals more prone to neurodegenerative diseases? Do protein quality control pathways fail over time? Accumulating evidence indicates that key proteostasis network pathways, including those involved in protein quality control, are compromised in older individuals. For example, the HSR is eroded as a function of aging,[6,88,150,151] suggesting that some arms of quality control become inoperative or less effective. Why might this be? Perhaps, epigenetic changes could lead to impaired HSR or UPR. For example, HSF1-binding sites in DNA appear to change with age.[152] In addition, accumulation of oxidative damage might overload quality control systems, making it easier for disease-related substrates to avoid detection. In this scenario, increased loads of proteotoxic proteins might increase the demand for quality control, which would be expected to increase the basal stress response and strain the ability to induce robust HSR. In turn, these chronic imbalances might decrease the efficiency of autophagy, induce lysosomal dysfunction, and reduce proteasome activity by shifting quality control responsibilities to these "arms" of the system. Despite this speculation, very little is known about the molecular mechanisms that link neurodegeneration to aging, and, similarly, our knowledge of the relationships between aging and protein quality control is still incomplete.

(2) How are certain disease-related substrates (e.g., tau, polyQ) able to avoid proper quality control? In some cases, the answer might be related to specific mutations that alter the molecular recognition between chaperones and the substrates. For example, the 42-amino acid version of Aβ is more prone to aggregation than the 40-amino acid form[153] and, concurrently, mutations that alter the ratio of Aβ42 to Aβ40 can impact disease (as

discussed in the chapter entitled Cerebral Amyloid Angiopathy). Another interesting case involves disease-associated mutations in the ER. Certain alleles of glucocerebrosidase (GCase) reduce the levels of functional enzyme by triggering premature degradation.[154] Interestingly, some of these GCase mutants do not appear to activate the UPR despite the fact that they are relatively defective for folding. Together, these observations suggest that GCase mutants may partially avoid normal quality control decisions, which leads to Gaucher disease.

In some cases, resistance of proteotoxic aggregates to normal clearance mechanisms might eventually lead to their accumulation. For example, some polyglutamine aggregates are thought to be relatively resistant to proteasomal degradation.[54] This scenario might lead to irreversible deposition, allowing the proteotoxic protein to accumulate unchecked. In turn, deposition of these aggregates has been shown to disrupt more global proteostasis by recruiting bystander proteins, especially intrinsically unfolded or metastable proteins.[155] Therefore, allowing substrates to evade degradation can have catastrophic consequences beyond just the intrinsic proteotoxicity of the aggregate.

We understand very little about how substrates are chosen for degradation, even under normal conditions. Thus, our ability to understand why some substrates accumulate in neurodegenerative disease is also limited. Perhaps these disease-related substrates share some unrecognized motif (e.g., high β-sheet content?) that is particularly troublesome for the quality control machinery. Perhaps the disease-related substrates are efficient at "hiding" their hydrophobic motifs, allowing them to appear normal and evade turnover. Do some proteins aggregate fast enough to evade quality control? For example, specific mutants of tau are especially prone to rapid aggregation and they are associated with FTDP17.[97] These types of questions will likely be of critical importance in the future of neurodegenerative disease research.

(3) Why are neurons sensitive to accumulation of misfolded proteins? Do different cells or tissues have different capacities for quality control? Do different cells have different proteotoxic protein loads? Do tissue-specific aging patterns alter the folding capacity of the brain? It seems likely that different tissues have distinct quality control capacities. For example, Hageman and Kampinga have demonstrated tissue- and organelle-specific differences in the expression levels of some molecular chaperones,[14,156] perhaps creating local differences in folding environments. In addition, postmitotic cells might be prone to accumulate some types of DNA damage as they age,[157–159] which could strain protein quality control systems. Still, these questions are under active study and it seems likely that multiple mechanisms will emerge.

(4) How do deficits in protein quality control contribute to neurodegenerative disease? It is thought that one central issue here is a balance between quality control and the proteotoxic load. In healthy individuals, balance appears to be maintained, with the capacity of the quality control systems capable of dealing with changes in protein loads. However, imbalance might be triggered by increases in the amount of misfolded protein combined with decreases in the capacity (e.g., perhaps aging-related?) of quality control systems. Interestingly, one would expect that the loss of balance would be different in different people (or tissues or cells) because they each have a different proteotoxic load, different accumulated amounts of DNA damage, and different levels of quality control activity. For example, some individuals may have a relatively "weak" oxidative stress response. These persons might be more susceptible to developing sporadic PD if they also carry other risk alleles or were exposed to high levels of environmental toxins. In this case, the capacity of quality control might be considered permissive for PD, but the same person with a "strong" oxidative stress response, all other things being equal, might not have developed disease.

Clearly, when pondering this question, it is critical to appreciate how the quality control components are linked within the proteostasis network.[32] For example, autophagy and proteasome pathways appear to be coregulated, allowing alternative shuttling of substrates into one or the other system.[160,161] Thus, a more useful way to view quality control with respect to neurodegenerative diseases is that pathology likely arises from global failures, not just problems in one "arm" (e.g., autophagy). In other words, a slow decline in total quality control capacity (defined as the additive effects of chaperones, HSR, UPR, autophagy, etc.) might lead to incremental accumulation of misfolded proteins. In turn, these initiating problems might nucleate the misfolding of disease-related targets and bystander proteins. In this scenario, proteostasis eventually collapses, causing rampant aggregation and, in theory, disease.

Another interesting way to view this question is that the quality control systems are not deteriorating *per se*, but that they might be attempting to "rescue" misfolded proteins because they are "misreading" the folding information. This well-intentioned attempt at rescuing a misfolded substrate might occur to the detriment of the health of the organism. In this case, mutations in the disease-related protein might evade even an intact system, potentially having relevance for early onset disease. Regardless of the mechanisms, it is likely that each of these possibilities occurs under some conditions and greater emphasis on integrated systems biology approaches might accelerate discovery in this area.

(5) Can protein quality control be a therapeutic target for neurodegenerative disease? One area of active investigation is the search for compounds that boost overall folding capacity through stimulating HSR or UPR pathways. For example, promising early stage efforts are focusing on activating HSF1.[4,162] Thus, one goal of these efforts is to counterbalance aging- or oxidation related decreases in folding capacity. Other efforts seek to control folding capacity by more specifically regulating the environment in specific organelles. For example, drugs directed to ER calcium channels can be used to increase calcium levels in the lumen, altering folding in this organelle and enhancing the levels of glycocerebrosidase mutants.[163] These studies suggest that a wide variety of potential drug targets exist in the broad realm of quality control pathways. The challenge will be how to put these networks together to predict off-target effects and identify the best (e.g., least toxic and most effective) targets for neurodegenerative disease.

One of the intellectually fascinating aspects of neurodegenerative diseases is that these disorders appear to be uniformly associated with some aspects of protein misfolding and accumulation. Thus, even though examples such as Alzheimer's and Huntington's diseases are clinically distinct and they manifest in different symptoms, they share some fundamental protein biochemistry. For example, fibrils formed by tau, Aβ, and all other amyloid proteins share a common core structure. These commonalities are exciting because they suggest related molecular mechanisms of formation. Further, as mentioned above, quality control needs to be somewhat nonselective in order to monitor the folding of an entire proteome. Can this same lack of selectivity be leveraged to impact AD, HD, PD, and other neurodegenerative disorders using a shared strategy? Clearly, more work is needed to answer these questions, but the basic biology and preliminary translational studies have provided tantalizing clues. For example, some evidence suggests that chemical targeting of chaperones can reduce protein aggregation in mouse models of both HD and AD.[88,105]

The theme in all of these early attempts at quality control-based drug discovery is to artificially boost or manipulate quality control. However, it is not yet clear whether specific substrates can be targeted for increased scrutiny. How do you get the chaperone system to reduce polyglutamine aggregation without impacting global proteostasis? What are the key drug targets to achieve selectivity? Addition insights and the answers to these questions are likely to emerge from ongoing studies into the basic biology of quality control and its relationships with aging and neurodegenerative disease.

References

1. Bonini NM. Chaperoning brain degeneration. *Proc Natl Acad Sci USA* 2002;**99**(Suppl. 4):16407–11.
2. Muchowski PJ. Protein misfolding, amyloid formation, and neurodegeneration: a critical role for molecular chaperones? *Neuron* 2002;**35**:9–12.
3. Muchowski PJ, Wacker JL. Modulation of neurodegeneration by molecular chaperones. *Nat Rev Neurosci* 2005;**6**:11–22.
4. Westerheide SD, Morimoto RI. Heat shock response modulators as therapeutic tools for diseases of protein conformation. *J Biol Chem* 2005;**280**:33097–100.
5. Balch WE, Morimoto RI, Dillin A, Kelly JW. Adapting proteostasis for disease intervention. *Science* 2008;**319**:916–9.
6. Douglas PM, Dillin A. Protein homeostasis and aging in neurodegeneration. *J Cell Biol* 2010;**190**:719–29.
7. Powers ET, Morimoto RI, Dillin A, Kelly JW, Balch WE. Biological and chemical approaches to diseases of proteostasis deficiency. *Annu Rev Biochem* 2009;**78**:959–91.
8. Bukau B, Weissman J, Horwich A. Molecular chaperones and protein quality control. *Cell* 2006;**125**:443–51.
9. Mayer MP. Gymnastics of molecular chaperones. *Mol Cell* 2010;**39**:321–31.
10. Liberek K, Lewandowska A, Zietkiewicz S. Chaperones in control of protein disaggregation. *EMBO J* 2008;**27**:328–35.
11. Sauer RT, Bolon DN, Burton BM, Burton RE, Flynn JM, Grant RA, et al. Sculpting the proteome with AAA(+) proteases and disassembly machines. *Cell* 2004;**119**:9–18.
12. Pratt WB, Morishima Y, Murphy M, Harrell M. Chaperoning of glucocorticoid receptors. *Handb Exp Pharmacol* 2006;**172**:111–38.
13. Hohfeld J, Cyr DM, Patterson C. From the cradle to the grave: molecular chaperones that may choose between folding and degradation. *EMBO Rep* 2001;**2**:885–90.
14. Hageman J, Kampinga HH. Computational analysis of the human HSPH/HSPA/DNAJ family and cloning of a human HSPH/HSPA/DNAJ expression library. *Cell Stress Chaperones* 2009;**14**:1–21.
15. McHaourab HS, Godar JA, Stewart PL. Structure and mechanism of protein stability sensors: chaperone activity of small heat shock proteins. *Biochemistry* 2009;**48**:3828–37.
16. Wegele H, Wandinger SK, Schmid AB, Reinstein J, Buchner J. Substrate transfer from the chaperone Hsp70 to Hsp90. *J Mol Biol* 2006;**356**:802–11.
17. Singh LR, Kruger WD. Functional rescue of mutant human cystathionine beta-synthase by manipulation of Hsp26 and Hsp70 levels in Saccharomyces cerevisiae. *J Biol Chem* 2009;**284**: 4238–45.
18. Hartl FU, Hayer-Hartl M. Converging concepts of protein folding in vitro and in vivo. *Nat Struct Mol Biol* 2009;**16**:574–81.
19. Rudiger S, Buchberger A, Bukau B. Interaction of Hsp70 chaperones with substrates. *Nat Struct Biol* 1997;**4**:342–9.
20. Pellecchia M, Montgomery DL, Stevens SY, Vander Kooi CW, Feng HP, Gierasch LM, et al. Structural insights into substrate binding by the molecular chaperone DnaK. *Nat Struct Biol* 2000;**7**:298–303.
21. Barkow SR, Levchenko I, Baker TA, Sauer RT. Polypeptide translocation by the AAA+ ClpXP protease machine. *Chem Biol* 2009;**16**:605–12.
22. Liu Q, Hendrickson WA. Insights into Hsp70 chaperone activity from a crystal structure of the yeast Hsp110 Sse1. *Cell* 2007;**131**:106–20.

23. Uversky VN. Flexible nets of malleable guardians: intrinsically disordered chaperones in neurodegenerative diseases. *Chem Rev* 2011;**111**:1134–66.

24. Anfinsen CB. Principles that govern the folding of protein chains. *Science* 1973;**181**:223–30.

25. Albanese V, Yam AY, Baughman J, Parnot C, Frydman J. Systems analyses reveal two chaperone networks with distinct functions in eukaryotic cells. *Cell* 2006;**124**:75–88.

26. Kramer G, Boehringer D, Ban N, Bukau B. The ribosome as a platform for co-translational processing, folding and targeting of newly synthesized proteins. *Nat Struct Mol Biol* 2009;**16**:589–97.

27. Cunningham CN, Krukenberg KA, Agard DA. Intra- and intermonomer interactions are required to synergistically facilitate ATP hydrolysis in Hsp90. *J Biol Chem* 2008;**283**:21170–8.

28. Rutherford SL, Lindquist S. Hsp90 as a capacitor for morphological evolution. *Nature* 1998;**396**:336–42.

29. Shi Y, Mosser DD, Morimoto RI. Molecular chaperones as HSF1-specific transcriptional repressors. *Genes Dev* 1998;**12**:654–66.

30. Anckar J, Sistonen L. Heat shock factor 1 as a coordinator of stress and developmental pathways. *Adv Exp Med Biol* 2007;**594**:78–88.

31. Voellmy R, Boellmann F. Chaperone regulation of the heat shock protein response. *Adv Exp Med Biol* 2007;**594**:89–99.

32. Prahlad V, Morimoto RI. Integrating the stress response: lessons for neurodegenerative diseases from C. elegans. *Trends Cell Biol* 2009;**19**:52–61.

33. Whitesell L, Lindquist S. Inhibiting the transcription factor HSF1 as an anticancer strategy. *Expert Opin Ther Targets* 2009;**13**:469–78.

34. Garrido C, Schmitt E, Cande C, Vahsen N, Parcellier A, Kroemer G. HSP27 and HSP70: potentially oncogenic apoptosis inhibitors. *Cell Cycle* 2003;**2**:579–84.

35. Mosser DD, Morimoto RI. Molecular chaperones and the stress of oncogenesis. *Oncogene* 2004;**23**:2907–18.

36. Maiuri MC, Zalckvar E, Kimchi A, Kroemer G. Self-eating and self-killing: crosstalk between autophagy and apoptosis. *Nat Rev Mol Cell Biol* 2007;**8**:741–52.

37. Brodsky JL. The protective and destructive roles played by molecular chaperones during ERAD (endoplasmic-reticulum-associated degradation). *Biochem J* 2007;**404**:353–63.

38. Zhang K, Kaufman RJ. The unfolded protein response: a stress signaling pathway critical for health and disease. *Neurology* 2006;**66**:S102–9.

39. Maattanen P, Gehring K, Bergeron JJ, Thomas DY. Protein quality control in the ER: the recognition of misfolded proteins. *Semin Cell Dev Biol* 2010;**21**:500–11.

40. Tu BP, Weissman JS. Oxidative protein folding in eukaryotes: mechanisms and consequences. *J Cell Biol* 2004;**164**:341–6.

41. Barik S. Immunophilins: for the love of proteins. *Cell Mol Life Sci* 2006;**63**:2889–900.

42. Germain D. Ubiquitin-dependent and -independent mitochondrial protein quality controls: implications in ageing and neurodegenerative diseases. *Mol Microbiol* 2008;**70**:1334–41.

43. Tatsuta T, Langer T. Quality control of mitochondria: protection against neurodegeneration and ageing. *EMBO J* 2008;**27**:306–14.

44. Friguet B, Bulteau AL, Petropoulos I. Mitochondrial protein quality control: implications in ageing. *Biotechnol J* 2008;**3**:757–64.

45. Bota DA, Davies KJ. Lon protease preferentially degrades oxidized mitochondrial aconitase by an ATP-stimulated mechanism. *Nat Cell Biol* 2002;**4**:674–80.

46. Lee CK, Klopp RG, Weindruch R, Prolla TA. Gene expression profile of aging and its retardation by caloric restriction. *Science* 1999;**285**:1390–3.

47. Menzies FM, Moreau K, Rubinsztein DC. Protein misfolding disorders and macroautophagy. *Curr Opin Cell Biol* 2010;**23**:190–7.

48. Rubinsztein DC, Gestwicki JE, Murphy LO, Klionsky DJ. Potential therapeutic applications of autophagy. *Nat Rev Drug Discov* 2007;**6**:304–12.
49. Yang Z, Klionsky DJ. Eaten alive: a history of macroautophagy. *Nat Cell Biol* 2010;**12**:814–22.
50. He C, Klionsky DJ. Regulation mechanisms and signaling pathways of autophagy. *Annu Rev Genet* 2009;**43**:67–93.
51. Koga H, Cuervo AM. Chaperone-mediated autophagy dysfunction in the pathogenesis of neurodegeneration. *Neurobiol Dis* 2010;**43**:29–37.
52. Mizushima N, Levine B, Cuervo AM, Klionsky DJ. Autophagy fights disease through cellular self-digestion. *Nature* 2008;**451**:1069–75.
53. Ravikumar B, Duden R, Rubinsztein DC. Aggregate-prone proteins with polyglutamine and polyalanine expansions are degraded by autophagy. *Hum Mol Genet* 2002;**11**:1107–17.
54. Waelter S, Boeddrich A, Lurz R, Scherzinger E, Lueder G, Lehrach H, et al. Accumulation of mutant huntingtin fragments in aggresome-like inclusion bodies as a result of insufficient protein degradation. *Mol Biol Cell* 2001;**12**:1393–407.
55. Rabinowitz JD, White E. Autophagy and metabolism. *Science* 2010;**330**:1344–8.
56. Martinez-Vicente M, Talloczy Z, Wong E, Tang G, Koga H, Kaushik S, et al. Cargo recognition failure is responsible for inefficient autophagy in Huntington's disease. *Nat Neurosci* 2010;**13**:567–76.
57. Pandey UB, Nie Z, Batlevi Y, McCray BA, Ritson GP, Nedelsky NB, et al. HDAC6 rescues neurodegeneration and provides an essential link between autophagy and the UPS. *Nature* 2007;**447**:859–63.
58. Gallastegui N, Groll M. The 26S proteasome: assembly and function of a destructive machine. *Trends Biochem Sci* 2010;**35**:634–42.
59. Glickman MH, Ciechanover A. The ubiquitin-proteasome proteolytic pathway: destruction for the sake of construction. *Physiol Rev* 2002;**82**:373–428.
60. Okiyoneda T, Barriere H, Bagdany M, Rabeh WM, Du K, Hohfeld J, et al. Peripheral protein quality control removes unfolded CFTR from the plasma membrane. *Science* 2010;**329**:805–10.
61. Deshaies RJ, Joazeiro CA. RING domain E3 ubiquitin ligases. *Annu Rev Biochem* 2009;**78**:399–434.
62. Varshavsky A. The N-end rule pathway of protein degradation. *Genes Cells* 1997;**2**:13–28.
63. Roychaudhuri R, Yang M, Hoshi MM, Teplow DB. Amyloid beta-protein assembly and Alzheimer disease. *J Biol Chem* 2009;**284**:4749–53.
64. Gatchel JR, Zoghbi HY. Diseases of unstable repeat expansion: mechanisms and common principles. *Nat Rev Genet* 2005;**6**:743–55.
65. Chen S, Ferrone FA, Wetzel R. Huntington's disease age-of-onset linked to polyglutamine aggregation nucleation. *Proc Natl Acad Sci USA* 2002;**99**:11884–9.
66. Bowman AB, Yoo SY, Dantuma NP, Zoghbi HY. Neuronal dysfunction in a polyglutamine disease model occurs in the absence of ubiquitin-proteasome system impairment and inversely correlates with the degree of nuclear inclusion formation. *Hum Mol Genet* 2005;**14**:679–91.
67. Ben-Zvi A, Miller EA, Morimoto RI. Collapse of proteostasis represents an early molecular event in Caenorhabditis elegans aging. *Proc Natl Acad Sci USA* 2009;**106**:14914–9.
68. Gidalevitz T, Ben-Zvi A, Ho KH, Brignull HR, Morimoto RI. Progressive disruption of cellular protein folding in models of polyglutamine diseases. *Science* 2006;**311**:1471–4.
69. Swayne LA, Braun JE. Aggregate-centered redistribution of proteins by mutant huntingtin. *Biochem Biophys Res Commun* 2007;**354**:39–44.
70. Duennwald ML, Jagadish S, Giorgini F, Muchowski PJ, Lindquist S. A network of protein interactions determines polyglutamine toxicity. *Proc Natl Acad Sci USA* 2006;**103**:11051–6.
71. Bilen J, Bonini NM. Genome-wide screen for modifiers of ataxin-3 neurodegeneration in Drosophila. *PLoS Genet* 2007;**3**:1950–64.

72. Cummings CJ, Sun Y, Opal P, Antalffy B, Mestril R, Orr HT, et al. Over-expression of inducible HSP70 chaperone suppresses neuropathology and improves motor function in SCA1 mice. *Hum Mol Genet* 2001;**10**:1511–8.

73. Krobitsch S, Lindquist S. Aggregation of huntingtin in yeast varies with the length of the polyglutamine expansion and the expression of chaperone proteins. *Proc Natl Acad Sci USA* 2000;**97**:1589–94.

74. Muchowski PJ, Schaffar G, Sittler A, Wanker EE, Hayer-Hartl MK, Hartl FU. Hsp70 and hsp40 chaperones can inhibit self-assembly of polyglutamine proteins into amyloid-like fibrils. *Proc Natl Acad Sci USA* 2000;**97**:7841–6.

75. Paul S. Polyglutamine-mediated neurodegeneration: use of chaperones as prevention strategy. *Biochemistry (Mosc)* 2007;**72**:359–66.

76. Wacker JL, Zareie MH, Fong H, Sarikaya M, Muchowski PJ. Hsp70 and Hsp40 attenuate formation of spherical and annular polyglutamine oligomers by partitioning monomer. *Nat Struct Mol Biol* 2004;**11**:1215–22.

77. Warrick JM, Chan HY, Gray-Board GL, Chai Y, Paulson HL, Bonini NM. Suppression of polyglutamine-mediated neurodegeneration in Drosophila by the molecular chaperone HSP70. *Nat Genet* 1999;**23**:425–8.

78. Jana NR, Tanaka M, Wang G, Nukina N. Polyglutamine length-dependent interaction of Hsp40 and Hsp70 family chaperones with truncated N-terminal huntingtin: their role in suppression of aggregation and cellular toxicity. *Hum Mol Genet* 2000;**9**:2009–18.

79. Robertson AL, Headey SJ, Saunders HM, Ecroyd H, Scanlon MJ, Carver JA, et al. Small heat-shock proteins interact with a flanking domain to suppress polyglutamine aggregation. *Proc Natl Acad Sci USA* 2010;**107**:10424–9.

80. Wyttenbach A, Sauvageot O, Carmichael J, Diaz-Latoud C, Arrigo AP, Rubinsztein DC. Heat shock protein 27 prevents cellular polyglutamine toxicity and suppresses the increase of reactive oxygen species caused by huntingtin. *Hum Mol Genet* 2002;**11**:1137–51.

81. Lotz GP, Legleiter J, Aron R, Mitchell EJ, Huang SY, Ng C, et al. Hsp70 and Hsp40 functionally interact with soluble mutant huntingtin oligomers in a classic ATP-dependent reaction cycle. *J Biol Chem* 2010;**285**:38183–93.

82. Gong WJ, Golic KG. Loss of Hsp70 in Drosophila is pleiotropic, with effects on thermotolerance, recovery from heat shock and neurodegeneration. *Genetics* 2006;**172**:275–86.

83. Wacker JL, Huang SY, Steele AD, Aron R, Lotz GP, Nguyen Q, et al. Loss of Hsp70 exacerbates pathogenesis but not levels of fibrillar aggregates in a mouse model of Huntington's disease. *J Neurosci* 2009;**29**:9104–14.

84. Howarth JL, Kelly S, Keasey MP, Glover CP, Lee YB, Mitrophanous K, et al. Hsp40 molecules that target to the ubiquitin-proteasome system decrease inclusion formation in models of polyglutamine disease. *Mol Ther* 2007;**15**:1100–5.

85. Iwata A, Christianson JC, Bucci M, Ellerby LM, Nukina N, Forno LS, et al. Increased susceptibility of cytoplasmic over nuclear polyglutamine aggregates to autophagic degradation. *Proc Natl Acad Sci USA* 2005;**102**:13135–40.

86. Fujikake N, Nagai Y, Popiel HA, Okamoto Y, Yamaguchi M, Toda T. Heat shock transcription factor 1-activating compounds suppress polyglutamine-induced neurodegeneration through induction of multiple molecular chaperones. *J Biol Chem* 2008;**283**:26188–97.

87. Fujimoto M, Takaki E, Hayashi T, Kitaura Y, Tanaka Y, Inouye S, et al. Active HSF1 significantly suppresses polyglutamine aggregate formation in cellular and mouse models. *J Biol Chem* 2005;**280**:34908–16.

88. Hay DG, Sathasivam K, Tobaben S, Stahl B, Marber M, Mestril R, et al. Progressive decrease in chaperone protein levels in a mouse model of Huntington's disease and induction of stress proteins as a therapeutic approach. *Hum Mol Genet* 2004;**13**:1389–405.

89. Katsuno M, Sang C, Adachi H, Minamiyama M, Waza M, Tanaka F, et al. Pharmacological induction of heat-shock proteins alleviates polyglutamine-mediated motor neuron disease. *Proc Natl Acad Sci USA* 2005;**102**:16801–6.

90. Waza M, Adachi H, Katsuno M, Minamiyama M, Sang C, Tanaka F, et al. 17-AAG, an Hsp90 inhibitor, ameliorates polyglutamine-mediated motor neuron degeneration. *Nat Med* 2005;**11**:1088–95.

91. Lim J, Hao T, Shaw C, Patel AJ, Szabo G, Rual JF, et al. A protein-protein interaction network for human inherited ataxias and disorders of Purkinje cell degeneration. *Cell* 2006;**125**:801–14.

92. Spires-Jones TL, Stoothoff WH, de Calignon A, Jones PB, Hyman BT. Tau pathophysiology in neurodegeneration: a tangled issue. *Trends Neurosci* 2009;**32**:150–9.

93. Wolfe MS. Tau mutations in neurodegenerative diseases. *J Biol Chem* 2009;**284**:6021–5.

94. Roberson ED, Scearce-Levie K, Palop JJ, Yan F, Cheng IH, Wu T, et al. Reducing endogenous tau ameliorates amyloid beta-induced deficits in an Alzheimer's disease mouse model. *Science* 2007;**316**:750–4.

95. Vossel KA, Zhang K, Brodbeck J, Daub AC, Sharma P, Finkbeiner S, et al. Tau reduction prevents Abeta-induced defects in axonal transport. *Science* 2010;**330**:198.

96. Hanger DP, Anderton BH, Noble W. Tau phosphorylation: the therapeutic challenge for neurodegenerative disease. *Trends Mol Med* 2009;**15**:112–9.

97. Brunden KR, Trojanowski JQ, Lee VM. Advances in tau-focused drug discovery for Alzheimer's disease and related tauopathies. *Nat Rev Drug Discov* 2009;**8**:783–93.

98. Maeda S, Sahara N, Saito Y, Murayama S, Ikai A, Takashima A. Increased levels of granular tau oligomers: an early sign of brain aging and Alzheimer's disease. *Neurosci Res* 2006;**54**:197–201.

99. Jeganathan S, von Bergen M, Mandelkow EM, Mandelkow E. The natively unfolded character of tau and its aggregation to Alzheimer-like paired helical filaments. *Biochemistry* 2008;**47**:10526–39.

100. Narayanan RL, Durr UH, Bibow S, Biernat J, Mandelkow E, Zweckstetter M. Automatic assignment of the intrinsically disordered protein Tau with 441-residues. *J Am Chem Soc* 2010;**132**:11906–7.

101. Sarkar M, Kuret J, Lee G. Two motifs within the tau microtubule-binding domain mediate its association with the hsc70 molecular chaperone. *J Neurosci Res* 2008;**86**:2763–73.

102. Dickey CA, Kamal A, Lundgren K, Klosak N, Bailey RM, Dunmore J, et al. The high-affinity HSP90-CHIP complex recognizes and selectively degrades phosphorylated tau client proteins. *J Clin Invest* 2007;**117**:648–58.

103. Dickey CA, Petrucelli L. Current strategies for the treatment of Alzheimer's disease and other tauopathies. *Expert Opin Ther Targets* 2006;**10**:665–76.

104. Jinwal UK, Miyata Y, Koren 3rd J, Jones JR, Trotter JH, Chang L, et al. Chemical manipulation of hsp70 ATPase activity regulates tau stability. *J Neurosci* 2009;**29**:12079–88.

105. O'Leary 3rd JC, Li Q, Marinec P, Blair LJ, Congdon EE, Johnson AG, et al. Phenothiazine-mediated rescue of cognition in tau transgenic mice requires neuroprotection and reduced soluble tau burden. *Mol Neurodegener* 2010;**5**:45.

106. Evans CG, Chang L, Gestwicki JE. Heat shock protein 70 (hsp70) as an emerging drug target. *J Med Chem* 2010;**53**:4585–602.

107. Koren 3rd J, Jinwal UK, Lee DC, Jones JR, Shults CL, Johnson AG, et al. Chaperone signalling complexes in Alzheimer's disease. *J Cell Mol Med* 2009;**13**:619–30.

108. Abisambra JF, Blair LJ, Hill SE, Jones JR, Kraft C, Rogers J, et al. Phosphorylation dynamics regulate Hsp27-mediated rescue of neuronal plasticity deficits in tau transgenic mice. *J Neurosci* 2010;**30**:15374–82.

109. Shimura H, Schwartz D, Gygi SP, Kosik KS. CHIP-Hsc70 complex ubiquitinates phosphory-lated tau and enhances cell survival. *J Biol Chem* 2004;**279**:4869–76.

110. Carrettiero DC, Hernandez I, Neveu P, Papagiannakopoulos T, Kosik KS. The cochaperone BAG2 sweeps paired helical filament-insoluble tau from the microtubule. *J Neurosci* 2009;**29**:2151–61.

111. Chambraud B, Sardin E, Giustiniani J, Dounane O, Schumacher M, Goedert M, et al. A role for FKBP52 in Tau protein function. *Proc Natl Acad Sci USA* 2010;**107**:2658–63.

112. Elliott E, Tsvetkov P, Ginzburg I. BAG-1 associates with Hsc70.Tau complex and regulates the proteasomal degradation of Tau protein. *J Biol Chem* 2007;**282**:37276–84.

113. Gong CX, Liu F, Wu G, Rossie S, Wegiel J, Li L, et al. Dephosphorylation of microtubule-associated protein tau by protein phosphatase 5. *J Neurochem* 2004;**88**:298–310.

114. Wang Y, Martinez-Vicente M, Kruger U, Kaushik S, Wong E, Mandelkow EM, et al. Tau fragmentation, aggregation and clearance: the dual role of lysosomal processing. *Hum Mol Genet* 2009;**18**:4153–70.

115. Wang Y, Martinez-Vicente M, Kruger U, Kaushik S, Wong E, Mandelkow EM, et al. Synergy and antagonism of macroautophagy and chaperone-mediated autophagy in a cell model of pathological tau aggregation. *Autophagy* 2010;**6**:182–3.

116. Evans CG, Wisén S, Gestwicki JE. Heat shock proteins 70 and 90 inhibit early stages of amyloid beta-(1-42) aggregation in vitro. *J Biol Chem* 2006;**281**:33182–91.

117. Fonte V, Kapulkin V, Taft A, Fluet A, Friedman D, Link CD. Interaction of intracellular beta amyloid peptide with chaperone proteins. *Proc Natl Acad Sci USA* 2002;**99**:9439–44.

118. Fonte V, Kipp DR, Yerg 3rd J, Merin D, Forrestal M, Wagner E, et al. Suppression of in vivo beta-amyloid peptide toxicity by overexpression of the HSP-16.2 small chaperone protein. *J Biol Chem* 2008;**283**:784–91.

119. Lee S, Carson K, Rice-Ficht A, Good T. Small heat shock proteins differentially affect Abeta aggregation and toxicity. *Biochem Biophys Res Commun* 2006;**347**:527–33.

120. Wu Y, Cao Z, Klein WL, Luo Y. Heat shock treatment reduces beta amyloid toxicity in vivo by diminishing oligomers. *Neurobiol Aging* 2010;**31**:1055–8.

121. Kumar P, Ambasta RK, Veereshwarayya V, Rosen KM, Kosik KS, Band H, et al. CHIP and HSPs interact with beta-APP in a proteasome-dependent manner and influence Abeta metabolism. *Hum Mol Genet* 2007;**16**:848–64.

122. Ansar S, Burlison JA, Hadden MK, Yu XM, Desino KE, Bean J, et al. A non-toxic Hsp90 inhibitor protects neurons from Abeta-induced toxicity. *Bioorg Med Chem Lett* 2007;**17**:1984–90.

123. Bandopadhyay R, de Belleroche J. Pathogenesis of Parkinson's disease: emerging role of molecular chaperones. *Trends Mol Med* 2009;**16**:27–36.

124. Zourlidou A, Payne Smith MD, Latchman DS. HSP27 but not HSP70 has a potent protective effect against alpha-synuclein-induced cell death in mammalian neuronal cells. *J Neurochem* 2004;**88**:1439–48.

125. Waudby CA, Knowles TP, Devlin GL, Skepper JN, Ecroyd H, Carver JA, et al. The interaction of alphaB-crystallin with mature alpha-synuclein amyloid fibrils inhibits their elongation. *Biophys J* 2010;**98**:843–51.

126. Zhou Y, Gu G, Goodlett DR, Zhang T, Pan C, Montine TJ, et al. Analysis of alpha-synuclein-associated proteins by quantitative proteomics. *J Biol Chem* 2004;**279**:39155–64.

127. Cuervo AM, Stefanis L, Fredenburg R, Lansbury PT, Sulzer D. Impaired degradation of mutant alpha-synuclein by chaperone-mediated autophagy. *Science* 2004;**305**:1292–5.

128. Dedmon MM, Christodoulou J, Wilson MR, Dobson CM. Heat shock protein 70 inhibits alpha-synuclein fibril formation via preferential binding to prefibrillar species. *J Biol Chem* 2005;**280**:14733–40.

129. Falsone SF, Kungl AJ, Rek A, Cappai R, Zangger K. The molecular chaperone Hsp90 modulates intermediate steps of amyloid assembly of the Parkinson-related protein alpha-synuclein. *J Biol Chem* 2009;**284**:31190–9.

130. Klucken J, Shin Y, Masliah E, Hyman BT, McLean PJ. Hsp70 reduces alpha-synuclein aggregation and toxicity. *J Biol Chem* 2004;**279**:25497–502.

131. Luk KC, Mills IP, Trojanowski JQ, Lee VM. Interactions between Hsp70 and the hydrophobic core of alpha-synuclein inhibit fibril assembly. *Biochemistry* 2008;**47**:12614–25.

132. Shimshek DR, Mueller M, Wiessner C, Schweizer T, van der Putten PH. The HSP70 molecular chaperone is not beneficial in a mouse model of alpha-synucleinopathy. *PLoS One* 2010;**5**:e10014.

133. Willingham S, Outeiro TF, DeVit MJ, Lindquist SL, Muchowski PJ. Yeast genes that enhance the toxicity of a mutant huntingtin fragment or alpha-synuclein. *Science* 2003;**302**:1769–72.

134. Flower TR, Chesnokova LS, Froelich CA, Dixon C, Witt SN. Heat shock prevents alpha-synuclein-induced apoptosis in a yeast model of Parkinson's disease. *J Mol Biol* 2005;**351**:1081–100.

135. Outeiro TF, Klucken J, Strathearn KE, Liu F, Nguyen P, Rochet JC, et al. Small heat shock proteins protect against alpha-synuclein-induced toxicity and aggregation. *Biochem Biophys Res Commun* 2006;**351**:631–8.

136. Auluck PK, Bonini NM. Pharmacological prevention of Parkinson disease in Drosophila. *Nat Med* 2002;**8**:1185–6.

137. Auluck PK, Meulener MC, Bonini NM. Mechanisms of suppression of {alpha}-synuclein neurotoxicity by geldanamycin in Drosophila. *J Biol Chem* 2005;**280**:2873–8.

138. Putcha P, Danzer KM, Kranich LR, Scott A, Silinski M, Mabbett S, et al. Brain-permeable small-molecule inhibitors of Hsp90 prevent alpha-synuclein oligomer formation and rescue alpha-synuclein-induced toxicity. *J Pharmacol Exp Ther* 2010;**332**:849–57.

139. Cohen FE, Prusiner SB. Pathologic conformations of prion proteins. *Annu Rev Biochem* 1998;**67**:793–819.

140. Rikhvanov EG, Romanova NV, Chernoff YO. Chaperone effects on prion and nonprion aggregates. *Prion* 2007;**1**:217–22.

141. Cashikar AG, Duennwald M, Lindquist SL. A chaperone pathway in protein disaggregation. Hsp26 alters the nature of protein aggregates to facilitate reactivation by Hsp104. *J Biol Chem* 2005;**280**:23869–75.

142. Shorter J, Lindquist S. Hsp104 catalyzes formation and elimination of self-replicating Sup35 prion conformers. *Science* 2004;**304**:1793–7.

143. Shorter J, Lindquist S. Destruction or potentiation of different prions catalyzed by similar Hsp104 remodeling activities. *Mol Cell* 2006;**23**:425–38.

144. Allen KD, Wegrzyn RD, Chernova TA, Muller S, Newnam GP, Winslett PA, et al. Hsp70 chaperones as modulators of prion life cycle: novel effects of Ssa and Ssb on the Saccharomyces cerevisiae prion [PSI+]. *Genetics* 2005;**169**:1227–42.

145. Douglas PM, Treusch S, Ren HY, Halfmann R, Duennwald ML, Lindquist S, et al. Chaperone-dependent amyloid assembly protects cells from prion toxicity. *Proc Natl Acad Sci USA* 2008;**105**:7206–11.

146. Mathur V, Hong JY, Liebman SW. Ssa1 overexpression and [PIN(+)] variants cure [PSI(+)] by dilution of aggregates. *J Mol Biol* 2009;**390**:155–67.

147. Steele AD, Hutter G, Jackson WS, Heppner FL, Borkowski AW, King OD, et al. Heat shock factor 1 regulates lifespan as distinct from disease onset in prion disease. *Proc Natl Acad Sci USA* 2008;**105**:13626–31.

148. DebBurman SK, Raymond GJ, Caughey B, Lindquist S. Chaperone-supervised conversion of prion protein to its protease-resistant form. *Proc Natl Acad Sci USA* 1997;**94**:13938–43.

149. Wilkins S, Choglay AA, Chapple JP, van der Spuy J, Rhie A, Birkett CR, et al. The binding of the molecular chaperone Hsc70 to the prion protein PrP is modulated by pH and copper. *Int J Biochem Cell Biol* 2010;**42**:1226–32.
150. Arosio B, Annoni G, Vergani C, Solano DC, Racchi M, Govoni S. Fibroblasts from Alzheimer's disease donors do not differ from controls in response to heat shock. *Neurosci Lett* 1998;**256**:25–8.
151. Kern A, Ackermann B, Clement AM, Duerk H, Behl C. HSF1-controlled and age-associated chaperone capacity in neurons and muscle cells of C elegans. *PLoS One* 2010;**5**:8568.
152. Anckar J, Sistonen L. Regulation of HSF1 function in the heat stress response: implications in aging and disease. *Annu Rev Biochem* 2011;**80**:1089–115.
153. Bitan G, Kirkitadze MD, Lomakin A, Vollers SS, Benedek GB, Teplow DB. Amyloid beta-protein (Abeta) assembly: Abeta 40 and Abeta 42 oligomerize through distinct pathways. *Proc Natl Acad Sci USA* 2003;**100**:330–5.
154. Sawkar AR, D'Haeze W, Kelly JW. Therapeutic strategies to ameliorate lysosomal storage disorders—a focus on Gaucher disease. *Cell Mol Life Sci* 2006;**63**:1179–92.
155. Olzscha H, Schermann SM, Woerner AC, Pinkert S, Hecht MH, Tartaglia GG, et al. Amyloid-like aggregates sequester numerous metastable proteins with essential cellular functions. *Cell* 2011;**144**:67–78.
156. Hageman J, Vos MJ, van Waarde MA, Kampinga HH. Comparison of intra-organellar chaperone capacity for dealing with stress-induced protein unfolding. *J Biol Chem* 2007;**282**:34334–45.
157. Intano GW, Cho EJ, McMahan CA, Walter CA. Age-related base excision repair activity in mouse brain and liver nuclear extracts. *J Gerontol A Biol Sci Med Sci* 2003;**58**:205–11.
158. Rao KS. DNA repair in aging rat neurons. *Neuroscience* 2007;**145**:1330–40.
159. Vyjayanti VN, Rao KS. DNA double strand break repair in brain: reduced NHEJ activity in aging rat neurons. *Neurosci Lett* 2006;**393**:18–22.
160. Kaganovich D, Kopito R, Frydman J. Misfolded proteins partition between two distinct quality control compartments. *Nature* 2008;**454**:1088–95.
161. Korolchuk VI, Menzies FM, Rubinsztein DC. Mechanisms of cross-talk between the ubiquitin-proteasome and autophagy-lysosome systems. *FEBS Lett* 2010;**584**:1393–8.
162. Neef DW, Turski ML, Thiele DJ. Modulation of heat shock transcription factor 1 as a therapeutic target for small molecule intervention in neurodegenerative disease. *PLoS Biol* 2010;**8**:e1000291.
163. Ong DS, Mu TW, Palmer AE, Kelly JW. Endoplasmic reticulum Ca2+ increases enhance mutant glucocerebrosidase proteostasis. *Nat Chem Biol* 2010;**6**:424–32.
164. Powers ET, Balch WE. Costly mistakes: translational infidelity and protein homeostasis. *Cell* 2008;**134**:204–6.

Biology of Mitochondria in Neurodegenerative Diseases

Lee J. Martin

Division of Neuropathology, Department of Pathology, The Pathobiology Graduate Program, Johns Hopkins University School of Medicine, Baltimore, Maryland, USA

Department of Neuroscience, Johns Hopkins University School of Medicine, Baltimore, Maryland, USA

Alzheimer's disease (AD), Parkinson's disease (PD), and amyotrophic lateral sclerosis (ALS) are the most common human adult-onset neurodegenerative diseases. They are characterized by prominent age-related neurodegeneration in selectively vulnerable neural systems. Some forms of AD, PD, and ALS are

Progress in Molecular Biology
and Translational Science, Vol. 107
DOI: 10.1016/B978-0-12-385883-2.00005-9

355

inherited, and genes causing these diseases have been identified. Nevertheless, the mechanisms of the neuronal degeneration in these familial diseases, and in the more common idiopathic (sporadic) diseases, are unresolved. Genetic, biochemical, and morphological analyses of human AD, PD, and ALS, as well as their cell and animal models, reveal that mitochondria could have roles in this neurodegeneration. The varied functions and properties of mitochondria might render subsets of selectively vulnerable neurons intrinsically susceptible to cellular aging and stress and the overlying genetic variations. In AD, alterations in enzymes involved in oxidative phosphorylation, oxidative damage, and mitochondrial binding of Aβ and amyloid precursor protein have been reported. In PD, mutations in mitochondrial proteins have been identified and mitochondrial DNA mutations have been found in neurons in the substantia nigra. In ALS, changes occur in mitochondrial respiratory chain enzymes and mitochondrial programmed cell death proteins. Transgenic mouse models of human neurodegenerative disease are beginning to reveal possible principles governing the biology of selective neuronal vulnerability that implicate mitochondria and the mitochondrial permeability transition pore. This chapter reviews several aspects of mitochondrial biology and how mitochondrial pathobiology might contribute to the mechanisms of neurodegeneration in AD, PD, and ALS.

Abbreviations: Aβ, amyloid β-protein; AD, Alzheimer's disease; AIF, apoptosis-inducing factor; ALS, amyotrophic lateral sclerosis; ANT, adenine nucleotide translocator; Apaf, apoptotic protease activating factor; APE, apurinic/apyrimidinic endonuclease (aka, HAP1); ApoE, apolipoprotein E; APP, amyloid precursor protein; Bax, Bcl-2-associated X protein; Bak1, Bcl-2-antagonist/killer 1; Bcl, B-cell lymphoma; BER, DNA base excision repair; CNS, central nervous system; Cu/ZnSOD, copper/zinc superoxide dismutase (also SOD1); CyPD, cyclophilin D; Caspase, cysteinyl aspartate-specific proteinase; DISC, death-inducing signaling complex; EM, electron microscopy; ER, endoplasmic reticulum; GPe, globus pallidus external; GPi, globus pallidus internal; HtrA2, high-temperature requirement protein A2; IAP, inhibitor of apoptosis protein; IMM, inner mitochondrial membrane; KA, kainic acid; LB, Lewy body; LRRK2, leucine-rich repeat kinase 2; Mfn, mitofusin; MnSOD, manganese SOD (also SOD2); mPT, mitochondrial permeability transition; mPTP, mitochondrial permeability transition pore; mSOD1, mutant SOD1; mtDNA, mitochondrial DNA; NAIP, neuronal apoptosis inhibitory protein; NFT, neurofibrillary tangle; NMDA, N-methyl-D-aspartate; NO, nitric oxide; NOS, nitric oxide synthase; NSC-34, neuroblastoma-spinal cord-34; $O_2^{\bullet-}$, superoxide radical; OGG1, 8-oxoguanine DNA glycosylase-1; OMM, outer mitochondrial membrane; ONOO⁻, peroxynitrite; OPA, optic atrophy type 1; PCD, programmed cell death; PEO, progressive external ophthalamoplegia; PD, Parkinson's disease; PINK1, phosphatase and tensin homolog-induced putative kinase-1; POLG, DNA polymerase γ; ROS, reactive oxygen species; Smac, second mitochondria-derived activator of caspases; SNpc, substantia nigra pars compacta; α-Syn, α-synuclein; Tg, transgenic; TIMM, translocase of inner mitochondrial membrane; TOMM, translocase of outer mitochondrial membrane; TSPO, translocator protein 18kDa (peripheral benzodiazepine receptor); TNF, tumor necrosis factor; TUNEL, terminal transferase-mediated biotin-dUTP nick-end labeling; UCHL1, ubiquitin carboxy-terminal hydrolyase-L1; VDAC, voltage-dependent anion channel

I. Introduction

An understanding of mitochondrial biology that is relevant to adult-onset neurodegenerative disorders has emerged from multiple disciplines. Mitochondria are dynamic multifunctional organelles.[1] In addition to their critical role in the production of ATP through the electron transport chain (Fig. 1), these organelles function in intracellular Ca^{2+} homeostasis, synthesis of steroid, heme and iron–sulfur clusters, heat production, and programmed cell death (PCD).[1,3,4] Mitochondria are sites of formation of reactive oxygen species (ROS), including superoxide anion $(O_2^{\bullet-})$[5] and the highly reactive hydroxyl radical ($^\bullet$OH) or its intermediates,[2] and reactive nitrogen species such as nitric oxide ($^\bullet$NO).[6] Mitochondrial stress and morphology have important relationships to various cell death mechanisms that can be apoptotic, necrotic, or apoptosis–necrosis hybrids that emerge along a cell death continuum.[7] Mitochondria have a variety of properties and functions (Fig. 1), discussed here, that might confer an intrinsic susceptibility of subsets of long-lived postmitotic cells, such as neurons, to aging and to stresses such as mutations and environmental toxins. Thus, it is reasonable that mitochondria could be involved in neurological disease. In fact, mitochondrial contribution to disease is found in disorders of acute interruptions in O_2 and substrate delivery to the brain; bioenergetic failure as seen in cerebral ischemia–reperfusion injury, trauma, and toxic exposures[1]; and neurodegenerative diseases.[7]

Several chronic neurodegenerative diseases are related causally to mitochondrial abnormalities. Optic atrophy type 1 (OPA1) is a hereditary optic neuropathy caused by mutations in the *OPA1* gene that encodes a mitochondrial dynamin-related GTPase that functions in maintenance of mitochondrial morphology, including fusion, and metabolism.[8] Charcot–Marie–Tooth disease, another autosomal-dominant neuropathy, is caused by mutations in the *mitofusin 2 (MFN2)* gene.[9] Leber's hereditary optic neuropathy, a neurodegenerative disease that causes optic nerve atrophy and blindness in young adults, is linked to at least 11 different missense mutations in mitochondrial DNA (mtDNA) genes that encode subunits in enzyme complexes I, III, and IV that function in oxidative phosphorylation.[10] Mutations in the *polymerase γ (POLG1)* gene, encoding the mtDNA polymerase catalytic subunit, are the most common causes of inherited mitochondrial disease in children and adults. These mutations are responsible for at least five phenotypes of neurodegenerative disease, including childhood myocerebrohepatopathy spectrum disorders, Alpers' syndrome, ataxia neuropathy spectrum disorders, myoclonus epilepsy myopathy sensory ataxia, and chronic progressive external ophthalamoplegia (PEO).[11] Mutations in the *ataxin-8* gene, encoding a protein called "twinkle" that functions as the replicative helicase for mtDNA, can cause infantile onset spinocerebellar ataxia and PEO.[12] Mitochondrial abnormalities caused by

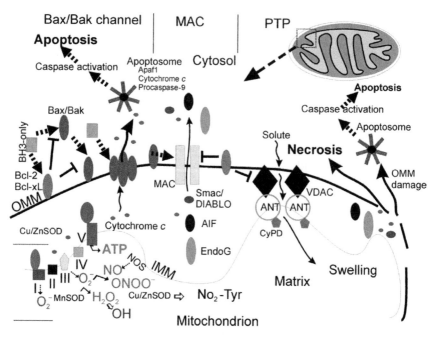

FIG. 1. Mitochondrial regulation of neuronal cell life and death in schematic representation (adapted from an earlier form Ref. 7). Mitochondria (upper right) are multifunctional organelles (see text). Oxygen- and proton pump-driven ATP production by the electron transport chain (lower left) is one function. The respiratory chain proteins (complex I–IV) establish an electrochemical gradient across the IMM by extruding protons out of the matrix into the intermembrane space, thereby creating an energy gradient that drives the production of ATP by complex V (lower left). Superoxide ($O_2^{\bullet-}$) is produced as a by-product in the process of electron transport and is converted to hydrogen peroxide (H_2O_2) by MnSOD (or Cu/ZnSOD in the intermembrane space). In pathological settings that can trigger cell aging and death, H_2O_2 can be converted to hydroxyl radical ($^\bullet$OH), or hydroxyl-like intermediates, and mitochondrial nitric oxide synthase (NOS) can produce nitric oxide (NO) that can combine with $O_2^{\bullet-}$ to form peroxynitrite (ONOO$^-$). Cu/ZnSOD can use ONOO$^-$ to catalyze the nitration (NO$_2$-Tyr) of mitochondrial protein tyrosine residues (bottom center) such as cyclophilin D (CyPD) and the adenine nucleotide translocator (ANT), which are core components of the mitochondrial permeability transition pore (PTP, another critical function of mitochondria). A third function of mitochondria is to regulate cell death. Bcl-2 family members regulate apoptosis by modulating the release of cytochrome c from mitochondria into the cytosol. Two models can account for this process, the Bax/Bak1 channel model and the mitochondrial apoptosis-induced channel (MAC). In the Bax/Bak1 channel model (left), Bax (Bcl-2-associated X protein) is a pro-apoptotic protein (Table I) found mostly in the cytosol in healthy mammalian cells but, after specific cell death-inducing stimuli, Bax undergoes a conformational change and translocates to the OMM, where it inserts. Bak1 (Bcl-2-antagonist/killer 1) is a similar pro-apoptotic protein localized mostly to the mitochondrial outer membrane. Bax/Bak1 monomers physically interact to form oligomeric or heteromeric channels that are permeable to cytochrome c. The formation of these channels is blocked by Bcl-2 and Bcl-x$_L$ at multiple sites. BH3-only members (Bad, Bid, Noxa, Puma) are pro-apoptotic and can modulate the conformation of Bax/Bak1 to sensitize this channel, possibly by exposing its membrane insertion domain (not shown). The MAC

mutations in the *adenine nucleotide translocator-1 (ANT1)* gene can also cause PEO.[13] Evidence for the involvement of mitochondria in other more common human adult-onset neurodegenerative disease is mostly circumstantial.[7]

This chapter summarizes several aspects of mitochondrial biology and the evidence for mitochondrial involvement in AD, PD, and ALS and some of their animal and cell models. In this regard, varying degrees of mitochondrial dysfunction and intrinsic mitochondrial-mediated cell death mechanisms could be critical determinants in the regulation of diseases and neurodegeneration, ranging along an apoptosis–necrosis cell death continuum.[7,14–16] Targeting mitochondrial properties, processes, or molecules, such as the mitochondrial permeability transition pore (mPTP)[17–20] (Fig. 2), could be important for developing new mechanism-based pharmacotherapies for these neurodegenerative diseases.

II. Some Aspects of Mitochondrial Biology Relevant to Neurodegeneration

A. Mitochondria and ROS

Mitochondria generate endogenous ROS as by-products of oxidative phosphorylation (Fig. 1).[4] Because many mitochondrial proteins possess iron–sulfur clusters for oxidation–reduction reactions, and because mtDNA lacks protective histones, these macromolecules are particularly vulnerable to ROS attack.[4] Electrons in the electron carriers, such as the unpaired electron of ubisemiquinone bound to coenzyme Q binding sites of complexes I–III, can be donated directly to O_2 to generate $O_2^{\bullet-}$.[4] $O_2^{\bullet-}$ does not easily pass through biological membranes and is inactivated in compartments where it is

---

could be a channel similar to the Bax/Bak1 channel, but it might also have additional components such as the voltage-dependent anion channel (VDAC). Released cytochrome c participates in the formation of the apoptosome, along with apoptotic protease activating factor 1 (Apaf1) and procaspase-9, in the cytosol that drives the activation of caspase-3. Second mitochondria-derived activator of caspases (Smac)/direct IAP-binding protein with low pI (DIABLO) are released into the cytosol to inactivate the anti-apoptotic actions of inhibitor of apoptosis proteins that inhibit caspases. The DNases apoptosis-inducing factor (AIF) and endonuclease G (EndoG) are released and translocate to the nucleus to stimulate DNA fragmentation. Another model (right) for mitochondrial-directed cell death involves the PTP. The PTP is a transmembrane channel formed by the interaction of ANT and VDAC at contact sites between the IMM and the OMM. CyPD, located in the matrix, can regulate the opening of the PTP by interacting with ANT. Opening of the PTP induces matrix swelling and OMM rupture, leading to release of cytochrome c and other apoptogenic proteins (AIF, EndoG). Certain Bcl-2 family members can modulate the activity of the PTP. (See Color Insert.)

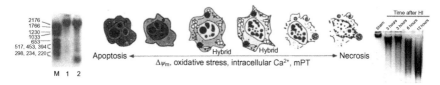

FIG. 2. The cell death continuum concept (modified from its original form Ref. 14). The concept as proposed in its original form envisions cell death as a spectrum. Apoptosis with internucleosomal fragmentation of genomic DNA (left) and necrosis with random digestion of genomic DNA (right) are at the extremes, and different syncretic hybrid forms are in between. The DNA gel at left shows a DNA fragmentation pattern typical of robust classical apoptosis (lane 2) and low amounts of apoptosis (lane 1) in developing rat brain (M is molecular weight markers in base pairs). The DNA gel at right shows a DNA fragmentation pattern typical of robust classical necrosis induced by brain hypoxia–ischemia (HI) and recovery of 3, 6, and 12 h. Only intact genomic DNA is seen in sham control brain. The syncretic forms of cell death depicted are predicted to manifest depending on the severities or amplitudes of the changes in mitochondrial membrane potential ($\Delta\psi_m$) oxidative stress, intracellular Ca^{2+} accumulation, and mitochondrial permeability transition pore (mPT) activation.

generated.[5] The mitochondrial matrix enzyme manganese superoxide dismutase (MnSOD or SOD2) or copper/zinc SOD (Cu/ZnSOD or SOD1) in the mitochondrial intermembrane space and cytosol convert $O_2^{\bullet-}$ to hydrogen peroxide (H_2O_2) in the reaction $O_2^{\bullet-}+O_2^{\bullet-}+2H^+\rightarrow H_2O_2+O_2$ (Fig. 1).[5] H_2O_2 is more stable than $O_2^{\bullet-}$ and can diffuse from mitochondria into the cytosol and nucleus. H_2O_2 is detoxified by glutathione peroxidase in mitochondria and the cytosol and by catalase in peroxisomes. In the presence of reduced transition metal (Fe^{2+}), H_2O_2 is converted to $^{\bullet}OH$.[2] $O_2^{\bullet-}$ also can react with $^{\bullet}NO$, which can be synthesized by three isoforms of nitric oxide synthase (NOS) enzymes,[6] to form the potent nucleophile oxidant and nitrating agent peroxynitrite ($ONOO^-$) (Fig. 1).[22] $ONOO^-$ or products of $ONOO^-$ can damage proteins by nitration.[22] $ONOO^-$ is also directly genotoxic to neurons by causing single- and double-strand breaks in DNA.[23] Mitochondria produce NO.[24,25] This reaction is catalyzed by a mitochondrial NOS (mtNOS) with similar cofactor and substrate requirements as constitutive NOS, but mtNOS can cross-react with antibodies to inducible NOS (NOS2).[24] The $^{\bullet}NO$ produced in mitochondria has direct actions in mitochondria.[24] $^{\bullet}NO$ at nanomolar concentrations can rapidly and reversibly inhibit mitochondrial respiration by nitration or nitrosylation.[25]

B. Mitochondria and Ca^{2+} Buffering

There is a 10,000-fold Ca^{2+} concentration gradient across the cell membrane, which separates an extracellular 1 mM Ca^{2+} concentration from an intracellular 50–100 nM Ca^{2+} concentration.[26] Mitochondria function in the regulation of cytoplasmic Ca^{2+} levels.[3] Utilizing specific transport systems, these organelles can move Ca^{2+} from the cytosol into their matrix by the Ca^{2+} uniporter and eject Ca^{2+} via the Na^+/Ca^{2+} exchanger.[3] Under conditions of elevated cytoplasmic Ca^{2+}, whenever the local free Ca^{2+} concentration rises above a set-point of $\approx 0.5\,\mu M$, mitochondria avidly accumulate Ca^{2+} to a fixed capacity.[3] The inner mitochondrial membrane (IMM) potential, $\Delta\Psi_m$, provides the driving force for the accumulation of Ca^{2+} into the matrix. Cytosolic Ca^{2+} concentration above set-point levels is believed to be achieved during tetanic stimulation and glutamate receptor activation.[3] In settings of excitotoxicity, resulting from excessive overstimulation of glutamate receptors, Ca^{2+} overload in neurons is significant. When mitochondria become overloaded with Ca^{2+}, they undergo mPT resulting in osmotic swelling and rupture of the outer mitochondrial membrane (OMM). Mitochondria within synapses appear to be more susceptible than nonsynaptic mitochondria to Ca^{2+} overload.[27]

C. Mitochondrial DNA

Each human cell contains hundreds of mitochondria and thousands of maternally inherited mtDNA copies residing in the matrix as double-stranded circular molecules of $\approx 16.5\,kb$ that contain 37 genes, all of which are transcribed.[4] mtDNA encodes 12S and 16S rRNAs and the 22 tRNAs required for mitochondrial protein synthesis occurring at mitochondrial ribosomes. mtDNA also encodes 13 proteins that are structural subunits of oxidative phosphorylation enzyme complexes, including 7 of the 46 proteins of complex I (NADH dehydrogenase), 1 of the 11 proteins of complex III (bc_1 complex), 3 of the 13 proteins of complex IV (cytochrome c oxidase), and 2 of the 16 proteins of complex V (ATP synthase).[4] Mitochondrial ROS can damage mtDNA, fostering the belief that mtDNA has a very high mutation rate.[4] Cells containing a mixed population of normal and mutant mtDNA are known as heteroplasmic. In postmitotic tissues, mutant mtDNA can be preferentially, clonally amplified. One type of mtDNA mutation, called the $\approx 5\,kb$ common mtDNA deletion (mtDNA4977), is found nonuniformly within different areas of the aging human brain.[28,29]

D. DNA Repair in Mitochondria

mtDNA sustains higher steady-state damage compared to nuclear DNA.[4] This greater lesion accumulation might be caused by greater local levels of ROS and lack of chromatin protection afforded by histones. Active DNA base excision repair (BER) proteins, all encoded by nuclear DNA, are present in mitochondria,[30] although at lower levels than in nuclei. Nuclei and mitochondria use variant proteins for BER.[31] Splice variants or truncation products of 8-oxoguanine DNA glycosylase-1 (OGG1), endonuclease III-like protein (NTH1), apurinic/apyrimidinic endonuclease (APE, also known as HAP1 and redox factor-1), and DNA ligase IIIβ are present in mitochondria.[31] Import mechanisms for OGG1 into mitochondria may undergo age-related perturbations.[32] DNA polymerase γ (POLG) is thought to be unique for mtDNA BER and replication. Human POLG is a nuclear-encoded gene product identified relatively recently,[33] and, since then, over 100 pathogenic mutations in POLG1 have been discovered to cause several neurological and nonneurological disorders.[11]

E. Mitochondrial Trafficking and Cytoskeletal Motor Proteins for Mitochondria

Various cargos are transported over short and long distances along microtubule tracks within neurons and their processes.[34] In axons, mitochondria move in both directions and can also be stationary for long periods.[35] Most of the mitochondria in axons move relatively rapidly at rates of 0.5–0.7 μm/s. Microtubules within axons are polarized, with their minus ends aligned toward the soma and their plus ends toward the synapse. Kinesin and dynein are directional molecular motors that are responsible for the fast transport of axonal mitochondria and vesicles.[34] Kinesin (a plus-end directed motor) moves mitochondria in the anterograde direction to the nerve terminal. Kinesin family members Kif1B and Kif5B are the motor proteins for anterograde movement of mitochondria.[36] Dynein (a minus-end directed motor) moves mitochondria in the retrograde direction to the soma. Cytoplasmic dynein is the primary motor for retrograde movement of mitochondria.[34] Mitochondrial movement may also occur on the actin cytoskeleton via myosin motors.[37]

The transport of mitochondria responds to physiological changes in the cells. Stimulation of glutamate receptors, tau phosphorylation, NO signaling, and intracellular Ca^{2+} and Zn^{2+} accumulation can induce changes in mitochondrial movement and shape, which might or might not be related to mPT.[38,39] In cultured developing peripheral neurons, mitochondria undergo net anterograde movement and then move retrogradely from the distal axon when growth cone activity ceases.[35] Moreover, in developing axons, NGF application causes mitochondria to accumulate at sites of neurotrophin stimulation through a mechanism involving the phosphoinositide 3 (PI3)–kinase pathway.[35] Mitochondrial movements in cultured neurons can be blocked by drugs that depolymerize microtubules (nocodazole) or aggregate actin filaments (cytochalasin D).[37] New data show that axonal mitochondrial transport and potential are correlated. Drugs that inhibit mitochondrial function can either block anterograde transport of mitochondria or stimulate retrograde transport of mitochondria.[40] Cell culture studies also show that NO and Ca^{2+} can inhibit mitochondrial movement by disrupting cytoskeletal structures and inhibiting ATP synthesis, and Zn^{2+} can inhibit mitochondrial movement without depolarizing mitochondria.[39] Other experiments on cultured neurons show that phosphorylated tau regulates mitochondrial anterograde transport and that blocking tau phosphorylation by inhibition of glycogen synthase kinase-3β decreases anterograde transport of mitochondria, causing them to cluster in the cell body.[41]

III. Mitochondria and Cell Death

Cells can die by several different processes.[7,42,43] These processes have been classified canonically into two distinct categories, apoptosis and necrosis. These forms of cellular degeneration were classified originally as different because they appeared different microscopically (Fig. 2). Apoptosis is an orderly and compartmental dismantling of single cells or groups of cells, while necrosis is a lytic destruction of individual or groups of cells. Apoptosis is an example of PCD, which is an ATP-driven (sometimes gene transcription-requiring) form of cell suicide often committed by demolition enzymes called caspases (cysteinyl aspartate-specific proteinases), but other apoptotic and nonapoptotic caspase-independent forms of PCD exist.[43] Apoptotic PCD is instrumental in developmental organogenesis and histogenesis and adult tissue homeostasis, functioning to eliminate excess cells.[44] In healthy people, estimates reveal that between 50–70 billion cells in adults and 20–30 billion cells in a child between the ages of 8 and 14 die each day due to apoptosis.[44]

A. Mitochondrial Regulation of Apoptosis

Apoptotic molecular networks are conserved in yeast, hydra, nematode, fruit fly, zebra fish, mouse, and human.[45] The current understanding of the molecular mechanisms of apoptosis in cells is built on studies by Robert Horvitz and colleagues on PCD in the nematode *Caenorhabditis elegans*.[46] They pioneered the understanding of the genetic control of developmental cell death by showing that it is regulated predominantly by three genes (*C. elegans death*, *ced-3*, *ced-4*, and *ced-9*).[46] This seminal work led to the identification of several families of apoptosis-regulation genes (Table I) in mammals, including the Bcl-2 family[47–49] and the caspase family of cysteine-containing, aspartate-specific proteases.[50] Other regulators of apoptotic cell death, most of which are mitochondrial proteins or influence mitochondria, are the p53 gene family, cell surface death receptors, cytochrome *c*, apoptosis-inducing factor (AIF), second mitochondrial derived activator of caspases (Smac), the inhibitor of apoptosis protein (IAP) family, and HtrA2/Omi.[51–57]

Mitochondria have been identified as critical for the apoptotic process. In the seminal work by Xiaodong Wang and colleagues it was discovered that the mitochondrion integrates death signals engaged by proteins in the Bcl-2 family

TABLE I
SOME MITOCHONDRIAL ASSOCIATED CELL DEATH PROTEINS AND THEIR ACTIONS

Protein	Function
Bcl-2[a]	Anti-apoptotic, blocks Bax/Bak channel formation
Bcl-X$_L$	Anti-apoptotic, blocks Bax/Bak channel formation
Bax[a]	Pro-apoptotic, forms pores for cytochrome c release
Bak[a]	Pro-apoptotic, forms pores for cytochrome c release
Bad	Pro-apoptotic, decoy for Bcl-2/Bcl-X$_L$ promoting Bax/Bak pore formation
Bid	Pro-apoptotic, decoy for Bcl-2/Bcl-X$_L$ promoting Bax/Bak pore formation
Noxa	Pro-apoptotic, decoy for Bcl-2/Bcl-X$_L$ promoting Bax/Bak pore formation
Puma	Pro-apoptotic, decoy for Bcl-2/Bcl-X$_L$ promoting Bax/Bak pore formation
p53[a]	Antagonizes activity of Bcl-2/Bcl-X$_L$, promotes Bax/Bak oligomerization
Cytochrome c	Activator of apoptosome
Smac/DIABLO	IAP inhibitor
AIF	Antioxidant flavoprotein/released from mitochondria to promote nuclear DNA fragmentation
Endonuclease G	Released from mitochondria to promote nuclear DNA fragmentation
HtrA2/Omi	IAP inhibitor
VDAC	mPTP component in outer mitochondrial membrane
ANT[b]	mPTP component in inner mitochondrial membrane
Cyclophilin D[b]	mPTP component in mitochondrial matrix
TSPO (peripheral benzodiazepine receptor)	Modulator of mPTP
Hexokinase	Modulator of VDAC

[a]Changes have been reported in human ALS.
[b]A reported target of oxidative modification in mouse ALS.

and releases pro-apoptotic molecules residing in the mitochondrial intermembrane space to activate caspases leading to internucleosomal cleavage of DNA (Figs. 1 and 3).[53,54] The endoplasmic reticulum (ER), which regulates intracellular Ca^{2+} levels, participates in a loop with mitochondria to modulate mitochondrial permeability and cytochrome c release through the actions of Bcl-2 protein family members (Fig. 1).[58]

Bcl-2 family members regulate apoptosis by modulating the release of cytochrome c from mitochondria into the cytosol (Table I). Different models can account for this process: the Bcl-2-associated X protein (Bax)/Bcl-2-antagonist/killer 1 (Bak1) channel model and the mitochondrial apoptosis-induced channel (MAC) (Fig. 1). The bcl-2 proto-oncogene family is a large group of apoptosis regulatory genes encoding about 20 different proteins (Table I). These proteins are defined by at least one of the four conserved B-cell lymphoma (Bcl) homology domains (BH1–BH4) in their amino acid sequence that function in protein–protein interactions.[47–49] Some of the proteins (e.g., Bcl-2, Bcl-x$_L$, and

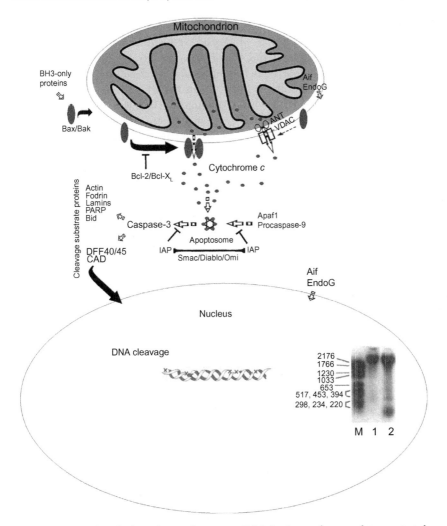

FIG. 3. Mitochrondrial regulation of apoptosis. Bcl-2 family members regulate apoptosis by modulating the release of cytochrome *c*. Bax and Bak are pro-apoptotic. They physically interact and form channels that are permeable to cytochrome *c*. BH3-only members (e.g., Bid, Noxa, Puma) are pro-apoptotic and can modulate the conformation of Bax. Bcl-2 and Bcl-X_L are anti-apoptotic and can block the function of Bax/Bak. The permeability transition pore (PTP), formed by the interaction of the adenine nucleotide translocator (ANT) and the voltage-dependent anion channel (VDAC) during the process of swelling, is a transmembrane channel that emerges at contact sites between the inner mitochondrial and the OMMs. The PTP has a role in regulating mitochondrial membrane potential and the release of cytochrome *c*. In the cytosol, cytochrome *c*, Apaf1, and procaspase-9 interact to form the apoptosome that drives the activation of caspase-3. The family of inhibitors of apoptosis (IAPs) blocks this process. The IAPs are inhibited by mitochondrially derived Smac, Diablo, and Omi. Caspase-3 cleaves many substrate proteins, some of which are

Mcl-1) have all four BH1–BH4 domains and are anti-apoptotic (Table I). Other proteins that are pro-apoptotic have BH1–BH3 sequences (e.g., Bax and Bak1) or only the BH3 domain (e.g., Bad, Bid, Bim, Bik, Noxa, and Puma) that contains the critical death domain (Table I). Bcl-x_L and Bax have α-helices resembling the pore-forming subunit of diphtheria toxin[59]; thus, Bcl-2 family members appear to function by conformation-induced insertion into the OMM to form channels or pores that can regulate the release of apoptogenic factors (Fig. 1).[60] Bcl-2 family members can form homo- or heterodimers and higher-order multimers with other family members.[47–49] Bax/Bak1 heterodimerization with either Bcl-2 or Bcl-x_L neutralizes their pro-apoptotic activity. When Bax and Bak1 are present in excess, the anti-apoptotic activity of Bcl-2 and Bcl-x_L is antagonized and apoptosis is promoted. In the Bax/Bak1 channel model (Fig. 1, left), after specific cell death-inducing stimuli, Bax undergoes a conformational change and translocates to the OMM where it inserts. Bak1 is a similar pro-apoptotic protein localized mostly to the OMM. Bax/Bak1 monomers physically interact to form oligomeric or hetero-meric membrane channels that are permeable to cytochrome c.[60] The formation of these channels is blocked by Bcl-2 and Bcl-x_L at multiple sites. BH3-only members (Bad, Bid, Noxa, Puma) are pro-apoptotic and can modulate the conformation of Bax/Bak1 to sensitize this channel, possibly by exposing its membrane insertion domain, or they serve as decoys for Bcl-x_L that allow Bax to form pores in the OMM.[61] Cells without *bax* and *bak* genes are resistant to mitochondrial cytochrome c release during apoptosis.[62] Release of cytochrome c from mitochondria (Fig. 1) can occur through mechanisms that involve the formation of membrane channels comprising Bax and the voltage-dependent anion channel (VDAC).[63] The MAC could be a channel similar to the Bax/Bak1 channel, but it might also have additional components, such as VDAC.

Released cytochrome c then triggers the assembly of the cytoplasmic apoptosome (Figs. 1 and 3). The apoptosome is a protein complex of apoptotic protease activating factor 1 (Apaf1), cytochrome c, and procaspase-9. This is the engine that drives caspase-3 activation in mammalian cells (Figs. 1 and 3).[53] Caspases are cysteine proteases that have a near-absolute substrate require-ment for aspartate in the P_1 position.[64] Fifteen *caspase* genes have been identified in mammals.[64] Caspases exist as constitutively expressed inactive pro-enzymes (30–50 kDa) in healthy cells. Their zymogens are found in

---

endonucleases that translocate to the nucleus to cleave DNA into internucleosomal fragments (see DNA gel at lower right, showing molecular weight standards [M] in base pairs, developing rat cerebral cortex showing very low DNA fragmentation [lane 1], and developing rat brainstem inferior colliculus undergoing considerable apoptosis [lane 2]). Aif and endonuclease G are mitochondrially released proteins with nuclease activity that can translocate to the nucleus. Genomic DNA (double helix in nucleus) is the site of action of nucleases that induce strand nicks (X in helix). See text for detailed descriptions.

different proportions in different subcellular locations. In Henrietta Lacks (HeLa) cervical epithelial carcinoma cells, most caspase-3 pro-enzyme is found in the cytoplasm, while only 10% is found in mitochondria.[65] Ninety percent of caspase-9 pro-enzyme is mitochondrial in rat heart and brain.[66]

So far, three caspase-related signaling pathways have been identified that can lead to apoptosis,[53,54,67,68] but cross talk among these pathways is possible. The intrinsic mitochondria-mediated pathway is controlled by Bcl-2 family proteins. It is regulated by cytochrome c release from mitochondria, promoting the activation of caspase-9 through Apaf1, and then caspase-3 activation (Figs. 1 and 3).[53,54] Apaf1 is a cytoplasmic protein that contains several copies of the WD-40 domain ($\approx$40-amino acid motifs often terminating in a Trp-Asp dipeptide), a caspase recruitment domain, and an ATPase domain. Upon binding cytochrome c and dATP, Apaf1 forms an oligomeric apoptosome. The apoptosome binds and cleaves caspase 9 preproprotein (Apaf3), releasing its mature, activated form. Activated caspase 9 cleaves pro-caspase-3. The extrinsic death receptor pathway involves the activation of cell-surface death receptors, including Fas and the tumor necrosis factor receptor, leading to the formation of the death-inducing signaling complex and caspase-8 activation, which, in turn, cleaves and activates downstream caspases such as caspase-3, -6, and -7.[68] Caspase-8 can also cleave Bid leading to the translocation, oligomerization, and insertion of Bax or Bak1 into the mitochondrial membrane.[68] Another pathway involves the activation of caspase-2 by DNA damage or ER stress as a premitochondrial signal.[69]

The activity of pro-apoptotic proteins is blocked to prevent untimely apoptosis in normal cells. Apoptosis can be antagonized by the IAP family in mammalian cells.[70–72] This family includes X chromosome-linked IAP (XIAP), IAP1, IAP2, neuronal apoptosis inhibitory protein, Survivin, Livin, and Apollon. These proteins are characterized by 1–3 baculoviral IAP repeat domains consisting of a zinc finger domain of ~70–80 amino acids.[71] Apollon is a huge (530kDa) protein that also has a ubiquitin-conjugating enzyme domain. The main identified anti-apoptotic function of IAPs is the suppression of caspase activity.[72] Procaspase-9 and -3 are major targets of several IAPs. IAPs reversibly interact directly with caspases to block substrate cleavage. Apollon also ubiquitinates and facilitates proteasomal degradation of active caspase-9 and second mitochondria-derived activator of caspases (Smac).[73]

Mitochondrial proteins exist that inhibit mammalian IAPs. A murine mitochondrial protein called Smac and its human ortholog DIABLO (for direct IAP-binding protein with low pI) inactivate the anti-apoptotic actions of IAPs and thus exert pro-apoptotic actions.[74,75] Smac/DIABLO are released into the cytosol to inactivate the anti-apoptotic actions of IAPs that inhibit caspases (Figs. 1 and 3). These IAP inhibitors are 23kDa mitochondrial proteins (derived from 29kDa precursor proteins processed in the mitochondria) that are released into the cytosol from the intermembrane space to sequester IAPs. High-temperature requirement protein A2 (HtrA2), also called Omi, is another mitochondrial serine protease that exerts pro-apoptotic activity by inhibiting IAPs.[76] HtrA2/Omi functions as a homotrimeric protein that cleaves IAPs irreversibly, thus facilitating caspase activity. The intrinsic mitochondrial-mediated cell death pathway is regulated by Smac and HtrA2/Omi.[76] Mutations in the *htra2* gene, identified as *PARK13* (Table II), have been linked to the development of PD,[77] but this linkage is controversial.[372]

B. Apoptosis-Inducing Factor

AIF is a mammalian mitochondrial protein identified as a flavoprotein oxidoreductase.[78] AIF has an N-terminal mitochondrial localization signal that is cleaved off to generate a mature protein of 57kDa after import into the mitochondrial intermembrane space. Under normal physiological conditions, AIF might function as a ROS scavenger targeting H_2O_2[55] or in redox cycling with nicotinamide adenine dinucleotide phosphate.[79] After some apoptotic stimuli, AIF is released from mitochondria (Figs. 1 and 3) and

TABLE II
MUTANT GENES LINKED TO FAMILIAL PD

Locus	Inheritance	Gene	Protein name/function
PARK1/4q21	Autosomal dominant	α-Syn	α-Syn/presynaptic maintenance?
PARK2/6q25.2-27	Autosomal recessive	parkin	Parkin/ubiquitin E3 ligase
PARK3/2p13	Autosomal dominant	?	?
PARK4/4p15	Autosomal dominant	α-Syn	α-Syn/presynaptic maintenance?
PARK5/4p14	Autosomal dominant	UCHL1	UCHL1/polyubiquitin hydrolase
PARK6/1p36	Autosomal recessive	PINK1	PTEN-induced putative kinase-1/ mitochondrial protein kinase
PARK7/1p36.33-36-12	Autosomal recessive	DJ-1	DJ-1/mitochondrial antioxidant, chaperone
PARK8/12q12	Autosomal dominant	LRRK2	Dardarin/multifunctional kinase/ GTPase
PARK9/1p36	Autosomal recessive	ATP13A2	Lysosomal type 5 P-ATPase
PARK10/1p32		?	?
PARK11/2q36-37	Autosomal dominant	GIGYF2?	Grb10-interacting GYP protein 2, modulates tyrosine kinase receptor signaling, including IGF-1
PARK12/Xq21-q25	X-linked	?	?
PARK13/2p12	Autosomal recessive susceptibility factor	Omi/ HtrA2	Omi/HtrA2, mitochondrial serine peptidase, inhibitor of IAPs
PARK14/22q13.1	Autosomal recessive	PLA2G6	Phospholipase A2 group VI
PARK15/22q12–q13	Autosomal recessive	FBXO7	F-box protein 7

translocates to the nucleus.[78] Overexpression of AIF in cultured cells induces cardinal features of apoptosis, including chromatin condensation, high-molecular-weight DNA fragmentation (Fig. 3), and loss of mitochondrial transmembrane potential.[78]

C. Cell Death by Necrosis and the Mitochondrial Permeability Transition Pore

Cell death caused by cytoplasmic swelling, nuclear dissolution (karyolysis), and lysis has been classified traditionally as necrosis (Fig. 2).[80,81] Cell necrosis (sometimes termed oncosis)[81] can result from rapid and severe failure to sustain cellular homeostasis, notably cell volume control.[82] The process of necrosis involves damage to the structural and functional integrity of the cell plasma membrane and associated enzymes (e.g., Na^+,K^+-ATPase), abrupt influx and overload of ions (e.g., Na^+ and Ca^{2+}) and H_2O, and rapid mitochondrial damage and bioenergetic collapse.[83–85] Metabolic inhibition, anoxia, and oxidative stress from ROS can trigger necrosis. Inhibitory cross talk between ion pumps causes pro-necrotic effects when Na^+,K^+-ATPase "steals" ATP from the plasma membrane Ca^{2+} ATPase, contributing to Ca^{2+} overload and mitochondrial damage.[86]

The morphology and some biochemical features of classic necrosis in neurons are distinctive (Fig. 2).[14,42] The main features are swelling and vacuolation/vesiculation of organelles (Fig. 2), destruction of membrane integrity, digestion of chromatin, and dissolution of the cell. The overall profile of the moribund cell is generally maintained as it degrades into the surrounding tissue parenchyma. The debris induces an inflammatory reaction in tissue. In necrosis, dying cells do not bud to form discrete membrane-bound fragments. The nuclear pyknosis and karyolysis appear as condensation of chromatin into many irregularly shaped small clumps, sharply contrasting with the formation of a few uniformly dense and regularly shaped chromatin aggregates that occur in apoptosis. In cells undergoing necrosis, genomic DNA is digested globally, because proteases that digest histones, which protect DNA, and DNases are coactivated to generate many randomly sized fragments seen as a DNA smear by gel electrophoresis (Fig. 2, right). These cytoplasmic and nuclear changes in pure necrosis are thought to be very diagnostic (Fig. 2).

Recent work has shown that cell necrosis in some settings is not as chaotic, random, and incomprehensible as envisioned originally but can involve the activation of specific signaling pathways to eventuate in cell death.[87–90] This idea is very important for developing new mechanism-based therapeutics to block cell necrosis.[16,17] For example, DNA damage can lead to poly(ADP-ribose) polymerase activation and ATP depletion, mitochondrial energetic collapse, and necrosis.[91] Other pathways for "programmed" necrosis involve death receptor signaling through NADPH oxidase, receptor-interacting protein 1, and mitochondrial permeability transition (mPT) (Figs. 1 and 2).[88–90,92,93]

mPT is a mitochondrial state in which the proton-motive force is disrupted reversibly or irreversibly.[18–20,92–95] Conditions of intramitochondrial Ca^{2+} overload, excessive oxidative stress, and decreased electrochemical gradient ($\Delta\Psi_m$), ADP, and ATP can favor mPT. This altered condition of mitochondria involves the mPTP that functions as a voltage, thiol, and Ca^{2+} sensor.[18–20,92–95] The mPTP is believed to be a poly-protein transmembrane channel (Fig. 1) formed at the contact sites between the OMM and the IMM. The collective components of the mPTP are still controversial. The VDAC (porin) in the OMM, the ANT (or solute carrier family 25) in the IMM, and cyclophilin D (CyPD) in the matrix at one time were believed to be the core components (Fig. 1 and Table I),[18–20,92–95] but these proteins now seem to be dispensable for mPTP formation. Other components or modulators of the mPTP appear to be the mitochondrial phosphate carrier, hexokinase, creatine kinase, translocator protein 18kDa (TSPO, or peripheral benzodiazepine receptor), and Bcl-2 family members (Table I).[95]

The VDAC family in human and mouse cells consists of three proteins of ≈31kDa (VDAC1–3) encoded by three different genes.[96] VDACs are the major transport proteins in the OMM, functioning in ATP rationing, Ca^{2+} homeostasis, oxidative stress response, and cell death.[96] Monomeric VDAC serves as the functional channel, although oligomerization of VDAC into dimers, tetramers, and higher-order multimers can occur and might function in cell death.[96] The VDAC adopts an open conformation at low or zero membrane potentials and a closed conformation at potentials above 30–40 mV.[96] In the open conformation, the VDAC makes the OMM permeable to most small hydrophilic molecules up to 1.3kDa for free exchange of respiratory chain substrates.[97] VDAC closure increases inward Ca^{2+} flux across the OMM[98,99] and causes oxidative stress.[99] Most data implicating VDAC opening or closing as an important regulator of cell death are based on in vitro conditions (cell culture and cell-free systems), while limited in vivo evidence is available.[100] VDAC1 binds Bak1, hexokinase, gelsolin, and ANT1/ANT2; VDAC2 binds Bak1, hexokinase, cytochrome c, glycerol kinase, and ANT1/ANT2; and VDAC3 binds glycerol kinase, CyPD, and ANT1-3.[96] In human tissues, VDAC1 and VDAC2 isoforms are expressed more abundantly than VDAC3. Highest levels are found in kidney, heart, skeletal muscle, and brain.[101] The effects of selective knockout of VDAC isoforms are not equivalent, implying different functions. Mice deficient in either VDAC1 or VDAC3 are viable,[102–104] but VDAC2 deficiency causes embryonic lethality.[105] Lack of both VDAC1 and VDAC3 causes growth retardation.[104] VDAC null mouse tissues exhibit deficits in mitochondrial respiration and abnormalities in mitochondrial ultrastructure.[102] However, mitochondria without VDAC1 have an intact mPT response.[106,107] VDAC2 deletion, but not lack of the more abundant

VDAC1, results in enhanced activation of the mitochondrial apoptosis pathway and enforced activation of Bak1 in mitochondria,[105] consistent with the idea that VDAC2 is a key inhibitor of Bak1-mediated apoptosis.[104] However, other data show that cells lacking individual VDACs or combinations of VDACs have normal death responses to Bax and Bid.[107] Recent work in yeast has revealed that SOD1 is necessary for proper functioning of VDAC. Specifically, SOD1 regulates VDAC channel activity and protein levels in mitochondria.[108]

The mitochondrial ANT family in humans consists of three members (ANT1–3, or solute carrier family 25, members 4–6) encoded by three different genes, but in mouse only two isoforms of the ANT are present.[109] The proteins are $\approx$33kDa in size and function as homodimers.[109] They are multipass membrane proteins, with odd-numbered transmembrane helices that mediate exchange of cytosolic ADP for mitochondrial ATP across the inner membrane utilizing the electrochemical gradient.[110] These helices have kinks because of proline residues.[110] ANT1 binds VDAC1, CyPD, Bax, twinkle (ataxin-8), and cyclophilin-40; ANT2 binds VDAC1–3 and cyclophilin-40; and ANT3 binds VDAC1, steroid sulfatase, and translocase of inner mitochondrial membrane-13 (TIMM13) and TIMM23.[109] The ANT isoforms are expressed differentially in tissue- and animal-specific patterns.[111] ANT1 is expressed highly in human and mouse heart and skeletal muscle. Human brain has low ANT1 mRNA but high ANT3 mRNA, while mouse brain has high ANT1 mRNA.[111] ANT2 mRNA is very low or not expressed in most adult human and mouse tissues, except kidney.[111] In tissue mitochondria in which more than one ANT isoform is expressed, it is ANT1 that binds preferentially to CyPD to form the mPTP at contact sites between the IMM and the OMM (Fig. 1).[112] It has been proposed that, in the presence of high mitochondrial Ca^{2+}, the binding of CyPD to proline residue 61 (Pro^{61}) in loop 1 of ANT1 results in a conformation that converts the ANT into a nonspecific pore.[109] Nonconditional ANT1 null mice are viable and grow normally but develop mitochondrial skeletal myopathy and cardiomyopathy.[110] Ablation of both ANT isoforms in mouse liver surprisingly did not change fundamentally mPT and cell death in hepatocytes,[113] and some ANT ligands induced mitochondrial dysfunction and cytochrome c release independent of mPT.[114] Clearly, the mechanisms of ANT-mediated cell death need further study.

CyPD (also named cyclophilin F, peptidyl prolyl isomerase F) is encoded by a single gene.[17,90] There is only one isoform of CyPD (EC 5.2.1.8, *ppif* gene product) in mouse and human. The $\approx$20kDa protein encoded by this gene is a member of the peptidyl-prolyl *cis–trans* isomerase (PPIase) family. PPIases catalyze the *cis–trans* isomerization of proline imidic peptide bonds in oligopeptides and accelerate the folding of proteins. CyPD binds ANT1.[109]

During normal mitochondrial function, the OMM and the IMM are separated by the intermembrane space, and the VDAC and the ANT do not interact.[93–95] Permeability transition is activated by the formation of the mPTP (Fig. 1). The IMM loses its integrity and the ANT changes its conformation from its native state into a nonselective pore.[109] This process is catalyzed by CyPD functioning as a protein *cis–trans* isomerase and chaperone.[17] The ANT and CyPD interact directly in this process.[115] The amount of CyPD (in heart mitochondria) is much lower than the ANT concentration (<5%); thus, under normal conditions, only a minor fraction of the ANT can be in a complex with CyPD.[115,116] When this occurs, small ions and metabolites permeate freely across the IMM and oxidation of metabolites by O_2 proceeds with electron flux not coupled to proton pumping, resulting in collapse of $\Delta\Psi_m$, dissipation of ATP production, elevated production of ROS, equilibration of ions between the matrix and the cytosol, increased matrix volume, and mitochondrial swelling.[94]

Very few studies have been published on the localizations of putative mPTP components in the mammalian central nervous system (CNS), and therefore details about cellular expression in different nervous system cell types are lacking. VDAC expression patterns are complicated by alternative splicing that generates two different VDAC1 mRNAs, three different VDAC2 mRNAs, and two different VDAC3 mRNAs.[96] Studies of nervous tissue have found VDAC in neurons and glial cells[117] and associated with mitochondria, the ER, and the plasma membrane.[118,119] Nonmitochondrial localizations of VDAC have been disputed.[120] Information on the cellular localizations of ANT in nervous tissue is scarce. ANT appears to be expressed in reactive astrocytes.[121] The few published studies on CyPD localization in mammalian CNS have found it enriched in subsets of neurons in adult rat brain, with some interneurons being positive[122] but astrocytes having relatively low levels.[123] In mouse spinal cord, putative components of the mPTP (VDAC, ANT, and CyPD) are enriched in motor neurons, as determined by immunohistochemistry.[124] The specific isoforms of ANT and VDAC in motor neurons have not been determined. CyPD, ANT, and VDAC have mitochondrial and nonmitochondrial localizations in motor neurons.[124] They are all nuclear-encoded mitochondrial-targeted proteins, thus a possible explanation for their nonmitochondrial localizations is that they are premitochondrial forms. Some cyclophilins are located in the cytoplasm,[125] such as cyclophilin A, but CyPD immunoreactivity is not observed in *ppif*[-/-] mice, demonstrating that the antibody is detecting only CyPD.[124] Spinal cord, brainstem, and forebrain had similar levels of CyPD, as well as similar levels of ANT and VDAC.[124] Thus, differences in the levels of individual mPTP components cannot explain the intrinsic differences in the sensitivity to Ca^{2+}-induced mPT seen in isolated mitochondria from spinal cord and brain.[126,127] Not all mitochondria within individual motor neurons contain CyPD, ANT, and VDAC.[124] This observation supports the idea that mitochondria in individual cells are not only heterogeneous in shape[128,129] but also in biochemical composition, metabolism,[130] and genetics.[4]

IV. Mitochondrial Autophagy

Autophagy is a mechanism whereby eukaryotic cells degrade their own cytoplasm and organelles.[131] Autophagy functions as a homeostatic nonlethal stress response mechanism for recycling proteins to protect cells from low supplies of nutrients and as a cell death mechanism.[131] Mitochondrial removal is mediated by autophagy. Nutrient and neurotrophin withdrawal can elicit mitochondrial autophagy in cultured neurons.[132] Degradation of mitochondria by autophagy can occur during death of neurons in culture.[132,133] In *in vivo* axotomy/target deprivation models, injured neurons can survive for months with a striking paucity of mitochondria within the cell body.[134,135] It is possible that some neurons might remove altered or injured mitochondria as part of their survival mechanisms either for eliminating toxic molecules or organelles or for scavenging nutrients. In contrast, neurons that die from axotomy accumulate mitochondria pre-apoptotically.[136,137] Spinal interneurons dying apoptotically after an excitotoxic insult also accumulate mitochondria pre-apoptotically.[138]

Autophagy is also called Type II PCD.[139] This degradation of organelles and long-lived proteins is carried out by the lysosomal system. A hallmark of autophagy thus is the accumulation of autophagic vacuoles of lysosomal origin. Autophagy has been seen in developmental and pathological conditions. For example, insect metamorphosis involves autophagy[43] and developing neurons can use autophagy as a PCD mechanism.[140,141] Degeneration of Purkinje neurons in the mouse mutant *Lurcher* appears to be a form of autophagy, thus linking excitotoxic constitutive activation of the GluRδ2 glutamate receptor to autophagic cell death.[142] However, loss of basal autophagic function in the CNS causes neurodegeneration in mice.[143,144] This finding could be a testimonial to the importance of Parkin, a ubiquitin kinase encoded by Parkinson's disease-related *PARK2*, which functions to promote autophagic turnover of mitochondria.[145]

The molecular controls of autophagy appear common in eukaryotic cells from yeast to human, and autophagy may have evolved before apoptosis.[146] However, much work has been done on yeast, while detailed work on autophagy in mammalian cells is emerging.[147] Double-membrane autophagosomes for sequestration of cytoplasmic components are derived from the ER or the plasma membrane. Tor kinase, PI3-kinase (a family of cysteine proteases called autophagins), and death-associated proteins function in autophagy.[148,149] Autophagic and apoptotic cell death pathways cross-talk. The product of the tumor suppressor gene Beclin1 (the human homolog of the yeast autophagy gene APG6) interacts with the anti-apoptosis regulator Bcl-2.[150] Autophagy can block apoptosis by sequestration of mitochondria. If the capacity for autophagy is reduced, stressed cells die by apoptosis, whereas inhibition or blockade of molecules that function in apoptosis can convert the cell death process into autophagy.[151] Thus, a continuum between autophagy and apoptosis could exist.

V. Mitochondrial Fission and Fusion

Mitochondria are not static organelles. They can undergo cycles of fusion and fission, and the balance of these events in part determines mitochondrial morphology. Mitochondrial fusion and fission events are controlled by several proteins, including dynamin-like proteins (Mfns and OPA1), dynamin-related protein 1 (Drp1), and mitochondrial fission protein 1 (Fis1).[9,152]

Mfn1, Mfn2, and OPA1 are mitochondrial GTPases essential for mitochondrial fusion.[9] Mfns and OPA1, residing in the OMM and intermembrane space, respectively, promote mitochondrial fusion. Mfn1 and Mfn2 can homodimerize and heterodimerize, but Mfns and OPA1 appear not to interact biochemically in mammalian cells. Mfns possess a C-terminal coiled-coil domain that mediates oligomerization between Mfn molecules on adjacent mitochondria. OPA1 forms oligomeric complexes consisting of membrane-bound and soluble forms of OPA1 and may form a diffusion barrier for proteins stored in mitochondrial cristae.[9] OPA1 also functions in apoptosis. Proteolytic processing in response to intrinsic apoptotic signals may lead to disassembly of OPA1 oligomers and release cytochrome c into the mitochondrial intermembrane space.

New mitochondria are generated by fission, which is controlled by at least two proteins (Fis1 and Drp1).[9] Fis1 is a single-pass transmembrane protein that is anchored to the OMM. Fis1 can weakly bind Drp1. In contrast, Drp1 is present mostly in the cytoplasm but a portion localizes to puncta on the mitochondria surface. It has been postulated that bound Drp1 at the mitochondrial surface couples GTP hydrolysis to mitochondrial membrane constriction and fission. Mitochondrial fission might even be an important part of the mechanisms driving cell death because inhibition of Drp1 function blocks apoptosis in nonneuronal cell lines. Under stress conditions, Drp1 translocates to mitochondria and localizes to scission foci.[153] Mitochondrial fission appears to involve Bax, because Bax translocates to scission foci as well.[153] It is believed that most mitochondriogenesis occurs in the perinuclear region.[154] NO can stimulate mitochondrial biogenesis.[155]

VI. Mitochondrial Involvement in Adult-Onset Neurodegenerative Diseases

A. Alzheimer's Disease

AD is the most common cause of dementia occurring in middle and late life.[156] Population-based surveys estimate that AD affects 7–10% of individuals >65 years of age and possibly 50–60% of people over 85 years of age.[157,158] AD now affects about 2% of the population, or about 4 million people in the United

States and ~35 million people worldwide.[159] The prevalence of AD is increasing proportionally to increased life expectancy, and estimates predict that the prevalence will reach ~107 million by 2050.[160]

Most cases of AD have late-onset and unknown etiologies and are termed "sporadic." However, some cases, particularly those with early onset, are familial and are inherited as autosomal-dominant disorders linked to mutations in the gene that encodes the amyloid precursor protein (APP)[161–164] or genes that encode presenilin proteins.[165,166] For late-onset sporadic cases, a variety of risk factors have been identified.[167] Aging is the strongest risk factor.[167] The *apolipoprotein E (APOE)* gene is a susceptibility locus, with the APOE ε4 allele showing dose-dependent contributions to AD risk.[168] Cardiovascular disease and head trauma are additional risk factors for AD.[156]

The dementia in AD is caused by severe atrophy of the forebrain. Neurons in the neocortex, hippocampus, and basal forebrain are selectively vulnerable in AD.[169–173] The numerous lesions that are formed in the brains of AD patients are termed senile plaques, which are abnormal extracellular deposits of amyloid β-protein (Aβ), and neurofibrillary tangles (NFTs), which are abnormal intracellular aggregates of protein containing hyperphosphorylated tau (a microtubule-associated protein).[174,175]

The mechanisms of neuronal degeneration AD are not known, and existing information is incomplete. Abnormal processing or modification of APP and the cytoskeletal protein tau are involved (Fig. 4).[176] Cortical and hippocampal neuronal degeneration could be the consequence of a combination of several mechanisms, including perturbations in protein metabolism, excitotoxicity, oxidative stress, mitochondrial perturbations, and inflammation. The possible specific mechanisms for neuronal degeneration in AD may involve dysfunction of N-methyl-D-aspartate receptors,[177,178] dysregulation of Ca^{2+} and mitochondrial homeostasis,[179,180] defects in synapses,[181–185] abnormalities in the metabolism of APP and presenilin proteins, toxic actions of Aβ,[185–187] and cytoskeletal pathology linked to abnormal phosphorylation.[188,189]

There are possible disease links between intraneuronal Aβ and mitochondria, suggesting an intracellular toxic activity of Aβ.[190–192] Importantly, APP possesses a targeting sequence for mitochondria.[190] When overexpressed in cultured cells, APP interacts with mitochondrial import proteins, can arrest mitochondrial import, and can result in bioenergetic deficits.[190] In postmortem human brain samples, APP variants were found to be associated with mitochondria from the AD brain, but not mitochondria from control brain,[21,191] and APP can interact with the translocase of the outer mitochondrial membrane (TOMM40) and TIMM23.[191] The human AD autopsy brain shows evidence for mitochondrial impairments.[180] High mitochondrial APP levels mirror abnormalities in respiratory chain subunit levels and activity and enhanced ROS production.[191] Aβ can interact with the mitochondrial matrix protein

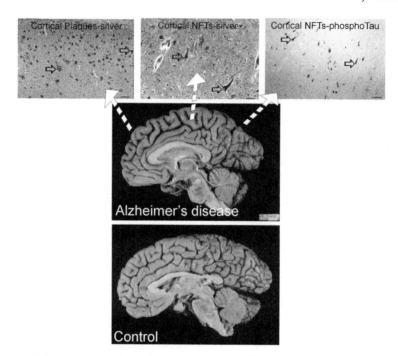

FIG. 4. Brain atrophy and neurodegeneration in AD. Midsagittal views (center pictures) of the brains from an 85-year-old individual with AD and an 86-year-old normal individual. The microscopic neuropathological hallmarks of AD are senile plaques (scale bar=200μm), and neurofibrillary tangles (NFTs, scale bar=50μm) The silver stain detects deposits and accumulations of abnormal proteins such as a amyloid beta protein-containing senile plaques (open arrows) in cerebral cortex (top left image) and NFTs in neurons and neuronal tombstones (open arrows, top middle image). Antibodies can be used to detect protein constituents of NFTs in neurons such a hyperphosphorylated tau (open arrows in top right image). These abnormalities are microscopic pathological entities associated with AD.

Aβ-binding alcohol dehydrogenase in human AD brain and is believed to participate in mitochondrial dysfunction and oxidative stress.[193] A possible intraneuronal Aβ-mitochondria link was shown by electron microscopy (EM) in aged nonhuman primate neocortex.[183]

In an APP transgenic (tg) mouse line (Tg2576), Aβ was also found to associate with mitochondria isolated from cerebral cortex.[192] It has also been reported that Aβ interacts with CyPD in mouse and human cerebral cortex mitochondria to potentiate synaptic stress.[194] Genetic deletion of the *ppif* gene (CyPD knockout) in human mutant APP tg mice (J-20 line) protected neurons from Aβ- and oxidative stress-induced cell death.[194] However, these abnormalities might not be related to mPTP-driven cell death because these mice, and most other mouse models of AD, show scant or modest evidence for neurodegeneration resulting in neuronal cell death, despite tremendous brain burdens of Aβ.[195]

B. Parkinson's Disease

PD is a chronically progressive, age-related, incapacitating movement disorder in humans. Estimates indicate that 4–6 million people have been diagnosed with PD and that ≈2% of the population will suffer from the disease at some time in life. The greatest prevalence occurs in the United States (100–250 cases per 100,000),[196] making PD as the second most common adult-onset neurodegenerative disease after AD. Most PD cases are sporadic (no known gene defects). However, some cases, particularly those with early onset, are familial (genetic) (Table II). Parkinsonism can be acquired from trauma and can be part of more global degenerative syndromes affecting multiple systems with features of multiple system atrophy, progressive supranuclear palsy, corticobasal degeneration, and dementia with Lewy bodies (LBs). Primary PD patients develop progressive resting tremor (4–7 Hz), rigidity (stiffness), bradykinesia (slowing of movement), akinesia, gait disturbance, and postural instability.[197] The disease progression is also associated with mood disturbances, dementia, sleep disturbances, and autonomic dysfunction.[197] There are currently no cures for PD. Medications and neurosurgery can relieve some of the symptoms.

A major neuropathological feature of PD is the degeneration of dopaminergic neurons in the substantia nigra pars compacta (SNpc) and in other brainstem regions, which causes the movement disorder.[198] However, PD should be regarded as a multiregional, multisystem neurodegenerative disorder in which the pathology appears in a regionally specific sequence, beginning in the dorsal motor nucleus of the vagus and olfactory bulbs and anterior nucleus followed by the locus coeruleus and then the SNpc, at which time (when ≈50–70% of SNpc neurons are lost) a clinical diagnosis of PD becomes possible.[199] The movement disorder in PD is thought to arise from reduced dopaminergic innervation of the striatum resulting from the loss of SNpc neurons.[198] The effect of reduced dopaminergic input is overactivity of striatal neurons that project to, and inhibit, neurons in external globus pallidus (GPe), thereby reducing the normal GPe inhibition of excitatory subthalamic neurons.[200] In addition, due to actions of dopamine on different dopamine receptor subtypes, there is also loss of normal dopaminergic excitation of striatal neurons that innervate the internal GP (GPi) and SN reticularis, causing increased γ-aminobutyric acidergic inhibition of thalamic nuclei that drive activation of cerebral cortex (Fig. 5).[198] The dopaminergic deficit in PD thus functionally translates to overactivity of the subthalamic nucleus and GPi.[198]

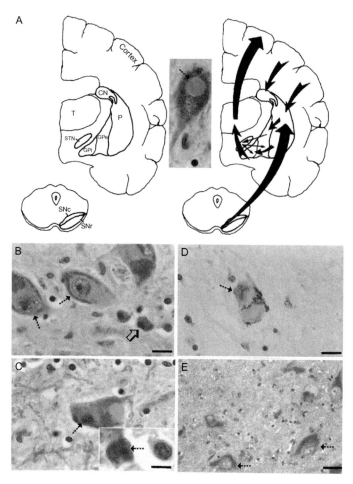

FIG. 5. Basal ganglia circuits in control of movement and SNpc neuron degeneration in people
with PD. (A) The basal ganglia are comprised (left panel) of the caudate nucleus (CN), putamen
(P), globus pallidus external (GPe) and internal (GPi) divisions, the subthalamic nucleus (STN), and
the substantia nigra compact (SNpc) and reticular (SNr) divisions. The cerebral cortex and
thalamus (T), although not part of the basal ganglia, participate in the connectivity loops (right
panel). The major excitatory input to the striatum (the caudate nucleus and putamen) is from the
cerebral cortex (top, right panel). The striatum, in turn, projects to the globus pallidus and the
substantia nigra reticular division. Striatal activity is modulated by an extensive dopaminergic input
from the SNpc. The major output of the basal ganglia is directed toward the thalamus, originating
from GPi and from the substantia nigra reticular division (not shown). The thalamic projection to
the cerebral cortex (premotor and supplementary motor areas) drives the activity of motor cortex
that executes somatic movements. Between the two panels is a SNpc neuron with a LB (arrow) seen
by hematoxylin–eosin (H&E) staining. (B and C) Degeneration of SNpc neurons in human PD.
H&E staining shows that the degeneration of pigmented SNpc neurons is characterized by
chromatolysis (B, hatched arrows) and nuclear condensation (C, hatched arrow), and severe

Another neuropathological feature of PD is eosinophilic proteinaceous intraneuronal or intraglial inclusions, known as LBs (Fig. 5A). LBs are comprised of a dense core of filamentous material enshrouded by filaments that are 10–20 nm in diameter that contain ubiquitin and α-synuclein (α-syn).[201] It is not clear if LBs are related causally to the disease process or are consequences of it.

The molecular pathogenesis of PD is still not understood. Epidemiological studies reveal several risk factors for developing idiopathic PD, in addition to aging. Pesticides now have been linked convincingly to the development of PD.[202] Herbicides, well water (contaminated with pesticides), and industrial chemicals are possible neurotoxic agents related to the development of PD.[203] It has been believed for over a decade that mitochondrial dysfunction is related to the pathogenesis of PD. Complex I activity was found to be reduced in the SNpc and platelets of PD cases, but changes in skeletal muscle remain contentious.[203] Complex I inhibitors, notably rotenone, cause damage to dopaminergic neurons in animal models.[204]

C. Gene Mutations that Cause Some Forms of PD

About 5–10% of PD patients suffer a genetic (familial) form of the disease.[203] Gene mutations with autosomal-dominant or autosomal-recessive inheritance patterns have been identified (Table II). PD-linked mutations occur in the genes encoding α-syn, Parkin, ubiquitin carboxy-terminal hydrolyase-L1 (UCHL1), phosphatase and tensin homolog-induced putative kinase-1 (PINK1), DJ-1, and leucine-rich repeat kinase-2 (LRRK2). In rare autosomal-dominant inherited forms of PD, missense mutations in the α-syn gene (*PARK1*) result in the amino acid substitutions Ala-53→Thr, Ala-30→Pro, or Glu-46→Lys. In addition, duplication and triplication mutations in the α-syn gene (*PARK4*) have been found.[205,206] A missense mutation in the UCHL1 gene (*PARK5*), resulting in the amino acid substitution Ile-93→Met, can also cause very rare autosomal-dominant PD.[207] Loss-of-function mutations due to large deletions

---

soma attrition (C, inset hatched arrow). The neuronal chromatolysis (in B) is indicated by the eccentrically placed nucleus, pale cytoplasm, and peripheral margination of the Nissl substance. Glial/macrophage-like cells (B, open arrow) are laden with phagocytosed cellular debris. The nucleus of SNpc neurons undergoes considerable condensation (C, hatched arrow) while the Nissl substance dissipates, but before appreciable somal shrinkage. The cell body of SNpc neurons then becomes attritional (C, inset, hatched arrow), resulting in residual neurons that are ≈10–20% their normal size. The cell shown is an atrophic neuron, rather than a debris-laden macrophage, because of the presence of a condensed nucleus with a single prominent nucleolus. This degeneration pattern could be indicative of autophagy. (D) The nuclear condensation stage of pigmented SNpc neuron degeneration is characterized by the appearance of DNA double-strand breaks as detected by TUNEL (arrow, brown staining). (E) SNpc neurons accumulate cleaved caspase-3 (arrows, brown staining). Scale bars: B, 20μm (same for C and D); C inset, 6μm; E, 45μm. (See Color Insert.)

and truncations, and also missense or nonsense mutations, in parkin (*PARK2*), PINK1 (*PARK6*), and DJ-1 (*PARK7*) cause autosomal-recessive PD.[208–211] The more commonly occurring autosomal-dominant, and possibly sporadic, PD are caused by several missense mutations in the LRRK2 gene (*PARK8*), resulting in amino acid substitutions Tyr-1654→Cys, Arg-1396→Gly, Tyr-1699→Cys, Arg-1441→Cys, Ile-1122→Val, or Ile-2020→Thr.[212,213] *PARK9* has been ascribed to a deletion mutation (cytosine at nucleotide position 3057) or a guanine-to-adenine transversion at a splice site of exon 13 in the ATP13A2 gene that encodes a predominantly neuronal P-type lysosomal ATPase.[214] Potential lysosomal dysfunction related to ATP13A2 mutant proteins might tie into PD etiology through abnormalities in autophagy. Mutations of genes at other PD loci are more controversial (Table II).

1. α-Synuclein

α-Syn is a relatively small (140 amino acids), very abundant (~1% of total protein) protein found in cells throughout the nervous system and is particularly enriched in axon terminals.[215–217] The functions of α-syn are not entirely understood. However, growing evidence supports a role for α-syn in neurotransmitter release. Mice without α-syn have no overt phenotype,[218] but neurons deficient in α-syn have a reduction in the reserve pool of synaptic vesicles needed for responses to tetanic stimulation and show defective mobilization of dopamine and glutamate.[217,219] Without α-syn, neurons have impaired, long-lasting enhancement of evoked and miniature neurotransmitter release.[220] α-Syn is highly mobile and rapidly dissociates from synaptic vesicle membranes after fusion in response to neuronal activity.[221] α-Syn appears to function as a molecular chaperone to assist cysteine-string protein-α in the folding and refolding of SNARE synaptic proteins.[222,373,374]

α-Syn is a soluble monomeric protein that can associate with mitochondrial membranes.[223] α-Syn can undergo α-helix→β-sheet conformational transition, which leads to fibril formation.[224] α-Syn is a major structural component of LBs, forming the ≈10-nm fibrils, but in most neurodegenerative diseases LBs are associated with accumulation of wild-type, not mutant, α-syn.[201] α-Syn mutations cause increased levels of protofibrils, possibly the more toxic form of the protein.[225] α-Syn protofibrils might also be toxic by making membranes of cells more porous.[226] Overexpression of human wild-type or mutant α-syn in cultured cells elevates the generation of intracellular ROS[227,228] and causes mitochondrial deficits.[227] Moreover, expression of mutant α-syn increases cytotoxicity to dopamine oxidation products.[229] Aggregation of wild-type and mutated α-syn is associated with enhanced cell death in cultured cells.[230] Nitration of α-syn, signifying the presence of potent reactive nitrogen species such as peroxynitrate (ONOO⁻), or its free-radical derivative nitrogen dioxide (·NO₂), is a major signature of human PD and other synuclinopathies and might be critical to the aggregation process.[231,232]

2. UCHL1 AND PARKIN

UCHL1 is a very abundant protein (~1–2% total soluble protein in brain) that functions in the formation and recycling of ubiquitin monomers for the ubiquitin–proteasome pathway.[233] This pathway is important for intracellular protein turnover and degradation and functions generally in quality control of proteins in cells to eliminate misfolded, mutated, and damaged proteins.[234] Ubiquitin is an abundant, small (≈8.5 kDa) protein that is attached covalently to lysine aliphatic chains in proteins to mark them for degradation carried out by the 26S proteasome. UCHL1 hydrolyses the C-terminus of fusion proteins containing polyubiquitin molecules and ribosomal protein, thereby generating ubiquitin monomers. *In vitro*, PD-linked mutant UCHL1 has reduced enzyme activity,[235] and inhibition of UCHL1 is associated with production of α-syn aggregates,[236] indicating that α-syn can be degraded by the proteasome.

The ubiquitination of proteins is catalyzed by the activities of three enzymes called ubiquitin-activating enzyme (E1), ubiquitin-conjugating enzyme (E2), and ubiquitin ligase (E3).[234] *PARK2*, encoding parkin, which is a ubiquitin E3 ligase, causes juvenile-onset recessive PD (before 40 years of age)[210] with relatively confined neuronal loss in the SNpc and locus coeruleus, but with an absence of LBs. Several substrates of parkin have been identified, including α-syn, synphilin-1, and other synaptic proteins.[237] Mutations in the *PARK2* result in loss of function of E3, thereby possibly causing some substrates of parkin to accumulate and aggregate within cells. One parkin mutation found in a Turkish patient (Gln-311→X), replacing a glutamine residue at position 311 with a stop codon, causes a C-terminal truncation of 155 amino acids of Parkin.[238]

3. PINK1

The *PARK6* locus contains the PINK1 gene.[209,211] *PARK6* kindreds have juvenile-onset PD and truncation mutations, missense mutations (His-271→Gln; Gly-309→Ala; Leu-347→Pro; Glu-417→Gly; Arg-246→X, where X is any other amino acid; Trp-437→X), or compound nonsense mutations (Gln-309→X/Arg-492→X).

PINK1 is a 581-amino acid protein (≈63 kDa) that contains a domain highly homologous to the serine/threonine protein kinases of the calcium/calmodulin family, and a mitochondrial targeting motif.[211] Thus, PINK1 is a mitochondrial kinase. It is processed at the N-terminus in a manner consistent with mitochondrial import, but the mature protein is also present in the cytosol.[239] Both human wild-type and mutant PINK1 localize to mitochondria.[240] Interestingly, most of the reported mutations are in the putative kinase domain. PINK1 is expressed in many adult human tissues.[241] In adult rodents, PINK1 is expressed throughout the brain.[242] It is unclear how PINK1 mutations cause the selective death of SNpc neurons in human PD (Fig. 5). PINK1 appears to function in mitochondrial trafficking by forming a multiprotein complex with the GTPase Miro and the adaptor protein Milton.[243] PINK1 may protect human dopaminergic neuroblastoma cells (SH-SY5Y) against mitochondrial malfunction under conditions of cell stress.[244] In rat neuroblastoma cells, mutant PINK1 can induce abnormalities in mitochondrial Ca^{2+} influx and aggravate the cytopathology caused by mutant α-syn in a mechanism that involves the mPTP (Fig. 1).[245]

4. DJ-1

The *PARK7* locus contains the DJ-1 gene.[208] *PARK7* kindreds can have homozygous deletion of a large region within the DJ-1 gene causing complete loss of DJ-1 expression or homozygous missense mutations in the DJ-1 gene resulting in single amino acid substitutions in the DJ-1 molecule (Met-26→Ile, Glu-64→Asp, Leu-166→Pro).[208]

DJ-1 is a small (189-amino acid, ∼20–25 kDa) protein with multiple apparent functions involving cellular transformation, male fertility, control of protein–RNA interaction, and oxidative stress response.[208] The protein exists *in vivo* as a dimer.[246] DJ-1 is expressed throughout the mouse nervous system,[247] where it might act as a neuroprotective intracellular redox sensor that can localize to the cytoplasmic side of mitochondria.[248] The localization of DJ-1 to mitochondria is associated with protective actions against some mitochondrial poisons.[248] Some DJ-1 mutant proteins have abnormalities in dimer formation and decreased stability.[249,250] It remains to be determined how mutated DJ-1 proteins trigger the degeneration of SNpc neurons in human PD (Fig. 5).

5. LRRK2

The *PARK8* locus contains the LRRK2 gene.[212,213] Mutations in this gene are, to date, the most common in both familial and sporadic PD. The LRRK2 protein is a large multidomain protein (2527 amino acids, 286 kDa), also called dardarin (derived from the Basque word dardara, meaning tremor), that is expressed throughout the body.[212,213] Currently, it is not evident how *PARK8* mutations relate to the selective death of neurons that causes human PD (Fig. 5). LRRK2 contains leucine-rich repeat domains, a Ras/small GTPase domain, a nonreceptor tyrosine kinase-like domain, and a WD-40 domain, consistent with the architecture of multifunctional Ras/GTPases of the Ras of complex family.[212,213] The presence of leucine-rich and WD40 domains suggests that LRRK2 is capable of multiple protein–protein interactions. The GTPase activity indicates that LRRK2 functions as a molecular switch, possibly involved in cytoskeleton organization and vesicle trafficking. The kinase domain may belong to the mitogen-activated protein kinase kinase kinase (MAPKKK or MEKK) family of kinases. Studies of rodent brain show little or no expression of LRRK2 in SNpc neurons.[251,252] However, expression of LRRK2 is high in dopamine-innervated brain regions.[251] Recent work shows that LRRK2 can influence mitochondrial- and death receptor-mediated cell death in cultured cells.[253,254] These findings might be hints that the target of SNpc neurons (i.e., the striatum) and SNpc neuron target deprivation are important to the understanding of pathogenic mechanisms of PD (Fig. 5).

D. PD α-Syn Transgenic Mice Develop Neuronal Mitochondrial Degeneration and Cell Death

Identification of human genes linked to familial PD by molecular genetics drives experimental work on the generation of animal and cell models of PD. Parkin[-/-] null mice have a normal lifespan, do not develop any major neurological abnormalities, have no loss of midbrain dopaminergic neurons, and do not form intracellular inclusions.[255,256] However, these mice exhibit modest evidence of dopaminergic presynaptic dysfunction in striatum and possible deficits in behavioral tests indicative of nigrostriatal dysfunction,[255] although this finding has not been confirmed in another mouse line.[256] Parkin[-/-] mice show decrease in proteins involved in mitochondrial oxidative phosphorylation and oxidative stress in ventral midbrain and exhibit reduced mitochondrial respiration in striatum, but they have no mitochondrial ultrastructural abnormalities.[257] Mice with null mutations in DJ-1 also have a normal lifespan and do not develop an overt phenotype or loss of dopaminergic neurons, but behavioral tests reveal age-dependent motor deficits and neurochemical assessments show altered striatal dopamine content.[258] DJ-1 null mice also show altered D2 dopamine receptor-mediated function.[259] In contrast, transgenic mice expressing the Parkin Q311X truncation mutation develop a progressive hypokinetic disorder, degeneration of SNpc neurons, and loss of striatal dopamine.[260] Thus, Parkin could be important for maintenance of mitochondrial function or mitochondrial turnover through autophagy and synaptic integrity distally within the SNpc neuron target region.

Several tg mouse lines have been made using a variety of different promoters to drive expression of human full-length wild-type or mutant α-syn.[230,261–266] Of these lines, mice expressing human A53T mutant α-syn have a shortened lifespan and develop a severe movement disorder and synucleinopathy.[230,263,266] It is noteworthy that there have been no reports of robust dopamine SNpc neuron degeneration in full-length α-syn tg mice or in any other tg or null mouse models of PD-linked genes. However, tg mice expressing a truncation mutant of human α-syn have an abnormal development-related loss of SNpc neurons.[267] Cell death mechanisms or thresholds for cell death activation in human and mouse brain dopamine neurons might differ.

α-Syn tg mice do develop extensive cell death and neuronal loss in other regions of brain and in spinal cord.[268] These tg mice express high levels of human wild-type or mutant (A53T and A30P) α-syn under the control of the mouse prion protein promoter.[263] Mice expressing A53T α-syn (lines G2-3 and H5), but not mice expressing wild-type (line I2-2) or A30P (line O2) α-syn, develop adult-onset progressive motor deficits, including reduced spontaneous activity with bradykinesia, mild ataxia, and dystonia at ≈10–15 months of age, followed by rapidly progressive paralysis and death.[263] A53T mice develop intraneuronal inclusions, mitochondrial degeneration, and cell death in neocortex, brainstem, and spinal cord.[268] Brainstem neurons and spinal motor neurons display a prominent chromatolysis reaction and axonal spheroids,[268] typical of that seen after axonal injury.[269] A53T mice form LB-like inclusions in neocortical and spinal motor neurons and have progressive profound loss (≈75%) of motor neurons that causes their paralysis.[268] Motor neuron loss in A53T mice has been shown by another group.[222]

Mitochondrial pathology in A53T mice involving mtDNA damage is seen frequently in the absence of nuclear DNA damage in large brainstem neurons and spinal motor neurons.[268] Subsets of mitochondria in brainstem and spinal cord cells in A53T mice appear dysmorphic, becoming shrunken, swollen, or vacuolated.[268] Human α-syn is found bound to some mitochondria in degenerating neurons in A53T mice.[268] Some abnormal intracellular inclusions in these cells are degenerating mitochondria. A mitochondrial defect in A53T mice is further indicated by biochemical evidence revealing loss of complex IV activity.[268]

The mechanisms for this mtDNA damage may be related to oxidative stress, which is suggested by evidence that mitochondrial associated metabolic proteins are oxidized in A30P mice.[270] α-Syn can generate H_2O_2[271] and $^\bullet OH$[229] in vitro upon incubation with Fe(II). Evidence for $ONOO^-$-mediated oxidative/nitrative stress in A53T mouse motor neurons has come from the observation of nitrated human synuclein.[268] Nitrated synuclein formed inclusions in motor neurons consistent with in vitro data showing that $ONOO^-$ promotes the formation of stable α-syn oligomers.[272,273] Our data showing mtDNA damage is in line with the presence of $ONOO^-$ or its derivatives near mitochondria, because $ONOO^-$ or products of $ONOO^-$ are directly genotoxic by causing single- and double-strand breaks in DNA.[23] Moreover, the loss of complex IV enzyme activity without a change in protein level[268] might be explained by inactivation of this mitochondrial enzyme by nitration. Overall, $ONOO^-$-mediated damage in mitochondria may be a key pathological mechanism leading to motor neuron degeneration in A53T mice.

The reasons for the vulnerability of mouse motor neurons to human A53T mutant α-syn are not clear. These mice express high levels of mRNA and protein for human α-syn in the forebrain, diencephalon, and midbrain,[263] but these regions are much less vulnerable than spinal cord. A53T α-syn causes axonopathy[230,266]; thus motor neuron vulnerability could be related to their long myelinated axons and interactions with oligodendrocytes and Schwann cells for myelin support. Motor neuron vulnerability could also be related to their unusual expression of inducible NO synthase (iNOS) in mitochondria.[274,275] Moreover, distal axonopathy and muscle disease may have roles in the pathogenesis in A53T mice.[268] Prominent skeletal muscle denervation occurs in α-syn tg mice.[266,268] This work is intriguing because the original goal was to develop a tg mouse model of PD, but our result is a profound mouse model of ALS. Thus, the mutant α-syn A53T tg mouse is a new model to study mechanisms of motor neuron degeneration and could provide insight into the selective vulnerability of motor neurons in age-related disorders and the possible roles of α-syn in synaptic maintenance and diseases of long-axon neurons.[268]

E. Amyotrophic Lateral Sclerosis

ALS is a progressive and severely disabling neurological disease in humans characterized by initial muscle weakness, then muscle atrophy, spasticity, and eventual paralysis and death, typically 3–5 years after symptoms begin.[276] The cause of the spasticity, paralysis, and death is progressive degeneration and elimination of upper motor neurons in cerebral cortex and lower motor neurons in brainstem and spinal cord (Fig. 6).[276,277] Degeneration and loss of spinal and neocortical interneurons also have been found in human ALS.[278,279] More than 5000 people in the United States are diagnosed with ALS each year (ALS Association, http://www.alsa.org), and, in parts of the United Kingdom, three people die every day from some form of motor neuron disease (http://www.mndassociation.org). Other than life support management, no effective treatments exist for ALS.[280]

It is still not understood why specific neuronal populations are selectively vulnerable in ALS, such as certain somatic motor neurons and interneurons.[276–279] The molecular pathogenesis of ALS is poorly understood, contributing to the lack of appropriate target identification and effective mechanism-based therapies to treat even the symptoms of this disease. As with other neurodegenerative diseases, both sporadic and familial forms of ALS occur. The majority of ALS is sporadic. Aging is a strong risk factor for ALS because the average age of onset is 55 years (ALS Association, www.alsa.org). Familial forms of ALS have autosomal-dominant or autosomal-recessive inheritance patterns and make up ≤10% of all ALS cases. ALS-linked

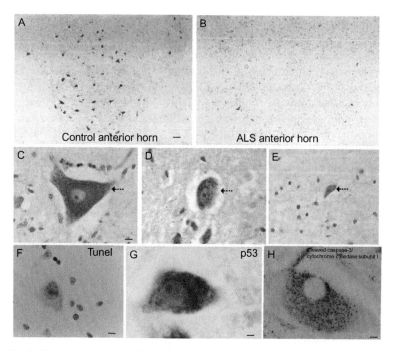

FIG. 6. Motor neurons in spinal cord degenerate in ALS. (A) In normal control individuals, the anterior horns of the spinal cord contain many large, multipolar motor neurons (dark cells). (B) In ALS cases, the anterior horn is depleted of large neurons (dark cells) and remaining neurons are atrophic. These attritional chromatolytic motor neurons display a dark condensed nucleus as seen microscopically. Scale bar in A=76 μm (same for B). (C, D, and E) Nissl staining shows that the degeneration of motor neurons in familial ALS is characterized by shrinkage and progressive condensation of the cytoplasm and nucleus. The motor neuron in (C) (arrow) appears normal. It has a large, multipolar cell body and a large nucleus containing a reticular network of chromatin and a large nucleolus. Scale bar=7 μm (same for D and E).The motor neuron in (D) (arrow) has undergone severe somatodentritic attrition. The motor neuron in (E) is at near end-stage degeneration (arrow). The cell has shrunken to ≈10% of its normal size and has become highly condensed. The cell in (E) is identified as a residual motor neuron based on the nucleus and nucleolus (seen as eccentrically placed darkly stained component to the lower left of cytoplasmic pale area) and residual large Nissl bodies. (F) Cell death assays (e.g., TUNEL) identify subsets of motor neurons in the process of DNA fragmentation. Nuclear DNA fragmentation (brown labeling) occurs in motor neurons as the nucleus condenses and the cell body shrinks. Motor neurons in the somato-dendritic attrition stage accumulate DNA double strand breaks. Scale bar=7 μm. (G) In individuals with ALS, p53 accumulates in the nucleus (brown labeling) of motor neurons. Scale bar=5 μm. (H) Degenerating motor neurons in human ALS are immunopositive for cleaved caspase-3 (black-dark green labeling) in the somatodendritic attrition stage. Around the nucleus (pale circle), motor neurons accumulate discrete mitochondria (brown-orange labeling, detected with antibody to cytochrome *c* oxidase subunit I) exhibiting little light microscopic evidence for swelling. Scale bar=5 μm. (See Color Insert.)

mutations occur in the genes (Table III) encoding SOD1 (*ALS1*), Alsin (*ALS2*), senataxin (*ALS4*), fused in sarcoma (FUS, *ALS6*), vesicle associated membrane protein (VAMP/synaptobrevin)-associated protein B (VAPB, *ALS8*), p150 dynactin, and TAR-DNA binding protein (TADBP or TDP43).[281–284] Most recently, variations in the phosphoinositide phosphatase *FIG4* gene have been found to cause ALS11.[285] Several other genes are believed to be susceptibility factors for ALS (Table III).

TABLE III
MUTANT/POLYMORPHIC GENES LINKED TO FAMILIAL ALS

Locus	Inheritance	Gene	Protein name/function
ALS1/21q22	Autosomal dominant (adult onset)	SOD1	Cu/Zn superoxide dismutase/ dismutation of superoxide
ALS2/2q33.2	Autosomal recessive (juvenile onset primary lateral sclerosis)	Alsin	Alsin/guanine exchange factor for RAB5A and Rac1
ALS4/9q34	Autosomal dominant (adult onset)	SETX	Senataxin/helicase, RNA processing
ALS6/16q12	Autosomal recessive (adult onset)	FUS	Fused in sarcoma, component of heterogeneous nuclear ribonuclear protein complex; RNA/DNA-binding protein
ALS8/20q13.33	Autosomal dominant	VAPB	VAMP-associated protein B/part of SNARE complex
2q13	Autosomal dominant (adult onset, atypical ALS)	DCTN1	Dynactin p150^{glued}/axonal transport, link between dynein and microtubule network
ALS10/1p36.22	Autosomal dominant	TARDBP	TAR DNA-binding protein, DNA- and RNA-binding protein, regulates RNA splicing
ALS11/6q21	Autosomal recessive	FIG4	FIG4 homolog, SAC1 lipid phosphatase domain containing; regulates phosphotidylinositol turnover
14q11.1–q11.2	Susceptibility factor	ANG	Angiogenin; angiogenesis; stimulates production of rRNA
22q12.2	Susceptibility factor	NEFH	Neurofilament, heavy polypeptide; neurofilament subunit
12q12–q13	Susceptibility factor	PRPH	Peripherin; intermediate filament formation
5q13	Susceptibility factor	SMN1/ SMN2	Survival motor neuron; RNA processing
7q36.6	Susceptibility factor?	DPP6	Dipeptidyl-peptidase 6; S9B serine protease, binds voltage-gated potassium channels

F. Mitochondrial Dysfunction in Human ALS

Human ALS is associated with mitochondrial abnormalities. Structural abnormalities in mitochondria are seen by EM in skeletal muscle, liver, spinal motor neurons, and motor cortex of ALS patients.[286,287] A mutation in cytochrome c oxidase subunit I was found in a patient with a motor neuron disease phenotype.[288] Another patient with motor neuron disease had a mutation in a mitochondrial tRNA gene.[289] One type of mtDNA mutation, called the common mtDNA deletion (mtDNA4977), is found nonuniformly within different human brain areas. The highest levels are detected in the striatum and SN.[28,29] However, no significant accumulation of the 5-kb common deletion in mtDNA has been found by single-cell analysis of motor neurons from sporadic ALS cases compared to controls.[290] This finding contrasts with the high levels of mtDNA deletions that accumulate in SNpc neurons in human PD.[291,292] Overall, there is a lack of strong direct evidence for mitochondrial involvement in human ALS, despite the existence of associational/correlative data from human, animal, or cell models.

Notwithstanding the lack of a clear seminal role for mitochondria in disease causation, intracellular Ca^{2+} abnormalities and excitotoxicity may link mitochondrial dysfunction and oxidative stress to ALS. Mitochondria regulate cytoplasmic Ca^{2+} levels.[1,3,293] EM on skeletal muscle biopsies of people with sporadic ALS shows changes indicative of elevated Ca^{2+} in motor neuron terminals, with some mitochondria showing an augmented Ca^{2+} signal.[294] Excitotoxicity has long been implicated in the pathogenesis of ALS[295] and is another possible mechanism of motor neuron damage in ALS.[216,276] While many drugs targeting excitotoxicity as a mechanism have failed in human ALS clinical trials, the anti-excitotoxic drug riluzole is currently approved by the Food and Drug Administration for ALS treatment.[280] Many sporadic ALS patients have reduced levels of synaptosomal high-affinity glutamate uptake[295] and astroglial glutamate transporter EAAT2 (excitatory amino acid transporter 2 or GLT1) in motor cortex and spinal cord.[296] Reductions in levels of activity of EAAT2 in spinal cord could increase the extracellular concentrations of glutamate at synapses on motor neurons. Motor neurons might be particularly sensitive to glutamate excitotoxicity because they have a low proportion of GluR2- or under-edited AMPA subtype glutamate receptor on their surfaces, predisposing these cells to risk of excess Ca^{2+} entry and mitochondrial perturbations.[297,298] Cell culture studies show that excess glutamate receptor activation in neurons can cause increased intracellular Ca^{2+}, mitochondrial ROS production, bioenergetic failure, and mitochondrial trafficking abnormalities.[299] Ca^{2+}-induced generation of ROS in brain mitochondria is mediated by mPT.[300] Motor neurons are particularly affected by inhibition of mitochondrial metabolism, which can cause elevated cytosolic Ca^{2+} levels, excitability, and oxidative stress.[301]

Markers of oxidative stress and ROS damage are elevated in postmortem human ALS tissues.[302] In sporadic and familial ALS, protein carbonyls are elevated in motor cortex.[303] Tyrosine nitration is elevated in human ALS nervous tissues.[304–306] Studies of respiratory chain enzyme activities are discrepant. Experiments have shown increases in complex I–III activities (Fig. 1) in vulnerable and nonvulnerable brain regions in patients with familial ALS mutant SOD1,[307] but experiments by others show decreased complex IV activity in spinal cord ventral horn[308] and skeletal muscle[309] of sporadic ALS cases. In sporadic ALS skeletal muscle, reductions in activity of respiratory chain complexes with subunits encoded by mtDNA are associated with reduced mtDNA content[309] and decreased NOS levels.[310] Alterations in skeletal muscle mitochondria are progressive[311] and could be intrinsic to skeletal muscle and not due merely to neurogenic atrophy, as commonly assumed.

G. Human ALS and Mitochondrial-Orchestrated PCD Involving p53

PCD appears to contribute to the selective degeneration of motor neurons in human ALS, albeit seemingly as a nonclassical form differing from apoptosis (Fig. 6).[312,313] Motor neurons appear to pass through sequential stages of chromatolysis, suggestive of initial axonal injury,[269] somatodendritic attrition without extensive cytoplasmic vacuolation, and then nuclear DNA fragmentation, nuclear and chromatin condensation, and cell death (Fig. 6).[312] Motor neurons in people who have died from sporadic ALS and familial ALS show the same type of degeneration.[312] This cell death in human motor neurons is identified by genomic DNA fragmentation (determined by DNA agarose gel electrophoresis and in situ DNA nick-end labeling) and cell loss and is associated with accumulation of perikaryal mitochondria, cytochrome c, and cleaved caspase-3.[312–314] p53 protein also increases in vulnerable CNS regions in people with ALS, and it accumulates specifically in the nucleus of lower and upper motor neurons with nuclear DNA damage.[315–317] This p53 is active functionally because it is phosphorylated at Ser392 and has increased DNA-binding activity.[315,316] However, the morphology of this cell death is distinct from classical apoptosis, despite the nuclear condensation.[14,312,316] Nevertheless, Bax and Bak1 protein levels are increased in mitochondria-enriched fractions of selectively vulnerable motor regions (spinal cord anterior horn and motor cortex gray matter), but not in regions unaffected by the disease (somatosensory cortex gray matter).[312] In marked contrast, Bcl-2 protein is depleted severely in mitochondria-enriched fractions of affected regions and is sequestered in the cytosol.[312] Although these Western blot results lacked direct specificity for motor neuron events,[312] subsequent immunohistochemistry (Fig. 6H)[313] and laser capture microdissection of motor neurons combined with mass spectroscopy-protein profiling have confirmed the presence of intact active caspase-3 in human ALS motor neurons.[314]

Our studies[312–316] support the concept of an aberrant reemergence of a mitochondrial-directed PCD mechanism, involving p53 activation and redistributions of mitochondrial cell death proteins, participating in the pathogenesis of motor neuron degeneration in human ALS (Figs. 1 and 6). The morphological and biochemical changes seen in human ALS are modeled robustly and faithfully at structural and molecular levels in axonal injury/target deprivation (axotomy) models of motor neuron degeneration in adult rodents,[317,318] but not in the current most commonly used human mutant SOD1 tg mouse models.[274,313] However, we have created recently a new tg mouse expressing human mutant SOD1 only in skeletal muscle and have found a motor neuron disease phenotype with morphological and biochemical changes very similar that those seen in human ALS motor neurons.[319]

H. Mitochondrial Pathobiology in Cell and Mouse Models of ALS

A common mutation in human *SOD1* that is linked to familial ALS (Table III) is G93A.[320] SOD1 is a metalloenzyme comprising 153 amino acids (≈ 16 kDa) that binds one copper ion and one zinc ion per subunit and is active as a noncovalently linked homodimer.[5,321] SOD1 is responsible, through catalytic dismutation, for the detoxification and maintenance of intracellular $O_2^{\bullet-}$ concentration in the low femtomolar range (Fig. 1).[5,321] SOD1 is ubiquitous (intracellular SOD concentrations are typically ≈ 10–$40\,\mu M$) in most tissues, possibly with highest levels in neurons.[322]

Cell culture experiments reveal mitochondrial dysfunction in the presence of human mutant SOD1 (mSOD1).[323] Expression of several mSOD1 variants increases mitochondrial $O_2^{\bullet-}$ levels and causes toxicity in rat primary embryonic motor neurons,[324] human neuroblastoma cells,[325] and mouse neuroblastoma-spinal cord (NSC)-34 cells, a hybrid cell line with some motor neuron-like characteristics produced by fusion of motor neuron-enriched embryonic mouse spinal cord cells with mouse neuroblastoma cells.[326] These responses can be attenuated by overexpression of manganese SOD.[325] ALS-mSOD1 variants, compared to human wild-type SOD1, associate more with mitochondria in NSC-34 cells and appear to form cross-linked oligomers that shift the mitochondrial GSH/GSSG ratio toward oxidation.[323]

Gurney *et al.* were the first to develop tg mice that express human mSOD1.[313,320,327,328] In these mice, human mSOD1 is expressed ubiquitously, driven by its endogenous promoter in a tissue/cell nonselective pattern, against a background of normal wild-type mouse SOD1.[327] Effects of this human mutant gene in mice are profound. Hemizygous tg mice expressing a high copy number of the G93A SOD1 mutant become completely paralyzed and die at ≈16–18 weeks of age.[327] G93A-mSOD1 mice with reduced transgene copy number have a much slower disease progression and die at ≈8–9 months of age.[124,327]

Spinal motor neurons and interneurons in mice expressing G93A[high]-mSOD1 undergo prominent degeneration. About 80% of lumbar motor neurons are eliminated by end-stage disease.[274,329] Subsets of spinal interneurons degenerate before motor neurons in G93A[high]-mSOD1 animals[274] and some are the glycinergic Renshaw cells.[329] Unlike the degeneration of motor neurons in human ALS,[312] motor neurons in G93A[high]-mSOD1 mice do not degenerate with a morphology resembling any form of apoptosis or hybrid apoptosis–necrosis.[14,124,274,313,316] The motor neuron degeneration more closely resembles a prolonged necrotic-like cell death process[124,274] involving early occurring mitochondrial damage, cellular swelling, and dissolution.[274,329–334] Biochemically, the death of motor neurons in G93A[high]-mSOD1 is characterized by cell body and mitochondrial swelling, as well as the formation of DNA single-strand breaks prior to double-strand breaks occurring in nuclear DNA and mtDNA.[274] The motor neuron death in G93A[high]-mSOD1 mice is independent of activation of caspases-1 and -3 and also appears to be independent of caspase-8 and AIF activation.[274] Indeed, caspase-dependent and p53-mediated apoptosis mechanisms might be blocked actively in G93A[high]-mSOD1 mouse motor neurons, possibly by upregulation of IAPs and changes in the nuclear import of proteins.[274] More work is needed on cell death and its mechanisms in G93A[low]-mSOD1 mice because these mice could be more relevant physiologically and preclinically to the human disease, compared to the G93A[high]-mSOD1 mouse.

Mitochondrial pathology has been implicated in the mechanisms of motor neuron degeneration in tg mSOD1 mouse models,[330–334] but until recently most evidence has been circumstantial. In different mSOD1 mouse models of ALS, mitochondria in spinal cord neurons exhibit structural pathology[274,329–334] and some of the mitochondrial degeneration occurs very early in the course of the disease.[274,330] Mitochondrial microvacuolar damage in motor neurons emerges, as seen by EM, by 4 weeks of age in G93A[high] mice.[274] It has been argued that mitochondrial damage in G93A[high]-mSOD1 mice is related to supranormal levels of SOD1 and might not be related causally to the disease process because tg mice expressing high levels of human wild-type SOD1 show some mitochondrial pathology.[333] However, mitochondrial abnormalities in motor neurons have been found histologically in G93A[low]-mSOD1 mice[334] and in mice with only skeletal muscle expression of human SOD1.[319] Thus, mitochondria could be primary sites of human SOD1 toxicity in tg mice irrespective of transgene copy number, tissue expression, and expression level of human SOD1, but direct, unequivocal causal relationships have been lacking.

Mutated and wild-type forms of human SOD1 could contribute to the development of ALS. Human mSOD1 proteins appear to acquire a toxic property or function, rather than having diminished $O_2^{\bullet-}$ scavenging activity.[282,335,336] Wild-type SOD1 can gain toxic properties through loss of Zn^{324} and oxidative modification.[337,338] A gain in aberrant oxidative chemistry appears to contribute to the mechanisms of mitochondriopathy in G93A[high] mice.[22,339] G93A-mSOD1 has enhanced free radical-generating capacity compared to wild-type enzyme[336] and can catalyze protein oxidation by hydroxyl-like intermediates and carbonate radical.[340] G93A[high] mice have increased protein carbonyl formation in total spinal cord tissue extracts at presymptomatic disease stages.[341] Protein carbonyl formation in mitochondrial membrane-enriched fractions of spinal cord is a robust signature of incipient disease.[341] A mass spectroscopy study of G93A[high] mice identified proteins in total spinal cord tissue extracts with greater than baseline carbonyl modification, including SOD1, translationally controlled tumor protein, and UCHL1.[342] Nitrated and aggregated cytochrome c oxidase subunit-I and α-syn accumulate in G93A[high] mouse spinal cord.[274] Nitrated MnSOD also accumulates in G93A[high] mouse spinal cord.[274]

Toxic properties of mSOD1 might also be mediated through protein binding or aggregation. Endogenous mouse SOD1[343] and human wild-type SOD1 and mSOD1[344] associate with mitochondria. Human SOD1 mutants associate with spinal cord mitochondria in mSOD1 mice and can bind Bcl-2,[345] thus potentially being decoys or dominant negative regulators of cell survival molecules (Fig. 1). It is not known whether this process occurs specifically in motor neurons. Binding of mSOD1 to mitochondria has been reported to be spinal cord-selective and age-dependent,[346] but this work also lacks cellular resolution. A recent *in vitro* study has shown that endogenous SOD1 in the mitochondrial intermembrane space controls cytochrome c-catalyzed peroxidation and that G93A-mSOD1 mediates greater ROS production in the intermembrane space compared to wild-type SOD1.[346] Human SOD1 mutants can also shift mitochondrial redox potential when expressed in cultured cells.[323] Nevertheless, the direct links between the physicochemical changes in wild-type and mutant SOD1 and the mitochondrial functional and structural changes associated with ALS and motor neuron degeneration remain uncertain.

EM studies of motor neurons in G93A[high] mice have shown that the OMM remains relatively intact to permit formation of mega-mitochondria.[124,274,329,332] Moreover, early in the disease of these mice, mitochondria in dendrites in spinal cord ventral horn undergo extensive cristae and matrix remodeling, while few mitochondria in motor neuron cell bodies show major structural changes.[124] Thus, the disease might start distally in mitochondria of motor neuron processes.[124] Another interpretation of ultrastructural findings is that the mSOD1 causes mitochondrial degeneration by inducing OMM extension and leakage and intermembrane space expansion.[347] Mechanisms for this damage could be related to mSOD1 gaining access to the mitochondrial intermembrane space.[343,344,347] This mitochondrial conformation seen by EM might favor the formation of the mPTP (Fig. 1). Indeed, we found evidence for increased contact sites between the OMM and the IMM in dendritic mitochondria in G93A[high] mice.[124] Another feature of motor neurons in young G93A[high] mice, before symptoms emerge, is the apparent fission of ultrastructurally normal mitochondria in cell bodies and fragmentation of abnormal mitochondria.[124] It is not clear whether the cristae and matrix remodeling and the apparent fragmentation and fission mitochondria are related or independent events, or whether these abnormalities interfere with mitochondrial trafficking. Nevertheless, morphological observations support the idea that mitochondria could be important for the pathobiology of mSOD1-induced toxicity of motor neurons in G93A[high] mice.

We have hypothesized that mitochondrial trafficking perturbations occur in motor neurons of mSOD1.[274] Some data support the novel idea that mitochondria might act as retrogradely transported couriers from distal regions (axon branches and dendrites) to the cell body of motor neurons in mSOD1 mice. G93A[high]-mSOD1 mouse motor neurons accumulate mitochondria from the axon terminals and generate higher levels of $O_2^{\bullet-}$, $^\bullet NO$, and $ONOO^-$ than motor neurons in tg mice expressing human wild-type SOD1.[274] This mitochondrial accumulation occurs at a time when motor neuron cell body volume is increasing, suggestive of ongoing abnormalities with ATP production or plasma membrane Na^+,K^+-ATPase.[274] G93A-mSOD1 perturbs anterograde axonal transport of mitochondria in cultured primary embryonic motor neurons,[348] making it possible that retrogradely transported mitochondria with toxic properties from the neuromuscular junction fail to be returned to distal processes.[274] Mitochondria with enhanced toxic potential from distal axons and terminals could therefore have a "Trojan horse" role in triggering degeneration of motor neurons in ALS via retrograde transport from diseased skeletal muscle.

Motor neurons in G93A[high]-mSOD1 mice also accumulate higher levels of intracellular Ca^{2+} than motor neurons in human wild-type SOD1 tg mice.[274] The intracellular Ca^{2+} signal in motor neurons is very compartmental and mitochondrial-like in its appearance.[274,349] Abnormal elevations of intracellular Ca^{2+} in G93A[high]-mSOD1 mouse motor neurons have been detected using a variety of Ca^{2+} detection methods.[350,351] Recent work on mouse neuromuscular junction preparations has shown that mitochondrial Ca^{2+} accumulation is accompanied by greater mitochondrial depolarization, specifically within motor neuron terminals of human mutant SOD1 tg mice.[352]

NO signaling mechanisms in mitochondria of ALS mice appear to be involved in pathogenesis. Motor neurons seem to be unique regarding NO production because they express constitutively low levels of iNOS.[274,275,349]

G93A[high]-mSOD1 mouse motor neurons accumulate nicotinamide adenine dinucleotide phosphate diaphorase and react with iNOS-specific antibodies.[274,275] iNOS also is upregulated aberrantly in human sporadic ALS motor neurons.[353] iNos (Nos2) gene knockout[274] and iNOS inhibition with 1400W[275] extend significantly the lifespan of G93A[high]-mSOD1 mice. Thus, mitochondrial oxidative stress, Ca^{2+} dysregulation, iNOS activation, protein nitration, and protein aggregation (not necessarily SOD1 aggregation) are all likely intrinsic, cell-autonomous mechanisms in the process of motor neuron degeneration caused by mSOD1 in mice.[275] The mechanistic basis for the differences between human ALS and mSOD1 mice, regarding cell death phenotype,[16,312,313,354] is not yet clear. The difference could be related to the extreme nonphysiological expression of toxic mSOD1, to fundamental differences in cell death mechanisms in human and mouse neurons,[355] or to tissue inflammation that drives motor neurons in mSOD1 tg mice to necrotic-like death according to the cell death matrix.[7,354] Another contributing factor to this difference between human and mouse motor neurons is that mitochondria are functionally diverse and have species-specific activities and molecular compositions, including the make-up of the mPTP (Fig. 1).[7,355] These facts raise questions about the relevance of previous tg mSOD1 mouse to human ALS. Therefore, we recently have created a tg mouse with restricted expression of human SOD1 in skeletal muscle that develop ALS with a motor neuron degeneration phenotype similar to that seen in human ALS.[319]

I. The MPTP Contributes to the Disease Mechanisms of ALS in Mice

Despite the implication of toxic effects of mSOD1 on mitochondria in mouse ALS, cause–effect relationships between abnormal functioning of mitochondria and initiation and progression of disease have been uncertain. These relationships need to be defined because this knowledge could lead to new mechanism-based treatments for ALS. One specific target of investigation for mitochondria in disease causality is the mPTP.

The mPTP was first implicated in mouse ALS pathogenesis using pharmacological approaches. Cyclosporine A treatment of G93A[high] mice, delivered into the cerebral ventricle or systemically to mice on a multiple drug resistance type 1a/b background (to inactive the blood–brain barrier), improved outcome modestly.[356–358] These studies were confounded by the immunosuppressant actions of cyclosporine A through calcineurin inhibition. Pharmacological studies using CyPD inhibitors devoid of effects on calcineurin need to be done on ALS mice. Another study showed that treatment with cholest-4-en-3-one oxime (TRO19622), a drug that binds VDAC and the 18-kDa translocator protein (TSPO, or peripheral benzodiazepine receptor), improved motor performance, delayed disease onset, and extended survival of G93A[high] mice.[359] However, another study using a different TSPO ligand (Ro-4864) did not show positive effects with G93A[high] mice.[360]

We identified CyPD and ANT as targets of nitration in ALS mice.[124] CyPD nitration is elevated in early to mid-symptomatic stages, but declines to baseline at end-stage disease.[124] ANT nitration is pertinent because it occurs in presymptomatic and symptomatic stages but not at end-stage disease or in tg mice expressing human wild-type SOD1.[124] The ANT is important in the context of age-related neurodegenerative disease because it undergoes carbonyl modification during aging, as seen in housefly flight muscle[361] and rat brain.[362] In vitro cell-free and cell experiments have shown that NO and $ONOO^-$ can act directly on the ANT to induce mitochondrial permeabilization in a cyclosporine A-sensitive manner.[363] Oxidative stress enhances the binding of CyPD to ANT.[364] Some SOD1 mutants are unstable and lose copper,[365] and interestingly, copper interactions or thiol modification of ANT can cause mPTP opening.[366,367] Together, these data and future work could reveal that oxidative and nitrative damage to proteins, some of which are core components of the mPTP, is targeted rather than stochastic and could impinge on the functioning of the mPTP.

We examined the role of CyPD in the process of motor neuron disease in ALS mice through gene ablation.[124] G93A[high]-mSOD1 mice without CyPD showed markedly delayed disease onset and lived significantly longer than tg mice with CyPD. The effect of CyPD deletion was much more prominent in female mice than in male mice.[124] Female mice showed positive effects with only haplo-deletion of CyPD. Ppif gene ablation in tg mice with much lower levels of human mSOD1 expression and a slower disease progression (G93A[low]-mSOD1 mice) also showed significantly delayed disease onset and lived significantly longer than tg mice with CyPD.[124] Thus, some form of mPTP pathobiology was occurring regardless of whether transgene expression of G93A is high or low. Nevertheless, most G93A-mSOD1 mice without CyPD eventually developed motor neuron disease and died. Other work on CyPD null mice has shown that high concentrations of Ca^{2+} (2mM) can still lead to mPTP activation without CyPD and that cell deaths caused by Bid, Bax, DNA damage, and TNF-α still occur without CyPD.[368] The effects of CyPD deficiency on motor neuron cell mechanisms thus need detailed examination, but the cell death phenotype might switch or convert to another form with the attenuation of mitochondrial swelling. A switch in cell death morphology and molecular mechanisms in motor neurons of mSOD1 mice without CyPD is an outcome consistent with the cell death matrix concept.[7,354]

VII. Reflections

Mitochondria have diverse functions and properties and could be critically important for the development of AD, PD and ALS. Structural and biochemical data from studies of human autopsy CNS as well as cell and animal models of these neurodegenerative disorders suggest that mitochondrial dysfunction is a trigger or propagator of neurodegeneration. Novel mechanisms for mitochondriopathy and neurodegeneration could involve the mPTP (Fig. 1). There is precedence for this suggestion in mouse models of AD,[194] multiple sclerosis,[369] stroke,[370] and ALS.[124] The mPTP actively participates in the mechanisms of motor neuron death in ALS mice in a gender-dependent pattern.[124] Thus, mPTP activation is a possible triggering event for motor neuron degeneration, and motor neuron selective vulnerability in ALS could be related to the amount, composition, and trafficking of mitochondria in these cells. The disappointing negative results of the recent minocycline clinical trial in ALS[371] should not be viewed as an outcome contradictory to the mPTP hypothesis in neurodegeneration, but instead as an example of the difficulty in adequately interpreting the *in vivo* effects of a multifunctional drug in a disease process and the importance of identifying and selecting appropriate dosing regimens. Basic biological considerations also may be relevant. For example, cell death and inflammatory mechanisms in human and mice are different,[313] so the effects of minocycline could differ. Further study of mitochondria and the mPTP in human and rodent neurons, glial cells, and skeletal myocytes may define new mechanisms of disease and lead to the identification of molecular mechanism-based therapies for treating PD and ALS.

Acknowledgments

This chapter is dedicated to Albert Lehninger for his seminal work on mitochondria done while Chair of the Department of Biological Chemistry at Johns Hopkins University School of Medicine. The author thanks all of the individuals in his lab for their hard work, particularly Yan Pan and Ann Price for data generated on human PD and ALS autopsy brain and spinal cord, the A53T α-syn tg mouse, and the G93A-mSOD1 tg mouse. This work was supported by grants from the U.S. Public Health Service, NIH-NINDS (NS034100, NS065895, NS052098) and NIH-NIA (AG016282).

References

1. Zorov DB, Isave NK, Plotnikov EY, Zorova LD, Stelmashook EV, Vasileva AK, et al. The mitochondrion as Janus Bifrons. *Biochemistry (Mosc)* 2007;**72**:1115–26.
2. Halliwell B. Role of free radicals in the neurodegenerative diseases. *Drugs Aging* 2001;**18**:685–716.

3. Nicholls DG. Mitochondrial function and dysfunction in the cell: its relevance to aging and aging-related disease. *Int J Biochem Cell Biol* 2002;**34**:1372–81.

4. Wallace DC. A mitochondrial paradigm of metabolic and degenerative diseases, aging, and cancer: a dawn of evolutionary medicine. *Annu Rev Genet* 2002;**39**:359–407.

5. Fridovich I. Superoxide radical and superoxide dismutases. *Annu Rev Biochem* 1995;**64**:97–112.

6. Mungrue IN, Bredt DS, Stewart DJ, Husain M. From molecules to mammals: what's NOS got to do with it? *Acta Physiol Scand* 2003;**179**.123–35.

7. Martin LJ. Mitochondrial and cell death mechanisms in neurodegenerative disease. *Pharmaceuticals* 2010;**3**:839–915.

8. Delettre C, Lenaers G, Pelloquin L, Belenguer P, Hamel CP. OPA1 (Kjer type) dominant optic atrophy: a novel mitochondrial disease. *Mol Genet Metab* 2002;**75**:97–107.

9. Chen H, Chan DC. Emerging functions of mammalian mitochondrial fusion and fission. *Hum Mol Genet* 2005;**14**:R283–9.

10. Brown MD, Voljavec AS, Lott MT, MacDolald I, Wallace DC. Leber's hereditary optic neuropathy: a model for mitochondrial neurodegenerative diseases. *FASEB J* 1992;**6**:2791–9.

11. Wong L-JC, Naviaux RK, Brunetti-Pierri N, Zhang Q, Schmitt ES, Truong C, et al. Molecular and clinical genetics of mitochondrial disease due to *POLG* mutations. *Hum Mutat* 2008;**29**: E150–72.

12. Goffart S, Cooper HM, Tyynismaa H, Wanrooij S, Suomalainen A, Spelbrink JN. Twinkle mutations associated with autosomal dominant progressive external opthalmoplegia lead to impaired helicase function and in vivo mtDNA replication stalling. *Hum Mol Genet* 2009;**18**:328–40.

13. Komaki H, Fukazawa T, Houzen H, Yoshida K, Nonaka I, Goto Y. A novel D104G mutation in the adenine nucleotide translocator 1 gene in autosomal dominant external opthalmoplegia patients with mitochondrial DNA with multiple deletions. *Ann Neurol* 2002;**51**:645–8.

14. Martin LJ, Al-Abdulla NA, Brambrink AM, Kirsch JR, Sieber FE, Portera-Cailliau C. Neurodegeneration in excitotoxicity, global cerebral ischemia, and target deprivation: a perspective on the contributions of apoptosis and necrosis. *Brain Res Bull* 1998;**46**:281–309.

15. Northington FJ, Graham EM, Martin LJ. Apoptosis in perinatal hypoxic-ischemic brain injury: how important is it and should it be inhibited? *Brain Res Rev* 2005;**50**:244–57.

16. Martin LJ. The mitochondrial permeability transition pore: a molecular target for amyotrophic lateral sclerosis. *Biochim Biophys Acta* 2010;**1802**:186–97.

17. Waldmeier PC, Zimmermann K, Qian T, Tintelnot-Blomley M, Lemasters JJ. Cyclophilin D as a drug target. *Curr Med Chem* 2003;**10**:1485–506.

18. Crompton M. Mitochondria and aging: a role for the permeability transition? *Aging Cell* 2004;**3**:3–6.

19. Halestrap AP. What is the mitochondrial permeability transition pore? *J Mol Cell Cardiol* 2009;**46**:821–31.

20. Bernardi P, Krauskopf A, Basso E, Petronilli V, Blalchy-Dyson E, Di Lisa F, et al. The mitochondrial permeability transition from *in vitro* artifact to disease target. *FEBS J* 2006;**273**:2077–99.

21. Caspersen C, Wang N, Yao J, Sosunov A, Chen X, Lustbader JW, et al. Mitochondrial Aβ: a potential focal point for neuronal metabolic dysfunction in Alzheimer's disease. *FASEB J* 2005;**19**:2040–1.

22. Beckman JS, Carson M, Smith CD, Koppenol WH. ALS, SOD and peroxynitrite. *Nature* 1993;**364**:548.

23. Martin LJ, Liu Z. DNA damage profiling in motor neurons: a single-cell analysis by comet assay. *Neurochem Res* 2002;**27**:1089–100.

24. Giulini C. Characterization and function of mitochondrial nitric-oxide synthase. *Free Radic Biol Med* 2003;**34**:397–408.

25. Brown GC, Borutaite V. Nitric oxide, cytochrome c, and mitochondria. *Biochem Soc Symp* 1999;**66**:17–25.

26. Choi DW. Cellular defences destroyed. *Nature* 2005;**433**:696–8.

27. Brown MR, Sullivan PG, Geddes JW. Synaptic mitochondria are more susceptible to Ca^{2+} overload than nonsynaptic mitochondria. *J Biol Chem* 2006;**281**:11658–68.

28. Soong NW, Hinton DR, Cortopassi G, Arnheim N. Mosaicism for a specific somatic mitochondrial DNA mutation in adult human brain. *Nat Genet* 1992;**2**:318–23.

29. Corral-Debrinski M, Horton T, Lott MT, Shoffner JM, Beal MF, Wallace DC. Mitochondrial DNA deletions in human brain: regional variability and increase with advanced age. *Nat Genet* 1992;**2**:324–9.

30. Driggers WJ, LeDoux SP, Wilson GL. Repair of oxidative damage within the mitochondrial DNA or RINr 38 cells. *J Biol Chem* 1993;**268**:22042–5.

31. Larsen NB, Rasmussen M, Rasmussen LJ. Nuclear and mitochondrial DNA repair: similar pathways? *Mitochondrion* 2005;**5**:89–108.

32. Szczesny B, Hazra TK, Papaconstantinou J, Mitra S, Boldogh I. Age-dependent deficiency in import of mitochondrial DNA glycosylases required for repair of oxidatively damaged bases. *Proc Natl Acad Sci USA* 2003;**100**:10670–5.

33. Popp PA, Copeland WC. Cloning and characterization of the human mitochondrial DNA polymerase, DNA polymerase gamma. *Genomics* 1996;**36**:449–58.

34. Gunawardena S, Goldstein LSB. Cargo-carrying motor vehicles on the neuronal highway: transport pathways and neurodegenerative disease. *J Neurobiol* 2003;**58**:258–71.

35. Chada SR, Hollenbeck PJ. Mitochondrial movement and positioning in axons: the role of growth factor signaling. *J Exp Biol* 2003;**206**:1985–92.

36. Tanaka Y, Kanai Y, Okada Y, Nonaka S, Takeda S, Harada A, et al. Targeted disruption of mouse conventional kinesin heavy chain, kif5B, results in abnormal perinuclear clustering of mitochondria. *Cell* 1998;**93**:1147–58.

37. Ligon LA, Steward O. Role of microtubules and actin filaments in the movement of mitochondria in the axons and dendrites of cultured hippocampal neurons. *J Comp Neurol* 2000;**427**:351–61.

38. Pintoul GL, Filiano AJ, Brocard JB, Kress GJ, Reynolds IJ. Glutamate decreases mitochondrial size and movement in primary forebrain neurons. *J Neurosci* 2003;**23**:7881–8.

39. Reynolds IJ, Malaiyandi LM, Coash M, Rintoul GL. Mitochondrial trafficking in neurons: a key variable in neurodegeneration? *J Bioeng Biomembr* 2004;**36**:283–6.

40. Miller KE, Sheetz MP. Axonal mitochondrial transport and potential are correlated. *J Cell Sci* 2004;**117**:2791–804.

41. Tatebayashi Y, Haque N, Tung YC, Iqbal K, Grundke-Iqbal I. Role of tau phosphorylation by glycogen synthase kinase-3β in the regulation of organelle transport. *J Cell Sci* 2004;**117**:1653–63.

42. Martin LJ. Neuronal cell death in nervous system development, disease, and injury. *Int J Mol Med* 2001;**7**:455–78.

43. Lockshin RA, Zakeri Z. Caspase-independent cell deaths. *Curr Opin Cell Biol* 2002;**14**:727–33.

44. Gilbert S. *Developmental biology*. Sunderland: Sinauer Associates; 2006.

45. Ameisen JC. On the origin, evolution, and nature of programmed cell death: a timeline of four billion years. *Cell Death Differ* 2002;**9**:367–93.

46. Metzstein MM, Stanfield GM, Horvitz HR. Genetics of programmed cell death in C. elegans: past, present and future. *Trends Genet* 1998;**14**:410–6.

47. Youle RJ, Strasser A. The Bcl-2 protein family: opposing activities that mediate cell death. *Nat Rev* 2008;**9**:47–59.

48. Merry DE, Korsmeyer SJ. Bcl-2 gene family in the nervous system. *Annu Rev Neurosci* 1997;**20**:245–67.

49. Cory S, Adams JM. The bcl-2 family: regulators of the cellular life-or-death switch. *Nat Rev* 2002;**2**:647–56.
50. Wolf BB, Green DR. Suicidal tendencies: apoptotic cell death by caspase family proteinases. *J Biol Chem* 1999;**274**:20049–52.
51. Nagata S. Fas ligand-induced apoptosis. *Annu Rev Genet* 1999;**33**:29–55.
52. Levrero M, De Laurenzi V, Costanzo A, Sabatini S, Gong J, Wang JYJ, et al. The p53/p63/p73 family of transcription factors: overlapping and distinct functions. *J Cell Sci* 2000;**113**:1661–70.
53. Li P, Nijhawan D, Budihardjo I, Srinivasula SM, Ahmad M, Alnemri ES, et al. Cytochrome c and dATP-dependent formation of Apaf-1/caspase-9 complex initiates an apoptotic protease cascade. *Cell* 1997;**91**:479–89.
54. Liu X, Kim CN, Yang J, Jemmerson R, Wang X. Induction of apoptotic program in cell-free extracts: requirement for dATP and cytochrome c. *Cell* 1996;**86**:147–57.
55. Klein JA, Longo-Guess CM, Rossmann MP, Seburn RE, Hurd RE, Frankel WN, et al. The harlequin mouse mutation downregulates apoptosis-inducing factor. *Nature* 2002;**419**:367–74.
56. Hegde R, Srinivasula SM, Zhang Z, Wassell R, Mukattash R, Cilentei L, et al. Identification of Omi/HtrA2 as a mitochondrial apoptotic serine protease that disrupts inhibitor of apoptosis protein-caspase interaction. *J Biol Chem* 2002;**277**:432–4538.
57. Liston P, Roy N, Tamai K, Lefebvre C, Baird S, Cherton-Horvat G, et al. Suppression of apoptosis in mammalian cells by NAIP and a related family of IAP genes. *Nature* 1996;**379**:349–53.
58. Scorrano L, Oakes SA, Opferman TJ, Cheng EH, Sorcinelli MD, Pozzan T, et al. Bax and Bak regulation of endoplasmic reticulum Ca^{2+}: a control point for apoptosis. *Science* 2003;**300**:135–9.
59. Muchmore SW, Sattler M, Liang H, Meadows RP, Harlan JE, Yoon HS, et al. X-ray and NMR structure of human Bcl-xL, an inhibitor of programmed cell death. *Nature* 1999;**381**:335–41.
60. Antonsson B, Conti F, Ciavatta A, Montessuit S, Lewis S, Martinou I, et al. Inhibition of bax channel-forming activity by bcl-2. *Science* 1997;**277**:370–2.
61. Letai A, Bassik MC, Walensky LD, Sorcinelli MD, Weiler S, Korsmeyer SJ. Distinct BH3 domains either sensitize or activate mitochondrial apoptosis, serving as prototype cancer therapeutics. *Cancer Cell* 2002;**2**:183–92.
62. Wei MC, Zong W-X, Cheng EH-Y, Lindsten T, Panoutsakopoulou V, Ross AJ, et al. Proapoptotic Bax and Bak: a requisite gateway to mitochondrial dysfunction and death. *Science* 2001;**292**:727–30.
63. Shimizu S, Ide T, Yanagida T, Tsujimoto Y. Electrophysiological study of a novel large pore formed by Bax and the voltage-dependent anion channel that is permeable to cytochrome c. *J Biol Chem* 2000;**275**:12321–5.
64. Chowdhury I, Tharakan B, Bhat GH. Caspases—an update. *Comp Biochem Physiol* 2008;**51**:10–27.
65. Mancini M, Nicholson DW, Roy S, Thornberry NA, Peterson EP, Casciola-Rosen LA, et al. The caspase-3 precursor has a cytosolic and mitochondrial distribution: implications for apoptotic signaling. *J Cell Biol* 1998;**140**:1485–95.
66. Krajewski S, Krajewska M, Ellerby LM, Welsh K, Xie Z, Deveraus QL, et al. Release of caspase-9 from mitochondria during neuronal apoptosis and cerebral ischemia. *Proc Natl Acad Sci USA* 1999;**96**:5752–7.
67. Liu X, Zou H, Slaughter C, Wang X. DFF, a heterodimeric protein that functions downstream of caspase-3 to trigger DNA fragmentation during apoptosis. *Cell* 1997;**89**:175–84.
68. Li H, Zhu H, Xu C-J, Yuan J. Cleavage of Bid by caspase 8 mediates the mitochondrial damage in the Fas pathway of apoptosis. *Cell* 1998;**94**:491–501.
69. Robertson JD, Enoksson M, Suomela M, Zhivotovsky B, Orrenius S. Caspase-2 acts upstream of mitochondria to promote cytochrome c release during etoposide-induced apoptosis. *J Biol Chem* 2002;**277**:29803–9.

70. LaCasse EC, Baird S, Korneluk RG, MacKenzie AE. The inhibitors of apoptosis (IAPs) and their emerging role in cancer. *Oncogene* 1998;**17**:3247–59.
71. Holcik M. The IAP proteins. *Trends Genet* 2002;**18**:537–8.
72. Deveraux QL, Roy N, Stennicke HR, Van Arsdale T, Zhou Q, Srinivasula SM, et al. IAPs block apoptotic events induced by caspase-8 and cytochrome c by direct inhibition of distinct caspases. *EMBO J* 1998;**17**:2215–23.
73. Hao Y, Sekine K, Kawabata A, Nakamura H, Ishioka T, Ohata H, et al. Apollon ubiquitinates SMAC and caspase-9, and has an essential cytoprotection function. *Nat Cell Biol* 2004;**6**:849–60.
74. Du C, Fang M, Li Y, Li L, Wang X. Smac, a mitochondrial protein that promotes cytochrome c-dependent caspase activation by eliminating IAP inhibition. *Cell* 2000;**102**:33–42.
75. Verhagen AM, Ekert PG, Pakusch M, Silke J, Connolly LM, Reid GE, et al. Identification of DIABLO, a mammalian protein that promotes apoptosis by binding to and antagonizing IAP proteins. *Cell* 2000;**102**:43–53.
76. Suzuki Y, Imai Y, Nakayama H, Takahashi K, Takio K, Takahashi R. A serine protease, HtrA2, is released from the mitochondria and interacts with XIAP, inducing cell death. *Mol Cell* 2001;**8**:613–21.
77. Bogaerts V, Nuytemans K, Reumers J, Pals R, Engelborghs S, Pickut B, et al. Genetic variability in the mitochondrial serine protease HTRA2 contributes to risk for Parkinson's disease. *Hum Mutat* 2008;**29**:832–40.
78. Susin SA, Lorenzo HK, Zamzami N, Marzo I, Snow BE, Brothers GM, et al. Molecular characterization of mitochondrial apoptosis-inducing factor. *Nature* 1999;**397**:441–6.
79. Mate MJ, Ortiz-Lombardia M, Boitel B, Haouz A, Tello D, Susin SA, et al. The crystal structure of the mouse apoptosis-inducing factor AIF. *Nat Struct Biol* 2002;**9**:442–6.
80. Trump BF, Berezesky IK. The role of altered $[Ca^{2+}]_i$ regulation in apoptosis, oncosis, and necrosis. *Biochim Biophys Acta* 1996;**1313**:173–8.
81. Majno G, Joris I. Apoptosis, oncosis, and necrosis. An overview of cell death. *Am J Pathol* 1995;**146**:3–15.
82. Trump BJ, Goldblatt PJ, Stowell RE. Studies on necrosis of mouse liver in vitro. Ultrastructural alterations in the mitochondria of hepatic parenchymal cells. *Lab Invest* 1964;**14**:343–71.
83. Bonfoco E, Krainc D, Ankarcrona M, Nicotera P, Lipton SA. Apoptosis and necrosis: two distinct events induced, respectively, by mild and intense insults with N-methyl-D-aspartate or nitric oxide/superoxide in cortical cell culture. *Proc Natl Acad Sci USA* 1995;**92**:7162–6.
84. Leist M, Single B, Castoldi AF, Kuhnles S, Nicotera P. Intracellular adenosine triphosphate (ATP) concentration: a switch in the decision between apoptosis and necrosis. *J Exp Med* 1997;**185**:1481–6.
85. Golden WC, Brambrink AM, Traystman RJ, Martin LJ. Failure to sustain recovery of Na, K ATPase function is a possible mechanism for striatal neurodegeneration in hypoxic-ischemic newborn piglets. *Mol Brain Res* 2001;**88**:94–102.
86. Castro J, Ruminot I, Porras OH, Flores CM, Hermosilla T, Verdugo E, et al. ATP steal between cation pumps: a mechanism linking Na^+ influx to the onset or necrotic Ca^{2+} overload. *Cell Death Differ* 2006;**13**:1675–85.
87. Proskuryakov SY, Konoplyannikov AG, Gabai VL. Necrosis: a specific form of programmed cell death. *Exp Cell Res* 2003;**283**:1–16.
88. Kim Y-S, Morgan MJ, Choksi S, Lu Z-G. TNF-induced activation of the Nox1 NADPH oxidase and its role in the induction of necrotic cell death. *Mol Cell* 2007;**26**:675–87.
89. Hitomi J, Christofferson DE, Ng A, Yao J, Degterev A, Xavier RJ, et al. Identification of a molecular signaling network that regulates a cellular necrotic cell death pathway. *Cell* 2008;**135**:1311–23.

90. Baines CP, Kaiser RA, Purcell NH, Blair HS, Osinska H, Hambleton MA, et al. Loss of cyclophilin D reveals a critical role for mitochondrial permeability transition in cell death. *Nature* 2005;**434**:658–62.

91. Ha HC, Snyder SH. Poly(ADP-ribose) polymerase-1 in the nervous system. *Neurobiol Dis* 2000;**7**:225–39.

92. Zoratti M, Szabo I. The mitochondrial permeability transition. *Biochem Biophys Acta* 1995;**1241**.139–76.

93. Crompton M. The mitochondrial permeability transition pore and its role in cell death. *Biochem J* 1999;**341**:233–49.

94. van Gurp M, Festjens N, van Loo G, Saelens X, Vandenabeele P. Mitochondrial intermembrane proteins in cell death. *Biochem Biophys Res Commun* 2003;**304**:487–97.

95. Leung AWC, Halestrap AP. Recent progress in elucidating the molecular mechanism of the mitochondrial permeability transition pore. *Biochim Biophys Acta* 2008;**1777**:946–52.

96. Shoshan-Barmatz V, Israelson A, Brdiczka D, Sheu SS. The voltage-dependent anion channel (VDAC): function in intracellular signaling, cell life and cell death. *Curr Pharm Des* 2006;**12**: 2249–70.

97. Rostovtseva TK, Tan W, Colombini M. On the role of VDAC in apoptosis: fact and fiction. *J Bioenerg Biomembr* 2005;**37**:129–42.

98. Tan W, Colombini M. VDAC closure increases calcium ion flux. *Biochim Biophys Acta* 2007;**1768**:2510–5.

99. Tikunov A, Johnson CB, Pediaditakis P, Markevich N, MacDonald JM, Lemasters JJ, et al. Closure of VDAC causes oxidative stress and accelerates the Ca^{2+}-induced mitochondrial permeability transition in rat liver mitochondria. *Arch Biochem Biophys* 2010;**495**:174–81.

100. Granville DJ, Gottlieb RA. The mitochondrial voltage-dependent anion channel (VDAC) as a therapeutic target for initiating cell death. *Curr Med Chem* 2003;**10**:1527–33.

101. Huizing M, Ruitenbeek W, van den Heuvel LP, Dolce V, Iacobazzi V, Smeitink JAM, et al. Human mitochondrial transmembrane metabolite carriers: tissue distribution and its implication for mitochondrial disorders. *J Bioenerg Biomembr* 1998;**30**:277–84.

102. Wu S, Sampson MJ, Decker WK, Craigen WJ. Each mammalian mitochondrial outer membrane porin protein is dispensable: effects on cellular respiration. *Biochem Biophys Acta* 1999;**1452**:68–78.

103. Anflous K, Armstrong DD, Craigen WJ. Altered sensitivity for ADP and maintenance of creatine-stimulated respiration in oxidative striated muscles from VDAC1-deficient mice. *J Biol Chem* 2001;**276**:1954–60.

104. Sampson MJ, Decker WK, Beaudet AL, Ruitenbeek W, Armstrong D, Hicks MJ, et al. Immotile sperm and infertility in mice lacking mitochondrial voltage-dependent anion channel type 3. *J Biol Chem* 2001;**276**:39206–12.

105. Cheng EH, Sheiko TV, Fisher JK, Craigen WJ, Korsemeyer SJ. VDAC2 inhibits Bak activation and mitochondrial apoptosis. *Science* 2003;**301**:513–7.

106. Chandra D, Choy G, Daniel PT, Tang DG. Bax-dependent regulation of Bak by voltage-dependent anion channel 2. *J Biol Chem* 2005;**280**:19051–61.

107. Baines CP, Kaiser RA, Sheiko T, Craigen WJ, Molkentin JD. Voltage-dependent anion channels are dispensable for mitochondrial-dependent cell death. *Nat Cell Biol* 2007;**9**:550–5.

108. Karachitos A, Galganska H, Wojtkowska M, Budzinska M, Stobienia O, Bartosz G, et al. Cu, Zn-superoxide dismutase is necessary for proper function of VDAC in Saccharomyces cerevisiae cells. *FEBS Lett* 2009;**583**:449–55.

109. Halestrap AP, Brenner C. The adenine nucleotide translocase: a central component of the mitochondrial permeability transition pore and key player in cell death. *Curr Med Chem* 2003;**10**:1507–25.

110. Graham BH, Waymire KG, Cottrell B, Trounce IA, MacGregor GR, Wallace DC. A mouse model for mitochondrial myopathy and cardiomyopathy resulting from a deficiency in the heart/muscle isoform of the adenine nucleotide translocator. *Nat Genet* 1997;**16**:226–34.
111. Stepien G, Torroni A, Chung AB, Hodge JA, Wallace DC. Differential expression of adenine nucleotide translocator isoforms in mammalian tissues and during muscle cell differentiation. *J Biol Chem* 1992;**267**:14592–7.
112. Vyssokikh MY, Katz A, Rueck A, Wuensch C, Dorner A, Zorov DB, et al. Adenine nucleotide translocator isoforms 1 and 2 are differently distributed in the mitochondrial inner membrane and have distinct affinities to cyclophilin D. *Biochem J* 2001;**358**:349–58.
113. Kikoszka JE, Waymire KG, Levy SE, Sligh JE, Cai J, Jones DP, et al. The ADP/ATP translocator is not essential for the mitochondrial permeability transition pore. *Nature* 2004;**427**:461–5.
114. Machida K, Hayashi Y, Osada H. A novel adenine nucleotide translocase inhibitor, MT-21, induces cytochrome c release through a mitochondrial permeability transition-independent mechanisms. *J Biol Chem* 2002;**277**:31243–8.
115. Woodfield K, Rück A, Brdiczka D, Halestrap AP. Direct demonstration of a specific interaction between cyclophilin-D and the adenine nucleotide translocase confirms their role in the mitochondrial permeability transition. *Biochem J* 1998;**336**:287–90.
116. Johnson N, Khan A, Virji S, Ward JM, Crompton M. Import and processing of heart mitochondrial cyclophilin D. *Eur J Biochem* 1999;**263**:353–9.
117. McEnery MW, Dawson TM, Verma A, Gurley D, Colombini M, Snyder SH. Mitochondrial voltage-dependent anion channel. *J Biol Chem* 1993;**268**:23289–96.
118. Shoshan-Barmatz V, Zalk R, Gincel D, Vardi N. Subcellular localization of VDAC in mitochondria and ER in the cerebellum. *Biochem Biophys Acta* 2004;**1657**:105–14.
119. Akanda N, Tofight R, Brask J, Tamm C, Elinder F, Ceccatelli S. Voltage-dependent anion channels (VDAC) in the plasma membrane play a critical role in apoptosis in differentiated hippocampal neurons but not in neural stem cells. *Cell Cycle* 2008;**7**:3225–34.
120. Yu WH, Wolfgang W, Forte M. Subcellular localization of human voltage-dependent anion channel isoforms. *J Biol Chem* 1995;**270**:13998–4006.
121. Buck CR, Jurynec MJ, Upta DE, Law AKT, Bilger J, Wallace DC, et al. Increased adenine nucleotide translocator 1 in reactive astrocytes facilitates glutamate transport. *Exp Neurol* 2003;**181**:149–58.
122. Hazelton JL, Petrasheuskaya M, Fiskum G, Kristian T. Cyclophilin D is expressed predominantly in mitochondria of γ-aminobutyric acidergic interneurons. *J Neurosci Res* 2009;**87**:1250–9.
123. Naga KK, Sullivan PG, Geddes JW. High cyclophilin D content of synaptic mitochondria results in increased vulnerability to permeability transition. *J Neurosci* 2007;**27**:7469–75.
124. Martin LJ, Gertz B, Pan Y, Price AC, Molkentin JD, Chang Q. The mitochondrial permeability transition pore in motor neurons: involvement in the pathobiology of ALS mice. *Exp Neurol* 2009;**218**:333–46.
125. Bose S, Freedman RB. Peptidyl prolyl cis-trans-isomerase activity associated with the lumen of the endoplasmic reticulum. *Biochem J* 1994;**300**:865–70.
126. Sullivan PG, Rabchevsky AG, Keller JN, Lovell M, Sodhi A, Hart RP, et al. Intrinsic differences in brain and spinal cord mitochondria: implications for therapeutic interventions. *J Comp Neurol* 2004;**474**:524–34.
127. Morota S, Hansson MJ, Ishii N, Kudo Y, Elmer E, Uchino H. Spinal cord mitochondria display lower calcium retention capacity compared with brain mitochondria without inherent differences in sensitivity to cyclophilin D inhibition. *J Neurochem* 2007;**103**:2066–76.
128. Collins TJ, Bootman MD. Mitochondria are morphologically heterogeneous within cells. *J Exp Biol* 2003;**206**:1993–2000.

129. Jensen RE. Control of mitochondrial shape. *Curr Opin Cell Biol* 2005;**17**:384–8.
130. Hamberger A, Blomstrand C, Lehninger AL. Comparative studies of mitochondria isolated from neuron-enriched and glia-enriched fractions of rabbit and beef brain. *J Cell Biol* 1970;**45**:221–34.
131. Klionsky DJ, Emr SD. Autophagy as a regulated pathway of cellular degradation. *Science* 2000;**290**:1717–21.
132. Tolkovsky AM, Xue L, Fletcher GC, Borutaite V. Mitochondrial disappearance from cells: a clue to the role of autophagy in programmed cell death and disease. *Biochimie* 2002;**84**: 233–40.
133. Gozuacik D, Kimchi A. Autophagy as a cell death and tumor suppressor mechanism. *Oncogene* 2004;**23**:2891–906.
134. Ginsberg SD, Martin LJ. Ultrastructural analysis of the progression of neurodegeneration in the septum following fimbria-fornix transaction. *Neuroscience* 1998;**86**:1259–72.
135. Ginsberg SD, Portera-Cailliau C, Martin LJ. Fimbria-fronix transection and excitotoxicity produce similar neurodegeneration in the septum. *Neuroscience* 1999;**88**: 1059–1071.
136. Martin LJ, Kaiser A, Price AC. Motor neuron degeneration after sciatic nerve avulsion in adult rat evolves with oxidative stress and is apoptosis. *J Neurobiol* 1999;**40**:185–201.
137. Al-Abdulla NA, Martin LJ. Apoptosis of retrogradely degenerating neurons occurs in association with accumulation of perikaryal mitochondria and oxidative damage to the nucleus. *Am J Pathol* 1998;**153**:447–56.
138. Martin LJ. An approach to experimental synaptic pathology using green fluorescent protein-transgenic mice and gene knockout mice: excitotoxic vulnerability of interneurons and motoneurons is associated with mitochondrial accumulation and mediated by the mitochondrial permeability transition pore. *Toxicol Pathol* 2011;**39**(1):220–33.
139. Clarke PGH. Developmental cell death: morphological diversity and multiple mechanisms. *Anat Embryol* 1990;**181**:195–213.
140. Schweichel JU, Merker HJ. The morphology of various types of cell death in prenatal tissues. *Teratology* 1973;**7**:253–66.
141. Xue LZ, Fletcher GC, Tolkovsky AM. Autophagy is activated by apoptotic signaling in sympathetic neurons: an alternative mechanism of death execution. *Mol Cell Neurosci* 1999;**14**:180–98.
142. Yue Z, Horton A, Bravin M, DeJager PL, Selimi F, Heintz N. A novel protein complex linking the δ2 glutamate receptor and autophagy: implications for neurodegeneration in Lurcher mice. *Neuron* 2002;**35**:921–33.
143. Hara T, Nakamura K, Matsui M, Yamamoto A, Nakahara Y, Suzuki-Migishima R, et al. Suppression of basal autophagy in neural cells causes neurodegenerative disease in mice. *Nature* 2006;**44**:885–9.
144. Komatsu M, Waguri S, Chiba T, Murata S, Iwata J, Tanida I, et al. Loss of autophagy in the central nervous system causes neurodegeneration in mice. *Nature* 2006;**44**:880–4.
145. Nakendra D, Tanaka A, Suen D-F, Youle RJ. Parkin is recruited selectively to impaired mitochondria and promotes their autophagy. *J Cell Biol* 2008;**183**:795–803.
146. Yuan J, Lipinski M, Degterev A. Diversity in the mechanisms of neuronal cell death. *Neuron* 2003;**40**:401–13.
147. Todde V, Veenhuis M, van der Klei IJ. Autophagy: principles and significance in health and disease. *Biochim Biophys Acta* 2009;**1792**:3–13.
148. Bursch W. The autophagosomal-lysosomal compartment in programmed cell death. *Cell Death Differ* 2001;**8**:569–81.
149. Inbal B, Bialik S, Sabanay I, Shani G, Kimchi A. DAP kinase and DRP-1 mediate membrane blebbing and the formation of autophagic vesicles during programmed cell death. *J Cell Biol* 2002;**157**:455–68.

150. Liange XH, Kleeman LK, Jiang HH, Gordon G, Goldman JE, Berry G, et al. Protection against fatal Sindbis virus encephalitis by beclin, a novel Bcl-2-interacting protein. *J Virol* 1998;**72**:8586–96.

151. Ogier-Denis E, Codogno P. Autophagy: a barrier or an adaptive response to cancer. *Biochim Biophys Acta* 2003;**1603**:113–28.

152. Varadi A, Johnson-Cadwell LI, Cirulli V, Yoon Y, Allan VJ, Rutter GA. Cytoplasmic dynein regulates the subcellular distribution of mitochondria by controlling the recruitment of the fission factor dynamin-related protein-1. *J Cell Sci* 2004;**117**:4389–400.

153. Karbowski M, Lee Y-L, Gaume B, Jeong S-Y, Frank S, Nechushtan A, et al. Spatial and temporal association of Bax with mitochondrial fission sites, Drp1, and Mfn2 during apoptosis. *J Cell Biol* 2002;**159**:931–8.

154. Davis AF, Clayton DA. In situ localization of mitochondrial DNA replication in intact mammalian cells. *J Cell Biol* 1996;**135**:883–93.

155. Nisoli E, Clementi E, Paolucci C, Cozzi V, Tonello C, Sciorati C, et al. Mitochondrial biogenesis in mammals: the role of endogenous nitric oxide. *Science* 2003;**299**:896–9.

156. Katzman R. Education and the prevalence of dementia and Alzheimer's disease. *Neurology* 1993;**43**:13–20.

157. Evans DA, Funkenstein HH, Albert MS, Scherr PA, Cook NR, Chown MJ, et al. Prevalence of Alzheimer's disease in a community population of older persons. Higher than previously reported. *JAMA* 1989;**262**:2551–6.

158. McKhann G, Drachman D, Folstein M, Katzman R, Price D, Stadlan EM. Clinical diagnosis of Alzheimer's disease: report of the NINCDS-ADRDA work group under the auspices of the Department of Health and Human Services task force on Alzheimer's disease. *Neurology* 1984;**34**:939–44.

159. Olshansky SJ, Carnes BA, Cassel CK. The aging of the human species. *Sci Am* 1993;**268**:46–52.

160. Minati L, Edginton T, Bruzzone MG, Giaccone G. Current concepts in Alzheimer's disease: a multidisciplinary review. *Am J Alz Dis Other Dem* 2009;**24**:95–121.

161. Chartier-Harlin M-C, Crawford F, Houlden H, Warren A, Hughes D, Fidani L, et al. Early-onset Alzheimer's disease caused by mutations at codon 717 of the β-amyloid precursor protein gene. *Nature* 1991;**353**:844–6.

162. Tilley L, Morgan K, Kalsheker N. Genetic risk factors for Alzheimer's disease. *J Clin Pathol Mol Pathol* 1998;**51**:293–304.

163. Goate A, Chartier-Harlin MC, Mullan M, Brown J, Crawford F, Fidani L, et al. Segregation of a missense mutation in the amyloid precursor protein gene with familial Alzheimer's disease. *Nature* 1991;**349**:704–6.

164. Naruse S, Igarashi S, Kobayashi H, Aoki K, Inuzuka T, Kaneko K, et al. Mis-sense mutation Val->Ile in exon 17 of amyloid precursor protein gene in Japanese familial Alzheimer's disease. *Lancet* 1991;**337**:978–9.

165. Campion D, Flaman JM, Brice A, Hannequin D, Dubois B, Martin C, et al. Mutations of the presenilin 1 gene in families with early-onset Alzheimer's disease. *Hum Mol Genet* 1995;**4**:2373–7.

166. Sherrington R, Rogaev EI, Liang Y, Rogaeva EA, Levesque G, Ikeda M, et al. Cloning of a gene bearing missense mutations in early-onset familial Alzheimer's disease. *Nature* 1995;**375**:754–60.

167. Kalaria RN. Dementia comes of age in the developing world. *Lancet* 2003;**361**:888–9.

168. Roses AD. Apolipoprotein E alleles as risk factors in Alzheimer's disease. *Annu Rev Med* 1996;**47**:387–400.

169. Whitehouse PJ, Price DL, Struble RG, Clark AW, Coyle JT, DeLong MR. Alzheimer's disease and senile dementia: loss of neurons in the basal forebrain. *Science* 1982;**215**:1237–9.

170. Gomez-Isla T, Price JL, McKeel Jr. DW, Morris JC, Growdon JH, Hyman BT. Profound loss of layer II entorhinal cortex neurons occurs in very mild Alzheimer's disease. *J Neurosci* 1996;**16**:4491–500.

171. Mouton PR, Martin LJ, Calhoun ME, Dal Forno G, Price DL. Cognitive decline strongly correlates with cortical atrophy in Alzheimer's disease. *Neurobiol Aging* 1998;**19**:371–7.

172. West MJ, Kawas CH, Martin LJ, Troncoso JC. The CA1 region of the human hippocampus is a hot spot in Alzheimer's disease. *Annu NY Acad Sci* 2000;**908**:255–9.

173. Pelvig DP, Pakkenberg H, Regeur L, Oster S, Pakkenberg B. Neocortical glial cell numbers in Alzheimer's disease. A stereological study. *Dement Geriatr Cogn Disord* 2003;**16**:212–9.

174. Troncoso JC, Cataldo AM, Nixon RA, Barnett JL, Lee MK, Checler F, et al. Neuropathology of preclinical and clinical late-onset Alzheimer's disease. *Ann Neurol* 1998;**43**:673–6.

175. Kosik KS, Joachim CL, Selkoe DJ. Microtubule-associated protein tau is a major antigenic component of paired helical filaments in Alzheimer's disease. *Proc Natl Acad Sci USA* 1986;**83**:4044–8.

176. Hardy J, Selkoe DJ. The amyloid hypothesis of Alzheimer's disease: progress and problems on the road to therapeutics. *Science* 2002;**297**:353–6.

177. Sze C-I, Bi H, Kleinschmidt-DeMasters BK, Filley CM, Martin LJ. N-Methyl-D-aspartate receptor subunit proteins and their phosphorylation status are altered selectively in Alzheimer's disease. *J Neurol Sci* 2001;**182**:151–9.

178. Kemp JA, McKernan RM. NMDA receptor pathways as drug targets. *Nat Neurosci* 2002;**5**:1039–42.

179. Mattson MP, Cheng B, Davis D, Bryant K, Lieberburg I, Rydel RE. β-Amyloid peptides destabilize calcium homeostasis and render human cortical neurons vulnerable to excitotoxicity. *J Neurosci* 1993;**12**:376–89.

180. Reddy PH, Beal MF. Amyloid beta, mitochondrial dysfunction and synaptic damage: implications for cognitive decline in aging and Alzheimer's disease. *Trends Mol Med* 2008;**14**:45–53.

181. DeKosky ST, Scheff SW. Synapse loss in frontal cortex biopsies in Alzheimer's disease: correlation with cognitive severity. *Ann Neurol* 1990;**27**:457–64.

182. Terry RD, Masliah E, Salmon DP, Butters N, DeTeresa R, Hill R, et al. Physical basis of cognitive alterations in Alzheimer's disease: synapse loss is the major correlate of cognitive impairment. *Ann Neurol* 1991;**30**:572–80.

183. Martin LJ, Pardo CA, Cork LC, Price DL. Synaptic pathology and glial responses to neuronal injury precede the formation of senile plaques and amyloid deposits in the aging cerebral cortex. *Am J Pathol* 1994;**145**:1358–81.

184. Sze C-I, Troncoso JC, Kawas C, Mouton P, Price DL, Martin LJ. Loss of the presynaptic vesicle protein synaptophysin in hippocampus correlates with cognitive decline in Alzheimer's disease. *J Neuropathol Exp Neurol* 1997;**56**:933–94.

185. Selkoe DJ. Alzheimer's disease is a synaptic failure. *Science* 2002;**298**:789–91.

186. Yankner BA, Dawes LR, Fisher S, Villa-Komaroff L, Oster-Granite ML, Neve RL. Neurotoxicity of a fragment of the amyloid precursor associated with Alzheimer's disease. *Science* 1989;**245**:417–20.

187. Younkin SG. Evidence that Abeta 42 is the real culprit in Alzheimer's disease. *Ann Neurol* 1995;**37**:287–8.

188. Fath T, Eidenmuller J, Brandt R. Tau-mediated cytotoxicity in a pseudohyperphosphorylation model of Alzheimer's disease. *J Neurosci* 2002;**22**:9733–41.

189. Rapoport M, Dawson HN, Binder LI, Vitek MP, Ferreira A. Tau is essential to β-amyloid-induced neurotoxicity. *Proc Natl Acad Sci USA* 2002;**99**:6364–9.

190. Anandatheerthavarada HK, Biswas G, Robin MA, Avadhani NG. Mitochondrial targeting and a novel transmembrane arrest of Alzheimer's amyloid precursor protein impairs mitochondrial function in neuronal cells. *J Cell Biol* 2003;**161**:41–54.

191. Devi L, Prabhu BM, Galati DF, Avadhani NG, Anandatheerthavarada HK. Accumulation of amyloid precursor protein in the mitochondrial import channels of human Alzheimer's disease brain is associated with mitochondrial dysfunction. *J Neurosci* 2006;**26**:9057–68.

192. Manczak M, Anekonda TS, Henson E, Park BS, Quinn J, Reddy PH. Mitochondria are a direct site of Aβ accumulation in Alzheimer's disease neurons: implications for free radical generation and oxidative damage in disease progression. *Hum Mol Genet* 2006;**15**:1437–49.

193. Lustbader JW, Cirilli M, Lin C, Xu HW, Takuma K, Wang N, et al. ABAD directly links Aβ to mitochondrial toxicity in Alzheimer's disease. *Science* 2004;**304**:448–52.

194. Du H, Guo L, Fang F, Chen D, Sosunov AA, McKhann GM, et al. Cyclophilin D deficiency attenuates mitochondrial and neuronal perturbation and ameliorates learning and memory in Alzheimer's disease. *Nat Med* 2008;**14**:1097–105.

195. Duyckaerts C, Potier MC, Delatour B. Alzheimer disease models and human neuropathology: similarities and differences. *Acta Neuropathol* 2008;**115**:5–38.

196. Van Den Eeden SK, Tanner CM, Bernstein AL, Fross RD, Leimpeter A, Bloch DA, et al. Incidence of Parkinson's disease: variations by age, gender and race ethnicity. *Am J Epidemol* 2003;**157**:1015–22.

197. Jankovic J. Parkinson's disease: clinical features and diagnosis. *J Neurol Neurosurg Psychiatry* 2008;**79**:368–76.

198. Lowe J, Lennox G, Leigh PN. Disorders of movement and system degeneration. In: Graham DI, Lantos PL, editors. *Greenfields neuropathology*. London: Arnold; 1997. pp. 281–366.

199. Braak H, Del Tredici K, Rüb U, de Vos RAI, Jansen Steur ENH, Braak E. Staging of brain pathology related to sporadic Parkinson's disease. *Neurobiol Aging* 2003;**24**:197–211.

200. Martin LJ. Neurodegenerative disorders of the human brain and spinal cord. In: Ramachandran VS, editor. *Encyclopedia of the human brain*. Amsterdam: Elsevier Science; 2002. pp. 441–63 Vol. 3.

201. Goedert M. Alpha-synuclein and neurodegenerative diseases. *Nat Rev Neurosci* 2001;**2**:492–501.

202. Ascherio A, Chen H, Weisskopf MG, O'Reilly E, McCullough ML, Calle EE, et al. Pesticide exposure and risk for Parkinson's disease. *Ann Neurol* 2006;**60**:197–203.

203. Schapira AHV. Etiology of Parkinson's disease. *Neurology* 2006;**66**(Suppl. 4):S10–23.

204. Shimohama S, Swada H, Kitamura Y, Taniguchi T. Disease model: Parkinson's disease. *Trends Mol Med* 2003;**9**:360–5.

205. Polymeropoulos MH, Lavedan C, Leroy E, Ide SE, Dehejia A, Dutra A, et al. Mutation in the α-synuclein gene identified in families with Parkinson's disease. *Science* 1997;**276**:2045–7.

206. Singleton AB, Farrer M, Johnson J, Singleton A, Hague S, Kachergus J, et al. Alpha-synuclein locus triplication causes Parkinson's disease. *Science* 2003;**302**:841.

207. Leroy E, Boyer R, Auburger G, Leube B, Ulm G, Mezey E, et al. The ubiquitin pathway in Parkinson's disease. *Nature* 1998;**395**:451–2.

208. Bonifati V, Rizzu P, van Baren MJ, Schaap O, Breedveld GJ, Krieger E, et al. Mutations in the DJ-1 gene associated with autosomal recessive early-onset parkinsonism. *Science* 2003;**299**:256–9.

209. Hatano Y, Li Y, Sato K, Asakawa S, Yamamura Y, Tomiyama H, et al. Novel PINK1 mutations in early-onset parkinsonism. *Ann Neurol* 2004;**56**:424–7.

210. Kitada T, Asakawa S, Hattori N, Matsumine H, Yamamura Y, Minoshima S, et al. Mutations in the parkin gene cause autosomal recessive juvenile parkinsonism. *Nature* 1998;**392**:605–8.

211. Valente EM, Abou-Sleiman PM, Caputo V, Muqit MM, Harvey K, Gispert A, et al. Hereditary early-onset Parkinson's disease caused by mutations in PINK1. *Science* 2004;**304**:1158–60.

212. Paisán-Ruíz C, Jain S, Whitney Evans E, Gilks WP, Simón J, van der Brug M, et al. Cloning of the gene containing mutations that cause PARK8-linked Parkinson's disease. *Neuron* 2004;**44**:595–600.

213. Zimprich A, Biskup S, Leitner P, Lichtner P, Farrer M, Lincoln S, et al. Mutations in LRRK2 cause autosomal-dominant parkinsonism with pleomorphic pathology. *Neuron* 2004;**44**:601–7.

214. Ramirez A, Heimbach A, Grundemann J, Stiller B, Hampshire D, Cid LP, et al. Hereditary parkinsonism with dementia is caused by mutations in ATP13A2, encoding a lysosomal type 5 P-type ATPase. *Nat Genet* 2006;**38**:1184–91.

215. Lesuisse C, Martin LJ. Long-term culture of mouse cortical neurons as a model for neuronal development, aging, and death. *J Neurobiol* 2002;**51**:9–23.

216. Maroteaux L, Campanelli JT, Scheller RH. Synuclein: a neuron-specific protein localized to the nucleus and presynaptic nerve terminals. *J Neurosci* 1998;**8**:2804–15.

217. Murphy DD, Rueter SM, Trojanowski JQ, Lee VMY. Synucleins are developmentally expressed, and α-synuclein regulates the size of the presynaptic vesicular pool in primary hippocampal neurons. *J Neurosci* 2000;**20**:3214–20.

218. Chandra S, Fornai F, Kwon HB, Yazdani U, Atasoy D, Liu X, et al. Double knockout mice for α- and β-synucleins: effect on synaptic functions. *Proc Natl Acad Sci USA* 2004;**101**:14966–71.

219. Gurevicine I, Gurevicius K, Tanila H. Role of α-synuclein in synaptic glutamate release. *Neurobiol Dis* 2007;**28**:83–9.

220. Liu S, Fa M, Ninan I, Trinchese F, Dauer W, Aranico O. α-Synuclein involvement in hippocampal synaptic plasticity: role of NO, cGMP, cGK and CAMKII. *Eur J Neurosci* 2007;**25**:3583–96.

221. Fortin DL, Nemani VM, Voglmaier SM, Anthony MD, Ryan TA, Edwards RH. Neural activity control the synaptic accumulation of α-synuclein. *J Neurosci* 2005;**25**:10913–21.

222. Chandra S, Gallardo G, Fernandez-Chacon R, Schluter OM, Sudholf TC. α-Synuclein cooperates with CSPα in preventing neurodegeneration. *Cell* 2005;**123**:383–96.

223. Gallardo G, Schluter OM, Sudhof TC. A molecular pathway of neurodegeneration linking α-synuclein to ApoE and Aβ peptides. *Nat Neurosci* 2008;**11**:301–8.

224. Serpell LC, Berriman J, Jakes M, Goedert M, Crowther RA. Fiber diffraction of synthetic alpha synuclein filaments shows amyloid-like cross-beta conformation. *Proc Natl Acad Sci USA* 2000;**97**:4897–902.

225. Conway KA, Lee SJ, Rochet JC, Ding TT, Williamson RE, Lansbury Jr. PT. Acceleration of oligomerization, not fibrilization, is a shared property of both alpha-synuclein mutations linked to early-onset Parkinson's disease: implications for pathogenesis and therapy. *Proc Natl Acad Sci USA* 2000;**97**:571–6.

226. Caughey B, Lansbury PT. Protofibrils, pores, fibril, and neurodegeneration: separating the responsible protein aggregates from the innocent bystanders. *Annu Rev Neurosci* 2003;**26**:267–98.

227. Hsu LJ, Sagara Y, Arroyo A, Rockenstein E, Sisk A, Mallory M, et al. Alpha-synuclein promotes mitochondrial deficit and oxidative stress. *Am J Pathol* 2000;**157**:401–10.

228. Junn E, Mouradian MM. Human α-synuclein over-expression increases intracellular reactive oxygen species levels and susceptibility to dopamine. *Neurosci Lett* 2002;**320**:146–50.

229. Tabrizi SJ, Orth M, Wilkinson JM, Taanman JW, Warner TT, Cooper JM, et al. Expression of mutant alpha-synuclein causes increased susceptibility to dopamine toxicity. *Hum Mol Genet* 2000;**9**:2683–9.

230. Giasson BI, Duda JE, Quinn SM, Zhang B, Trojanowski JQ, Lee WM-Y. Neuronal α-synucleinopathy with severe movement disorder in mice expressing A53T human α-synuclein. *Neuron* 2002;**34**:521–33.

231. Giasson BI, Duda JE, Murray IVJ, Chen Q, Souza JM, Hurtig HI, et al. Oxidative damage linked to neurodegeneration by selective alpha-synuclein nitration in synucleinopathy lesions. *Science* 2000;**290**:985–9.

232. Ischiropoulos H. Oxidative modification of alpha-synuclein. *Annu NY Acad Sci* 2003;**991**:93–100.

233. Wilkinson KD, Lee KM, Deshpande S, Duerken-Hughes P, Boss JM, Pohl J. The neuron-specific protein PGP 9.5 is a ubiquitin carboxyl terminal hydrolase. *Science* 1989;**246**:670–3.

234. Hershko A, Ciechanover A. The ubiquitin system. *Annu Rev Biochem* 1998;**67**:425–79.

235. Lansbury Jr. PT, Brice A. Genetics of Parkinson's disease and biochemical studies of implicated gene products. *Curr Opin Cell Biol* 2002;**14**:653–60.

236. McNaught KS, Mytilineou C, Jnobaptiste R, Yabut J, Shahidharan P, Jennert P, et al. Impairment of the ubiquitin-proteasome system causes dopaminergic cell death and inclusion body formation in ventral mesencephalic cultures. *J Neurochem* 2002;**81**:301–6.

237. Imai Y, Takahashi R. How do Parkin mutations result in neurodegeneration? *Curr Opin Neurobiol* 2004;**14**:384–9.

238. Hattori N, Matsumine H, Asakawa S, Kitada T, Yoshino H, Elibol B, et al. Point mutations (Thr240Arg and Gln311Stop) in the Parkin gene. *Biochem Biophys Res Commun* 1998;**249**:754–8.

239. Beilina A, Van Der Brug M, Ahmad R, Kesavapany S, Miller DW, Petsko GA, et al. Mutations in PTEN-induced putative kinase 1 associated with recessive parkinsonism have differential effects on protein stability. *Proc Natl Acad Sci USA* 2005;**102**:5703–8.

240. Silvestri L, Caputo V, Bellacchio E, Atorino L, Dallapiccola B, Valente EM, et al. Mitochondrial import and enzymatic activity of PINK1 mutants associated to recessive parkinsonism. *Hum Mol Genet* 2005;**14**:3477–92.

241. Unoki M, Nakamura Y. Growth-suppressive effects of BPOZ and RGR2, two genes involved in the PTEN signaling pathway. *Oncogene* 2001;**20**:4457–65.

242. Taymans J-M, Van den Haute C, Baekelandt V. Distribution of PINK1 and LRRK2 in rat and mouse brain. *J Neurochem* 2006;**98**:951–61.

243. Weihofen A, Thomas KJ, Ostazewski BL, Cooksen MR, Selkoe DJ. Pink1 forms a multiprotein complex with Miro and Milton, linking Pink1 function to mitochondrial trafficking. *Biochemistry* 2009;**48**:2045–52.

244. Deng H, Jankovic J, Guo Y, Xie W, Le W. Small interfering RNA targeting the PINK1 induces apoptosis in dopaminergic cells SH-SY5Y. *Biochem Biophys Res Commun* 2005;**337**:1133–8.

245. Marongiu R, Spencer B, Crews L, Adame A, Patrick C, Trejo M, et al. Mutant Pink1 induces mitochondrial dysfunction in a neuronal cell model of Parkinson's disease by disturbing calcium flux. *J Neurochem* 2009;**108**:1561–74.

246. Wilson MA, Collins JL, Hod Y, Ringe D, Petsko GA. The 1.1-A resolution crystal structure of DJ-1, the protein mutated in autosomal recessive early onset Parkinson's disease. *Proc Natl Acad Sci USA* 2003;**100**:9256–61.

247. Shang H, Lang D, Jean-Marc B, Kaelin-Lang A. Localization of DJ-1 mRNA in the mouse brain. *Neurosci Lett* 2004;**367**:273–7.

248. Canet-Aviles RM, Wilson MA, Miller DW, Ahmad R, McLendon C, Bandyopadhyay S, et al. The Parkinson's disease protein DJ-1 is neuroprotective due to cysteine-sulfinic acid-driven mitochondrial localization. *Proc Natl Acad Sci USA* 2004;**101**:9103–8.

249. Miller DW, Ahmad R, Hague S, Baptista MJ, Canet-Aviles R, McLendon C, et al. L166P mutant DJ-1, causative for recessive Parkinson's disease, is degraded through the ubiquitin-proteasome system. *J Biol Chem* 2003;**278**:36588–95.

250. Takahashi-Niki K, Niki T, Taira T, Iguchi-Ariga SM, Ariga H. Reduced anti-oxidative stress activities of DJ-1 mutants found in Parkinson's disease patients. *Biochem Biophys Res Commun* 2004;**320**:389–97.

251. Galter D, Westerlund M, Carmine A, Lindqvist E, Sydow O, Olson L. LRRK2 expression linked to dopamine-innervated areas. *Ann Neurol* 2006;**59**:714–9.

252. Melrose H, Lincoln S, Tyndall G, Dickson D, Farrer M. Anatomical localization of leucine-rich repeat kinase 2 mouse brain. *Neuroscience* 2006;**139**:791–4.

253. Iaccarino C, Crosio C, Vitale C, Sanna G, Carri MT, Barone P. Apoptotic mechanisms in mutant LRRK2-mediated cell death. *Hum Mol Genet* 2007;**16**:1319–26.
254. Ho CC-Y, Rideout HJ, Ribe E, Troy CM, Dauer WT. The Parkinson's disease protein leucine-rich repeat kinase 2 transduces death signals via Fas-associated protein with death domain and caspase-8 in a cellular model of neurodegeneration. *J Neurosci* 2009;**29**:1011–6.
255. Goldberg MS, Fleming SM, Palacino JJ, Capedam C, Lam HA, Bhatnagar A, et al. Parkin-deficient mice exhibit nigrostriatal deficits but not loss of dopaminergic neurons. *J Biol Chem* 2003;**278**:43628–35.
256. Perez FA, Palmiter RD. Parkin-deficient mice are not a robust model of parkinsonism. *Proc Natl Acad Sci USA* 2005;**102**:2174–9.
257. Palacino JJ, Sagi D, Goldberg MS, Krauss S, Motz C, Wacker M, et al. Mitochondrial dysfunction and oxidative damage in parkin-deficient mice. *J Biol Chem* 2004;**279**:18614–22.
258. Chen L, Cagniard B, Mathews T, Jones S, Koh HC, Ding Y, et al. Age-dependent motor deficits and dopaminergic dysfunction in DJ-1 null mice. *J Biol Chem* 2005;**280**:21418–26.
259. Goldberg MS, Pisani A, Haburcak M, Vortherms TA, Kitada Y, Costa C, et al. Nigrostriatal dopaminergic deficits and hypokinesia caused by inactivation of the familial parkinsonism-linked gene DJ-1. *Neuron* 2005;**45**:489–96.
260. Lu X-H, Fleming SM, Meurers B, Ackerson LC, Mortazavi F, Lo V, et al. Bacterial artificial chromosome transgenic mice expressing a truncated mutant Parkin exhibit age-dependent hypokinetic motor deficits, dopaminergic neuron degeneration, and accumulation of protein-ase K-resistant α-synuclein. *J Neurosci* 2009;**29**:1962–76.
261. Gispert S, Del Turco D, Garrett L, Chen A, Bernard DJ, Hamm-Clement J, et al. Transgenic mice expressing mutant A53T human alpha-synuclein show neuronal dysfunction in the absence or aggregate formation. *Mol Cell Neurosci* 2003;**24**:419–29.
262. Kahle PJ, Neumann M, Ozmen L, Muller V, Jacobsen H, Schindzielorz A, et al. Subcellular localization of wild-type and Parkinson's disease-associated mutant alpha-synuclein in human and transgenic mouse brain. *J Neurosci* 2000;**20**:6365–73.
263. Lee MK, Stirling W, Xu Y, Xu X, Qui D, Mandir AS, et al. Human α-synuclein-harboring familial Parkinson's disease-linked Ala-53→Thr mutation causes neurodegenerative disease with α-synuclein aggregation in transgenic mice. *Proc Natl Acad Sci USA* 2002;**99**:8968–73.
264. Masliah E, Rockenstein E, Veinbergs I, Mallory M, Hashimoto M, Takeda A, et al. Dopami-nergic loss and inclusion body formation in α-synuclein mice: implications for neurodegener-ative disorders. *Science* 2000;**287**:1265–9.
265. Richfield EK, Thiruchelvam MJ, Cory-Slechta DA, Wuetzer C, Gainetdinov RR, Caron MG, et al. Behavioral and neurochemical effects of wild-type and mutated human alpha-synuclein in transgenic mice. *Exp Neurol* 2002;**175**:35–48.
266. van der Putten H, Wiederhold K-H, Probst A, Barbieri S, Mistl C, Danner S, et al. Neuropa-thology in mice expressing human α-synuclein. *J Neurosci* 2000;**20**:6021–9.
267. Wakamatsu M, Ishii A, Iwata S, Sakagami J, Ukai Y, Ono M, et al. Selective loss of nigral dopamine neurons induced by overexpression of truncated human α-synuclein. *Neurobiol Aging* 2008;**29**:547–85.
268. Martin LJ, Pan Y, Price AC, Sterling W, Copeland NG, Jenkins NA, et al. Parkinson's disease α-synuclein transgenic mice develop neuronal mitochondrial degeneration and cell death. *J Neurosci* 2006;**26**:41–50.
269. Lieberman AR. The axon reaction: a review of the principal features of perikaryal responses to axon injury. *Int Rev Neurobiol* 1971;**14**:49–124.
270. Poon HF, Frasier M, Shreve N, Calabrese V, Wolozin B, Butterfield DA. Mitochondrial associated metabolic proteins are selectively oxidized in A30P α-synuclein transgenic mice—a model of familial Parkinson's disease. *Neurobiol Dis* 2005;**18**:492–8.

271. Turnbull S, Tabner BJ, El-Agnaf OM, Moore S, Davies Y, Allsop D. Alpha-synuclein implicated in Parkinson's disease catalyses the formation of hydrogen peroxide in vitro. *Free Radic Biol Med* 2001;**30**:1163–70.

272. Paxinou E, Chen Q, Weisse M, Giasson BI, Norris EH, Rueter SM, et al. Induction of alpha-synuclein aggregation by intracellular nitrative insult. *J Neurosci* 2001;**21**:8053–61.

273. Souza JM, Giasson BI, Chen Q, Lee VM-Y, Ischiropoulos H. Dityrosine cross-linking promotes formation of stable α-synuclein polymers. *J Biol Chem* 2000;**265**:18344–9.

274. Martin LJ, Liu Z, Chen K, Price AC, Pan Y, Swaby JA, et al. Motor neuron degeneration in amyotrophic lateral sclerosis mutant superoxide dismutase-1 transgenic mice: mechanisms of mitochondriopathy and cell death. *J Comp Neurol* 2007;**500**:20–46.

275. Chen K, Northington FJ, Martin LJ. Inducible nitric oxide synthase is present in motor neuron mitochondria and Schwann cells and contributes to disease mechanisms in ALS mice. *Brain Struct Funct* 2010;**214**:219–34.

276. Rowland LP, Shneider NA. Amyotrophic lateral sclerosis. *N Engl J Med* 2001;**344**:1688–700.

277. Sathasivam S, Ince PG, Shaw PJ. Apoptosis in amyotrophic lateral sclerosis: a review of the evidence. *Neuropathol Appl Neurobiol* 2001;**27**:257–74.

278. Stephens B, Guiloff RJ, Navarrete R, Newman P, Nikhar N, Lewis P. Widespread loss of neuronal populations in spinal ventral horn in sporadic motor neuron disease. A morphometric study. *J Neurol Sci* 2006;**244**:41–58.

279. Maekawa S, Al-Sarraj S, Kibble M, Landau S, Parnavelas J, Cotter D, et al. Cortical selective vulnerability in motor neurons disease: a morphometric study. *Brain* 2004;**127**:1237–51.

280. Zoccolella S, Santamato A, Lamberti P. Current and emerging treatments for amyotrophic lateral sclerosis. *Neuropsychiatr Dis Treat* 2009;**5**:577–95.

281. Kabashi E, Valdmains PN, Dion P, Spiegelman D, McConkey BJ, Vande Velde C, et al. TARDBP mutations in individuals with sporadic and familial amyotrophic lateral sclerosis. *Nat Genet* 2008;**40**:572–4.

282. Deng H-X, Hentati A, Tainer JA, Iqbal Z, Cayabyab A, Hung W-Y, et al. Amyotrophic lateral sclerosis and structural defects in Cu, Zn superoxide dismutase. *Science* 1993;**261**:1047–51.

283. Vance C, Rogelj B, Hortobagyi T, de Vos KJ, Nishimura AL, Sreedharan J, et al. Mutations in FUS, an RNA processing protein, cause familial amyotrophic lateral sclerosis type 6. *Science* 2009;**323**:1208–11.

284. Schymick JC, Talbot K, Traynor GJ. Genetics of amyotrophic lateral sclerosis. *Hum Mol Genet* 2007;**16**:R233–42.

285. Chow CY, Lander JE, Bergren SK, Sapp PC, Grant AE, Jones JM, et al. Deleterious variants of FIG4, a phosphoinositade phosphatase, in patients with ALS. *Am J Hum Genet* 2009;**84**:85–8.

286. Sasaki S, Iwata M. Ultrastructural changes of synapses of Betz cell in patients with amyotrophic lateral sclerosis. *Neurosci Lett* 1999;**268**:29–32.

287. Menzies FM, Ince PG, Shaw PJ. Mitochondrial involvement in amyotrophic lateral sclerosis. *Neurochem Int* 2002;**40**:543–51.

288. Comi GP, Bordoni A, Salani S, Franeschina L, Sciacco M, Prelle A, et al. Cytochrome c oxidase subunit I microdeletion in a patient with motor neuron disease. *Ann Neurol* 1998;**43**:110–6.

289. Borthwick GM, Taylo RW, Walls TJ, Tonska K, Taylor GA, Shaw PJ, et al. Motor neuron disease in a patient with a mitochondrial tRNAIle mutation. *Ann Neurol* 2006;**59**:570–4.

290. Mawrin C, Kirches E, Krause G, Wiedemann FR, Vorwerk CK, Bogerts B, et al. Single-cell analysis of mtDNA levels in sporadic amyotrophic lateral sclerosis. *Neuroreport* 2004;**15**:939–43.

291. Bender A, Krishnan KJ, Morris CM, Taylor GA, Reve AK, Perry RP, et al. High levels of mitochondrial DNA deletions in substantia nigra neurons in aging and Parkinson's disease. *Nat Genet* 2006;**38**:515–7.

292. Kraytsberg Y, Kudryavtseva E, McKee AC, Geula C, Kowall NW, Khrapko K. Mitochondrial DNA deletions are abundant and cause functional impairment in aged human substantia nigra neurons. *Nat Genet* 2006;**38**:518–20.

293. Babcock D, Hille B. Mitochondrial oversight of cellular Ca^{2+} signaling. *Curr Opin Neurobiol* 1998;**8**:398–404.

294. Siklos L, Engelhardt J, Harat Y, Smith RG, Joo F, Appel SH. Ultrastructural evidence for altered calcium in motor nerve terminals in amyotrophic lateral sclerosis. *Ann Neurol* 1996;**39**:203–16.

295. Rothstein JD, Martin LJ, Kuncl RW. Decreased glutamate transport by brain and spinal cord in amyotrophic lateral sclerosis. *N Engl J Med* 1992;**326**:1464–8.

296. Rothstein JD, Van Kammen M, Levey AI, Martin LJ, Kuncl RW. Selective loss of glial glutamate transporter GLT-1 in amyotrophic lateral sclerosis. *Ann Neurol* 1995;**38**:73–84.

297. Heath PR, Tomkins J, Ince PG, Shaw PJ. Quantitative assessment of AMPA receptor mRNA in human spinal motor neurons isolated by laser capture microdissection. *Neuroreport* 2002;**13**:1753–7.

298. Kwak S, Kawahara Y. Deficient RNA editing of GluR2 and neuronal death in amyotrophic lateral sclerosis. *J Mol Med* 2005;**83**:110–20.

299. Chang DTW, Reynolds IJ. Mitochondrial trafficking and morphology in healthy and injured neurons. *Prog Brain Res* 2006;**80**:241–68.

300. Hansson MJ, Mansson R, Morota S, Uchino H, Kallur T, Sumi T, et al. Calcium-induced generation of reactive oxygen species in brain mitochondria is mediated by permeability transition. *Free Radic Biol Med* 2008;**45**:284–94.

301. Bergmann F, Keller BU. Impact of mitochondrial inhibition on excitability and cytosolic Ca^{2+} levels in brainstem motoneurones. *J Physiol* 2004;**555**:45–59.

302. Beal MF. Oxidatively modified protein in aging and disease. *Free Radic Biol Med* 2002;**32**:797–803.

303. Ferrante RJ, Browne SE, Shinobu LA, Bowling AC, Baik MJ, MacGarvey U, et al. Evidence of increased oxidative damage in both sporadic and familial amyotrophic lateral sclerosis. *J Neurochem* 1997;**69**:2064–74.

304. Abe K, Pan L-H, Watanabe M, Kato T, Itoyama Y. Induction of nitrotyrosine-like immunoreactivity in the lower motor neuron of amyotrophic lateral sclerosis. *Neurosci Lett* 1995;**199**:152–4.

305. Beal MF, Ferrante RJ, Browne SE, Matthews RT, Kowall NW, Brown Jr. RH. Increased 3-nitrotyrosine in both sporadic and familial amyotrophic lateral sclerosis. *Ann Neurol* 1997;**42**:644–54.

306. Sasaki S, Warita H, Abe K, Iwata M. Inducible nitric oxide synthase (iNOS) and nitrotyrosine immunoreactivity in the spinal cords of transgenic mice with mutant SOD1 gene. *J Neuropathol Exp Neurol* 2001;**60**:839–46.

307. Browne SE, Bowling AC, Baik MJ, Gurney M, Brown Jr. RH, Beal MF. Metabolic dysfunction in familial, but not sporadic, amyotrophic lateral sclerosis. *J Neurochem* 1998;**71**:281–7.

308. Borthwick GM, Johnson MA, Ince PG, Shaw PJ, Turnbull DM. Mitochondrial enzyme activity in amyotrophic lateral sclerosis: implications for the role of mitochondria in neuronal cell death. *Ann Neurol* 1999;**46**:787–90.

309. Vielhaber S, Kunz D, Winkler K, Wiedemann FR, Kirches E, Feistner H, et al. Mitochondrial DNA abnormalities in skeletal muscle of patients with sporadic amyotrophic lateral sclerosis. *Brain* 2000;**123**:1339–48.

310. Soraru G, Vergani L, Fedrizzi L, D'Ascenzo C, Polo A, Bernazzi B, et al. Activities of mitochondrial complexes correlate with nNOS amount in muscle from ALS patients. *Neuropathol Appl Neurobiol* 2007;**33**:204–11.

311. Echaniz-Laguna A, Zoll J, Ponsot E, N'Guessan B, Tranchant C, Loeffler J-P, et al. Muscular mitochondrial function in amyotrophic lateral sclerosis is progressively altered as the disease develops: a temporal study in man. *Exp Neurol* 2006;**198**:25–30.

312. Martin LJ. Neuronal death in amyotrophic lateral sclerosis is apoptosis: possible contribution of a programmed cell death mechanism. *J Neuropathol Exp Neurol* 1999;**58**:459–71.

313. Martin LJ, Liu Z. Opportunities for neuroprotection in ALS using cell death mechanism rationales. *Drug Discov Today* 2004;**1**:135–43.

314. Ginsberg SD, Hemby SE, Mufson EJ, Martin LJ. Cell and tissue microdissection in combination with genomic and proteomic profiling. In: Zaborszky L, Wouterlood FG, Lanciego JL, editors. *Neuroanatomical tract-tracing 3, molecules, neurons, and systems.* New York: Springer; 2006. pp. 109–41.

315. Martin LJ. p53 is abnormally elevated and active in the CNS of patients with amyotrophic lateral sclerosis. *Neurobiol Dis* 2000;**7**:613–22.

316. Martin LJ. The apoptosis-necrosis cell death continuum in CNS development, injury and disease: contributions and mechanisms. In: Lo EH, Marwah J, editors. *Neuroprotection.* Scottsdale, AZ: Prominent Press; 2002. pp. 379–412.

317. Martin LJ, Price AC, Kaiser A, Shaikh AY, Liu Z. Mechanisms for neuronal degeneration in amyotrophic lateral sclerosis and in models of motor neuron death. *Int J Mol Med* 2000;**5**:3–13.

318. Liu Z, Martin LJ. Motor neurons rapidly accumulate DNA single-strand breaks after *in vitro* exposure to nitric oxide and peroxynitrite and in vivo axotomy. *J Comp Neurol* 2001;**432**:35–60.

319. Wong M, Martin LJ. Skeletal muscle-restricted expression of human SOD1 causes motor neuron degeneration in transgenic mice. *Hum Mol Genet* 2010;**9**:2284–302.

320. Turner BJ, Talbot K. Transgenics, toxicity and therapeutics in rodent models of mutant SOD1-mediated familial ALS. *Prog Neurobiol* 2008;**85**:94–134.

321. McCord JM, Fridovich I. Superoxide dismutase, an enzymic function for erythrocuprein (hemocuprein). *J Biol Chem* 1969;**244**:6049–55.

322. Rakhit R, Crow JP, Lepock JR, Kondejewski LH, Cashman NR, Chakrabartty A. Monomeric Cu, Zn-superoxide dismutase is a common misfolding intermediate in the oxidation models of sporadic and familial amyotrophic sclerosis. *J Biol Chem* 2004;**279**:15499–504.

323. Ferri A, Cozzolino M, Crosio C, Nencini M, Casciati A, Gralla EB, et al. Familial ALS-superoxide dismutases associate with mitochondria and shift their redox potentials. *Proc Natl Acad Sci USA* 2006;**103**:13860–5.

324. Estévez AG, Crow JP, Sampson JB, Reiter C, Zhuang Y, Richardson GJ, et al. Induction of nitric oxide-dependent apoptosis in motor neurons by zinc-deficient superoxide dismutase. *Science* 1999;**286**:2498–500.

325. Flanagan SW, Anderson RD, Ross MA, Oberley LW. Overexpression of manganese superoxide dismutase attenuates neuronal death in human cells expressing mutant (G37R) Cu/Zn-superoxide dismutase. *J Neurochem* 2002;**81**:170–7.

326. Bilsland LG, Nirmalananthan N, Yip J, Greensmith L, Duhcen MR. Expression of mutant SOD1G93A in astrocytes induces functional deficits in motoneuron mitochondria. *J Neurochem* 2008;**107**:1271–83.

327. Gurney ME, Pu H, Chiu AY, Dal Canto MC, Polchow CY, Alexander DD, et al. Motor neuron degeneration in mice that express a human Cu, Zn superoxide dismutase mutation. *Science* 1994;**264**:1772–5.

328. Dal Canto MC, Gurney ME. Development of central nervous system pathology in a murine transgenic model of human amyotrophic lateral sclerosis. *Am J Pathol* 1994;**145**:1271–9.

329. Chang Q, Martin LJ. Glycinergic innervation of motoneurons is deficient in amyotrophic lateral sclerosis mice: a confocal quantitative analysis. *Am J Pathol* 2009;**174**:574–85.

330. Bendotti C, Calvaresi N, Chiveri L, Prelle A, Moggio M, Braga M, et al. Early vacuolization and mitochondrial damage in motor neurons of FALS mice are not associated with apoptosis or with changes in cytochrome oxidase histochemical reactivity. *J Neurol Sci* 2001;**191**:25–33.

331. Wong PC, Pardo CA, Borchelt DR, Lee MK, Copeland NG, Jenkins NA, et al. An adverse property of a familial ALS-linked SOD1 mutation causes motor neuron disease characterized by vacuolar degeneration of mitochondria. *Neuron* 1995;**14**:1105–16.

332. Kong J, Xu Z. Massive mitochondrial degeneration in motor neurons triggers the onset of amyotrophic lateral sclerosis in mice expressing a mutant SOD1. *J Neurosci* 1998;**18**:3241–50.

333. Jaarsma D, Rognoni F, van Duijn W, Verspaget HW, Haasdijk ED, Holstege JC. CuZn superoxide dismutase (SOD1) accumulates in vacuolated mitochondria in transgenic mice expressing amyotrophic lateral sclerosis-linked SOD1 mutations. *Acta Neuropathol* 2001;**102**:293–305.

334. Sasaki S, Warita H, Murakami T, Abe K, Iwata M. Ultrastructural study of mitochondria in the spinal cord of transgenic mice with a G93A mutant SOD1 gene. *Acta Neuropathol* 2004;**107**:461–74.

335. Borchelt DR, Lee MK, Slunt HH, Guarnieri M, Xu Z-S, Wong PC, et al. Superoxide dismutase 1 with mutations linked to familial amyotrophic lateral sclerosis possesses significant activity. *Proc Natl Acad Sci USA* 1994;**91**:8292–6.

336. Yim MB, Kang J-H, Yim H-S, Kwak H-S, Chock PB, Stadtman ER. A gain-of-function of an amyotrophic lateral sclerosis-associated Cu, Zn-superoxide dismutase mutant: an enhancement of free radical formation due to a decrease in Km for hydrogen peroxide. *Proc Natl Acad Sci USA* 1996;**93**:5709–14.

337. Kabashi E, Valdmanis PN, Dion P, Rouleau GA. Oxidized/misfolded superoxide dismutase-1: the cause of all amyotrophic lateral sclerosis? *Ann Neurol* 2007;**62**:553–9.

338. Ezzi SA, Urushitani M, Julien J-P. Wild-type superoxide dismutase acquires binding and toxic properties of ALS-linked mutant forms through oxidation. *J Neurochem* 2007;**102**:170–8.

339. Liochev SI, Fridovich I. Mutant Cu, Zn superoxide dismutases and familial amyotrophic lateral sclerosis: evaluation of oxidative hypotheses. *Free Radic Biol Med* 2003;**34**:1383–9.

340. Pacher P, Beckman JS, Liaudet L. Nitric oxide and peroxynitrite in health and disease. *Physiol Rev* 2007;**87**:315–424.

341. Andrus PK, Fleck TJ, Gurney ME, Hall ED. Protein oxidative damage in a transgenic mouse model of familial amyotrophic lateral sclerosis. *J Neurochem* 1998;**71**:2041–8.

342. Poon HF, Hensley K, Thongboonkerd V, Merchant ML, Lynn BC, Pierce WM, et al. Redox proteomics analysis of oxidatively modified proteins in G93A-SOD1 transgenic mice—a model of familial amyotrophic lateral sclerosis. *Free Radic Biol Med* 2005;**39**:435–62.

343. Okado-Matsumoto A, Fridovich I. Subcellular distribution of superoxide (SOD) in rat liver. *J Biol Chem* 2001;**276**:38388–93.

344. Higgins CMJ, Jung C, Ding H, Xu Z. Mutant Cu Zn Superoxide dismutase that causes motoneuron degeneration is present in mitochondria in the CNS. *J Neurosci* 2002;**22**(RC215):1–6.

345. Pasinelli P, Belford ME, Lennon N, Bacskai BJ, Hyman BT, Trotti D, et al. Amyotrophic lateral sclerosis-associated SOD1 mutant protein bind and aggregate with Bcl-2 in spinal cord mitochondria. *Neuron* 2004;**43**:19–30.

346. Goldsteins G, Keksa-Goldsteine V, Ahtiniemi T, Jaronen M, Arens E, Akerman K, et al. Deleterious role of superoxide dismutase in the mitochondrial intermembrane space. *J Biol Chem* 2008;**283**:8446–52.

347. Higgins CM, Jung C, Xu Z. ALS-associated mutant SOD1G93A causes mitochondrial vacuolation by expansion of the intermembrane space and by involvement of SOD1 aggregation and peroxisomes. *BMC Neurosci* 2003;**4**:16.

348. De Vos KJ, Chapman AL, Tennant ME, Manser C, Tudor EL, Lau K-F, et al. Familial amyotrophic lateral sclerosis-linked SOD1 mutants perturb fast axonal transport to reduce axonal mitochondrial content. *Hum Mol Genet* 2007;**16**:2720–8.

349. Martin LJ, Chen K, Liu Z. Adult motor neuron apoptosis is mediated by nitric oxide and Fas death receptor linked by DNA damage and p53 activation. *J Neurosci* 2005;**25**:6449–59.

350. Siklos L, Engelhardt JI, Alexianu ME, Gurney ME, Siddique T, Appel SH. Intracellular calcium parallels motoneuron degeneration in SOD-1 mutant mice. *J Neuropathol Exp Neurol* 1998;**57**:571–87.

351. Jaiswal MK, Keller BU. Cu/Zn superoxide dismutase typical for familial amyotrophic lateral sclerosis increases the vulnerability of mitochondria and perturbs Ca^{2+} homeostasis in $SOD1^{G93A}$ mice. *Mol Pharmacol* 2009;**75**:478–89.

352. Nguyen KT, Garcia-Chacon LE, Barrett JN, Barrett EF, David G. The ψ_m depolarization that accompanies mitochondrial Ca^{2+} uptake is greater in mutant SOD1 than in wild-type mouse motor terminals. *Proc Natl Acad Sci USA* 2009;**106**:2007–11.

353. Sasaki S, Shibata N, Komori T, Iwata M. iNOS and nitrotyrosine immunoreactivity in amyotrophic lateral sclerosis. *Neurosci Lett* 2000;**291**:44–8.

354. Martin LJ. Olesoxime, a cholesterol-like neuroprotectant for the potential treatment of amyotrophic lateral sclerosis. *IDrugs* 2010;**13**:568–80.

355. Kunz WS. Different metabolic properties of mitochondrial oxidative phosphorylation in different cell types—important implications for mitochondrial cytopathies. *Exp Physiol* 2003;**88**(1):149–54.

356. Keep M, Elmér E, Fong KSK, Csiszar K. Intrathecal cyclosporin prolongs survival of late-stage ALS mice. *Brain Res* 2001;**894**:27–331.

357. Karlsson J, Fong KS, Hansson MJ, Elmer E, Csiszar K, Keep MF. Life span extension and reduced neuronal death after weekly intraventricular cyclosporine injections in the G93A transgenic mouse model of amyotrophic lateral sclerosis. *J Neurosurg* 2004;**101**:128–37.

358. Kirkinezos IG, Hernandez D, Bradley WG, Moraes CT. An ALS mouse model with a permeable blood-brain barrier benefits from systemic cyclosporine A treatment. *J Neurochem* 2004;**88**:821–6.

359. Bordet T, Buisson B, Michaud M, Drouot C, Galea P, Delaage P, et al. Identification and characterization of Cholest-4-en-3-one, oxime (TRO19622), a novel drug candidate for amyotrophic lateral sclerosis. *J Pharmacol Exp Ther* 2007;**322**:709–20.

360. Mills C, Makwana M, Wallace A, Benn S, Schmidt H, Tegeder I, et al. Ro5-4864 promotes neonatal motor neuron survival and nerve regeneration in adult rats. *Eur J Neurosci* 2008;**27**:937–46.

361. Yan L-J, Sohal RS. Mitochondrial adenine nucleotide translocase is modified oxidatively during aging. *Proc Natl Acad Sci USA* 1998;**95**:12896–901.

362. Prokai L, Yan L-J, Vera-Serrano JL, Stevens Jr. SM, Forster MJ. Mass spectrometry-based survey of age-associated protein carbonylation in rat brain mitochondria. *J Mass Spectrom* 2007;**42**:1583–9.

363. Vieira HLA, Belzacq A-S, Haouzu D, Bernassola F, Cohen I, Jacotot E, et al. The adenine nucleotide translocator: a target of nitric oxide, peroxynitrite, and 4-hydroxynonenal. *Oncogene* 2001;**20**:4305–16.

364. McStay GP, Clarke SJ, Halestrap AP. Role of critical thiol groups on the matrix surface of the adenine nucleotide translocase in the mechanism of the mitochondrial permeability transition pore. *Biochem J* 2002;**367**:541–8.

365. Trumbull KA, Beckman JS. A role for copper in the toxicity of zinc-deficient superoxide dismutase to motor neurons in amyotrophic lateral sclerosis. *Antioxid Redox Signal* 2009;**11**:1627–39.

366. Costantini P, Belzacq A-S, Vieira HLA, Larochette N, de Pablo MA, Zamzami N, et al. Oxidation of a critical thiol residue of the adenine nucleotide translocator enforces Bcl-2-independent permeability transition pore opening and apoptosis. *Oncogene* 2000;**19**:307–14.

367. García N, Martínez-Abundis E, Pavón N, Correa F, Chávez E. Copper induces permeability transition through its interaction with the adenine nucleotide translocase. *Cell Biol Int* 2007;**31**:893–9.

368. Grimm S, Brdiczka D. The permeability transition pore in cell death. *Apoptosis* 2007;**12**:841–55.
369. Forte M, Gold BG, Marracci G, Chaudhary P, Basso E, Johnsen D, et al. Cyclophilin D inactivation protects axons in experimental autoimmune encephalomyelitis, an animal model of multiple sclerosis. *Proc Natl Acad Sci USA* 2007;**104**:7558–63.
370. Schinzel AC, Takeuchi O, Huang Z, Fisher JK, Zhou Z, Rubens J, et al. Cyclophilin D is a component of mitochondrial permeability transition and mediates neuronal cell death after focal cerebral ischemia. *Proc Natl Acad Sci USA* 2005;**102**:12005–10.
371. Gordon PH, Moore DH, Miller RG, Florence JM, Verheijde JL, Doorish C, et al. Efficacy of minocycline in patients with amyotrophic lateral sclerosis: a phase III randomized trial. *Lancet Neurol* 2007;**6**:1045–53.
372. Simon-Sanchez J, Singleton AB. Sequencing analysis of OMI/HTRA2 shows previously reported pathogenic mutations in neurologically normal controls. *Hum Mol Genet* 2008;**17**:1988–93.
373. Burré J, Sharma M, Tsetsenis T, Buchman V, Etherton M, Sudhof TC. α-Synulcein promotes SNARE-complex assembly in vivo and in vitro. *Science* 2010;**329**(5999):1663–7.
374. Garcia-Reitbock P, Anichtchik O, Bellucci A, Iovino M, Ballini C, Fineberg E, Ghetti B, et al. SNARE protein redistribution and synaptic failure in a transgenic mouse model of Parkinson's disease. *Brain* 2010;**133**:2032–44.

Fungal Prions

GEMMA L. STANIFORTH AND
MICK F. TUITE

Kent Fungal Group, School of Biosciences,
University of Kent, Canterbury, Kent,
United Kingdom

For both mammalian and fungal prion proteins, conformational templating drives the phenomenon of protein-only infectivity. The conformational conversion of a protein to its transmissible prion state is associated with changes to host cellular physiology. In mammals, this change is synonymous with disease, whereas in fungi no notable detrimental effect on the host is typically observed. Instead, fungal prions can serve as epigenetic regulators of inheritance in the form of partial loss-of-function phenotypes. In the presence of environmental challenges, the prion state [PRION⁺], with its resource for phenotypic plasticity, can be associated with a growth advantage. The growing number of yeast proteins that can switch to a heritable [PRION⁺] form represents diverse and metabolically penetrating cellular functions, suggesting that the [PRION⁺] state in yeast is a functional one, albeit rarely found in nature. In this chapter, we introduce the biochemical and genetic properties of fungal prions, many of which are shared by the mammalian prion protein PrP, and then outline the major contributions that studies on fungal prions have made to prion biology.

I. Introduction

Proteins that have a propensity to form aggregates in the human brain have a reputation as harbingers of disease, as exemplified by the α-synuclein protein in Parkinson's disease, $A\beta_{42}$ in Alzheimer's disease, and the huntingtin protein with an expanded polyglutamine tract in Huntington's disease.[1] As discussed elsewhere in this volume, the deposited protein aggregates associated with these increasingly common neurodegenerative disorders are typically amyloid in nature. However, one class of neurodegenerative diseases aligned to this family of catastrophic protein misfolding and aggregation diseases is the prion diseases (transmissible spongiform encephalopathies, TSE) typified by Creutzfeldt–Jakob disease (CJD) in humans and scrapie and bovine spongiform encephalopathy (BSE) in animals.[2] In mammalian prion diseases, the protein implicated, prion (PrP), shares with α-synuclein and $A\beta_{42}$ a propensity to aggregate and form amyloid, but in addition, misfolded forms of PrP are infectious—they can be transmitted between individuals. Whether the mechanism of cell-to-cell spread within the CNS via conformational templating is also unique to prions, or shared by proteins such as α-synuclein and $A\beta_{42}$, is presently the subject of some debate.[3,4]

PrP was identified by Prusiner and colleagues as a highly aggregated protein associated with amyloid plaques in the brains of individuals suffering from CJD and other TSEs.[5,6] While the focus on prions quite rightly was on their potential as disease-causing agents, the subsequent discovery of prions in the budding yeast *Saccharomyces cerevisiae*[7] and the filamentous fungus *Podospora anserina*[8] suggested that the term "prion" should not necessarily be synonymous with "disease-causing agent." Most of the described fungal prions do not negatively impact cell growth, at least under standard laboratory conditions. Indeed, under conditions of stress, certain prions in their aggregated state may actually be beneficial to the host cell.[9–12] Thus, one of the most intriguing aspects of yeast prion biology is the pleiotropic effects prions can have on host cell physiology, and with an increasing number of prion proteins found to occupy central positions in many regulatory processes,[13,14] the most debated question now emerging in this field is whether fungal prions represent a functional or a diseased state.[14,15]

Fungal prions have proven to be an invaluable tool with which to establish the generic principles not only of prion propagation and transmission[16] but also of proteinopathies in general, especially through the study of amyloid formation and the mechanisms of amyloid-related toxicity.[17,18] As a starting point, we will present evidence that yeast prions do share many of the properties associated with the mammalian prion. The focus will then shift to examining how the study of yeast prions has identified key sequence features and cellular factors integral to effective prion propagation and transmission. This will be followed by a discussion of the emerging role of yeast prions as epigenetic elements of inheritance and the apparent functional significance of prions in adaptation and survivability. Finally, we consider the utility of yeast prion proteins in identifying therapeutic molecules and intervention points that can be extrapolated to mammalian prions and potentially other protein-misfolding diseases.

A note on nomenclature: Despite fungal and mammalian prion proteins showing significant sequence or functional differences, both within and between species, in both fungi and mammals the term "prion" describes a protein that is capable of existing in at least one, but often more, self-propagating conformational states. The accepted standard nomenclature used to differentiate between this prion state and the normal non-prion state of the protein will be applied throughout this chapter. Specifically, in fungi the soluble state of a prion protein with its native fold is conveyed within brackets by italicized lower case text and a minus sign ($[prion^-]$). Similarly, the insoluble state of the prion protein, associated with the prion-specific conformation, is written within brackets in italicized upper case text with a positive sign ($[PRION^+]$). Historical precedence will, however, prevail where the nomenclature first ascribed to a fungal prion (e.g., $[URE3]$, [Het-s]) does not necessarily precisely fit the standard nomenclature.

II. Establishing the Existence of Prions in Fungi

A. Defining the Criteria—What Constitutes a Prion?

Until Prusiner's seminal paper in 1982,[5] scrapie, a long known disease of sheep, was the prototype for a class of chronic neurodegenerative diseases deemed to be caused by "slow viruses." Prusiner and his colleagues demonstrated that, instead of being a nucleic acid-based virion or other microorganism, the infectious scrapie agent was a novel, proteinaceous particle that consistently copurified with scrapie infectivity. Prusiner coined the portmanteau "prion," from "proteinaceous infectious," to describe the molecular nature of this disease-causing agent and identified PrP as the protein that formed the scrapie prion, PrP^{Sc}. In 1994, Wickner[7] was the first to propose that the self-templating properties attributed to PrP^{Sc} during infection could also account for the unusual behavior of two genetic elements in *S. cerevisiae*: $[URE3]$, which modified nitrogen utilization[7,19,20]; and $[PSI^+]$, which had an associated defect in translation termination.[21] In the case of $[URE3]$, the underlying prion protein was identified as Ure2,[7] a regulator of transcription of nitrogen utilization genes.[22,23] The protein associated with the $[PSI^+]$ prion was Sup35, a protein that mediated translation termination.[7,24] This radical and seamless integration of the "prion hypothesis" into yeast genetics, to explain the non-Mendelian behavior of the [URE3] and [PSI⁺] determinants at meiosis, provided a new approach to studying prion biology in a model system that was significantly safer and more tractable than mammalian prion systems.

Wickner proposed a number of criteria that could be used to distinguish yeast prion protein determinants of phenotype from those arising as a consequence of inheritance of a cytoplasmic nucleic acid-based determinant. Specifically, Wickner proposed the following:

(1) The [PRION⁺]-associated phenotype can also arise by mutation in the gene encoding the prion protein.
(2) Non-DNA active chemical agents can reversibly eliminate the [PRION⁺] state, and the resulting [prion⁻] cells can return to the [PRION⁺] state by cytoplasmic transfer of the determinant from the [PRION⁺] strain.
(3) Overexpression of the prion protein in a [prion⁻] cell should greatly elevate the frequency of *de novo* appearance of the corresponding [PRION⁺] state.
(4) Expression of the nuclear gene encoding the prion protein must be essential to the establishment and maintenance of the [PRION⁺] state.

Each criterion must be met before a yeast protein is considered a *bona fide* prion protein, as was so elegantly demonstrated in Wickner's initial study for Ure2/[URE3].[7] Subsequently, a range of other biochemical and genetic criteria have been applied to confirm whether or not a given protein can form a stable, transmissible prion. Of these, the three most important diagnostic criteria are as follows:

(1) The putative prion protein forms sodium dodecyl sulphate (SDS)-resistant aggregates in [PRION⁺] but not [prion⁻] cells.
(2) Incubation of a purified form of the putative prion protein can form amyloid-like polymers *in vitro* by a process of self-seeded polymerization.
(3) The resulting amyloid form generated *in vitro* can introduce the [PRION⁺] state into a [prion⁻] cell by transformation.

To illustrate how Wickner's criteria can apply to a given yeast prion, we will use the [PSI⁺] prion as an exemplar.

B. Applying the Criteria to the [PSI⁺] Prion

The [PSI⁺] genetic determinant was originally identified in a study of the genetic control of nonsense suppression in yeast.[21,25] The [PSI⁺]-associated phenotype was "allosuppression," that is, showing enhanced efficiency of suppression of a nonsense mutation. Cox had chosen a visual assay to detect and quantify nonsense suppression based on an ochre (UAA) allele of the *ADE2* gene (designated *ade2-1*). Cells carrying this allele form red colonies on agar plates and the cells are unable to utilize adenine (Ade⁻). In otherwise wild-type *ade2-1* cells, even when they also carry a weak nonsense suppressor tRNA mutation (*SUQ5*—a tRNA^{Ser}), no suppression of the red Ade⁻ phenotype of the *ade2-1* carrying cells is observed. However, Cox identified a rare Ade⁺ revertant that gave a white Ade⁺ phenotype and he designated this mutant [*PSI⁺*]. Surprisingly, when the [*PSI⁺*] mutant was crossed to a normal [*psi⁻*] *SUQ5 ade2-1* mutant, not only was the resulting diploid [*PSI⁺*] (indicating dominance of the trait), but also all haploid progeny obtained following meiosis were [*PSI⁺*], indicative of non-Mendelian inheritance of the underlying "mutation." Subsequently, it emerged that the [*PSI⁺*] determinant alone led to a defect in translation termination that allowed near-cognate tRNAs to translate a stop codon.[26]

The nature of the [*PSI⁺*] determinant remained a mystery for 30 years. The prime candidates were a range of cytoplasmically located nucleic acid determinants, including mitochondrial DNA, 2-μm plasmid DNA, and virus-like dsRNA genomes. Each of these nucleic acid replicons was systematically ruled out as the genetic determinant of [*PSI⁺*].[27] The first clues to the molecular nature of the [*PSI⁺*] determinant came from studies that established a connection between [*PSI⁺*] and the cytoplasmic protein Sup35. This connection emerged from the finding that Sup35 overexpression (a) caused growth defects in a [*PSI⁺*] but not in a [*psi⁻*] strain[28,29]; (b) resulted in a nonsense suppression phenotype in [*psi⁻*] cells[28]; and (c) induced the *de novo* appearance of [*PSI⁺*] in a [*psi⁻*] cell.[30] As a consequence of Wickner's findings on the prion-like nature of the [*URE3*] determinant, both Wickner[7] and Cox[24] suggested that [*PSI⁺*] could be the prion form of Sup35.

The primary function of Sup35 (also known as eRF3) is the formation of a translation termination complex with binding partner Sup45/eRF1 upon arrival of an in-frame stop codon (UAA, UAG, or UGA) in the ribosomal A-site.[31,32] Specifically, when Sup45 is accommodated by the ribosome in the presence of a stop codon, Sup35 binds to Sup45 and, by so doing, facilitates the correct positioning of this essential release factor in order that it can promote efficient peptidyl-tRNA cleavage.[32,33] This in turn requires activation of the Sup35 GTPase domain located in the C-terminal half of the protein.[34] In this way, Sup35 promotes effective positioning of Sup45 within the ribosomal A-site, allowing efficient translation termination.

Conformational conversion of Sup35 to its [*PSI⁺*] prion conformation, and its subsequent aggregation, leads to a reduction in the functional pool of Sup35 molecules available for translation termination,[35,36] thus readily explaining the [*PSI⁺*]-associated allosuppressor phenotype first identified by Cox.[21] This phenotype is now widely used to assay for the presence of [PSI^+] in cells either using the original $SUQ5$ $ade2-1$ markers[21] or, more recently, a different nonsense allele ($ade1-14$)[37] that gives the same red/white Ade$^-$/Ade$^+$ phenotypes in response to [PSI^+] but does not require the presence of a secondary nonsense suppressor tRNA to enhance the phenotype.[38]

The [PSI^+] prion satisfies Wickner's original genetic criteria and the subsequent biochemical criteria:

(a) Mutations in the $SUP35$ gene lead to the same phenotype as [PSI^+]. Such mutations were originally mapped to the $SAL3$[39] and the $SUP2$[40] genes and both genes are allelic with $SUP35$.

(b) Overexpression of the $SUP35$ gene in a [psi^-] cell induces the formation of the [PSI^+] determinant that shows non-Mendelian inheritance.[30]

(c) A number of chemical agents cause loss of the [PSI^+] determinant from growing cells at a frequency considerably higher than one would expect for a DNA-active mutagen.[41,42] These include guanidine hydrochloride (GdnHCl),[41] high salt concentrations,[43] and methanol.[42] The [psi^-] cells so generated can be restored to [PSI^+] either through a genetic backcross to a [PSI^+] cell, by the process of cytoduction, which involves cytoplasmic transfer without nuclear fusion, or by $de\ novo$ induction when Sup35 is overexpressed.[30]

(d) Sup35 forms high-molecular weight aggregates in [PSI^+] cells as has been demonstrated by subcellular fractionation, sucrose gradient analysis, and fusion of Sup35 to GFP.[35,36] The Sup35 aggregates found in [PSI^+] cells are SDS-resistant unlike nontransmissible and disordered protein aggregates.[44]

(e) Recombinantly produced Sup35 protein, or Sup35 N-terminal fragments, forms high-molecular-weight polymers $in\ vitro$.[45-47] This spontaneous conformational rearrangement does not require additional proteins or cofactors and results, after a lag of up to 90h, in the formation of amyloid fibrils. The lag phase can be considerably shortened by the addition of either sonicated preformed Sup35 fibrils or using extracts from [PSI^+] cells.[45,48,49]

(f) The amyloid fibrils of Sup35 produced $in\ vitro$ can stably convert a [psi^-] cell to [PSI^+] with high efficiency when added to spheroplasts generated from the former[50,51] (Fig. 1). This is also the case for whole or fractionated extracts prepared from [PSI^+] cells. In each case, sonication increases the efficiency of conversion as would be expected by an end-catalyzed seeding mechanism.

The only prion criterion that cannot be tested directly for [PSI^+] is the requirement for expression of the $SUP35$ gene for continued propagation of the [PSI^+], because $SUP35$ is an essential gene.[40]

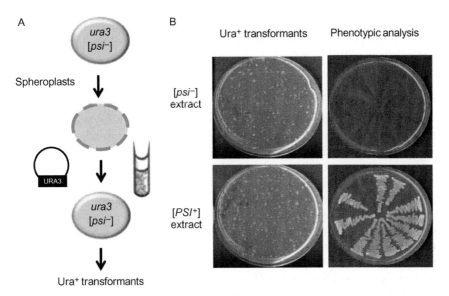

FIG. 1. Infectious yeast prions. (A) Scheme outlining the experimental strategy for introducing the [*PSI*⁺] prion into [*psi*⁻] cells using a plasmid-mediated transformation protocol, as devised by Tanaka *et al.*[50] (B) Results of an analysis of whole-cell extracts prepared from [*PSI*⁺] and [*psi*⁻] cells. The primary transformants were selected using the plasmid-borne *URA3* marker, giving rise to Ura⁺ cells (left) and the phenotypic analysis of 16 independent transformants from each of these plates (right). Those [*psi*⁻] cells that acquired the [*PSI*⁺] prion show a change from red to white colony color due to suppression of the *ade1-14* allele. (See Color Insert.)

In part through the ease by which it can be assayed, coupled with the relative mitotic stability of the prion, [*PSI*⁺] has emerged as the most extensively studied yeast prion. Increasingly sophisticated biochemical and biophysical techniques have been used in these studies.[38,52] Importantly, what we have learned from these studies is generally applicable to other yeast prions, as will be described below.

III. The Prions of Yeast and Their Cellular Roles

The recent development of a wide range of systems biology tools and bioinformatics coupled with the increasing availability of biological and technical high-throughput materials has resulted in the identification and validation of seven yeast prions, including [*PSI*⁺]. This figure is likely to be an underestimate and it is anticipated that the number will rise to more than 20.[53]

The following provides a brief introduction to the most widely studied of the fungal prion proteins to illustrate the range of cellular processes that can be potentially modified by prions.

A. [URE3] and the Regulation of Nitrogen Catabolism

The [URE3] prion alters the way the host cell can utilize poor nitrogen sources especially when in the presence of rich nitrogen sources such as ammonia.[7,19,20] Ure2, the protein that gives rise to the [URE3] prion, is an integral component of the nitrogen metabolism network in yeast.[54] Yeast cells are capable of sensing both the quality and quantity of various nitrogen sources in their environment and the metabolism of the various nitrogen sources available requires input from 90 or more different genes. All of these genes are under the control of only four transcription factors: Gln3, Gat1, Dal80, and Gzf3.[55] An elaborate network of negative and positive feedback loops and phosphorylation/dephosphorylation cycles finely coordinate relevant subsets of these genes, and the Ure2 conformational switch impacts this regulation.[56]

Ure2 acts as a cytoplasmic anchor for the GATA-binding transcription factor, Gln3, when preferred sources of nitrogen such as glutamine or ammonia are available.[22,23,54] When only poor nitrogen sources are available, the compartmentalization of Gln3 is lost following its dissociation from Ure2, and nuclear-localized Gln3 is able to activate transcription of the nitrogen catabolite repression genes. The stabilization of general amino acid permeases occurs in poor nitrogen conditions, as does the turnover of more specific amino acid permeases, and an upregulation of autophagy places an emphasis on the internal supply of nitrogen for sustenance.[57] The conversion of Ure2 to its [URE3] prion conformation results in persistent nuclear localization of Gln3 and subsequent lack of nitrogen discrimination. Thus, a [URE3] cell will metabolize a poor nitrogen source such as ureidosuccinate (USA) in the presence of an otherwise preferred nitrogen source such as ammonia.[7,19]

B. [PIN⁺] and the Unknown Function of Rnq1

Perhaps the most intriguing of all yeast prions is [PIN⁺], the discovery of which emerged from studies on the ability to reintroduce [PSI⁺] into [psi⁻] cells by the overexpression of Sup35. Two distinct [psi⁻] strains were identified: those in which it was possible to induce [PSI⁺] by Sup35 overexpression and those where no such induction was seen.[58,59] The difference between these two strains was the presence of a cytoplasmic genetic element in those [psi⁻] strains in which [PSI⁺] could be induced. This element was called [PIN⁺], an abbreviation for [PSI⁺] inducibility ([PIN⁺]), and like [PSI⁺], the [PIN⁺] determinant showed non-Mendelian inheritance.[58]

Rnq1, a protein of unknown function but rich in N (Asn) and Q (Gln) residues, was identified as the [PIN⁺] prion determinant because deletion of the RNQ1 gene led to loss of [PIN⁺].[60,61] Furthermore, Rnq1 was shown to exist as a high-molecular weight insoluble form in [PIN⁺] strains, but was soluble in a [pin⁻] strain.[60,62] Definitive proof that Rnq1 was the determinant of [PIN⁺] was obtained when in vitro generated, amyloid-like fibrils formed from Rnq1 were able to induce the [PIN⁺] state when transformed into a [pin⁻] yeast strain.[63] Yet, even though it is widely considered that the [PIN⁺] determinant is Rnq1, the original studies of Derkatch et al.[60] showed that several other proteins with a high proportion of Gln and Asn residues also can act as [PIN⁺] determinants.

Why is [PIN⁺] required for de novo formation of [PSI⁺], and perhaps other prions, in yeast? The prion form of Rnq1 most likely acts as a nucleating factor for Sup35 because Rnq1 fibrils can increase the in vitro polymerization rate of Sup35.[64] This nucleating activity is not, however, restricted to Sup35. For example, aggregation and toxicity of the polyglutamine-expanded huntingtin protein in yeast are dependent on [PIN⁺] being present.[61,65] Yet, despite intense analysis, the cellular function of Rnq1 remains to be identified. Rnq1 is not an essential cytoplasmic protein, and to date the only phenotype associated with Rnq1 is its ability to enhance the de novo formation of other yeast prions, or the aggregation of other aggregation-prone proteins, when in its [PRION⁺] conformation.

C. Prion-Forming Transcription Factors in Yeast

Yeast prions represent a unique class of epigenetic element that can transmit "genetic" information from mother to daughter cell or to meiotic progeny. Acquisition of the prion can transiently alter cell phenotype without imparting any underlying permanent change in DNA sequence. Other more commonly cited examples of epigenetic regulatory mechanisms in yeast, and other eukaryotes, include those that remodel nuclear chromatin to facilitate regulation of gene transcription.[66] Chromatin remodeling is carried out by macromolecular assemblies such as the SWI/SNF (SWItch/Sucrose Non-Fermentable) complex which includes the Swi1 protein. The discovery that Swi1 is able to spontaneously convert to a heritable prion form called [SWI⁺] provided the first clear link between these two distinct modes of epigenetic inheritance.[67,68]

The conversion of Swi1 to [SWI⁺] is accompanied by a partial loss-of-function phenotype, as is usually observed when other yeast prion proteins convert to their prion form. This has important ramifications for the host, not least because in S. cerevisiae the SWI/SNF chromatin remodeling complex regulates the transcriptional behavior of nearly 6% of all genes.[69] The formation of the [SWI^+] prion should therefore lead to a global change in gene expression that would affect a diverse range of processes, including cell growth, stress response, and DNA replication.[67]

An additional link between global regulation of gene expression and yeast prions came with the discovery that the Cyc8 protein, which forms a corepressor complex with the Tup1 protein, can form the heritable [OCT^+] prion.[70] As with Swi1, the "prionization" of Cyc8 is accompanied by a partial loss-of-function phenotype that leads to the incomplete derepression of Cyc8-Tup1 targeted genes. These include genes involved in mating-type DNA repair and glucose repression.[71]

Apart from Ure2, there are two other yeast prion proteins that function as transcription factors and whose activity is regulated by a prion-based conformational switch. The first of these, Sfp1, is a transcription factor that regulates the expression of nearly 10% of yeast genes, including those involved in ribosome biogenesis and the regulation of cell size.[72,73] Sfp1 is the protein determinant of the [ISP^+] prion, which forms very efficiently following Sfp1 overexpression at a spontaneous frequency significantly higher than that observed for other yeast prions (10^{-4} per cell per generation compared to 10^{-6} per cell per generation for [PSI^+]).[74] [ISP^+] was first detected through its ability to reverse a nonsense suppression phenotype associated with a nuclear *sup35* gene mutation.[75] Sfp1 apparently shuttles between the cytoplasm and nucleus, but in an [ISP^+] strain, Sfp1 forms aggregates in the nucleus.[74] [ISP^+] is thus an exclusively nuclear prion, and so it remains unclear how prionization of Sfp1 could impact translation termination.

The other transcription factor known to form a prion is Mot3, whose prion-like properties emerged from a rigorous and systematic approach to prion protein identification in *S. cerevisiae*.[53] Mot3 also acts as a global regulator of gene expression, including those genes involved in the biosynthesis of ergosterol. The partial loss of function of Mot3 in its [$MOT3^+$] prion form leads to changes to the cell wall which can confer a growth advantage to [$MOT3^+$] cells under conditions requiring increased cell wall resilience.[53]

D. Other Yeast Prions and Prion-Like Determinants

The results of combined genetic, bioinformatic, and biochemical screening approaches to identify new yeast prions have revealed that many proteins possess an array of prion-like features, though few possess a "complete set."[53,76,77] These quasi-prions present opportunities to understand the mechanistic subtleties that result in *bona fide* prion proteins, but they also challenge our basic assumptions about what defines a prion protein. This in turn has important implications for any search for new prions outside the fungal kingdom.

One example is the atypical prion-like element [GAR^+]. [GAR^+] cells can utilize alternative carbon sources in the presence of glucose.[78] Ordinarily, glucose is the favored carbon source for yeast, with even small amounts of glucose preventing the metabolism of different carbon sources, such as glycerol. The [GAR^+] prion appears to be an oligomeric complex composed of the Pma1 and Std1 proteins, though neither is known to be amyloid-forming. Pma1 is a plasma-membrane ATPase pump,[79] whereas Std1 is involved in the control of glucose-regulated gene expression.[80] Consistent with its other prion-like properties, [GAR^+] is cytoplasmically inherited and forms *de novo* at a high frequency following overexpression of either Pma1 or Std1.[78] The ability of a prion-like system to alter carbon source utilization strategies again implies a role for prions as useful epigenetic switches rather than as disease-causing entities.

While in most cases the conversion of yeast prion proteins to their [$PRION^+$] state is associated with a partial loss-of-function phenotype, one example currently exists in yeast in which [$PRION^+$] formation demonstrates a gain-of-function phenotype, namely, vacuolar protease PrB. PrB is required to be in its [$PRION^+$] state to activate its own inactive precursor. The [β] prion formed by PrB is reversibly curable and its *de novo* appearance is increased in frequency by overexpression of the encoding gene *PRB1*.[81] PrB has a well-established role as a serine protease involved in protein degradation in the vacuole, and the presence of [β] confers a growth advantage during conditions of starvation.[81] Yet, although [β] does satisfy the majority of Wickner's genetic criteria to be classified as a prion, it does not show the biochemical characteristics expected. This fact highlights the importance of carefully evaluating all of the criteria before applying the term "prion" to a non-Mendelian genetic element. This is further illustrated by the atypical properties of the [GAR^+] prion.

E. [HET-s] and Vegetative Incompatibility in *P. anserina*

Although the focus has been on prions in one fungal species, *S. cerevisiae*, one other fungal prion has also been identified and verified, namely, the [Het-s] prion that controls vegetative incompatibility in the filamentous fungus *P. anserina*.[8] The [Het-s] prion state represents an example of a gain of function compared with cells expressing the soluble form of the underlying protein HET-s. The [Het-s] prion regulates the fate of the vegetative heterokaryon that arises when hyphae from two different fungal strains come into contact and fuse, resulting in the mixing of genetically different nuclei within the cytoplasm of a single hyphal cell (Fig. 2). Genetic differences between individual strains at several different *het* (heterokaryon incompatibility) loci can lead to death of the heterokaryon, a phenomenon known as vegetative incompatibility.[82] One such locus, *het-s*, encodes the prion-forming protein HET-s and the conformational

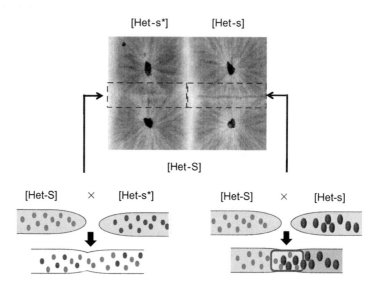

FIG. 2. Vegetative incompatibility in the fungus *Podospora anserina* is controlled by the [Het-s] prion. The consequences of hyphal fusion between a [Het-S] strain of *P. anserina* expressing the soluble form of the Het-S protein and either a [Het-s°] strain expressing the soluble form of the related Het-s protein or a [Het-s] strain containing the prion form of the Het-s protein. The distinct band of cell death that is the hallmark of vegetative incompatibility is indicated in the upper panel. (Fungal images kindly provided by Sven Saupe, IBGC, Bordeaux). (See Color Insert.)

change in HET-s to the [*PRION*⁺] state. Vegetative incompatibility results when the HET-s protein is in the prion form [Het-s], whereas the heterokaryon survives if the HET-s protein is in the non-prion form [Het-s°].

The evidence that [Het-s] is a prion is overwhelming,[83] but unlike its yeast counterparts, the propagation of the [Het-s] prion is not strictly dependent on chaperone (Hsp104)-mediated fragmentation (see Section IV.B) although loss of Hsp104 function leads to a detectable impairment in propagation.[84] Furthermore, the prion-forming domain (PrD) of HET-s does not share the conserved sequence features shared by the yeast prions[85] (see Section IV.A), yet there is no doubt that [Het-s] is a transmissible fungal amyloid.

IV. Propagating the [*PRION*⁺] State

The [*PRION*⁺] form of most fungal prion proteins can be propagated efficiently over many cell generations in the absence of any selective pressure. Yeast cells grow and divide under optimal laboratory conditions (2% glucose, 30°C and vigorous aeration) every 90–120min although this doubling time may be significantly extended in a natural environment. Nevertheless, this contrasts sharply with mammalian prions, which effectively propagate in nondividing neuronal cells.

Fungal prions are propagated by a mechanism that ensures that sufficient new "prion seeds" are generated within a single cell cycle in order to be passed on to daughter cells prior to cell separation. This mechanism is similar to that which propagates mammalian prions[16] and consists of three discrete sequential stages (Fig. 3):

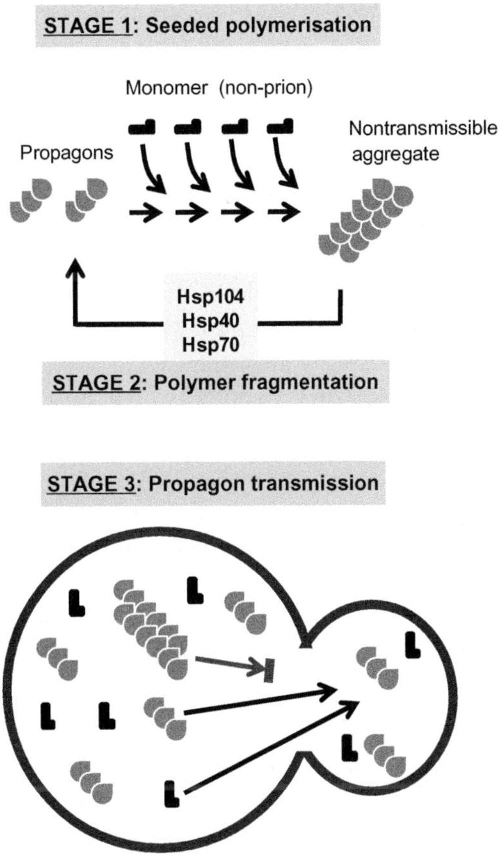

FIG. 3. The yeast prion propagation cycle. The generation and transmission of propagons occur in three stages. While the molecular nature of the propagon has not been established, it is assumed that it is a small oligomer generated by chaperone-mediated fragmentation of high-molecular weight prion polymers (aggregates) that are too large to pass from mother to daughter cell. (For color version of this figure, the reader is referred to the Web version of this chapter.)

Stage 1: *Seeded polymerization.* The soluble prion protein monomers interact with existing prions and are converted to the $[PRION^+]$ state, leading to the formation of high-molecular weight amyloid aggregates.

Stage 2: *Polymer fragmentation.* The growing amyloid fibrils are fragmented to produce lower molecular weight polymers that can seed further rounds of polymerization of existing of newly synthesized soluble protein molecules into polymers. The term "propagon" is used to describe such "prion seeds."[86]

Stage 3: *Propagon transmission.* The propagons are transferred in sufficient numbers to daughter cells via cytoplasmic transfer to ensure continued prion propagation.

The elucidation of these steps in yeast prion propagation has been achieved through the use of a variety of experimental approaches that fully exploit the yeast experimentalist's toolbox. Perhaps the two most successful approaches have been (a) to identify mutants that block prion propagation *in vivo* and (b) to re-create the generation of infectious prion polymers by *in vitro* polymerization of prion protein.[38,52] These approaches have not only helped establish the key amino acid sequence and secondary structural features of the prion protein that are required for the protein to be able to form a stable and transmissible $[PRION^+]$ form but, in addition, also have revealed the roles played by a number of cellular factors.

The most prominent of the cellular factors that participate in prion propagation are the various molecular chaperones that ensure that Stage 2 of the mechanism, polymer fragmentation, occurs in the time frame of minutes. Of particular importance is the heat-inducible molecular chaperone Hsp104 (heat shock protein of 104,000 Da)[87] which is required for the propagation of all verified yeast prions.[88,89] The role of Hsp104 in propagation of the [Het-s] prion is less well defined,[84] and to date, no cellular factors have been identified that are essential for mammalian prion propagation, although at least one factor (protein x) has been implicated in early studies on PrP.[90]

A. The Role of the Prion-Forming Domain

Fungal prion proteins contain a region at either their N- or C-terminus that is essential for the conformational change leading to the prion form. These regions constitute the PrD, are dispensable for the function of that protein, and do not participate significantly in organizing the native protein fold.[91,92] The PrD also is active when fused to a protein that is not naturally prion forming, for example, rat glucocorticoid receptor[93] or green fluorescent protein.[35,94] The most characteristic sequence feature of the yeast PrDs is their relatively high proportion of Asn and Gln residues and the relative lack of charged and hydrophobic amino acids.[53,76,95] Various studies have suggested that the PrD defines a highly flexible domain separated from the overall protein fold.[96]

Increasingly sophisticated algorithms have been developed in an attempt to guide a search for new yeast prions. Early iterations of these algorithms identified Rnq1 as a prion-forming protein.[97] The most recent iteration, using an approach based on the hidden Markov model, predicts that nearly 3% of the approximately 6200 proteins constituting the yeast proteome have PrDs.[53] A systematic genetic and biochemical analysis of the top-scoring 100 proteins (which included the yeast prions already identified) confirmed at least 19 additional potential prions.[53] This analysis also suggested that PrDs that showed the highest propensity to drive the formation of SDS-resistant aggregates had a greater proportion of Asn residues. While it remains to be seen if the rules that define a yeast PrD can be used to interrogate other fungal proteomes, or even those of other kingdoms, for as-yet-unidentified prions, Asn/Gln-rich regions are found in a wide range of proteins in eukaryotes, although they are rare in mesophilic bacteria and almost completely absent from thermophilic bacteria and Archaebacteria.[76] However, the presence of a Gln/Asn-rich region is not sufficient to confer prion properties to a yeast protein[98]

Studies on the N-terminal Sup35-PrD have been particularly informative with respect to the role of specific sequences and structures within the yeast PrD and their contribution to the *de novo* formation and continued propagation of the [*PSI*+] prion from. The first mutant identified in *SUP35* that blocked the propagation of [*PSI*+] was the *PNM2* ([*PSI*+] No More) mutant. This dominant negative mutant contains a single amino acid substitution in the N-terminal region of Sup35 (G58D)[99] in one of the five copies of a slightly degenerated oligopeptide present between residues 41 and 97 (Fig. 4). The Sup35 repeats show a striking four out of eight match with the sequence of the octarepeats found in mammalian PrP. Whether the existence *per se* of such repeats is a critical functional feature of Sup35-PrD remains a subject of debate. For example, deletions of one or more of the repeats can lead to a significant defect in [*PSI*+] propagation,[101-104] yet Ross et al.[105] have reported that the amino acid sequence encompassing the Sup35 repeats can be scrambled without impairing the protein's prion-like behavior. Few yeast PrDs contain oligopeptide repeats, yet the fact that single amino acid substitutions in the Sup35-PrD repeats can impair propagation[99,106] and that the spacing of Pro residues in the region of the PrD encompassing the repeats make an important contribution to the prion properties of Sup35[53] means that it is still unclear as to what the primary and secondary structure features of a yeast PrD are that are key for prion propagation *in vivo*.

The prion properties of Sup35 are exclusively conferred by the N-terminal region between residues 1 and 97,[103,104,107] although the adjacent and highly charged M region controls solubility of Sup35 and indirectly can impact prion

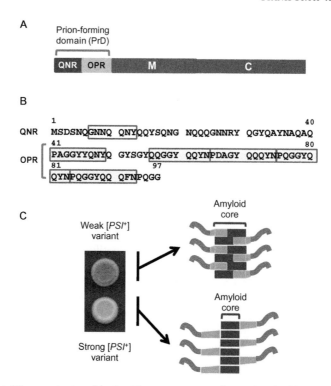

Fig. 4. The organization of the Sup35 protein in prion polymers (amyloid fibrils) in variants of the [*PSI*⁺] prion. (A) The proposed three-domain organization of Sup35 indicating the location of the prion-forming domain (PrD), the highly charged M domain, and the C domain that defines the functional region of the protein. The PrD is made up of two regions: QNR, a region rich in Asn (Q) and Gln (N) residues; and OPR, the oligopeptide repeat-containing region. (B) The sequence of residues 1–97 of the Sup35-PrD, indicating the location of the QNR and OPR regions and the location of the five imperfect oligopeptide repeats (boxed). (C) Weak and strong variants of the [*PSI*⁺] prion as defined by their different efficiencies of suppression of the *ade1-14* allele (left) and the extent of protein–protein interaction between the PrD of Sup35 molecules, as defined by Toyama *et al.*[100] (See Color Insert.)

stability.[108] PrDs do not necessarily have to reside at the N-terminus. For example, the complex Rnq1–PrD is located in the C-terminal half of that protein.[97,109,110] The Sup35-PrD has two functionally distinct regions (Fig. 4B). The first 40 residues, which are particularly enriched in Asn and Gln residues, and hence is referred to as the "QN-rich (QNR) region," likely represent the region through which protein–protein contact is made during seeded polymerization. This conclusion is based on two sets of observations: First, single amino acid substitutions introduced into this region can lead to loss of [*PSI*⁺], but not because of a defect in propagation, but rather due to inhibition of Sup35 polymerization.[48] Consequently, these mutants, termed *ASU* (antisuppressor) mutants, have a much higher than normal level of Sup35, which leads to increased termination efficiency. This results in loss of the [*PSI*⁺] phenotype, but these cells can still transfer [*PSI*⁺] propagons to ASU⁺ daughter cells. The second line of evidence emerged from the analysis of the inability of related Sup35 sequences to be seeded by *S. cerevisiae* Sup35 propagons,[111,112] a phenomenon akin to the prion species barrier in mammals.[113] The QNR region contains amino acid sequence information that defines the species barrier by blocking heterotypic protein–protein interactions. This region also contains a highly amyloidogenic peptide Gly-Asn-Asn-Gln-Gln-Asn-Tyr that has been widely used in defining the generic structures of amyloids.[114]

In addition to being required to maintain the prion fold of the respective yeast prion protein, overexpression of the respective PrD is also a more potent inducer of the [*PRION*⁺] state than is observed for overexpression of the corresponding full-length protein. For example, the minimal Ure2-PrD corresponds to the N-terminal 65 amino acids,[115] and overexpression of this fragment induces the *de novo* appearance of [*URE3*] several thousand fold above the spontaneous rate of *de novo* formation of the prion. This is compared to an increase in [*URE3*] appearance of only some 20- to 200-fold induced by overexpression of full-length Ure2.[115,116] Consequently, considering the PrD a discrete unstructured module that is required to promote the formation of the range of disordered molten globule-like states necessary for nucleating prion protein polymerization[96,117] may be an oversimplification. This suggestion is supported in part by the identification of mutations outside the Ure2-PrD that increase the frequency of the [*PRION*⁺] state.[118]

While Gln/Asn richness is a general characteristic of yeast PrDs, this does not hold true for the HET-s prion protein in *P. anserina* (see Section III.E). The ability of the non-Gln/Asn-rich HET-s prion to propagate its prion fold in both *P. anserina*[85,119] and in *S. cerevisiae*[120] indicates that QN richness is not an essential property of *S. cerevisiae* or other fungal prions. Consequently, the current bioinformatics tools used to identify yeast prions through signature QNR regions may be overlooking a separate, perhaps nonconventional, class of prion proteins. Furthermore, the fact that the propagation of HET-s as a prion in *S. cerevisiae* is enhanced by the presence of the [*PIN*⁺] prion[120] suggests that coexistence of QNR and non-QNR prion or aggregation-prone proteins may positively modulate each other's propagation, or that the presence of QNR aggregates modifies proteostasis, creating a more aggregation-conducive environment.

B. The Role of Cellular Factors

Seeded polymerization resulting in the formation of the diagnostic amyloid-like fibrillar polymers is not sufficient for the continued propagation of the [*PRION*⁺] state. This can be achieved only if cells are able to pass on these polymers to daughter cells where they can continue to seed polymerization (Fig. 3). Although we know very little about the molecular mass or composition of any yeast propagon, we do know that there is a physical limit to the size of prion polymer that can be transmitted to daughter cells, at least for the [PSI⁺] prion.[121] Consequently, the maintenance of a prion in a population of growing cells relies on a mechanism for transfer of appropriately sized propagons from mother to daughter cells. In yeast, this outcome is achieved through the action of molecular chaperones.

The primary role of molecular chaperones in the cell is to ensure that proteins are correctly folded as they are released from the ribosome. However, chaperones also are important in the refolding of misfolded proteins in cells exposed to stresses, such as heat shock and oxidative stress, that compromise protein folding and stability.[122] One outcome of protein misfolding is the formation of amorphous protein aggregates that must be disaggregated in order to make them accessible to the refolding chaperones. The molecular chaperones that mediate this disaggregation are also responsible for the continued propagation of yeast prions through their disaggregation function.

The first chaperone implicated in fungal prion propagation was Hsp104. Deletion or overexpression of the nonessential HSP104 gene in nonstressed [PSI⁺] cells resulted in the rapid appearance of prion-free [psi⁻] cells indicative of a block in one or other of the stages of prion propagation.[37] While the defect caused by Hsp104 overexpression is restricted to relatively few yeast prions, all, bar the recently described [ISP⁺] prion,[74,75] can propagate only in cells expressing Hsp104, which provides further support for the prion hypothesis to explain the behavior of [PSI⁺] and other yeast prions.

The explanation for why yeast prion propagation depends on Hsp104 is relatively simple—Hsp104 functions as a "disaggregase" that fragments prion fibrils into the smaller fibrils or oligomers which constitute the propagons that then can be transmitted to the daughter cell.[37,45,123] As would be expected, deletion of the HSP104 gene or impairment of the chaperone's function leads to the formation of much larger polymers that become physically trapped in the mother cell, thus limiting their transmissibility to daughter cells.[121,123,124] Similarly, overexpression of Hsp104 is believed to "over-fragment" the fibrils such that they can be efficiently resolubilized, restoring the [psi⁻] state. However, this has not been formally demonstrated and there are data that suggest otherwise.[125,126]

The only native yeast prions described that does not require Hsp104 for its propagation is [ISP⁺].[75] This atypical property might be due to the nuclear localization of the Sfp1/[ISP⁺] prion aggregates, whereas most other Hsp104-dependent yeast prions, including the nuclear residing chromatin remodeling factor Swi1,[68] form aggregates predominantly in the cytoplasm. The nuclear aggregates of Sfp1 may therefore have established dependency on an alternative chaperone, possibly one that shuttles between the nucleus and the cytoplasm. This would present a novel means of regulating [ISP⁺] maintenance.

Conditions that relocalize the chaperone predominantly to the cytoplasm or the nucleus would result in the loss or the maintenance of the $[ISP^+]$ prion, respectively. There is evidence, however, that Hsp104 can shuttle between the nucleus and cytoplasm.[127]

One important experimental tool that can be used to probe the role of Hsp104 in yeast prion propagation is the use of the chaotropic agent guanidine hydrochloride. GdnHCl inactivates Hsp104 *in vivo*[128–130] by inhibiting the chaperone's ATPase activity. This leads to a block in the fragmentation of the prion polymers, resulting in an increase in the size of the prion polymer and leading to a physical block in transfer to the daughter cells and hence generation of $[prion^-]$ cells.[123] The GdnHCl-mediated inhibition of prion propagation takes effect within one cell generation and is reversible, allowing the experimenter to block and restore prion propagation at will.[131–133] Further proof that the propagation of yeast prions requires the ATPase activity of Hsp104 comes from the demonstration that ATPase-negative mutants of Hsp104 are unable to maintain the $[PSI^+]$ prion state.[37]

Chaperones other than Hsp104 also have critical or influential roles in the yeast prion propagation cycle and ensure that the prion polymers are efficiently fragmented under normal (nonstress) physiological conditions. In particular, members of the Hsp70 chaperone family and their obligate co-chaperones, the Hsp40 chaperones, play crucial roles, the most critical being the Hsp40 chaperone Sis1. Sis1 is required for $[PSI^+]$, $[PIN^+]$, and $[URE3]$ propagation,[134] but the essential nature of the *SIS1* gene precludes the use of a *sis1Δ* strain to elucidate its role. This role has been demonstrated by engineering reduced cellular levels of Sis1 and revealing prion propagation defects.[134,135] The role of Sis1 in conjunction with Ssa1, a member of the Hsp70 family, is to tag the prion polymers and recruit Hsp104.[134,135] *In vitro* fragmentation studies using Sup35 fibrils support such a mechanism.[136] Ssa1 (and its close homologue, Ssa2) also contributes to the prion propagation cycle,[89] but because of the functional redundancy within the Ssa/Hsp70 family and the loss of viability when all four Ssa homologues are deleted, whether yeast prion propagation is strictly dependent on Ssa/Hsp70s has not been formally demonstrated.

C. The Role of the Cytoskeleton

While the seeded polymerization step of the yeast prion protein propagation mechanism (Fig. 3) can be readily re-created *in vitro*, there is increasing evidence that *in vivo* some form of scaffold is necessary for the polymers to form. This was revealed by the finding that the polymerization mechanisms employed for the ordered assembly of prion polymers and key elements of the cytoskeleton have overlapping cellular requirements. For example, Sla1 and a number of other proteins that form the cortical actin cytoskeleton of yeast interact directly with the Sup35-PrD.[137,138] In so doing, they play a role in the *de novo* generation of new $[PSI^+]$ prions rather than in the continued propagation of the $[PRION^+]$ state.[137,138] Treatment of yeast cells with the actin depolymerizing drug latrunculin A can eliminate the $[PSI^+]$ prion, consistent with a role for the actin cytoskeleton in yeast prion propagation, although yeast prion aggregates do not appear to colocalize with actin patches.[139] The actin cytoskeleton may therefore provide a scaffold that facilitates prion protein polymerization and, by exploiting the actin cytoskeleton to do so, mimics aggresome formation in mammalian cells.[138]

D. Cell-to-Cell Transmission of Prions

Continued propagation of a yeast prion requires the transmission of one or more propagons to the daughter cell prior to cytokinesis, and any defect in propagon transmission leads to the appearance of prion-free daughter cells. Whether or not an active prion transmission mechanism exists in S. cerevisiae remains to be established, but a number of "active" mother-to-daughter delivery systems are used in yeast ensuring efficient transfer of organelles such as mitochondria. These organelle transfer mechanisms typically involve a cytoskeletal track, most likely in the form of actin filaments coupled with one or more molecular motors.[140]

In order to fully evaluate the need for an active prion transmission mechanism in vivo, it is important to establish experimentally n_0, the number of propagons in an individual $[PRION^+]$ cell. Direct visualization of yeast prion protein–GFP fusion proteins reveals a number of specific GFP aggregates (foci) in the $[PRION^+]$ cell,[35,141–143] but it remains unclear whether these foci represent individual propagons. However, n_0 can be indirectly estimated by exploiting the Hsp104 ATPase inhibitor GdnHCl (see Section IV.B). Inhibiting the fragmentation activity of Hsp104 by GdnHCl ensures that no new propagons are generated, and those propagons present at the time of GdnHCl addition are subsequently diluted out by cell division[132,133] (Fig. 5). Monitoring the loss of $[PSI^+]$ from GdnHCl-treated cells has led to the development of stochastic models that can predict n_0, which typically indicates between 200 and 400 propagons per cell.[131,132,144] This method also allows an estimation of the relative number of propagons transmitted to the daughter cell at cytokinesis.[131] The value so derived, 0.4, is very similar to the relative volumetric differences between mother and daughter cell at division, a result consistent with propagon transfer being via simple cytoplasmic distribution and with no active transmission mechanism.

V. Yeast Prion Variants and Phenotypic Variability

One of the most remarkable properties of the mammalian prion protein PrP is that it can generate a number of distinct conformational variants of infectious PrP^{Sc}, each of which is associated with a characteristic neuropathology

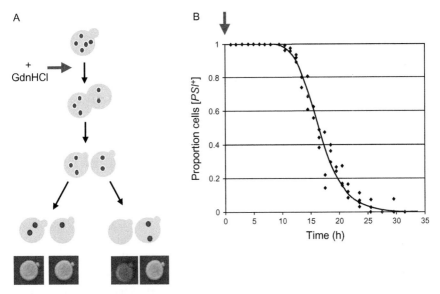

FIG. 5. How the number of propagons in a [*PSI*⁺] cell can be determined experimentally. (A) The mechanism by which guanidine hydrochloride (GdnHCl) eliminates the [*PSI*⁺] prion from growing cells. Inhibition of the chaperone Hsp104 results in a failure to generate new propagons (five are shown in the original [*PSI*⁺] cell) and the propagons remaining are diluted out by cell division.[133] Daughter cells that fail to inherit a single propagon become [*psi*⁻]. (B) The kinetics of elimination of the [*PSI*⁺] prion over time following the addition of GdnHCl at *t*=0. (See Color Insert.)

but which have no variation in the primary structure.[145] These prion strains have also been described for the yeast prions [*PSI*⁺],[59] [*URE3*],[146] and [*PIN*⁺],[62,147] although in yeast they are referred to as "variants" since the term "strain" is typically used to indicate cells with different nuclear genotypes. As with mammalian prion strains, the only underlying difference between the yeast prion variants is the conformation of the prion protein.[50,100,147–149] The ability to generate prion variants reflects the conformational plasticity of prion proteins and their ability to exist in conformationally stable forms following prion conversion. However, the yeast prion variants do display differences in (a) the relative levels of soluble and aggregated forms of the prion protein; (b) the extent of loss of native function; and (c) the hierarchical dominance of the inheritance of one prion variant over another.[62,141] Yeast prion variants may also differ in their relative mitotic stabilities, the latter reflecting the number of propagons present in the cell, which in turn reflects the efficiency with which the prion aggregates are fragmented by Hsp104 and its co-chaperones.[149]

The most actively studied variants in yeast are of the [*PSI*⁺] prion. When [*PSI*⁺] cells arise *de novo* from a [*psi*⁻] strain, they are typically classified into one of two types based on the efficiency with which the reporter nonsense allele (usually *ade1-14*) is suppressed.[58,59] The "strong" variants show a high level of nonsense suppression and generate white colonies, whereas the "weak" variants, which contain a significantly higher proportion of the soluble, and hence functional, form of Sup35, give rise to pink colonies[150] (Fig. 4C). In practice, there are likely to be many different [*PSI*⁺] variants, each representing subtly different conformations of Sup35, which in turn affect the relative fragility of the prion polymer and hence the ability to generate new propagons.[142,149] Mutations within Sup35 can also modify the spectrum of new [*PSI*⁺] variants that arise *de novo*.[151,152]

[*PSI*⁺] variants are self-perpetuating amyloid conformations which retain their physical properties both *in vivo* and *in vitro*. The ability to generate different conformations of Sup35 *in vitro* by altering the temperature at which polymerization takes place and showing that the different conformers formed give rise to distinct [*PSI*⁺] variants has provided formal proof of this hypothesis.[50] Furthermore, the availability of a ready means of generating conformationally distinct, yet biologically relevant, amyloid structures has facilitated studies into the structural differences between [*PSI*⁺] variants[100,148,149] and the role of nonnative interactions in generating the conformational diversity.[153]

What has emerged from these studies is a demonstration of the importance of intermolecular contacts within the Sup35-PrD in the formation of the amyloid core of the prion polymers in the different variants (Fig. 4C). Biophysical studies on the polymers reveal that those associated with strong [*PSI*⁺] variants are more readily fragmented than those from weak [*PSI*⁺] variants.[149] This observation provides an explanation of the relative mitotic instability of weak [*PSI*⁺] variants in comparison with strong variants, namely a reduction in the number of propagons (n_0) available to be passed on to daughter cells.[50,142]

In contrast to the yeast prion variants, which have relatively subtle effects on cell phenotype, the mammalian PrP prion strains present a number of distinguishable and stable phenotypes that are also differentially toxic. This would indicate that, for both yeast and mammalian prions, there exists a strong correlation between the final prion protein conformation and the observed phenotype. This makes questionable one of the most striking differences between mammalian and yeast prions—the phenotypic relationship between a [*PRION*⁺] state and the loss of the gene encoding the respective prion. In yeast, conversion of a prion protein to the [*PRION*⁺] state is similar, albeit not as severe, as a deletion of the gene encoding the prion protein. For example, a complete loss of function of Sup35 results in a loss of viability, yet [*PSI*⁺] cells show no clear growth difference compared to [*psi*⁻] cells.[131] Conversely, in mammals, while certain

PrP polymorphisms are associated with familial forms of prion disease, animals lacking the PrP gene ($Prnp^{-/-}$) show no disease phenotype.[154] Thus PrP appears to acquire a toxic gain of function in switching to the PrP^{Sc} form. The existence of stable conformational variants of both yeast and mammalian prions, each with unique phenotypic characteristics, suggests this overarching discrepancy between yeast and mammalian prions, where the former causes a loss-of-function while the latter a gain-of-function phenotype, may be superficial and explained instead by the remarkable differences in cellular context and environment in which the yeast and mammalian prion proteins appear.

VI. Fungal Prions: Friend or Foe?

The context of their initial identification implicated prions as causative agents of disease, specifically neurodegenerative diseases of mammals. The subsequent identification of prion proteins in yeast, and the absence of any significant growth defect in their presence, suggested that the expected correlation between prions and toxicity may not necessarily apply to all fungal prions. Moreover, fungal prions can rightly be considered epigenetic determinants of phenotype. With an expanding repertoire of yeast proteins being accorded "prion-forming status," the role of these epigenetic regulators has risen from biological anomaly to one where they can clearly have a significant impact on cell survival.[13,14]

A. Fungal Prions as Disease-Causing Agents

Studies so far have demonstrated a notable absence of yeast prions in natural isolates of S. cerevisiae, [PIN+] being the exception.[155] Such a lack of "native" prions in yeast could indicate that yeast prions are not of benefit to the species in the environmental niche(s) in which it has had to evolve and may represent a form of sexually transmitted disease in this microorganism.[156] Further support for this claim is the apparent lack of conservation in the prion-forming ability of close orthologues of Ure2[156–158] and the existence of a significant number of wild-type strains of yeast that have mutations in the Sup35-PrD which prevent the possibility of switching to the [PSI+] state.[155]

Similar to other nonchromosomal genetic elements found in yeast, for example, mitochondrial DNA, prions are transmitted horizontally via cytoplasmic mixing during mating and like the aforementioned elements, segregate 4+:0− in a non-Mendelian fashion during meiosis. Consequently, one might expect prions to be relatively prolific in wild populations and the apparent absence, or low frequency, of prion appearance in wild yeast strains may argue that prions have a net deleterious effect on their host.[156]

B. Fungal Prions as Agents of Adaptation and Survival

Given the above discussion, one cannot escape the fact that some fungal prions do confer a selective advantage on the host cell under certain growth conditions.[9–12] In the case of the [*PSI*⁺] prion, the decrease in stop codon recognition that accompanies Sup35 conversion to [*PSI*⁺] would be expected to lead to the production of C-terminally extended polypeptides. However, there is little evidence that such stop codon readthrough occurs at a significant level across the proteome, or if it does, that it significantly affects viability or growth competitiveness.[131] [*PSI*⁺]-mediated translation of stop codons at the end of open reading frames provides the potential to expose previously hidden genetic variation encoded by sequences 3′ of the stop codon. Translation readthrough would provide a novel means of exploring genetic space and may be the source of the [*PSI*⁺] survival advantage seen under conditions of stress, since it may allow for a greater probability of stress adaptation than can occur in a [*psi*⁻] cell.[12]

Yeast prions, however, cannot be considered completely benign. A number of conditions have been engineered in which the presence of a prion becomes toxic to the host. For example, overexpression of the Rnq1 protein in cells carrying the prion form of Rnq1 (i.e., [*PIN*⁺]) results in cell lethality that is not observed with otherwise isogenic [*pin*⁻] cells[159] (Fig. 6). The overexpression of Sup35 also shows toxicity in [*PSI*⁺], but not [*psi*⁻], cells, although this likely reflects the titration of the essential functional partner of Sup35, namely Sup45 (eRF1).[160,161]

It is intuitive that the more genetically and phenotypically heterogenous a given population of cells, or the greater their potential for phenotypic flexibility, the greater are the chances that one cell among the population will adapt successfully to a novel environmental challenge. The presence of prion proteins

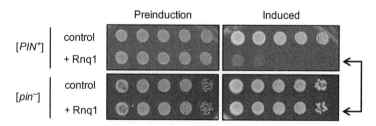

Fɪɢ. 6. The prion ([*PIN*⁺])-mediated killing of yeast in cells engineered to overexpress the prion protein Rnq1. An otherwise isogenic pair of [*PIN*⁺]/[*pin*⁻] yeast strains transformed either with a control plasmid (control) or a plasmid in which *RNQ1* gene expression is under the control of a galactose-inducible promoter (+Rnq1) are shown. Both the [*PIN*⁺] and [*pin*⁻] strains are viable prior to inductions, but when *RNQ1* expression is induced, the [*PIN*⁺], but not the [*pin*⁻], cells lose viability. Each row shows a series of dilutions in cell number of the respective strains. (See Color Insert.)

could plausibly introduce such a resource for phenotypic diversity in yeast and other fungi. The *de novo* appearance of [*PSI*$^+$], for example, is triggered by a number of environmentally induced physiological changes, including oxidative stress,[162,163] though the impact on cell phenotype depends on the strain background. Furthermore, *de novo* generation of [*PSI*$^+$] and other prions can lead to different prion variants emerging in the same population, each variant having a subtly different effect on host cell phenotype (see Section V). Some of these variants may be detrimental to the host,[164] but more likely is the possibility that prion variants act as "bet-hedging" devices that contribute to short-term adaptation to new, perhaps hostile, environments and perhaps facilitate the evolution of new traits required for survival.[14]

C. Fungal Prions as Epigenetic Determinants of Phenotype

To function as an epigenetic regulator of gene expression requires a means both of replication and of inheritance. Yeast prion proteins achieve these two essential events as follows: first, replication proceeds through conformational change from a [*prion*$^-$] to a [*PRION*$^+$] state, because any prion protein that undergoes this conformational change to the [*PRION*$^+$] state can induce the same conformational change in homologous prion protein molecules in the [*prion*$^-$] state. This behavior effectively defines the [*PRION*$^+$] protein as "infectious." Second, the inheritance or transmission of the prion protein in the [*PRION*$^+$] state from mother cell to daughter cell can occur during mitosis and meiosis by distribution of the "infected" cytoplasm between the mother and daughter cells or the four spores, respectively. New prion protein synthesized by either the mother or the daughter cell(s) is equally susceptible to conversion by existing [*PRION*$^+$] molecules, thus replenishing the pool of [*PRION*$^+$] proteins that might otherwise be diluted out through continued cell division. It is for these reasons that yeast harboring a prion protein in the [*PRION*$^+$] state display changes in gene expression with a non-Mendelian dominant inheritance pattern via the infected host cell lineage.

Presently, limitations in our understanding of the *raison d'être* for prions in the fungal cell, be they disease-causing or conduits for adaptation, mean that we must accept the possibility that any protein could evolve prionic properties. However, gene ontology (GO) analysis of the 29 yeast proteins that each exhibit a convincing array of biochemical and genetic signatures indicative of a *bona fide* prion[53] reveals an enrichment of certain cellular functions, processes, and localizations (Fig. 7). Specifically, GO terms such as response to chemical stimulus, transcription, RNA metabolic process, transcription regulator activity, DNA binding, and localization to cell cortex, cellular bud, or nucleus are especially frequent when one analyses *bona fide* prions.

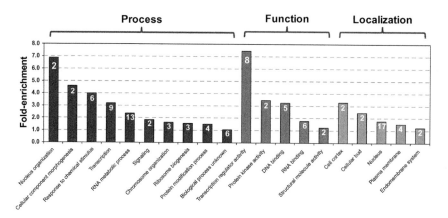

Fig. 7. Identifying bias in yeast prion protein identity. A gene ontology (GO) analysis of 29 yeast proteins,[53] each exhibiting multiple prion properties reveals an enrichment of some GO function, process, and localization terms relative to their genome frequency (fold-enrichment). The number within each bar of the chart reflects the actual number of proteins, in the group of 29, assigned to each factor. (For color version of this figure, the reader is referred to the Web version of this chapter.)

When prion-mediated regulation of certain activities occurs, beyond the normal distribution of those activities encoded by the genome, there is an implied evolutionary or functional significance to this correlation. However, the enrichment of transcription factor activities within the prion protein group may be a natural reflection of the observed evolutionary differences between hubs of generic protein interactions and transcription factor interactions.[165] Many protein–protein interaction networks are well conserved and represent the plethora of activities required for continued viability, with proteins that participate in multiple interactions or those proteins under tight regulatory control experiencing stronger functional constraints. While transcription factors organize the activity of proteins, the protein–protein interactions shift the functional constraint on transcription factors to generic proteins, and this is possibly skewed further by the transient nature of transcription factor regulation.[165] The overrepresentation of transcription factors with prion properties may merely be a consequence of the enhanced evolutionary capacity of transcription factors relative to generic proteins. However, regulation by transcription factors must reflect both the environment and the needs of the cell. Rewiring of the network by changes to transcription factor hubs would facilitate significant phenotypic diversity and allow for increased ecological adaptiveness. It is easy to conceive the contribution of prion-mediated regulation of transcription factors in this scenario.

Finally, the existence of prion variants (see Section V) adds a further layer of complexity to the inheritance of epigenetic information, which through its encryption in the overall conformation of the prion protein can be as variable as the number of stable conformations. In the case of the $[PSI^+]$ variants, the different conformational forms of Sup35 form polymers that show different levels of fragmentation mediated by Hsp104,[50,100,149] and as a result, different levels of the soluble and functional Sup35 that control the efficiency of translation termination.

Despite the opposing views on the short-term and long-term impact of fungal prions, it is becoming apparent that yeast prion proteins participate at many levels of cellular control, with transcriptional processes increasingly overrepresented.[53] The question then becomes whether prions represent an additional epigenetic level of control over cellular processes, superimposed upon a backbone of cellular coordination determined by nucleic acid. One might expect that, were this to be the case, there would also exist some degree of coordination within this novel mechanism for control, for example, interactions between the different prion proteins that maintain them in a "cell optimal" arrangement. We propose that Rnq1, the determinant of the $[PIN^+]$ prion, could serve as such a "central communicator" or a molecular pivot point, the point at which cellular information converges and is conveyed in a structural manner to the regulatory level of prion proteins, for example, conversion of Rnq1 to the $[PIN^+]$ state and the subsequent induction of other yeast prions and their own antagonistic interactions. The low rate of spontaneous conversion to the $[PRION^+]$ state, for example,[166] may indicate that, while Rnq1 conversion (or an equivalent in this role) initiates the cascade of epigenetic events, subsequent prion conversion events are still tolerably rare enough to allow the cell time to navigate the environmental condition that triggered Rnq1 conversion with the existing repertoire of proteins and functions. This theory would require an ability to lose the $[PIN^+]$ prion faster than the rate at which $[PSI^+]$, for example, would appear in the $[PIN^+]$ background. Intriguingly, a recent proteomics analysis of Rnq1 has suggested that many central processes are altered as a consequence of changes to Rnq1 abundance (G. Staniforth, K. Rumpel, and M.F. Tuite, unpublished data). This suggests that Rnq1 might be adequately positioned to translate changes in cellular physiology to an epigenetic signal. Rnq1 conversion to the $[PIN^+]$ prion might therefore be likened to switching on a prion-based epigenetic overlay on cellular physiology from a red stop light to amber.

VII. Yeast Prions for Therapeutic Discovery

In addition to its utility for probing the mechanism by which prions are propagated, yeast has also proven effective in screening for chemical agents that interfere with prion formation and propagation. There are several possible targets for therapeutics that could effectively block or remove prions from infected cells and these fall into two classes: (1) targets that would lead to a block in the formation of the infectious prion form(s) and (2) those that result in upregulation of the cellular pathway(s) that are responsible for clearing cells of such aggregates, for example, the ubiquitin–proteasome system and the molecular chaperone network.[167]

Identifying broad-spectrum inhibitors of prion formation or propagation also presents a number of conceptual challenges. For example, both mammalian and yeast prions do not possess single stable folds but rather populate a substantial volume of conformational space, that is, they exist in a highly dynamic structural equilibrium. Some of these conformers may constitute drug-resistant forms that may be selected from the ensemble of structures present.[168,169] Finding compounds that can act on such a "moving target" is a challenge. Furthermore, there remain concerns that the amyloid polymers are in fact a benign species and that the amyloid-associated toxicity is associated with low molecular weight species in the fibrillization pathway.[170] Nevertheless, progress has been made in using yeast to identify novel antiprion compounds.

To date, no efficient chemical treatment for arresting or reversing the progression of prion diseases in mammals has emerged, although some compounds can promote PrP^{Sc} clearance from scrapie-infected ScN2a neuroblastoma cells. These include the phenothiazine derivative, chlorpromazine, an antipsychotic drug, and quinacrine, an antimalaria drug.[171,172] Yet, while there are aspects of prion propagation in yeast that differ from what is seen with mammalian prions, for example, chaperone-mediated fragmentation of the prion polymers, some compounds can antagonize both yeast and mammalian prions. This gives confidence that yeast can be used as an effective screening tool for novel antiprion or even antiamyloid inhibitors.

The red/white Ade$^-$/Ade$^+$ phenotypic screen used to identify prion-free [psi$^-$] cells provides a simple visual screen for agents active against this yeast prion.[173] A similar phenotypic screen has been developed for the [URE3] prion[146] and exploited in a drug discovery screen.[173,174] The added advantage of using yeast-based assays is that they can be used in the high-throughput mode and are significantly cheaper for drug discovery and validation than the use of cultured mammalian cells.

A number of compounds that interfere specifically with yeast prion propagation have proven useful in experiments probing the underlying mechanism of propagation. The most important of such inhibitors is the chaotropic agent GdnHCl (see Section IV), although there are no reports of PrP^{Sc} clearance from cultured animal cells grown in the presence of GdnHCl. The latter is not surprising given that the primary target of GdnHCl is the chaperone Hsp104 and that guanidinium ions bind selectively to the chaperone and perturb the crucial ATPase activity that is required for efficient protein disaggregation and prion propagation.[129] The specificity of GdnHCl for this particular AAA+ ATPase is surprising, but other cellular targets cannot be completely ruled out.[175]

GdnHCl also has facilitated the search for other prion "curing" compounds. Bach et al.[174] carried out a screen of some 880 compounds to search for those that could eliminate the [PSI+] prion using the red/white colony color screen. However, this assay required the presence of subcuring levels of GdnHCl (0.1–0.2 mM) to allow the detection of positive hits. Why there should be synergy with subeffective amounts of GdnHCl remains to be explained, although we have identified several compounds (e.g., quinoline) that show such synergism in [PSI+] elimination but which apparently act by increasing the uptake of GdnHCl into the cell and triggering the observed loss of [PSI+] (C. Kyprianidou and M.F. Tuite, unpublished). That said, the two compounds known to eliminate PrP^{Sc} from infected ScN2a cells, namely quinacrine and chlorpromazine,[171,172] also eliminate both the [PSI+] and the [URE3] prions in this screening assay. Importantly, the screen by Bach et al.[173] also identified a novel class of antiprion compounds, the kastellpaolitines, which also proved effective at eliminating PrP^{Sc} in the ScN2a cell-based assay.[173,175]

Two of the most active compounds identified in the yeast-based screen carried out by Bach et al.[173,174] were the phenanthridine derivative 6-aminophenanthridine (6AP) and Guanabenz (GA), a drug widely used to treat hypertension. Both compounds inhibit yeast prion propagation and are active against mammalian prions in a mouse model for prion propagation. The mode of action of these compounds is the same. They show ribosomal RNA-dependent binding to the ribosome leading to a defect in a novel ribosome-associated protein folding activity.[176] Whether the elimination of yeast and mammalian prions by 6AP and GA is a direct or an indirect effect remains to be established, although neither compound binds directly to PrP or Sup35.

There also has been recent success in identifying small molecules that act either to inhibit prion formation in yeast[177] or lead to the elimination of the amyloid fibrils from the cell.[169,178] Studies using Sup35 or mouse PrP have led to the identification of a number of prion protein aggregation inhibitors, including tetraiodophenolphthalein (TIPT), one of a number of compounds identified that are able to form colloidal particles and also inhibit amyloid polymerization in general. TIPT and a number of other "chemical aggregators," including rotterlin and baicalein, also inhibit (in some cases, almost totally) the infectivity of sonicated Sup35 polymers when added to [psi−] spheroplasts.[177] Some small molecules, for example, epigallocatechin-3-gallate (EGCG), which is a known inhibitor of amyloid formation, can also eliminate preformed Sup35 prion aggregates from [PSI+] cells in vivo, but not all conformational variants of the prion are equally susceptible[169,178] and may therefore represent drug-resistant forms. The finding that another compound with a different mode of action, namely, 4,5-bis-(4-methoxyanilino)phthalimide (DAPH-12), can eliminate the EGCG-resistant forms of the $[PSI^+]$ prion[169] points to the use of combined therapeutic strategies for dealing with this problem.

Studies on antiprion compounds in yeast have identified a number of new antiprion drugs with previously unsuspected modes of action, and although none of these is as yet licensed for use in treating human prion diseases, the progress made so far bodes well for future antiprion drug discovery programs. In spite of these successes, further studies are needed to develop effective chemical therapies applicable to human TSEs and, more widely, to treat and eradicate human amyloidoses.

VIII. Reflections

Since their existence was first postulated,[7] fungal prions have provided us with exciting new insights into many aspects of prion biology. The ready availability of an increasing repertoire of prions, the ease with which a wide range of techniques can be applied to interrogate their structure and behavior, and the lack of biohazard concerns make the fungal prion a very attractive model.

In studying fungal prions, it is important to recognize the translatability of findings in a simple eukaryote to a fuller understanding of the mammalian prion. The underlying mechanism by which the fungal $[PRION^+]$ state is propagated shares many facets with the mechanism that generates new PrP^{Sc} prions.

In spite of the immense progress made in the understanding of prions and their impact on host phenotype,[13] many important questions remain to be answered, not only relating to the mechanism by which prions are propagated in both dividing and nondividing cells but also to the very nature of the propagon and how it relates to various physical forms of the corresponding prion protein. For example, the assumption continues to be made, without verification, that a yeast prion protein aggregate is amyloid in form in the living cell. There is no doubt that yeast proteins can form amyloid fibrils *in vitro*, and much of the high-resolution structural details of amyloids have come from studying peptides of Sup35[114] and Het-S.[179]

Attempts to study the ultrastructure of yeast prions *in vivo* have been limited. In one study, Speransky *et al.*[180] showed, that in [URE3] cells in which Ure2 levels have been artificially elevated, the prion protein forms a distinctive filamentous network in the cytoplasm. Subsequent studies on $[PSI^+]$ cells have demonstrated, using Sup35NM–GFP fusions, that this prion protein also forms fibrillar cytoplasmic structures in $[PSI^+]$, but not $[psi^-]$, cells.[94,181] Whether these fibrils are amyloid in nature was not proven in any of these studies, although the Sup35NM–GFP fibrils observed *in vivo* showed similar morphology to the amyloid fibrils formed *in vitro*.[181] Kimura *et al*.[182] also have shown that Sup35NM–GFP aggregates formed *in vivo* in $[PSI^+]$ cells stain with thioflavin S, an amyloid-specific dye.

While the controversy surrounding the very existence of protein-only infectious or epigenetic agents has waned, there remains considerable uncertainty about the molecular nature of the infectious form of the prion protein.[2] For example, the presence of aggregates of PrP and of yeast prion proteins is strongly predictive of the prion state, with both oligomeric intermediates and larger polymers (or fibrils) having been postulated as the infectious form. In yeast, the transmission of GFP-labeled prion aggregates from mother to daughter cell can be observed readily,[183,184] and in mammals, there is a large body of evidence that shows such aggregates as amyloid deposits in the brain of patient with prion disease.[2] For the mammalian TSEs in particular, it remains uncertain whether the structural form that transmits the disease is identical to the pathogenic entity. Recently, results have emerged that indicate that, for PrP at least, the infectivity and toxicity of the prion can be uncoupled.[185]

What has emerged over that past decade is that, in fungi, prions can have a wide-ranging, often beneficial, impact on host cell behavior. In addition to being exploited as a model to gain insights into rare infectious entities of humans, fungal prions can provide tremendously valuable insights into a form of genetic inheritance which involves the transfer of phenotype-defining information without the requirement for genome alteration. The next big challenge will be to establish whether these "protein genes" are restricted to a select group of fungi or are widely exploited throughout the microbial, plant, and animal kingdoms.

ACKNOWLEDGMENTS

Work on yeast prions in the authors' laboratory has been supported by funds from The Wellcome Trust (project no. 081881) and the Biotechnology and Biological Sciences Research Council (BBSRC). G. L. S. was supported by a PhD studentship from the BBSRC.

REFERENCES

1. Chiti F, Dobson C. Protein misfolding, functional amyloid, and human disease. *Annu Rev Biochem* 2006;**75**:333–66.
2. Aguzzi A, Calella AM. Prions: protein aggregation and infectious diseases. *Physiol Rev* 2009;**89**:1105–52.
3. Brudin P, Melki R, Kopito R. Prion-like transmission of protein aggregates in neurodegenerative diseases. *Nat Rev Mol Cell Biol* 2010;**11**:301–7.
4. Aguzzi A, Rajendran L. The transcellular spread of cytosolic amyloids, prions, and prionoids. *Neuron* 2009;**64**:783–90.

5. Prusiner SB. Novel proteinaceous infectious particles cause scrapie. *Science* 1982;**216**:136–44.

6. Prusiner SB. Prions. *Proc Natl Acad Sci USA* 1998;**95**:13363–83.

7. Wickner RB. [*URE3*] as an altered URE2 protein: evidence for a prion analog in *Saccharomyces cerevisiae. Science* 1994;**264**:566–9.

8. Coustou V, Deleu C, Saupe S, Begueret J. The protein product of the het-s heterokaryon incompatibility gene of the fungus *Podospora anserina* behaves as a prion analog. *Proc Natl Acad Sci USA* 1997;**94**:9773–8.

9. Eaglestone SS, Cox BS, Tuite MF. Translation termination efficiency can be regulated in *Saccharomyces cerevisiae* by environmental stress through a prion-mediated mechanism. *EMBO J* 1999;**18**:1974–81.

10. Namy O, Galopier A, Martini C, Matsufuji S, Fabret C, Rousset JP. Epigenetic control of polyamines by the prion [*PSI*⁺]. *Nat Cell Biol* 2008;**10**:1069–75.

11. True HL, Berlin I, Lindquist SL. Epigenetic regulation of translation reveals hidden genetic variation to produce complex traits. *Nature* 2004;**431**:184–7.

12. True HL, Lindquist SL. A yeast prion provides a mechanism for genetic variation and phenotypic diversity. *Nature* 2000;**407**:477–83.

13. Tuite MF, Serio TR. The prion hypothesis: from biological anomaly to basic regulatory mechanism. *Nat Rev Mol Cell Biol* 2010;**11**:823–33.

14. Halfmann R, Alberti S, Lindquist S. Prions, protein homeostasis, and phenotypic diversity. *Trends Cell Biol* 2010;**20**:125–33.

15. Wickner RB, Shewmaker F, Edskes H, Kryndushkin D, Nemecek J, McGlinchey R, et al. Prion amyloid structure explains templating: how proteins can be genes. *FEMS Yeast Res* 2010;**16**:980–91.

16. Tuite MF, Koloteva-Levin N. Propagating prions in fungi and mammals. *Mol Cell* 2004;**14**:541–52.

17. Braun RJ, Buttner S, Ring J, Kroemer G, Madeo F. Nervous yeast: modeling neurotoxic cell death. *Trends Biochem Sci* 2009;**35**:135–44.

18. Tenreiro S, Outerio TF. Simple is good: yeast models of neurodegeneration. *FEMS Yeast Res* 2010;**10**:970–9.

19. Aigle M, Lacroute F. Genetical aspects of [*URE3*], a non-Mendelian, cytoplasmically-inherited mutation in yeast. *Mol Gen Genet* 1975;**136**:327–35.

20. Lacroute F. Non-Mendelian mutation allowing ureidosuccinic acid uptake in yeast. *J Bacteriol* 1971;**106**:519–22.

21. Cox BS. [*PSI*], a cytoplasmic suppressor of super-suppressors in yeast. *Heredity* 1965;**20**:505–21.

22. Courchesne WE, Magasanik B. Regulation of nitrogen assimilation in *Saccharomyces cerevisiae*: roles of the URE2 and GLN3 genes. *J Bacteriol* 1988;**170**:708–13.

23. Cunningham TS, Andhare R, Cooper TG. Nitrogen catabolite repression of *DAL80* expression depends on the relative levels of Gat1p and Ure2p production in *Saccharomyces cerevisiae. J Biol Chem* 2000;**275**:14408–14.

24. Cox BS. Cytoplasmic inheritance. Prion-like factors in yeast. *Curr Biol* 1994;**4**:744–8.

25. Tuite MF, Cox BS. The [*PSI*⁺] prion of yeast: a problem of inheritance. *Methods* 2006;**39**:9–22.

26. Liebman S, Sherman F. Extrachromosomal psi+ determinant suppresses nonsense mutations in yeast. *J Bacteriol* 1979;**139**:1068–71.

27. Tuite MF, Lund PM, Futcher AB, Dobson MJ, Cox BS, McLaughlin CS. Relationship of the [psi] factor with other plasmids of *Saccharomyces cerevisiae. Plasmid* 1982;**8**:103–11.

28. Chernoff YO, Ingevechtomov SG, Derkach IL, Ptyushkina MV, Tarunina OV, Dagkesamanskaya AR, et al. Dosage-dependent translational suppression in yeast *Saccharomyces cerevisiae. Yeast* 1992;**8**:489–99.

29. Dagkesamanskaya AR, Ter-Avanesyan MD. Interaction of the yeast omnipotent suppressors *SUP1(SUP45)* and *SUP2(SUP35)* with non-mendelian factors. *Genetics* 1991;**128**:513–20.

30. Chernoff YO, Derkach IL, Inge-Vechtomov SG. Multicopy *SUP35* gene induces *de-novo* appearance of psi-like factors in the yeast *Saccharomyces cerevisiae*. *Curr Genet* 1993;**24**:268–70.

31. Stansfield I, Jones KM, Kushnirov VV, Dagkesamanskaya AR, Poznyakovski AI, Paushkin SV, et al. The products of the *SUP45* (eRF1) and *SUP35* genes interact to mediate translation termination in *Saccharomyces cerevisiae*. *EMBO J* 1995;**14**:4365–73.

32. Zhouravleva G, Frolova L, Le Goff X, Le Guellec R, Inge-Vechtomov S, Kisselev L, et al. Termination of translation in eukaryotes is governed by two interacting polypeptide chain release factors, eRF1 and eRF3. *EMBO J* 1995;**14**:4065–72.

33. Cheng Z, Saito K, Pisarev AV, Wada M, Pisareva VP, Pestova TV, et al. Structural insights into eRF3 and stop codon recognition by eRF1. *Genes Dev* 2009;**23**:1106–18.

34. Salas-Marco J, Bedwell DM. GTP hydrolysis by eRF3 facilitates stop codon decoding during eukaryotic translation termination. *Mol Cell Biol* 2004;**24**:7769–78.

35. Patino MM, Liu JJ, Glover JR, Lindquist S. Support for the prion hypothesis for inheritance of a phenotypic trait in yeast. *Science* 1996;**273**:622–6.

36. Paushkin SV, Kushnirov VV, Smirnov VN, Ter-Avanesyan MD. Propagation of the yeast prion-like [*psi*+] determinant is mediated by oligomerization of the *SUP35*-encoded polypeptide chain release factor. *EMBO J* 1996;**15**:3127–34.

37. Chernoff YO, Lindquist SL, Ono B, Inge-Vechtomov SG, Liebman SW. Role of the chaperone protein Hsp104 in propagation of the yeast prion-like factor [psi+]. *Science* 1995;**268**:880–4.

38. Tuite M, Byrne L, Josse L, Ness F, Koloteva-Levine N, Cox B. Yeast prions and their analysis in vivo. *Methods Microbiol* 2007;**36**:491–526.

39. Crouzet M, Tuite MF. Genetic control of translational fidelity in yeast: molecular cloning and analysis of the allosuppressor gene *SAL3*. *Mol Gen Genet* 1987;**210**:581–3.

40. Kushnirov VV, Teravanesyan MD, Telckov MV, Surguchov AP, Smirnov VN, Ingevechtomov SG. Nucleotide sequence of the Sup2 (Sup35) Gene of *Saccharomyces cerevisiae*. *Gene* 1988;**66**:45–54.

41. Tuite MF, Mundy CR, Cox BS. Agents that cause a high frequency of genetic change from [psi+] to [psi-] in *Saccharomyces cerevisiae*. *Genetics* 1981;**98**:691–711.

42. Cox BS, Tuite MF, McLaughlin CS. The psi factor of yeast: a problem in inheritance. *Yeast* 1988;**4**:159–78.

43. Singh A, Helms C, Sherman F. Mutation of the non-Mendelian suppressor, Psi+, in yeast by hypertonic media. *Proc Natl Acad Sci USA* 1979;**76**:1952–6.

44. Salnikova AB, Kryndushkin DS, Smirnov VN, Kushnirov VV, Ter-Avanesyan MD. Nonsense suppression in yeast cells overproducing Sup35 (eRF3) is caused by its non-heritable amyloids. *J Biol Chem* 2005;**280**:8808–12.

45. Glover JR, Kowal AS, Schirmer EC, Patino MM, Liu JJ, Lindquist S. Self-seeded fibers formed by Sup35, the protein determinant of *PSI*⁺, a heritable prion-like factor of S. cerevisiae. *Cell* 1997;**89**:811–9.

46. King CY, Tittmann P, Gross H, Gebert R, Aebi M, Wuthrich K. Prion-inducing domain 2-114 of yeast Sup35 protein transforms in vitro into amyloid-like filaments. *Proc Natl Acad Sci USA* 1997;**94**:6618–22.

47. Krzewska J, Melki R. Molecular chaperones and the assembly of the prion Sup35p, an *in vitro* study. *EMBO J* 2006;**25**:822–33.

48. DePace AH, Santoso A, Hillner P, Weissman JS. A critical role for amino-terminal glutamine/asparagine repeats in the formation and propagation of a yeast prion. *Cell* 1998;**93**:1241–52.

49. Paushkin SV, Kushnirov VV, Smirnov VN, Ter-Avanesyan MD. *In vitro* propagation of the prion-like state of yeast Sup35 protein. *Science* 1997;**277**:381–3.

50. Tanaka M, Chien P, Naber N, Cooke R, Weissman JS. Conformational variations in an infectious protein determine prion strain differences. *Nature* 2004;**428**:323–8.

51. Tanaka M, Weissman JS. An efficient protein transformation protocol for introducing prions into yeast. *Methods Enzymol* 2006;**412**:185–200.

52. Alberti S, Halfmann R, Lindquist S. Biochemical, cell biological, and genetic assays to analyze amyloid and prion aggregation in yeast. *Methods Enzymol* 2010;**470**:709–34.

53. Alberti S, Halfmann R, King O, Kapila A, Lindquist S. A systematic survey identifies prions and illuminates sequence features of prionogenic proteins. *Cell* 2009;**137**:146–58.

54. Coschigano PW, Magasanik B. The *URE2* gene product of *Saccharomyces cerevisiae* plays an important role in the cellular response to the nitrogen source and has homology to glutathione s-transferases. *Mol Cell Biol* 1991;**11**:822–32.

55. Cooper TG. Transmitting the signal of excess nitrogen in *Saccharomyces cerevisiae* from the Tor proteins to the GATA factors: connecting the dots. *FEMS Microbiol Rev* 2002; **26**:223–38.

56. Hofman-Bang J. Nitrogen catabolite repression in *Saccharomyces cerevisiae*. *Mol Biotechnol* 1999;**12**:35–73.

57. Tsukada M, Ohsumi Y. Isolation and characterization of autophagy-defective mutants of *Saccharomyces cerevisiae*. *FEBS Lett* 1993;**333**:169–74.

58. Derkatch IL, Bradley ME, Zhou P, Chernoff YO, Liebman SW. Genetic and environmental factors affecting the *de novo* appearance of the [*PSI*$^+$] prion in *Saccharomyces cerevisiae*. *Genetics* 1997;**147**:507–19.

59. Derkatch IL, Chernoff YO, Kushnirov VV, Inge-Vechtomov SG, Liebman SW. Genesis and variability of [*PSI*] prion factors in *Saccharomyces cerevisiae*. *Genetics* 1996;**144**:1375–86.

60. Derkatch IL, Bradley ME, Hong JY, Liebman SW. Prions affect the appearance of other prions: the story of [*PIN*$^+$]. *Cell* 2001;**106**:171–82.

61. Osherovich LZ, Weissman JS. Multiple Gln/Asn-rich prion domains confer susceptibility to induction of the yeast [*PSI*$^+$] prion. *Cell* 2001;**106**:183–94.

62. Bradley ME, Edskes HK, Hong JY, Wickner RB, Liebman SW. Interactions among prions and prion "strains" in yeast. *Proc Natl Acad Sci USA* 2002;**99**:16392–9.

63. Patel BK, Liebman SW. "Prion-proof" for [*PIN*$^+$]: infection with *in vitro*-made amyloid aggregates of Rnq1p-(132-405) induces [*PIN*$^+$]. *J Mol Biol* 2007;**365**:773–82.

64. Derkatch IL, Uptain SM, Outeiro TF, Krishnan R, Lindquist SL, Liebman SW. Effects of Q/N-rich, polyQ, and non-polyQ amyloids on the *de novo* formation of the [*PSI*$^+$] prion in yeast and aggregation of Sup35 *in vitro*. *Proc Natl Acad Sci USA* 2004;**101**:12934–9.

65. Meriin AB, Zhang X, He X, Newnam GP, Chernoff YO, Sherman MY. Huntington toxicity in yeast model depends on polyglutamine aggregation mediated by a prion-like protein Rnq1. *J Cell Biol* 2002;**157**:997–1004.

66. Bonasio R, Reinberg D. Molecular signals of epigenetic states. *Science* 2010;**330**:612–6.

67. Crow E, Du Z, Li L. New insights into prion biology from the novel [*SWI*$^+$] system. *Prion* 2009;**2**:141–4.

68. Du Z, Park KW, Yu H, Fan Q, Li L. Newly identified prion linked to the chromatin-remodeling factor Swi1 in *Saccharomyces cerevisiae*. *Nat Genet* 2008;**40**:460–5.

69. Sudarsanam P, Iyer VR, Brown PO, Winston F. Whole-genome expression analysis of *snf/swi* mutants of *Saccharomyces cerevisiae*. *Proc Natl Acad Sci USA* 2000;**97**:3364–9.

70. Patel BK, Gavin-Smyth J, Liebman SW. The yeast global transcription co-respressor protein Cyc8 can propagate as a prion. *Nat Cell Biol* 2009;**11**:344–9.

71. Smith RL, Johnson AD. Turning genes off by Ssn6-Tup1: a conserved system of transcriptional repression in eukaryotes. *Trends Biochem Sci* 2000;**25**:325–30.

72. Jorgensen P, Nishikawa JL, Breitkreutz BJ, Tyers M. Systematic identification of pathways that couple cell growth and division in yeast. *Science* 2002;**297**:395–400.

73. Jorgensen P, Rupes I, Sharom JR, Schneper L, Broach JR, Tyers M. A dynamic transcriptional network communicates growth potential to ribosome synthesis and critical cell size. *Genes Dev* 2004;**18**:2491–505.

74. Rogoza T, Goginashvili A, Rodionova S, Ivanov M, Viktorovskaya O, Rubel A, et al. Non-Mendelian determinant [*ISP*⁺] in yeast is a nuclear-residing prion form of the global transcriptional regulator Sfp1. *Proc Natl Acad Sci USA* 2010;**107**:10573–7.

75. Volkov KV, Aksenova AY, Soom MJ, Osipov KV, Svitin AV, Kurischko C, et al. Novel non-Mendelian determinant involved in the control of translation accuracy in *Saccharomyces cerevisiae*. *Genetics* 2002;**160**:25–36.

76. Michelitsch MD, Weissman JS. A census of glutamine/asparagine-rich regions: implications for their conserved function and the prediction of novel prions. *Proc Natl Acad Sci USA* 2000;**97**:11910–5.

77. Nemecek J, Nakayashiki T, Wickner RB. A prion of yeast metacaspase homolog (Mca1p) detected by a genetic screen. *Proc Natl Acad Sci USA* 2009;**106**:1892–6.

78. Brown JC, Lindquist S. A heritable switch in carbon source utilization driven by an unusual yeast prion. *Genes Dev* 2009;**23**:2320–32.

79. Serrano R, Kielland-Brandt MC, Fink GR. Yeast plasma membrane ATPase is essential for growth and has homology with (Na⁺ + K⁺), K⁺- and Ca2⁺-ATPases. *Nature* 1986; **319**:689–93.

80. Kaniak A, Xue Z, Macool D, Kim JH, Johnston M. Regulatory network connecting two glucose signal transduction pathways in *Saccharomyces cerevisiae*. *Eukaryot Cell* 2004;**3**:221–31.

81. Roberts BT, Wickner RB. Heritable activity: a prion that propagates by covalent autoactivation. *Genes Dev* 2003;**17**:2083–7.

82. Saupe SJ. Molecular genetics of heterokaryon incompatibility in filamentous ascomycetes. *Microbiol Mol Biol Rev* 2000;**64**:489–502.

83. Saupe SJ. A short history of small s: a prion of the fungus *Podospora anserina*. *Prion* 2007;**1**:110–5.

84. Malato L, Dos Reis S, Benkemoun L, Sabate R, Saupe SJ. Role of Hsp104 in the propagation and inheritance of the [Het-s] prion. *Mol Biol Cell* 2007;**18**:4803–12.

85. Balguerie A, Dos Reis S, Ritter C, Chaignepain S, Coulary-Salin B, Forge V, et al. Domain organization and structure-function relationship of the HET-s prion protein of *Podospora anserina*. *EMBO J* 2003;**22**:2071–81.

86. Cox BS, Ness F, Tuite MF. Analysis of the generation and segregation of propagons: entities that propagate the [*PSI*⁺] prion in yeast. *Genetics* 2003;**165**:23–33.

87. Sanchez Y, Taulien J, Borkovich KA, Lindquist S. Hsp104 is required for tolerance to many forms of stress. *EMBO J* 1992;**11**:2357–64.

88. Grimminger-Marquardt V, Lashuel HA. Structure and function of the molecular chaperone Hsp104 from yeast. *Biopolymers* 2010;**93**:252–76.

89. Jones GW, Tuite MF. Chaperoning prions: the cellular machinery for propagating an infectious protein? *Bioessays* 2005;**27**:823–32.

90. Telling GC, Scott M, Mastrianni J, Gabizon R, Torchia M, Cohen FE, et al. Prion propagation in mice expressing human and chimeric PrP transgenes implicates the interaction of cellular PrP with another protein. *Cell* 1995;**83**:79–90.

91. Tuite MF. Yeast prions and their prion-forming domain. *Cell* 2000;**100**:289–92.

92. Ross ED, Minton A, Wickner RB. Prion domains: sequences, structures and interactions. *Nat Cell Biol* 2005;**7**:1039–44.

93. Li L, Lindquist S. Creating a protein-based element of inheritance. *Science* 2000;**287**:661–4.

94. Tyedmers J, Treusch S, Dong J, McCaffery JM, Bevis B, Lindquist S. Prion induction involves an ancient system for the sequestration of aggregated proteins and heritable changes in prion fragmentation. *Proc Natl Acad Sci USA* 2010;**107**:8633–8.

95. Harrison LB, Yu Z, Stajich JE, Dietrich FS, Harrison PM. Evolution of budding yeast prion-determinant sequences across diverse fungi. *J Mol Biol* 2007;**368**:273–82.
96. Scheibel T, Lindquist SL. The role of conformational flexibility in prion propagation and maintenance for Sup35p. *Nat Struct Biol* 2001;**8**:958–62.
97. Sondheimer N, Lindquist S. Rnq1: an epigenetic modifier of protein function in yeast. *Mol Cell* 2000;**5**:163–72.
98. Toombs JA, McCarty BR, Ross ED. Compositional determinants of prion formation in yeast. *Mol Cell Biol* 2010;**30**:319–32.
99. Doel SM, McCready SJ, Nierras CR, Cox BS. The dominant *PNM2⁻* mutation which eliminates the psi factor of *Saccharomyces cerevisiae* is the result of a missense mutation in the *SUP35* gene. *Genetics* 1994;**137**:659–70.
100. Toyama BH, Kelly MJ, Gross JD, Weissman JS. The structural basis of yeast prion strain variants. *Nature* 2007;**449**:233–7.
101. Borchsenius AS, Wegrzyn RD, Newnam GP, Inge-Vechtomov SG, Chernoff YO. Yeast prion protein derivative defective in aggregate shearing and production of new 'seeds'. *EMBO J* 2001;**20**:6683–91.
102. Liu JJ, Lindquist S. Oligopeptide-repeat expansions modulate 'protein-only' inheritance in yeast. *Nature* 1999;**400**:573–6.
103. Parham SN, Resende CG, Tuite MF. Oligopeptide repeats in the yeast protein Sup35p stabilize intermolecular prion interactions. *EMBO J* 2001;**20**:2111–9.
104. Ter-Avanesyan MD, Dagkesamanskaya AR, Kushnirov VV, Smirnov VN. The *SUP35* omnipotent suppressor gene is involved in the maintenance of the non-Mendelian determinant [psi+] in the yeast *Saccharomyces cerevisiae*. *Genetics* 1994;**137**:671–6.
105. Ross E, Baxa U, Wickner R. Scrambled prion domains form prions and amyloid. *Mol Cell Biol* 2005;**24**:7206–13.
106. Osherovich LZ, Cox BS, Tuite MF, Weissman JS. Dissection and design of yeast prions. *PLoS Biol* 2004;**2**:442–51.
107. Ter-Avanesyan MD, Kushnirov VV, Dagkesamanskaya AR, Didichenko SA, Chernoff YO, Inge-Vechtomov SG, et al. Deletion analysis of the *SUP35* gene of the yeast *Saccharomyces cerevisiae* reveals two non-overlapping functional regions in the encoded protein. *Mol Microbiol* 1993;**7**:683–92.
108. Liu JJ, Sondheimer N, Lindquist SL. Changes in the middle region of Sup35 profoundly alter the nature of epigenetic inheritance for the yeast prion [*PSI⁺*]. *Proc Natl Acad Sci USA* 2002;**99**:16446–53.
109. Kadnar ML, Articov G, Derkatch IL. Distinct type of transmission barrier revealed by study of multiple prion determinants of Rnq1. *PLoS Genet* 2010;**6**:e1000824.
110. Vitrenko YA, Pavon ME, Stone SI, Liebman SW. Propagation of the [*PIN⁺*] prion by fragments of Rnq1 fused to GFP. *Curr Genet* 2007;**51**:309–19.
111. Chernoff YO, Galkin AP, Lewitin E, Chernova TA, Newnam GP, Belenkiy SM. Evolutionary conservation of prion-forming abilities of the yeast Sup35 protein. *Mol Microbiol* 2000;**35**:865–76.
112. Kushnirov VV, Kochneva-Pervukhova N, Chechenova MB, Frolova NS, Ter-Avanesyan MD. Prion properties of the Sup35 protein of yeast *Pichia methanolica*. *EMBO J* 2000;**19**:324–31.
113. Santoso A, Chien P, Osherovich LZ, Weissman JS. Molecular basis of a yeast prion species barrier. *Cell* 2000;**100**:277–88.
114. Sawaya M, Sambashivan S, Nelson R, Ivanova M, Sievers S, Apostol M, et al. Atomic structures of amyloid cross-beta spines reveal varied steric zippers. *Nature* 2007;**447**:453–7.
115. Masison DC, Wickner RB. Prion-inducing domain of yeast Ure2p and protease resistance of Ure2p in prion-containing cells. *Science* 1995;**270**:93–5.

116. Masison DC, Maddelein ML, Wickner RB. The prion model for [URE3] of yeast: sponta-neous generation and requirements for propagation. *Proc Natl Acad Sci USA* 1997; **94**:12503–8.

117. Serio TR, Cashikar AG, Kowal AS, Sawicki GJ, Moslehi JJ, Serpell L, et al. Nucleated conformational conversion and the replication of conformational information by a prion determinant. *Science* 2000;**289**:1317–21.

118. Maddelein ML, Wickner RB. Two prion-inducing regions of Ure2p are non-overlapping. *Mol Cell Biol* 1999;**19**:4516–24.

119. Ritter C, Maddelein ML, Siemer AB, Luhrs T, Ernst M, Meier BH, et al. Correlation of structural elements and infectivity of the HET-s prion. *Nature* 2005;**435**:844–8.

120. Taneja V, Maddelein ML, Talarek N, Saupe SJ, Liebman SW. A non-Q/N-rich prion domain of a foreign prion, [Het-s], can propagate as a prion in yeast. *Mol Cell* 2007;**27**:67–77.

121. Derdowski A, Sindi SS, Klaips CL, DiSalvo S, Serio TR. A size threshold limits prion transmission and establishes phenotypic diversity. *Science* 2010;**330**:680–3.

122. Broadley SA, Hartl FU. The role of molecular chaperones in human misfolding diseases. *FEBS Lett* 2009;**583**:2647–53.

123. Satpute-Krishnan P, Langseth SX, Serio TR. Hsp104-dependent remodeling of prion com-plexes mediates protein-only inheritance. *PLoS Biol* 2007;**5**:e24.

124. Wegrzyn RD, Bapat K, Newnam GP, Zink AD, Chernoff YO. Mechanism of prion loss after Hsp104 inactivation in yeast. *Mol Cell Biol* 2001;**21**:4656–69.

125. Moosavi B, Wongwigkarn J, Tuite MF. Hsp70/Hsp90 co-chaperones are required for efficient Hsp104-mediated elimination of the yeast [PSI+] prion but not for prion propagation. *Yeast* 2010;**27**:167–79.

126. Reidy M, Masison D. Sti1 regulation of Hsp70 and Hsp90 is critical for curing of *Saccharo-myces cerevisiae* [PSI+] prions by Hsp104. *Mol Cell Biol* 2010;**30**:3542–52.

127. Tkach JM, Glover JR. Nucleocytoplasmic trafficking of the molecular chaperone Hsp104 in unstressed and heat-shocked cells. *Traffic* 2008;**9**:39–56.

128. Ferreira PC, Ness F, Edwards SR, Cox BS, Tuite MF. The elimination of the yeast [PSI+] prion by guanidine hydrochloride is the result of Hsp104 inactivation. *Mol Microbiol* 2001;**40**:1357–69.

129. Grimminger V, Richter K, Imhof A, Buchner J, Walter S. The prion curing agent guanidinium chloride specifically inhibits ATP hydrolysis by Hsp104. *J Biol Chem* 2004;**279**:7378–83.

130. Jung GM, Masison DC. Guanidine hydrochloride inhibits Hsp104 activity *in vivo*: a possible explanation for its effect in curing yeast prions. *Curr Microbiol* 2001;**43**:7–10.

131. Byrne LJ, Cole DJ, Cox BS, Ridout MS, Morgan BJ, Tuite MF. The number and transmission of [PSI+] prion seeds (propagons) in the yeast *Saccharomyces cerevisiae*. *PLoS One* 2009;**4**: e4670.

132. Byrne LJ, Cox BS, Cole DJ, Ridout MS, Morgan BJ, Tuite MF. Cell division is essential for elimination of the yeast [PSI+] prion by guanidine hydrochloride. *Proc Natl Acad Sci USA* 2007;**104**:11688–93.

133. Eaglestone SS, Ruddock LW, Cox BS, Tuite MF. Guanidine hydrochloride blocks a critical step in the propagation of the prion-like determinant [PSI+] of *Saccharomyces cerevisiae*. *Proc Natl Acad Sci USA* 2000;**97**:240–4.

134. Higurashi T, Hines JK, Sahi C, Aron R, Craig EA. Specificity of the J-protein Sis1 in the propagation of 3 yeast prions. *Proc Natl Acad Sci USA* 2008;**105**:16596–601.

135. Tipton KA, Verges KJ, Weissman JS. In vivo monitoring of the prion replication cycle reveals a critical role for Sis1 in delivering substrates to Hsp104. *Mol Cell* 2008;**32**:584–91.

136. Shorter J, Lindquist S. Hsp104, Hsp70 and Hsp40 interplay regulates formation, growth and elimination of Sup35 prions. *EMBO J* 2008;**27**:2712–24.

137. Bailleul PA, Newnam GP, Steenbergen JN, Chernoff YO. Genetic study of interactions between the cytoskeletal assembly protein sla1 and prion-forming domain of the release factor Sup35 (eRF3) in *Saccharomyces cerevisiae*. *Genetics* 1999;**153**:81–94.

138. Ganusova EE, Ozolins LN, Bhagat S, Newnam GP, Wegrzyn RD, Sherman MY, et al. Modulation of prion formation, aggregation, and toxicity by the actin cytoskeleton in yeast. *Mol Cell Biol* 2006;**26**:617–29.

139. Bailleul-Winslett PA, Newnam GP, Wegrzyn RD, Chernoff YO. An antiprion effect of the anticytoskeletal drug latrunculin A in yeast. *Gene Expr* 2000;**9**:145–56.

140. Catlett NL, Weisman LS. Divide and multiply: organelle partitioning in yeast. *Curr Opin Cell Biol* 2000;**12**:509–16.

141. Bradley ME, Liebman SW. Destabilizing interactions among [*PSI*⁺] and [*PIN*⁺] yeast prion variants. *Genetics* 2003;**165**:1675–85.

142. Cox B, Byrne L, Tuite MF. Prion stability. *Prion* 2007;**1**:170–8.

143. Zhou P, Derkatch IL, Liebman SW. The relationship between visible intracellular aggregates that appear after overexpression of Sup35 and the yeast prion-like elements [*PSI*⁺] and [*PIN*⁺]. *Mol Microbiol* 2001;**39**:37–46.

144. Cole DJ, Morgan BJ, Ridout MS, Byrne LJ, Tuite MF. Estimating the number of prions in yeast cells. *Math Med Biol* 2004;**21**:369–95.

145. Bruce ME. Scrapie strain variation and mutation. *Br Med Bull* 1993;**49**:822–38.

146. Schlumpberger M, Prusiner SB, Herskowitz I. Induction of distinct [*URE3*] yeast prion strains. *Mol Cell Biol* 2001;**21**:7035–46.

147. Kalastavadi T, True HL. Analysis of the [*RNQ*⁺] prion reveals stability of amyloid fibers as the key determinant of yeast prion variant propagation. *J Biol Chem* 2010;**285**:20748–55.

148. Krishnan R, Lindquist SL. Structural insights into a yeast prion illuminate nucleation and strain diversity. *Nature* 2005;**435**:765–72.

149. Tanaka M, Collins SR, Toyama BH, Weissman JS. The physical basis of how prion conformations determine strain phenotypes. *Nature* 2006;**442**:585–9.

150. Uptain SM, Sawicki GJ, Caughey B, Lindquist S. Strains of [*PSI*⁺] are distinguished by their efficiencies of prion-mediated conformational conversion. *EMBO J* 2001;**20**:6236–45.

151. King CY. Supporting the structural basis of prion strains: induction and identification of [*PSI*] variants. *J Mol Biol* 2001;**307**:1247–60.

152. Chang HY, Lin JY, Lee HC, Wang HL, King CY. Strain-specific sequences required for yeast [*PSI*⁺] prion propagation. *Proc Natl Acad Sci USA* 2008;**105**:13345–50.

153. Ohhashi Y, Ito K, Toyama BH, Weissman JS, Tanaka M. Differences in prion strain conformations result from non-native interactions in a nucleus. *Nat Chem Biol* 2010;**6**:225–30.

154. Weissmann C, Büeler H, Fischer M, Sailer A, Aguzzi A, Aguet M. PrP-deficient mice are resistant to scrapie. *Ann N Y Acad Sci* 1994;**724**:235–40.

155. Resende CG, Outeiro TF, Sands L, Lindquist S, Tuite MF. Prion protein gene polymorphisms in *Saccharomyces cerevisiae*. *Mol Microbiol* 2003;**49**:1005–17.

156. Nakayashiki T, Kurtzman CP, Edskes HK, Wickner RB. Yeast prions [*URE3*] and [*PSI*⁺] are diseases. *Proc Natl Acad Sci USA* 2005;**102**:10575–80.

157. Edskes HK, McCann LM, Hebert AM, Wickner RB. Prion variants and species barriers among Saccharomyces Ure2 proteins. *Genetics* 2009;**181**:1159–67.

158. Talarek N, Maillet L, Cullin C, Aigle M. The [*URE3*] prion is not conserved among *Saccharomyces* species. *Genetics* 2005;**171**:23–34.

159. Douglas PM, Treusch S, Ren HY, Halfmann R, Duennwald ML, Lindquist S, et al. Chaperone-dependent amyloid assembly protects cells from prion toxicity. *Proc Natl Acad Sci USA* 2008;**105**:7206–11.

160. Vishveshwara N, Bradley ME, Liebman SW. Sequestration of essential proteins causes prion associated toxicity in yeast. *Mol Microbiol* 2009;**73**:1101–14.

161. Derkatch IL, Bradley ME, Liebman SW. Overexpression of the *SUP45* gene encoding a Sup35p-binding protein inhibits the induction of the *de novo* appearance of the [*PSI*⁺] prion. *Proc Natl Acad Sci USA* 1998;**95**:2400–5.

162. Sideri TC, Stojanovski K, Tuite MF, Grant CM. Ribosome-associated peroxiredoxins suppress oxidative stress-induced *de novo* formation of the [*PSI*⁺] prion in yeast. *Proc Natl Acad Sci USA* 2010;**107**:6394–9.

163. Tyedmers J, Madariaga ML, Lindquist SL. Prion switching in response to environmental stress. *PLoS Biol* 2008;**6**:2605–13.

164. McGlinchey RP, Kryndushkin D, Wickner RB. Suicidal [*PSI*⁺] is a lethal yeast prion. *Proc Natl Acad Sci USA* 2011;**108**:5337–41.

165. Wang Y, Franzosa EA, Zhang XS, Xia Y. Protein evolution in yeast transcription factor subnetworks. *Nucleic Acids Res* 2010;**38**:5959–69.

166. Lancaster AK, Bardill JP, True HL, Masel J. The spontaneous appearance rate of the yeast prion [*PSI*⁺] and its implications for the evolution of the evolvability properties of the [*PSI*⁺] system. *Genetics* 2010;**184**:393–400.

167. Tyedmers J, Mogk A, Bukau B. Cellular strategies for controlling protein aggregation. *Nat Rev Mol Cell Biol* 2010;**11**:777–88.

168. Duennwald ML, Shorter J. Countering amyloid polymorphism and drug resistance with minimal drug cocktails. *Prion* 2010;**4**:244–51.

169. Roberts BE, Duennwald ML, Wang H, Chung C, Lopreiato NP, Sweeny EA, et al. A synergistic small-molecule combination directly eradicates diverse prion strain structures. *Nat Chem Biol* 2009;**5**:936–46.

170. Haass C, Selkoe DJ. Soluble protein oligomers in neurodegeneration: lessons from the Alzheimer's amyloid beta-peptide. *Nat Rev Mol Cell Biol* 2007;**8**:101–12.

171. Doh-Ura K, Iwaki T, Caughey B. Lysosomotropic agents and cysteine protease inhibitors inhibit scrapie-associated prion protein accumulation. *J Virol* 2000;**74**:4894–7.

172. Kocisko DA, Baron GS, Rubenstein R, Chen J, Kuizon S, Caughey B. New inhibitors of scrapie-associated prion protein formation in a library of 2000 drugs and natural products. *J Virol* 2003;**77**:10288–96.

173. Bach S, Tribouillard D, Talarek N, Desban N, Gug F, Galons H, et al. A yeast-based assay to isolate drugs active against mammalian prions. *Methods* 2006;**39**:72–7.

174. Bach S, Talarek N, Andrieu T, Vierfond JM, Mettey Y, Galons H, et al. Isolation of drugs active against mammalian prions using a yeast-based screening assay. *Nat. Biotechnol.* 2003;**21**:1075–81.

175. Tribouillard D, Bach S, Gug F, Desban N, Beringue V, Andrieu T, et al. Using budding yeast to screen for anti-prion drugs. *Biotechnol J* 2006;**1**:58–67.

176. Tribouillard-Tanvier D, Dos Reis S, Gug F, Voisset C, Beringue V, Sabate R, et al. Protein folding activity of ribosomal RNA is a selective target of two unrelated antiprion drugs. *PLoS One* 2008;**3**:e2174.

177. Feng BY, Toyama BH, Wille H, Colby DW, Collins SR, May BC, et al. Small-molecule aggregates inhibit amyloid polymerization. *Nat Chem Biol* 2008;**4**:197–9.

178. Wang H, Duennwald ML, Roberts BE, Rozeboom LM, Zhang YL, Steele AD, et al. Direct and selective elimination of specific prions and amyloids by 4,5-dianilinophthalimide and analogs. *Proc Natl Acad Sci USA* 2008;**105**:7159–64.

179. Wasmer C, Lange A, Van Melckebeke H, Siemer A, Riek R, Meier B. Amyloid fibrils of the HET-s(218-289) prion form a beta solenoid with a triangular hydrophobic core. *Science* 2008;**319**:1523–6.

180. Speransky VV, Taylor KL, Edskes HK, Wickner RB, Steven AC. Prion filament networks in [*URE3*] cells of *Saccharomyces cerevisiae*. *J Cell Biol* 2001;**153**:1327–36.

181. Kawai-Noma S, Pack CG, Kojidani T, Asakawa H, Hiraoka Y, Kinjo M, et al. *In vivo* evidence for the fibrillar structures of Sup35 prions in yeast cells. *J Cell Biol* 2010;**190**:223–31.

182. Kimura Y, Koitabashi S, Fujita T. Analysis of yeast prion aggregates with amyloid-staining compound in vivo. *Cell Struct Funct* 2003;**28**:187–93.
183. Taguchi H, Kawai-Noma S. Amyloid oligomers: diffuse oligomer-based transmission of yeast prions. *FEBS J* 2010;**277**:1359–68.
184. Satpute-Krishnan P, Serio TR. Prion protein remodelling confers an immediate phenotypic switch. *Nature* 2005;**437**:262–5.
185. Sandberg MK, Al-Doujaily H, Sharps B, Clarke AR, Collinge J. Prion propagation and toxicity in vivo occur in two distinct mechanistic phases. *Nature* 2011;**470**:540–2.

Index

Note: Page numbers followed by "*f*" indicate figures, and "*t*" indicate tables.

N

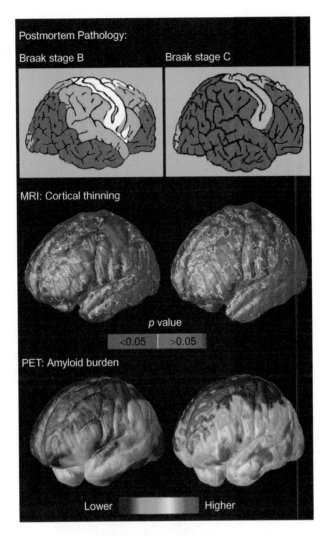

Thompson and Vinters, Fig. 3.

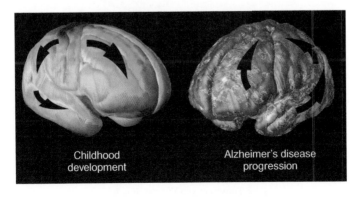

Thompson and Vinters, Fig. 4.

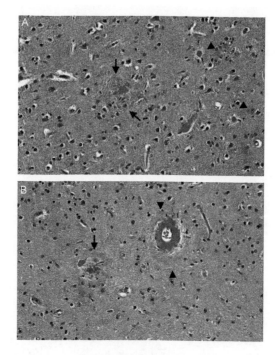

Thompson and Vinters, Fig. 5.

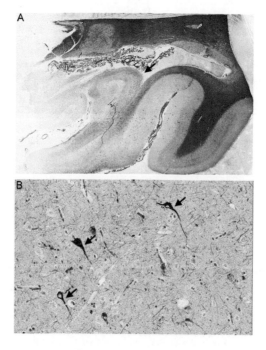

Thompson and Vinters, Fig. 6.

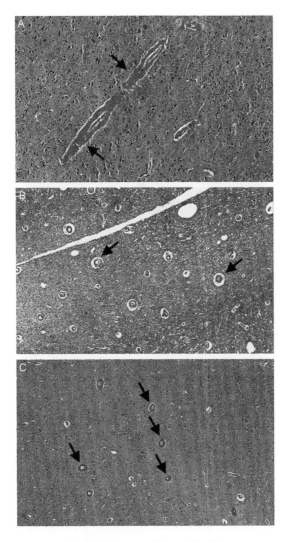

THOMPSON AND VINTERS, FIG. 7.

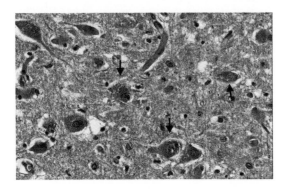

THOMPSON AND VINTERS, FIG. 8.

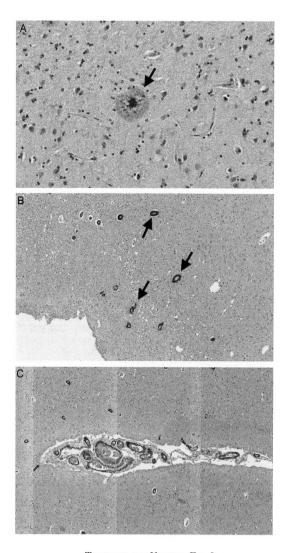

THOMPSON AND VINTERS, FIG. 9.

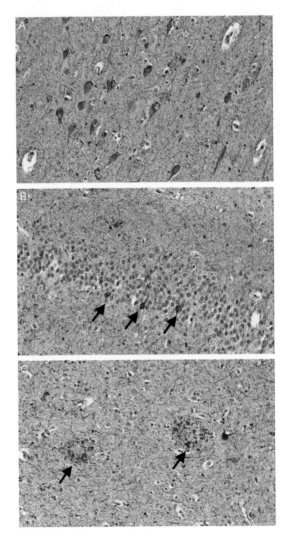

THOMPSON AND VINTERS, FIG. 10.

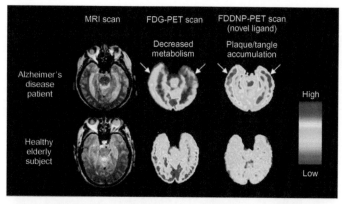

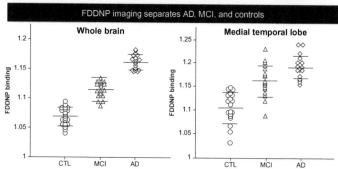

THOMPSON AND VINTERS, FIG. 11.

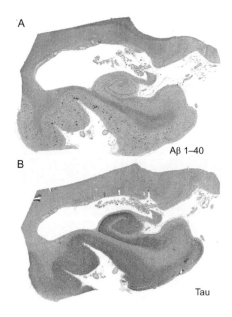

THOMPSON AND VINTERS, FIG. 12.

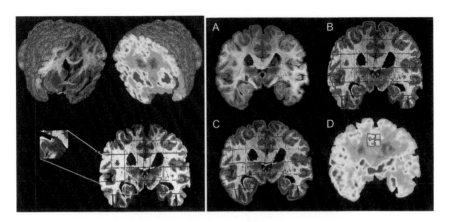

THOMPSON AND VINTERS, FIG. 13.

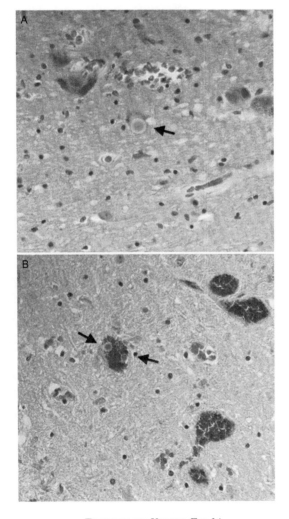

THOMPSON AND VINTERS, FIG. 14.

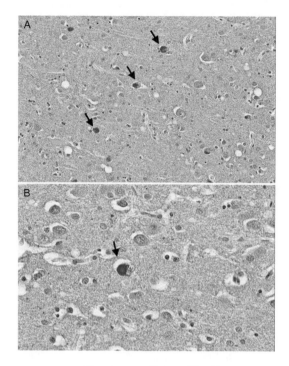

THOMPSON AND VINTERS, FIG. 15.

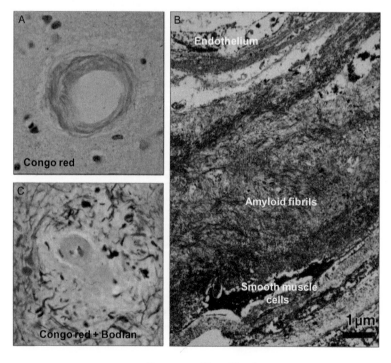

YAMADA AND NAIKI, FIG. 1.

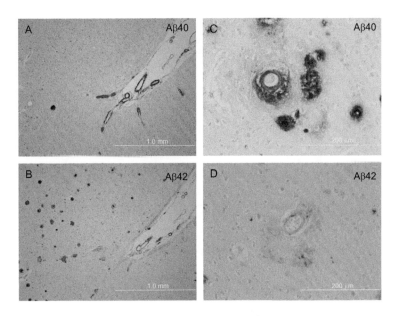

Yamada and Naiki, Fig. 3.

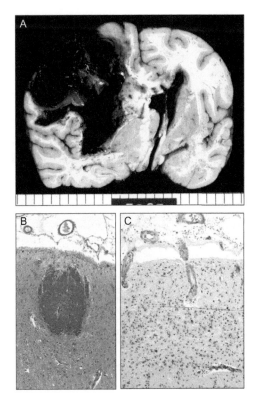

Yamada and Naiki, Fig. 5.

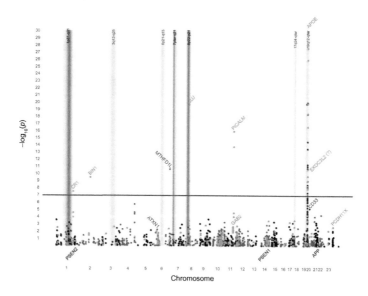

BERTRAM AND TANZI, FIG. 1.

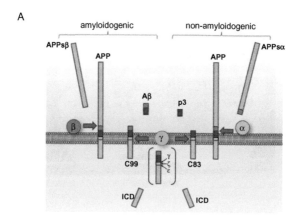

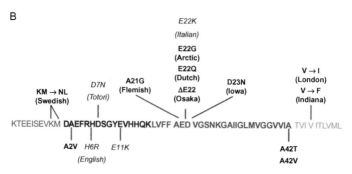

WALSH AND TEPLOW, FIG. 1.

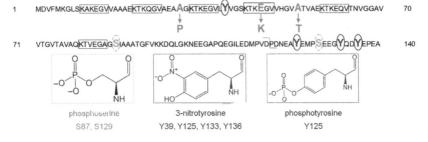

1   MDVFMKGLSKAKEGVVAAAEKTKQGVAEAAGKTKEGVLYVGSKTKEGVVHGVATVAEKTKEQVTNVGGAV   70

                               P   K   T

71   VTGVTAVAQKTVEGAGSIAAATGFVKKDQLGKNEEGAPQEGILEDMPVDPDNEAYEMPSEEGYQDYEPEA   140

phosphoserine
S87, S129

3-nitrotyrosine
Y39, Y125, Y133, Y136

phosphotyrosine
Y125

ROCHET *ET AL.*, FIG. 1.

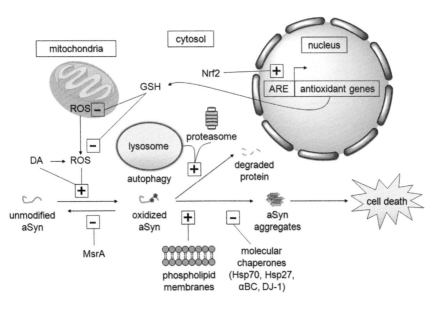

ROCHET *ET AL.*, FIG. 3.

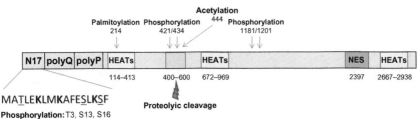

MATLEKLMKAFESLKSF

Phosphorylation: T3, S13, S16
Ubiquitination/sumoyalation: K6, K9, K15
NES: L4, L7, F11, L14

ZHENG AND DIAMOND, FIG. 1.

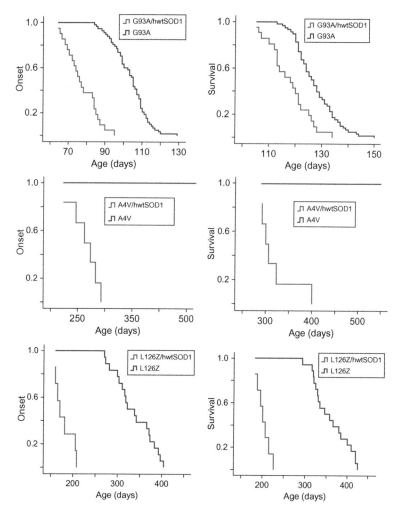

REDLER AND DOKHOLYAN, FIG. 5.

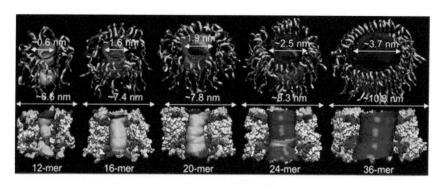

BRUCE L. KAGAN, FIG. 1.

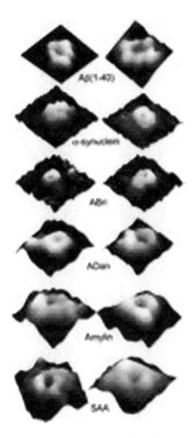

BRUCE L. KAGAN, FIG. 2.

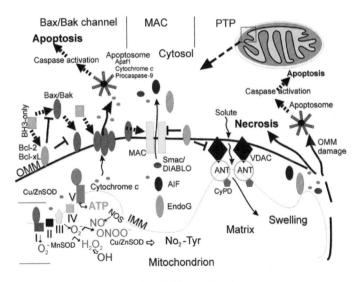

LEE J. MARTIN, FIG. 1.

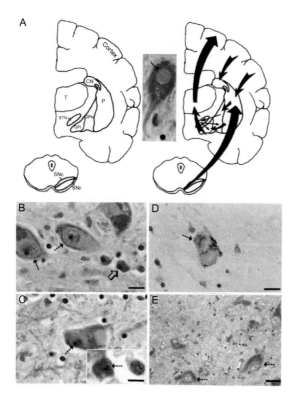

Lee J. Martin, Fig. 5.

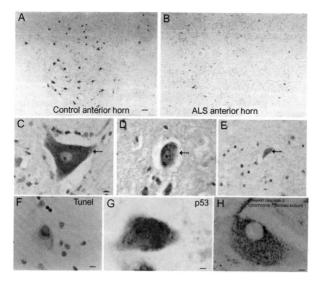

Lee J. Martin, Fig. 6.

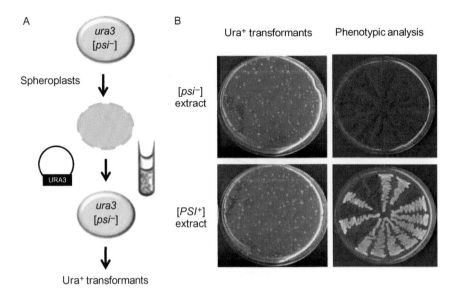

STANIFORTH AND TUITE, FIG. 1.

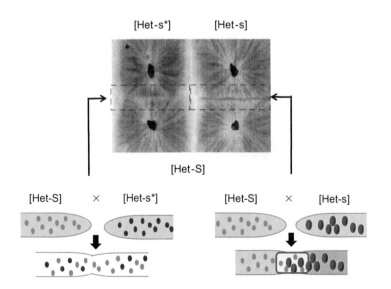

STANIFORTH AND TUITE, FIG. 2.

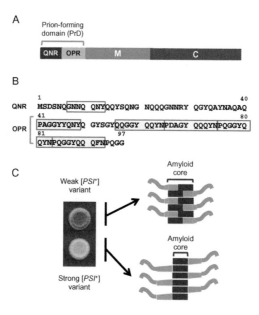

STANIFORTH AND TUITE, FIG. 4.

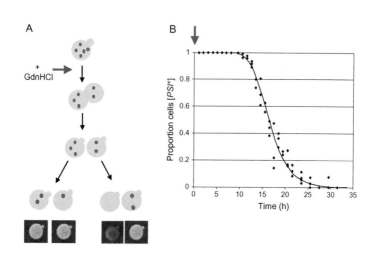

STANIFORTH AND TUITE, FIG. 5.

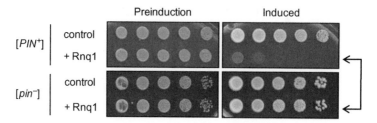

STANIFORTH AND TUITE, FIG. 6.

Printed and bound by CPI Group (UK) Ltd, Croydon, CR0 4YY

08/05/2025

01864953-0005